Lecture Notes in Mathematics

Volume 2288

This series reports on new developments in all areas of mathematics and their applications - quickly, informally and at a high level. Mathematical texts analysing new developments in modelling and numerical simulation are welcome. The type of material considered for publication includes:

1. Research monographs
2. Lectures on a new field or presentations of a new angle in a classical field
3. Summer schools and intensive courses on topics of current research.

Texts which are out of print but still in demand may also be considered if they fall within these categories. The timeliness of a manuscript is sometimes more important than its form, which may be preliminary or tentative.

Titles from this series are indexed by Scopus, Web of Science, Mathematical Reviews, and zbMATH.

More information about this series at http://www.springer.com/series/304

Alberto Arabia

Equivariant Poincaré Duality on G-Manifolds

Equivariant Gysin Morphism and Equivariant Euler Classes

Springer

Alberto Arabia
IMJ-PRG, CNRS
Paris Diderot University
Paris Cedex 13
France

ISSN 0075-8434 ISSN 1617-9692 (electronic)
Lecture Notes in Mathematics
ISBN 978-3-030-70439-1 ISBN 978-3-030-70440-7 (eBook)
https://doi.org/10.1007/978-3-030-70440-7

Mathematics Subject Classification: 55, 55M05, 57R91, 57S15, 55N30, 14F05, 20Cxx

This Springer imprint is published by the registered company Springer Nature Switzerland AG.
The registered company address is: Gewerbestrasse 11, 6330 Cham, Switzerland

Preface

This is an introduction to equivariant Poincaré duality of oriented G-manifolds, for a compact Lie group G, and to equivariant Gysin morphisms for both proper and non-proper maps, with applications as: the equivariant Gysin exact sequence, the equivariant Lefschetz fixed point theorem, the equivariant Thom isomorphism, and the equivariant Thom and Euler classes.

For the most part, we have chosen to focus on smooth manifolds and de Rham (equivariant) cohomology, thereby restricting coefficients to the field of real numbers $\mathbb{R}$. However, we also give a more general and parallel approach, based on the Grothendieck-Verdier's formalism, which allows us to simultaneously treat equivariant and nonequivariant theories as well as substituting the field of coefficients.

The material is organized into three parts:

- The first part, which consists of Chaps. 2 and 3, is about nonequivariant de Rham cohomology. Chapter 2 summarizes familiar results and constructions related to Poincaré duality and Gysin morphisms, which we extended, in Chap. 3, to fiber bundles following Grothendieck's *relative point of view*. In doing this, we were motivated by the fact that, for a G-manifold M, Poincaré duality of its homotopy quotient M_G, *relative to the classifying space* $\mathbb{B}G$, is the precise analogue to G-equivariant Poincaré duality of M. This approach thus opens the way to a general definition of equivariant cohomology, not only beyond manifolds but also over arbitrary coefficient rings.
- The second part consists of Chaps. 4 to 7 devoted to equivariant *de Rham* cohomology of manifolds. Chapter 4 reviews the origins of equivariant cohomology, recalls standard definitions and constructions in equivariant de Rham cohomology, and ends with the equivariant de Rham comparison theorems that identify equivariant de Rham and singular cohomologies (4.8). Chapter 5 is devoted to equivariant de Rham Poincaré duality, which we apply in Chap. 6 to define equivariant de Rham Gysin morphisms and Euler classes, which are localized in Chap. 7.
- The third part, Chap. 8, explains the approach of equivariant cohomology within the framework of Grothendieck-Verdier duality, whose great advantage is its

enormous flexibility, in particular, the fact that the coefficients field $\Bbbk$ is irrelevant. Chapter 8 thus presents an alternative approach for dealing with equivariant cohomology, equivariant Poincaré duality and equivariant Gysin morphisms for manifolds, allowing the results of previous chapters to be extended to arbitrary coefficient fields.

We assume the reader to be familiar with algebraic topology techniques as well as with sheaf cohomology and with derived categories as they appear in Grothendieck-Verdier's duality theory, for example, as explained in Iversen [58], Kashiwara-Schapira [61, 62], and Weibel [95]. However, for readers with little experience in these subjects we added Appendix A, which gives an overview of Derived Categories and Derived Functors. Beyond recalling the usual terminology and definitions, this appendix describes the common underlying approach for dealing with duality in derived categories of dg-modules over dg-algebras, as it occurs throughout the book.

We also assume that the reader knows the basics of equivariant cohomology, both as the ordinary cohomology of the homotopy quotient of G-spaces and as the cohomology of the equivariant de Rham complex in the case of G-manifolds. References for these topics are, for ordinary equivariant cohomology: W. Hsiang [55] and Allday-Puppe [2], and for de Rham equivariant cohomology: Guillemin and Sternberg [50] and Loring Tu [91].

A more specialized reference in equivariant Poincaré duality is the article of Brylinski [24] on G-equivariant Poincaré duality, which, written for stratified spaces and for intersection homology, uses a modified version of the Cartan model relative to a Thom-Mather stratification. When the space is smooth, the model coincides with the usual Cartan model, which is the starting point of this book.

Other specialized references are articles by Allday, Franz, and Puppe [2–5], which address G-equivariant Poincaré duality for $G := T$ a torus and coefficients in arbitrary fields using what they call *the singular Cartan Model*. Equivariant Gysin morphisms, Thom isomorphism, and Euler classes can also be found in Allday et al. [4], Allday-Puppe [2], and Kawakubo [63].

Many people have contributed to the development of equivariant cohomology and equivariant Poincaré duality in significant ways since the 1990s by pursuing other routes than those that we will travel. I apologize in advance to each of them for not having been able to refer to their work as it deserves.

The following signs have been used in the book:

$\square$ to indicate the end of the proof of a theorem, proposition, etc.,
$\boxdot$ to indicate the end of a sketch of proof,
$\boxminus$ to indicate the end of a lemma within a proof,
to indicate a helpful hint.

Acknowledgements

To Matthias Franz for his remarks on the subject of reflexivity of equivariant cohomology; to Bernhard Keller for enlightening discussions on dg-algebras; to Bruno Klingler both for his helpful remarks and for directing me to valuable references on classification spaces; to the referees whose valuable reviews made this a much better book; to Loïc Merel, Director of the Mathematical Institute of Jussieu (IMJ), who made the institute's research resources available to me; to Frances Cowell, whose careful proofreading and her wise comments have helped me improve this work; to Ute McCrory, editor at Springer, for her constant encouragement, her gentle prodding and, especially, her remarkable patience, and, of course, to SPRINGER for offering me the opportunity to publish this book.

Paris, France
January 2021

Alberto Arabia

Contents

Chapter 1
Introduction

1.1 Equivariant de Rham Poincaré Duality

Given a compact Lie group G, we prove G-equivariant Poincaré duality for an oriented G-manifold M following J.-L. Brylinski's approach for the equivariant intersection cohomology of G-pseudomanifolds [24]. We will therefore work in the category $\mathrm{DGM}(\Omega_G)$ of Ω_G-differential graded modules (Ω_G-dgm), where Ω_G denotes the ring $S(\mathfrak{g}^\vee)^G$ of G-invariant real polynomial functions on $\mathfrak{g} := \mathrm{Lie}(G)$ endowed with the grading that doubles polynomial degrees. By setting its differential to zero, Ω_G becomes a differential graded ring; which coincides with its cohomology ring, traditionally denoted by H_G.

The complexes of G-equivariant differential forms $\Omega_G(M)$ and $\Omega_{G,\mathrm{c}}(M)$ (4.4.4), which respectively compute the equivariant de Rham cohomologies $H_G(M)$ and $H_{G,\mathrm{c}}(M)$, belong to the category $\mathrm{DGM}(\Omega_G)$.

When M is an oriented G-manifold of dimension d_M, ordinary *Poincaré pairing* naturally extends to equivariant differential forms as an Ω_G-pairing:

$$\langle \cdot, \cdot \rangle_{M,G} : \Omega_G(M) \times \Omega_{G,\mathrm{c}}(M) \to \Omega_G\,, \quad \langle \alpha, \beta \rangle_{M,G} := \int_M \alpha \wedge \beta\,,$$

giving rise to the *G-equivariant Poincaré adjunctions*, the morphisms in $\mathrm{DGM}(\Omega_G)$:

$$\begin{cases} I\!D_{G,M} : \ \Omega_G(M)[d_M] \to \mathbf{Hom}^\bullet_{\Omega_G}(\Omega_{G,\mathrm{c}}(M), \Omega_G) \\ I\!D'_{G,M} : \Omega_{G,\mathrm{c}}(M)[d_M] \to \mathbf{Hom}^\bullet_{\Omega_G}(\Omega_G(M), \Omega_G) \end{cases} \tag{1.1}$$

defined respectively by:

$$I\!D_{G,M}(\alpha) : \beta \mapsto \langle \alpha, \beta \rangle_{M,G} \quad \text{and} \quad I\!D'_{G,M}(\beta) : \alpha \mapsto \langle \alpha, \beta \rangle_{M,G}\,,$$

for which we have the following equivariant analogue to Poincaré duality theorem.

A. Arabia, *Equivariant Poincaré Duality on G-Manifolds*, Lecture Notes in Mathematics 2288, https://doi.org/10.1007/978-3-030-70440-7_1

Theorem (5.6.2.1) *Let G be a compact Lie group, and let M an oriented G-manifold of dimension d_M.*

1. *The following left Poincaré adjunction is an injective quasi-isomorphism,*

$$\mathbb{D}_{G,M} : \Omega_G(M)[d_M] \longrightarrow \mathbf{Hom}^{\bullet}_{\Omega_G}(\Omega_{G,c}(M), \Omega_G)\,. \tag{1.2}$$

2. *The morphism $\mathbb{D}_{G,M}$ induces* the Poincaré duality morphism in G-equivariant cohomology (see 5.4.7.2–(1))

$$D_{G,M} : H_G(M)[d_M] \longrightarrow \mathbf{Hom}^{\bullet}_{H_G}(H_{G,c}(M), H_G)$$

which is an isomorphism if $\mathrm{Ext}^i_{H_G}(H_{G,c}(M), H_G) = 0$ for all $i > 0$, for example if $H_{G,c}(M)$ is a free H_G-module.

3. *If G is connected, then there exist spectral sequences converging to $H_G(M)[d_M]$*

$$\begin{cases} \mathbb{E}_2^{p,q}(M) = \ (\mathrm{Ext}^p_{H_G}(H_{G,c}(M), H_G))^q \ \Rightarrow H_G^{d_M+p+q}(M) \\ \mathbb{F}_2^{p,q}(M) = H_G^p \otimes_{\mathbb{R}} \mathbf{Hom}^{\bullet}_{\mathbb{R}}(H_c^{-q}(M), \mathbb{R}) \Rightarrow H_G^{d_M+p+q}(M) \end{cases}$$

where, in the first, q refers to the graded vector space grading.

4. *If, in addition, M is of finite type, the right Poincaré adjunction*

$$\mathbb{D}'_{G,M} : \Omega_{G,c}(M)[d_M] \longrightarrow \mathbf{Hom}^{\bullet}_{\Omega_G}(\Omega_G(M), \Omega_G) \tag{1.3}$$

is an injective quasi-isomorphism, and mutatis mutandis *for (2) and (3).*

1.2 Equivariant de Rham Gysin Morphisms

As we will show *equivariant Poincaré duality* is better approached in the derived category $\mathcal{D}(\mathrm{DGM}(\Omega_G))$ of $\mathrm{DGM}(\Omega_G)$ (5.4.6, A.2.5) by a systematic use of the *derived* duality functor

$$\mathbb{R}\,\mathbf{Hom}^{\bullet}_{\Omega_G}(\_, \Omega_G) : \mathcal{D}(\mathrm{DGM}(\Omega_G)) \rightsquigarrow \mathcal{D}(\mathrm{DGM}(\Omega_G))\,. \tag{1.4}$$

Indeed, although in the statements of theorem 5.6.2.1 we still use nonderived $\mathbf{Hom}^{\bullet}$, possible since both $\Omega_G(M_G)$ and $\Omega_{G,c}(M_G)$ are free Ω_G-graded modules (*cf*. proof of 5.6.2.1-(1)), the right framework to deal with Gysin morphisms is derived category, the reason being that the general definition of these requires inverting quasi-isomorphisms.

Let $f : M \to N$ be a G-equivariant map between oriented G-manifolds of dimensions d_M and d_N respectively.

When f is proper, the pullback $f^* : \Omega_{G,c}(N) \to \Omega_{G,c}(M)$ induces, through the duality functor (1.4), a morphism in $\mathcal{D}(\mathrm{DGM}(\Omega_G))$, which we denote:

$$f_\vee : I\!R\,\mathbf{Hom}^\bullet_{\Omega_G}(\Omega_{G,c}(M), \Omega_G) \to I\!R\,\mathbf{Hom}^\bullet_{\Omega_G}(\Omega_{G,c}(N), \Omega_G)\,.$$

We therefore have in $\mathcal{D}(\mathrm{DGM}(\Omega_G))$ the diagram:

$$\begin{array}{ccc} I\!R\,\mathbf{Hom}^\bullet_{\Omega_G}(\Omega_{G,c}(M), \Omega_G) & \xrightarrow{f_\vee} & I\!R\,\mathbf{Hom}^\bullet_{\Omega_G}(\Omega_{G,c}(N), \Omega_G) \\ {\scriptstyle I\!D_{G,M}}\uparrow{\scriptstyle \simeq} & & {\scriptstyle I\!D_{G,N}}\uparrow{\scriptstyle \simeq} \\ \Omega_G(M)[d_M] & \xrightarrow{f_*} & \Omega_G(N)[d_N] \end{array} \tag{1.5}$$

by which, we define *the Gysin morphism f_* for the proper f* as

$$f_* := I\!D_{G,N}{}^{-1} \circ f_\vee \circ I\!D_{G,M} : \Omega_G(M)[d_M] \to \Omega_G(N)[d_N]\,,$$

thanks to equivariant Poincaré duality (5.6.2.1–(1)).

Now if $g : N \to N'$ is also G-equivariant and proper between oriented G-manifolds, we immediately get

$$\begin{aligned}(g \circ f)_* &:= I\!D_{G,N'}{}^{-1} \circ (g \circ f)_\vee \circ I\!D_{G,M} \\ &= (I\!D_{G,N'}{}^{-1} \circ g_\vee \circ I\!D_{G,N}) \circ (I\!D_{G,N}{}^{-1} \circ f_\vee \circ I\!D_{G,M}) = g_* \circ f_*\,.\end{aligned}$$

Whence, the *Gysin functor for proper maps*

$$(\_)_* : G\text{-}\mathrm{Man}_{\mathrm{pr}} \rightsquigarrow \mathcal{D}(\mathrm{DGM}(\Omega_G))\,, \quad M \rightsquigarrow \Omega_G(M)[d_M]\,, \quad f \rightsquigarrow f_*\,,$$

where G-$\mathrm{Man}_{\mathrm{pr}}$ is the category of G-manifolds and G-equivariant *proper* maps.

When f is not proper, we can proceed in the same way if $\dim(H_c(N)) < \infty$, since in that case the morphism $I\!D'_{G,N}$ (1.3) is also an isomorphism (5.6.2.1–(3)). We can then consider the pullback $f^* : \Omega_G(N) \to \Omega_G(M)$ with no condition on supports and replace $I\!D_{G,M}$ with $I\!D'_{G,M}$ in diagram (1.5).

The definition of *the Gysin morphism $f_!$ for arbitrary f* is then

$$f_! := (I\!D'_{G,N})^{-1} \circ (f^*)^\vee \circ I\!D'_{G,M} : \Omega_{G,c}(M)[d_M] \to \Omega_{G,c}(N)[d_N]\,.$$

Whence, the *Gysin functor for arbitrary maps*

$$(\_)_! : G\text{-}\mathrm{Man} \rightsquigarrow \mathcal{D}(\mathrm{DGM}(\Omega_G))\,, \quad M \rightsquigarrow \Omega_{G,c}(M)[d_M]\,, \quad f \rightsquigarrow f_!\,,$$

where G-Man is the category of G-manifolds and G-equivariant maps.

1.3 Adjunction Properties of Gysin Morphisms

By taking cohomology, Gysin morphisms induce morphisms of graded H_G-modules which we denote by the same notations:

$$\begin{cases} f_* = H_G(M)[d_M] \to H_G(N)[d_N], \\ f_! = H_{G,c}(M)[d_M] \to H_{G,c}(N)[d_N]. \end{cases}$$

The commutativity of diagram (1.5) then amounts to the equalities

$$\begin{cases} \langle \alpha, f^*(\beta) \rangle_{M,G} = \langle f_*(\alpha), \beta \rangle_{N,G}, \\ \langle f^*(\alpha), \beta \rangle_{M,G} = \langle \alpha, f_!(\beta) \rangle_{N,G}. \end{cases} \tag{1.6}$$

for all $\alpha \in H_G(M)$ and all $\beta \in H_{G,c}(N)$, showing that Gysin morphisms are Poincaré adjoints to pullbacks.

This is often the departure point for introducing Gysin morphisms, but to solve the adjunction equalities (1.6) in the category $\mathrm{DGM}(\Omega_G)$ generally demands additional constraints in the map f, constraints that are no more needed in $\mathcal{D}(\mathrm{DGM}(\Omega_G))$. In other words, the Gysin morphisms f_* and $f_!$ cannot always be defined in $\mathrm{DGM}(\Omega_G)$.

In the classical approach, given a map $f : M \to N$, one has explicit definitions of Gysin morphism for compact supports in the category $\mathrm{DGM}(\Omega_G)$ for both the projection $p : M \times N \to N$ and the graph embedding $\mathrm{Gr}(f) : M \to M \times N$. These allow defining $f_! := p_! \circ \mathrm{Gr}(f)_!$. But although this is a necessary condition to functoriality, it does not justify it, something which would still have to be proven.

Working in derived category $\mathcal{D}(\mathrm{DGM}(\Omega_G))$ simplifies things.

- It allows an a priori justification for the functoriality of Gysin morphisms.
- Given a closed inclusion of oriented manifolds $N \subseteq M$, we can simply dualize the exact triangle in $\mathcal{D}(\mathrm{DGM}(\Omega_G))$

  $$\Omega_{G,c}(M \setminus N) \to \Omega_{G,c}(M) \to \Omega_{G,c}(N) \to,$$

  applying the duality functor (1.4), immediately getting, thanks to equivariant Poincaré duality, the Gysin exact triangle in $\mathcal{D}(\mathrm{DGM}(\Omega_G))$:

  $$\Omega_G(N)[d_N] \to \Omega_G(M)[d_M] \to \Omega_G(M \setminus N)[d_M] \to,$$

  whence, *the Gysin exact sequence of equivariant cohomology*:

  $$\to H^i_G(N) \to H^{i+d_M-d_N}_G(M) \to H^{i+d_M-d_N}_G(M \setminus N) \to .$$

- In the same easy way, if $E \to B$ is a vector bundle with fibers of dimension n and oriented base space, the zero section map $\iota : B \to E$, being proper, automatically induces the Gysin morphism $\iota_* : \Omega_G(B) \to \Omega_G(E)[n]$ in $\mathcal{D}(\mathrm{DGM}(\Omega_G))$. We deduce the *equivariant Thom morphism* of H_G-graded modules $\iota_* : H^i_G(B) \simeq H^{i+n}_G(E)$, the *Thom equivariant class* then being $\iota_*(1) \in H^n_G(E)$.
  In a more refined approach of Poincaré duality of E *relative to the base space* B developed in Chap. 3, we see in the same easy way that the Thom morphism establishes an isomorphism $\iota_* : H^i_G(B) \simeq H^{i+n}_{G,B}(E)$, where $H_{G,B}(\_)$ denotes the *equivariant cohomology with supports in B*.
- The Equivariant Euler class for a closed embedding $i : N \subseteq M$ of oriented manifolds, is simply $\mathrm{Eu}_G(N, M) := i^*i_*(1)$ where $i^*i_* : H_G(N) \to H_G(N)$ is the push-pull operator.

1.4 Equivariant Cohomology Viewed as a Relative Cohomology Theory

Chapter 3 is a transition chapter between nonequivariant and equivariant de Rham cohomologies, in which we make explicit, in the category of fiber bundles (E, B, π, M), the Grothendieck-Verdier formalism of cohomology and of Poincaré duality of the total space E *relative to the base space* B. The classic setting, thereafter referred to as *absolute*, corresponds to the case where $B := \{\bullet\}$.

The change from the absolute to the relative point of view amounts to the following replacements:

$$\begin{aligned}
\text{absolute cohomology} &\leftrightarrow \text{relative cohomology}\\
\mathbb{R} := \Omega(\{\bullet\}) &\leftrightarrow \underline{\underline{\Omega}}(B) \simeq \underline{\underline{\mathbb{R}}}_B\\
\mathcal{D}(\mathrm{DGM}(\mathbb{R})) &\leftrightarrow \mathcal{D}(\mathrm{DGM}(\underline{\underline{\Omega}}(B), d)) \sim \mathcal{D}(B; \mathbb{R})\\
\Omega(M) &\leftrightarrow \pi_*\, \underline{\underline{\Omega}}(E) \simeq I\!R\, \pi_* \underline{\underline{\mathbb{R}}}_E\\
\Omega_c(M) &\leftrightarrow \pi_!\, \underline{\underline{\Omega}}(E) \simeq I\!R\, \pi_! \underline{\underline{\mathbb{R}}}_E\\
\mathbf{Hom}^\bullet_{\mathbb{R}}(_, \mathbb{R}) &\leftrightarrow I\!R\, \underline{\mathbf{Hom}}^\bullet_{(\underline{\underline{\Omega}}_B, d)}(_, \underline{\underline{\Omega}}_B) \simeq I\!R\, \underline{\mathbf{Hom}}^\bullet_{\underline{\underline{\mathbb{R}}}_B}(_, \underline{\underline{\mathbb{R}}}_B)
\end{aligned}$$

where $\underline{\underline{\Omega}}(\_)$ denotes the sheaf of differential forms, and $\mathcal{D}(B; \mathbb{R})$ is the derived category of sheaves of $\mathbb{R}$-vector spaces on B.

The relative Poincaré duality, also known as the Poincaré-Verdier Duality, states the existence of an isomorphism in the *derived category* of sheaves over B.

Theorem (3.3.3.1) *Let (E, B, π, M) be an oriented fiber bundle with fiber M a manifold of dimension d_M.*

1. The following morphisms induced by the left Poincaré adjunction,

$$\begin{array}{ccc} \pi_*\underline{\underline{\Omega}}_E[d_M] & \xrightarrow[\text{q.i.}]{\mathbb{D}_{B,M}} & \mathbb{R}\,\underline{\underline{\boldsymbol{Hom}}}^\bullet_{(\underline{\underline{\Omega}}_B,d)}(\pi_!\underline{\underline{\Omega}}_E,\underline{\underline{\Omega}}_B) \\ & \text{Grothendieck-Verdier Duality} \searrow^{\text{q.i.}} & \downarrow \text{q.i.} \\ & & \mathbb{R}\,\underline{\underline{\boldsymbol{Hom}}}^\bullet_{\underline{\underline{\mathbb{R}}}_B}(\pi_!\underline{\underline{\Omega}}_E,\underline{\underline{\mathbb{R}}}_B) \end{array} \tag{1.7}$$

are isomorphisms in $\mathcal{D}(B;\mathbb{R})$.

4. If M is of finite type, the previous statements remain true if we swap terms $\pi_!\underline{\underline{\Omega}}_E \leftrightarrow \pi_*\underline{\underline{\Omega}}_E$ and $\Omega_{\mathrm{cv}}(E) \leftrightarrow \Omega(E)$. The right *Poincaré adjunction:*

$$\pi_!\underline{\underline{\Omega}}_E[d_M] \xrightarrow[\text{q.i.}]{\mathbb{D}'_{B,M}} \mathbb{R}\,\underline{\underline{\boldsymbol{Hom}}}^\bullet_{(\underline{\underline{\Omega}}_B,d)}(\pi_*\underline{\underline{\Omega}}_E,\underline{\underline{\Omega}}_B) \tag{1.8}$$

is therefore a quasi-isomorphism in $\mathcal{D}(B;\mathbb{R})$ too.

This theorem is the basis of all versions of Poincaré Duality. Indeed, when $B := \{\bullet\}$, we have the classical (absolute) Poincaré Duality; when B is the classifying space $\mathbb{B}G$ and that we consider the fiber bundle $(M_G, \mathbb{B}G, \pi, M)$, where $\pi : M_G := \mathbb{E}G \times_G M \to \mathbb{B}G$ is the Borel construction for an oriented G-manifold M, the cohomology of M_G relative to $\mathbb{B}G$ is precisely the singular equivariant cohomology of M with coefficients in $\mathbb{R}$. The equivariant comparison de Rham theorems, proved in Sect. 4.8, establish the existence of canonical isomorphisms

$$H_G(M) \simeq H(M_G;\mathbb{R}) \quad \text{and} \quad H_{G,\mathrm{c}}(M) \simeq H_{\mathrm{cv}}(M_G;\mathbb{R}),$$

where $H(M_G;\mathbb{R}) := H(\mathbb{B}G;\pi_*\underline{\underline{\Omega}}_{M_G})$ and $H_{\mathrm{cv}}(M_G;\mathbb{R}) := H(\mathbb{B}G;\pi_!\underline{\underline{\Omega}}_{M_G})$,

1.5 Equivariant Poincaré Duality over Arbitrary Fields

As we recalled in the Preface, Grothendieck-Verdier's formalism is especially valuable because of its generality: not only does it allow topological spaces to be much more general than simple manifolds, but also coefficients rings can be arbitrary. Nevertheless, in this book we limited generality to fiber bundles (E, B, π, M) where the fiber M is a manifold and the base space B is a mild topological space (*cf*. Sect. 1.7 in this chapter). We also limited the coefficients ring to be a field $\Bbbk$, albeit of arbitrary characteristic. Poincaré-Verdier Duality is then relative to B with coefficients in $\Bbbk$. In Chap. 8, we extend Theorem 3.3.3.1 to this more general framework, thus extending equivariant Poincaré duality and Gysin morphisms to arbitrary coefficients fields.

1.6 Conditions on the Group G

In the literature on G-equivariant cohomology, the most frequent assumption about group G is that it is a compact *connected* Lie group whose action on manifolds is differentiable.

Compactness and differentiability hypotheses seem unavoidable when one needs, as we do, G-averaging operators on G-manifolds, which allow us to prove the following fundamental (and constantly used) facts for G-manifolds.

- Existence of proper invariant functions (2.5.4.2).
- Existence of G-invariant Riemannian metrics.
- Existence of G-Slices (4.5.3.1, 7.5.1);
- Existence of G-invariant tubular neighborhoods around G-stable submanifolds, essential to the proof of Localization formulas (7.7)
- Existence of quotient manifolds for free actions.
- Equivariant de Rham theorems, which, for G-manifolds M, N, identify the cohomology of the Cartan Models $\Omega_G(M)$, $\Omega_{G,c}(M)$ and $\Omega_{G,N}(M)$ with the corresponding cohomologies for the homotopy quotients M_G, N_G (4.8).

Connectedness of G is less necessary and can (and will) frequently be avoided.

This hypothesis is usually invoked to warrant

- *homotopic triviality* of the action of G on G-spaces. But connectedness can be unnecessarily restrictive, for example, when, G is connected and we are interested in the K-equivariant cohomology $H_K(M)$ of a G-manifold, where K is any closed subgroup of G, connected or not (see 4.4.6.3).
- the action of G *preserves the orientation* of a G-manifold, which is not always true when G is not connected. But here, rather than excluding non-connectedness, we call *oriented G-manifold* a G-manifold in which the action of every $g \in G$ preserves orientation (see 5.5.1)
- *simple connectedness* of the classifying space $\mathbb{B}G$. This is a useful property when dealing with local systems on $\mathbb{B}G$ since, in that case, local systems are trivial and the description of Leray spectral sequences converging to $H_G(M)$ and $H_{G,c}(M)$ is greatly simplified. We can however still avoid connectedness of G altogether and get around nontrivial local systems by including in our considerations the finite group G/G_0 where G_0 denotes the connected component of the identity element $e \in G$, and by replacing $H_G(M)$ and $H_{G,c}(M)$ respectively with $H_{G_0}(M)^{G/G_0}$ and $H_{G_0,c}(M)^{G/G_0}$. (We will then need to add the hypothesis that the cardinality of G/G_0 be prime to char($\mathbb{k}$) when introducing equivariant cohomology over fields of positive characteristic.)
- *use of Cartan-Weil theory* as described in Cartan's Seminar (4.1.1). Indeed, the theory concerns the category of $\mathfrak{g}$-differential graded algebras, where $\mathfrak{g} :=$ Lie(G_0). But, again, instead of restricting to connected G, we extend Cartan-Weil theory in such a way that it incorporates the action of the finite group G/G_0, which we do in 4.4.5, Proposition 4.4.5.2, beyond which we need no longer assume G to be connected to apply this theory (see 4.4.4.2).

Convention For these reasons, in this book, and otherwise explicitly stated, the group G is only assumed to be a *compact Lie group whose action on manifolds is differentiable*.

1.7 Conditions on Topological Spaces

All topological spaces are assumed to be *Hausdorff*, *locally contractible*, *paracompact* and *perfectly normal*.[1] We call such spaces *mild spaces* (*cf*. Appendix B.1). The most immediate examples of which are: manifolds and inductive limits of such, as are the classifying spaces $\mathbb{B}G$, and, more generally, open subspaces of CW-complexes.

Working with mild spaces is especially pleasant in so far as Alexander-Spanier, Singular, Čech, de Rham (for manifolds) and Sheaf cohomology are all canonically isomorphic to each other.[2] This flexibility allows us to approach Poincaré Duality from different points of view and, in particular, through Grothendieck-Verdier formalism. Notice that, in contrast to frequent practice, we do not require the property of being locally compact in the definition of mild spaces. The reason being that we need to apply Poincaré-Verdier Duality to fiber bundles (E, B, π, M) where, while the fiber space M is a manifold, the base space B can fail to be so, in which case being a mild space is a sufficient condition for the theory to work.

[1] A topological space X is '*normal*' if for every pair of disjoint closed subspaces F_1, F_2, there exist disjoint neighborhoods $V_i \supseteq F_i$, and it is '*perfectly normal*' if, in addition, there exist continuous functions $f : X \to \mathbb{R}_{\geq 0}$ verifying $f|_{V_1} = 1$ and $f|_{V_2} = 0$.

[2] See Bredon [20], ch. III, Comparison with other cohomology theories.

Chapter 2
Nonequivariant Background

2.1 Category of Cochain Complexes

We begin recalling standard terminology and notations.

2.1.1 *Fields in Use*

We denote by $\Bbbk$ an arbitrary field, and by $\mathrm{Vec}(\Bbbk)$ the category of $\Bbbk$-vector spaces and $\Bbbk$-linear maps. Later, as we consider de Rham cohomology on manifolds, the field $\Bbbk$ will be the field of real numbers $\mathbb{R}$.

2.1.2 *Vector Spaces and Pairings*

Expressions such as *vector space, linear, bilinear,...* always refer to the field $\Bbbk$, unless otherwise stated.

The *dual* of the vector space V is denoted by $V^\vee := \mathrm{Hom}_\Bbbk(V, \Bbbk)$.

A bilinear map $b : V \times W \to \Bbbk$, also called *a pairing*, gives rise to the linear maps $\gamma_b : V \to W^\vee$, $\gamma_b(v)(w) := b(v, w)$, and $\rho_b : W \to V^\vee$, $\rho_b(w)(v) := b(v, w)$, respectively called the *left and right adjunctions associated with b*. One says that *b is a nondegenerate pairing* whenever the adjunctions are injective, and that *b is a perfect pairing* whenever they are bijective. For example, the canonical pairing $V^\vee \times V \to \Bbbk$, $(\lambda, v) \mapsto \lambda(v)$, is always nondegenerate and is perfect if and only if V is finite dimensional.

A. Arabia, *Equivariant Poincaré Duality on G-Manifolds*, Lecture Notes in Mathematics 2288, https://doi.org/10.1007/978-3-030-70440-7_2

2.1.3 The Category $\mathrm{GV}(\Bbbk)$ of Graded Spaces

A *graded (vector) space* is a family $\boldsymbol{V} := \{V^m\}_{m\in\mathbb{Z}}$ of vector spaces indexed by the set of integers. Given graded spaces $\boldsymbol{V}$ and $\boldsymbol{W}$, a *graded homomorphism* $\boldsymbol{\alpha} : \boldsymbol{V} \to \boldsymbol{W}$ *of degree* $d =: \deg(\boldsymbol{\alpha})$ is a family $\boldsymbol{\alpha} := \{\alpha_m : V^m \to W^{m+d}\}_{m\in\mathbb{Z}}$ of linear maps. The composition of graded homomorphisms is defined degree by degree, i.e. $\boldsymbol{\beta} \circ \boldsymbol{\alpha} = \{\beta_{m+d} \circ \alpha_m\}_{m\in\mathbb{Z}}$, and we have $\deg(\boldsymbol{\alpha} \circ \boldsymbol{\beta}) = \deg(\boldsymbol{\alpha}) + \deg(\boldsymbol{\beta})$.

We denote by $\mathrm{Homgr}^d_{\Bbbk}(\boldsymbol{V}, \boldsymbol{W})$ the vector space of graded homomorphisms of degree d. When $d = 0$, we simply write $\mathrm{Homgr}_{\Bbbk}(\boldsymbol{V}, \boldsymbol{W})$ for $\mathrm{Homgr}^0_{\Bbbk}(\boldsymbol{V}, \boldsymbol{W})$.

The *graded space of graded homomorphisms* from $\boldsymbol{V}$ to $\boldsymbol{W}$ is defined as

$$\mathrm{Homgr}^*_{\Bbbk}(\boldsymbol{V}, \boldsymbol{W}) := \big\{\mathrm{Homgr}^d_{\Bbbk}(\boldsymbol{V}, \boldsymbol{W})\big\}_{d\in\mathbb{Z}}\,. \tag{2.1}$$

The *category* $\mathrm{GV}(\Bbbk)$ *of graded spaces*, also denoted by $\mathrm{Vec}(\Bbbk)^{\mathbb{Z}}$, is the category whose objects are graded spaces and whose *morphisms* are graded homomorphisms of degree 0. Given $\boldsymbol{V}, \boldsymbol{W} \in \mathrm{GV}(\Bbbk)$, we denote by $\mathrm{Mor}_{\mathrm{GV}(\Bbbk)}(\boldsymbol{V}, \boldsymbol{W})$ the set of morphisms from $\boldsymbol{V}$ to $\boldsymbol{W}$. Note that the three notations $\mathrm{Mor}_{\mathrm{GV}(\Bbbk)}(\_, \_)$, $\mathrm{Homgr}_{\Bbbk}(\_, \_)$ and $\mathrm{Homgr}^0_{\Bbbk}(\_, \_)$ denote the same set.

Comment 2.1.3.1 It is worth mentioning that since a graded space is simply a family of vector spaces $\boldsymbol{V} := \{V^m\}_{m\in\mathbb{Z}}$, it underlies the direct sum $\bigoplus_{m\in\mathbb{Z}} V^m$ and the direct product $\prod_{m\in\mathbb{Z}} V^m$. Nevertheless, it would be a mistake to identify graded spaces with any kind of sum or product. In categories of sheaves, for example, the distinction between a graded sheaf $\{\mathcal{F}^i\}_{i\in\mathbb{Z}}$ and a direct-sum sheaf $\bigoplus_{i\in\mathbb{Z}} \mathcal{F}^i$ is crucial since the inclusion $\bigoplus_{i\in\mathbb{Z}} \Gamma(X; \mathcal{F}^i) \subseteq \Gamma(X; \bigoplus_{i\in\mathbb{Z}} \mathcal{F}^i)$ is generally strict (*cf*. fn. ([6]), p. 305).

2.1.4 The Subcategories of Bounded Graded Spaces

A graded space $\boldsymbol{V} := \{V^m\}_{m\in\mathbb{Z}}$ is said to be *bounded*, *bounded below (by ℓ)*, *bounded above (by ℓ)*, if $V^m = 0$ respectively for $|m| \gg 0$, $m \ll 0$ ($m < \ell$), $m \gg 0$ ($m > \ell$). We denote by $\mathrm{GV}^b(\Bbbk)$, $\mathrm{GV}^+(\Bbbk)$, $\mathrm{GV}^{\geq\ell}(\Bbbk)$, $\mathrm{GV}^-(\Bbbk)$ and $\mathrm{GV}^{\leq\ell}(\Bbbk)$ the full subcategories of $\mathrm{GV}(\Bbbk)$ whose objects are respectively the bounded, bounded below and bounded above graded spaces. The notation $\mathrm{GV}^*(\Bbbk)$ will be used in statements concerning all of these categories.

2.1.5 Graded Algebras over Fields

A *graded* $\Bbbk$*-algebra* is a graded space $\boldsymbol{A} \in \mathrm{GV}(\Bbbk)$ together with a *multiplication* operation, i.e. a family of $\Bbbk$-bilinear maps $\{\cdot : A^a \times A^b \to A^{a+b}\}_{a,b\in\mathbb{Z}}$, such that the triple $(\boldsymbol{A}, 0, +, \cdot)$ verifies the axioms of a $\Bbbk$-algebra.

The graded algebra is said to be

- *positively graded*: if $A^m = 0$ for all $m < 0$,
- *evenly graded*: if $A^m = 0$, for all odd m,
- *anticommutative*:[1] if $a_1 \cdot a_2 := (-1)^{d_1 d_2} a_2 \cdot a_1$, for all $a_1 \in A^{d_1}, a_2 \in A^{d_2}$.

A *morphism of graded (resp. unital) algebras* $\boldsymbol{\alpha} : (\boldsymbol{A}, 0, +, \cdot) \to (\boldsymbol{B}, 0, +, \cdot)$ is a morphism of graded spaces $\boldsymbol{\alpha} \in \mathrm{Homgr}_{\mathbb{k}}(\boldsymbol{A}, \boldsymbol{B})$ which is compatible with multiplication, i.e. $\boldsymbol{\alpha}(x \cdot y) = \boldsymbol{\alpha}(x) \cdot \boldsymbol{\alpha}(y)$ (resp. $\boldsymbol{\alpha}(\mathbf{1}_{\boldsymbol{A}}) = \mathbf{1}_{\boldsymbol{B}}$).

Examples 2.1.5.1

1. For any $\boldsymbol{V} \in \mathrm{GV}(\mathbb{k})$, the space $(\mathrm{Endgr}^*_{\mathbb{k}}(\boldsymbol{V}), 0, +, \mathrm{id}, \circ)$ of graded endomorphisms (*cf*. Sect. 2.1.3–(2.1)) is a noncommutative graded algebra.
2. The algebra $\Omega_G := S(\mathfrak{g}^\vee)^{\mathfrak{g}}$ of $\mathfrak{g}$-invariant polynomial functions over a real Lie algebra $\mathfrak{g}$ equipped with the grading that doubles the polynomial grading is an evenly positive graded commutative $\mathbb{R}$-algebra.
3. Given a manifold M, the algebras of differential forms $(\Omega(M), d)$ and $(\Omega_c(M), d)$ are positively graded anticommutative $\mathbb{R}$-algebras. Notice that $\Omega_c(M)$ does not have an identity element if M is not compact.

Exercise 2.1.5.2 Let $\overline{\mathrm{GV}}(\mathbb{k})$ be the category of graded spaces where the sets of morphisms are the graded spaces $\mathrm{Mor}_{\overline{\mathrm{GV}}(\mathbb{k})}(\boldsymbol{V}, \boldsymbol{W}) := \mathrm{Homgr}^*_{\mathbb{k}}(\boldsymbol{V}, \boldsymbol{W})$, where the composition rule is the coordinate-wise composition of linear maps.

Denote by $\overline{\mathrm{GV}}^b(\mathbb{k})$ the full subcategory of $\overline{\mathrm{GV}}(\mathbb{k})$ of bounded graded spaces. Given $\boldsymbol{V} := \{V^m\}_{m\in\mathbb{Z}}$, let $\oplus\boldsymbol{V} := \bigoplus_{m\in\mathbb{Z}} V^m$ and $\Pi\boldsymbol{V} := \prod_{m\in\mathbb{Z}} V^m$.

1. Show the correspondence

$$\oplus : \overline{\mathrm{GV}}(\mathbb{k}) \rightsquigarrow \mathrm{Vec}(\mathbb{k}), \quad \begin{cases} \boldsymbol{V} \rightsquigarrow \oplus\boldsymbol{V} \\ (\boldsymbol{\alpha} : \boldsymbol{V} \to \boldsymbol{W}) \rightsquigarrow (\oplus\boldsymbol{\alpha} : \oplus\boldsymbol{V} \to \oplus\boldsymbol{W}) \end{cases}$$

   where $\oplus\boldsymbol{\alpha}(v_m)_{m\in\mathbb{Z}} := \sum_{m\in\mathbb{Z}} \alpha_m(v_m)$, is a faithful functor,[2] and conclude that $\mathrm{GV}(\mathbb{k})$ and $\overline{\mathrm{GV}}(\mathbb{k})$ can be viewed as subcategories of $\mathrm{Vec}(\mathbb{k})$.
2. Same as (1), replacing $\oplus$ by Π.

---

[1] This is the terminology used in the *Colloque de Topologie sur les Espaces Fibrés de Bruxelles (1950)* [32], notably in Cartan's talks [25, 26]. Although the terminology is still widely used, it is being progressively replaced by *graded commutative*, especially in the theory of dg-algebras, see Stacks Project [87].

[2] A functor $F : C_1 \to C_2$ is said to be *faithful* (resp. *full*) if, for all pair of objects $X, Y \in \mathrm{Ob}(C_1)$, the map $F_{X,Y} : \mathrm{Mor}_{C_1}(X, Y) \to \mathrm{Mor}_{C_2}(F(X), F(Y))$ is injective (resp. surjective).

3. Let $\boldsymbol{V}, \boldsymbol{W} \in \mathrm{GM}(\Bbbk)$. Given a $\Bbbk$-linear map $\lambda : \oplus\boldsymbol{V} \to \oplus\boldsymbol{W}$ define, for all $k, d \in \mathbb{Z}$, the map $\lambda_{k,d} \in \mathrm{Hom}_{\Bbbk}(V^k, W^{k+d})$ as the composition

$$V^k \xrightarrow{\iota_k} \oplus_{m\in\mathbb{Z}} V^m \xrightarrow{\lambda} \oplus_{m\in\mathbb{Z}} W^m \xrightarrow{p_{k+d}} W^{k+d}$$
$$\lambda_{k,d}$$

where ι_k denotes the canonical injection of the k-th coordinate space, and p_{k+d} the canonical projection onto the $(k+d)$-th coordinate space.
The family $\lambda_d := \{\lambda_{m,d}\}_{m\in\mathbb{Z}}$ belongs to $\mathrm{Homgr}^d_{\Bbbk}(\boldsymbol{V}, \boldsymbol{W})$, hence the map:

$$\mathrm{Hom}_{\Bbbk}\left(\oplus\boldsymbol{V}, \oplus\boldsymbol{W}\right) \xrightarrow{\Phi_{\boldsymbol{V},\boldsymbol{W}}} \mathrm{Homgr}^*_{\Bbbk}(\boldsymbol{V}, \boldsymbol{W}), \quad \lambda \mapsto \{\lambda_d\}_{d\in\mathbb{Z}}.$$

- Show that $\Phi_{\boldsymbol{V},\boldsymbol{W}}$ is a natural *injective* homomorphism.
- Same question replacing $\oplus$ with Π. (☞, p. 329)

4. Give the conditions of surjectivity for $\Phi_{\boldsymbol{V},\boldsymbol{W}}$. (☞, p. 329)
   a. Show that $\overline{\mathrm{GV}^b}(\Bbbk)$ is equivalent to the category whose objects are the finite direct sums of $\Bbbk$-vector spaces, and whose morphisms are the sets of all $\Bbbk$-linear maps.
   b. Show that $\overline{\mathrm{GV}}(\Bbbk)$ is not equivalent by $\oplus$ (resp. Π) to the category of countable direct sums (resp. products) of $\Bbbk$-vector spaces and $\Bbbk$-linear maps.

2.1.6 *The Category* DGV($\Bbbk$) *of Differential Graded Vector Spaces*

A *differential graded (vector) space* $(\boldsymbol{V}, \boldsymbol{d})$ (also called *complex of vector spaces*), is a graded space $\boldsymbol{V}$ together with a graded endomorphism $\boldsymbol{d} \in \mathrm{Endgr}^1_{\Bbbk}(\boldsymbol{V})$ such that $\boldsymbol{d}^2 = 0$, called *the differential.*

- Differential graded spaces are usually represented as cochain complexes:

$$(\boldsymbol{V}, \boldsymbol{d}) := \left(\cdots \xrightarrow{d_{m-2}} V^{m-1} \xrightarrow{d_{m-1}} V^m \xrightarrow{d_m} V^{m+1} \xrightarrow{d_{m+1}} \cdots\right).$$

- A *morphism of complexes* $\boldsymbol{\alpha} : (\boldsymbol{V}, \mathbf{d}) \to (\boldsymbol{V}', \mathbf{d}')$ is a graded homomorphism of degree 0 such that $\boldsymbol{\alpha} \circ \mathbf{d} = \mathbf{d}' \circ \boldsymbol{\alpha}$.

$$\begin{array}{ccccccccc}
(\boldsymbol{V}, \mathbf{d}) := (\cdots & \xrightarrow{d_{m-2}} & V^{m-1} & \xrightarrow{d_{m-1}} & V^m & \xrightarrow{d_m} & V^{m+1} & \xrightarrow{d_{m+1}} & \cdots) \\
\boldsymbol{\alpha}\downarrow & & \downarrow{\alpha_{m-1}} & & \downarrow{\alpha_m} & & \downarrow{\alpha_{m+1}} & & \\
(\boldsymbol{V}', \mathbf{d}') := (\cdots & \xrightarrow{d'_{m-2}} & V'^{m-1} & \xrightarrow{d'_{m-1}} & V'^m & \xrightarrow{d'_m} & V'^{m+1} & \xrightarrow{d'_{m+1}} & \cdots)
\end{array}$$

- The complexes and their morphisms constitute the *category* $\mathrm{DGV}(\Bbbk)$ *of differential graded vector spaces*. Also denoted as $C(\mathrm{Vec}(\Bbbk))$ and called *the category of complexes of* $\Bbbk$*-vector spaces*.
- The *cohomology* of a complex $(\boldsymbol{V}, \mathbf{d})$ is the graded space denoted by $\boldsymbol{h}(\boldsymbol{V}, \mathbf{d})$, or simply $\boldsymbol{h}(\boldsymbol{V})$, whose i-th term is $h^i(\boldsymbol{V}, \mathbf{d}) := \ker(d_i)/\operatorname{im}(d_{i-1})$.
- A morphism of complexes $\boldsymbol{\alpha} : (\boldsymbol{V}, \mathbf{d}) \to (\boldsymbol{V}', \mathbf{d}')$ induces a morphism between the graded spaces of cohomologies denoted by

$$\boldsymbol{h}(\boldsymbol{\alpha}) : \boldsymbol{h}(\boldsymbol{V}, \mathbf{d}) \to \boldsymbol{h}(\boldsymbol{V}', \mathbf{d}') .$$

  The morphism $\boldsymbol{\alpha}$ is called *a quasi-isomorphism, a quasi-injection or a quasi-surjection*, if $\boldsymbol{h}(\boldsymbol{\alpha})$ is an isomorphism, an injection or a surjection respectively.
- The correspondence which associates $\boldsymbol{V} \rightsquigarrow \boldsymbol{h}(\boldsymbol{V})$ and $\boldsymbol{\alpha} \rightsquigarrow \boldsymbol{h}(\boldsymbol{\alpha})$ constitutes the (covariant) cohomology functor which we denote by

$$\boldsymbol{h} : \mathrm{DGV}(\Bbbk) \rightsquigarrow \mathrm{GV}(\Bbbk) .$$

Exercise 2.1.6.1 Let Ab and Ab$'$ be abelian categories. Show that every additive functor $F : \mathrm{Ab} \to \mathrm{Ab}'$ is exact if and only if Ab is *split*. [3] Apply to $\mathrm{Vec}(\Bbbk)$ and $\mathrm{GV}(\Bbbk)$. Explain the exactness of bifunctors 2.1.8.1–(2.8). (♝, p. 329)

2.1.7 The Shift Functors

Let $s \in \mathbb{Z}$.

S-1. Given $V \in \mathrm{Vec}(\Bbbk)$, we denote by $V[s]$ the graded space with $V[s]^{-s} := V$ and $V[s]^m := 0$ for $s + m \neq 0$. Given a $\Bbbk$-linear map $\alpha : V \to V'$, we denote by $\alpha[s] : V[s] \to V'[s]$ the morphism of graded spaces with $\alpha[s]_{-s} := \alpha$ and $\alpha[s]_m := 0$ for $m \neq -s$. The correspondences $V \rightsquigarrow V[s]$ and $V \rightsquigarrow (V[s], 0)$,

[3] In an abelian category Ab, a morphism $f : X \to Y$ is said to be *split*, if there are isomorphisms $X \sim K \oplus L$ and $L \oplus M \sim Y$ through which f reads $(k, l) \mapsto (l, 0)$. The category Ab is called *split*, if all its morphisms are split.

together with $\alpha \rightsquigarrow \alpha[s]$, define respectively the *shift* functors

$$[s] : \mathrm{Vec}(\Bbbk) \to \mathrm{GV}(\Bbbk) \quad \text{and} \quad [s] : \mathrm{Vec}(\Bbbk) \to \mathrm{DGV}(\Bbbk)\,. \tag{2.2}$$

S-2. Given $\boldsymbol{V} := \{V^m\}_{m\in\mathbb{Z}} \in \mathrm{GV}(\Bbbk)$, we denote by $\boldsymbol{V}[s]$ the graded space with $\boldsymbol{V}[s]^m := V^{s+m}$. Given a morphism of graded spaces $\boldsymbol{\alpha} : \boldsymbol{V} \to \boldsymbol{W}$, we denote by $\boldsymbol{\alpha}[s] : \boldsymbol{V}[s] \to \boldsymbol{W}[s]$ the morphism with $\boldsymbol{\alpha}[s]_m := \alpha_{s+m}$.
The correspondence $\boldsymbol{V} \rightsquigarrow \boldsymbol{V}[s]$ and $\boldsymbol{\alpha} \rightsquigarrow \boldsymbol{\alpha}[s]$ defines the *shift* functor

$$[s] : \mathrm{GV}(\Bbbk) \rightsquigarrow \mathrm{GV}(\Bbbk)\,. \tag{2.3}$$

S-3. Given $(\boldsymbol{V}, \boldsymbol{d}) \in \mathrm{DGV}(\Bbbk)$, we set

$$(\boldsymbol{V}, \boldsymbol{d})[s] := \big(\boldsymbol{V}[s], (-1)^s \boldsymbol{d}[s]\big)\,.$$

The correspondence $(\boldsymbol{V}, \boldsymbol{d}) \rightsquigarrow (\boldsymbol{V}, \boldsymbol{d})[s]$, and $\boldsymbol{\alpha} \rightsquigarrow \boldsymbol{\alpha}[s]$ is the *shift* functor

$$[s] : \mathrm{DGV}(\Bbbk) \rightsquigarrow \mathrm{DGV}(\Bbbk)\,. \tag{2.4}$$

All these functors are covariant additive and exact.[4]

2.1.8 *The Functors* $\mathbf{Hom}^{\bullet}_{\Bbbk}(\_, \_)$ *and* $(\_ \otimes_{\Bbbk} \_)^{\bullet}$

Given two complexes $\boldsymbol{V} := (\boldsymbol{V}, \boldsymbol{d})$ and $\boldsymbol{V}' := (\boldsymbol{V}', \boldsymbol{d}')$ in $\mathrm{DGV}(\Bbbk)$, we recall the definition of the complexes

$$\big(\mathbf{Hom}^{\bullet}_{\Bbbk}(\boldsymbol{V}, \boldsymbol{V}'), \boldsymbol{D}\big) \quad \text{and} \quad \big((\boldsymbol{V} \otimes_{\Bbbk} \boldsymbol{V}')^{\bullet}, \boldsymbol{\Delta}\big)\,. \tag{2.5}$$

As graded spaces they are[5]

$$m \in \mathbb{Z} \mapsto \begin{cases} \mathbf{Hom}^{m}_{\Bbbk}(\boldsymbol{V}, \boldsymbol{V}') := \mathrm{Homgr}_{\Bbbk}(\boldsymbol{V}, \boldsymbol{V}'[m]) = \mathrm{Homgr}^{m}_{\Bbbk}(\boldsymbol{V}, \boldsymbol{V}')\,, \\ (\boldsymbol{V} \otimes_{\Bbbk} \boldsymbol{V}')^m \ := \bigoplus_{b+a=m} V^a \otimes_{\Bbbk} V'^b\,. \end{cases} \tag{2.6}$$

[4] In an abelian category Ab the sets of morphisms $\mathrm{Mor}_{\mathrm{Ab}}(X, Y)$ are abelian groups. A covariant functor between abelian categories $F : \mathrm{Ab} \to \mathrm{Ab}'$ is then said to be *additive* if, for all $X, Y \in \mathrm{Ob}(\mathrm{Ab})$, the map $F_{X,Y} : \mathrm{Mor}_{\mathrm{Ab}}(X, Y) \to \mathrm{Mor}_{\mathrm{Ab}'}(F(X), F(Y))$ is a group homomorphism. Likewise for contravariant functors. The functor is then said to be exact, left exact or right exact, if it transforms every short exact sequence of Ab in respectively an exact, left exact or right exact sequence of Ab′, and this regardless of whether F is covariant or contravariant.

[5] These notations are standard and will be used in several different categories, in the present case, we get two equivalent notations $\mathbf{Hom}^{\bullet}_{\Bbbk}(\_, \_) = \mathrm{Homgr}^{*}_{\Bbbk}(\_, \_)$.

The differentials are[6]

$$\begin{cases} \boldsymbol{D}_m(\boldsymbol{\alpha}) := \boldsymbol{d}' \circ \boldsymbol{\alpha} - (-1)^m \boldsymbol{\alpha} \circ \boldsymbol{d}, \\ \boldsymbol{\Delta}_m(v \otimes v') := \boldsymbol{d}(v) \otimes v' + (-1)^a v \otimes \boldsymbol{d}'(v'), \end{cases} \tag{2.7}$$

where $v \otimes v' \in V^a \otimes V'^b$ and $a + b = m$.

Definition 2.1.8.1 The previous constructions are natural w.r.t. each entry, and define the following two bifunctors:

$$\begin{cases} \mathbf{Hom}^\bullet_{\Bbbk}(\_, \_) : \mathrm{DGV}(\Bbbk) \times \mathrm{DGV}(\Bbbk) \rightsquigarrow \mathrm{DGV}(\Bbbk), \\ (\_ \otimes_{\Bbbk} \_)^\bullet : \mathrm{DGV}(\Bbbk) \times \mathrm{DGV}(\Bbbk) \rightsquigarrow \mathrm{DGV}(\Bbbk). \end{cases} \tag{2.8}$$

The functor 'Hom$^\bullet$' is contravariant and exact in the first entry, and covariant and exact in the second entry. The functor '$\otimes$' is covariant and exact in each entry.

Exercise 2.1.8.2 Show that the natural morphism of GV($\Bbbk$)

$$\Psi : \boldsymbol{V}' \otimes_{\Bbbk} \mathbf{Hom}^\bullet_{\Bbbk}(\boldsymbol{V}, \Bbbk) \to \mathbf{Hom}^\bullet_{\Bbbk}(\boldsymbol{V}, \boldsymbol{V}')$$

defined, for all homogeneous $\boldsymbol{\lambda} \in \mathbf{Hom}^\bullet_{\Bbbk}(\boldsymbol{V}, \Bbbk)$ and $v' \in \boldsymbol{V}'$, by

$$\Psi(v' \otimes \boldsymbol{\lambda}) := \big(v \mapsto \boldsymbol{\lambda}(v)\, v'\big),$$

is a morphism of DGV($\Bbbk$). (p. 331)

Exercise 2.1.8.3 Check that the following complexes coincide as graded spaces but not as complexes even though they are naturally isomorphic. (p. 331)

$$\mathbf{Hom}^\bullet_{\Bbbk}(\boldsymbol{V}[s], \boldsymbol{W}[t]) \simeq \mathbf{Hom}^\bullet_{\Bbbk}(\boldsymbol{V}, \boldsymbol{W})[t - s],$$

$$\boldsymbol{V}[s] \otimes \boldsymbol{W}[t] \simeq (\boldsymbol{V} \otimes \boldsymbol{W})[s + t].$$

2.1.9 On the Koszul Sign Rule

This is the name given to the mnemonic determining signs when manipulating differential graded objects,[7] as we did in the previous paragraphs. Essentially, it says that when transposing two consecutive homogeneous objects x and y respectively of degrees $[x]$ and $[y]$, the sign $(-1)^{[x][y]}$ must be added, e.g. $xy = (-1)^{[x][y]} yx$.

[6] See Stacks Project [88] §15.67 Hom complexes, and [87] §12. Tensor product, p. 15.

[7] See Stacks Project [86] §68 Sign rules, p. 167, for a commented full list of these rules.

In the case of the differential $\boldsymbol{D}$ in (2.7), if $\boldsymbol{\alpha} \in \mathbf{Hom}^\bullet_{\Bbbk}(\boldsymbol{V}, \boldsymbol{V'})$ and $v \in \boldsymbol{V}$ are homogeneous, and if we write $\boldsymbol{\alpha}(v) \leftrightarrow \boldsymbol{\alpha}\, v$, we see that the rule gives

$$\boldsymbol{d}'(\boldsymbol{\alpha}\, v) = \boldsymbol{D\alpha}\, v + (-1)^{[\boldsymbol{\alpha}][\boldsymbol{D}]}\boldsymbol{\alpha}\, \boldsymbol{d}v\,,$$

where the sign $(-1)^{[\boldsymbol{\alpha}][\boldsymbol{d}']}$ corresponds to the transposition $\boldsymbol{D\alpha} \leftrightarrow \boldsymbol{\alpha d}$.

2.1.10 *The Functor* $\mathbf{Hom}^\bullet_{\Bbbk}(\_, \boldsymbol{W})$

Let $(\boldsymbol{W}, \boldsymbol{d})$ be a complex.

Given a morphism of complexes $\boldsymbol{\alpha} : (\boldsymbol{V'}, \boldsymbol{d}) \to (\boldsymbol{V}, \boldsymbol{d})$, the map

$$\boldsymbol{\alpha}^* : \mathbf{Hom}^m_{\Bbbk}(\boldsymbol{V}, \boldsymbol{W}) \to \mathbf{Hom}^m_{\Bbbk}(\boldsymbol{V'}, \boldsymbol{W})\,, \quad \boldsymbol{\beta} \mapsto \boldsymbol{\beta} \circ \boldsymbol{\alpha}\,,$$

is well-defined for all $m \in \mathbb{Z}$ and commutes with differentials:

$$\begin{aligned} \boldsymbol{D}(\boldsymbol{\alpha}^*(\boldsymbol{\beta})) &= \boldsymbol{d} \circ (\boldsymbol{\beta} \circ \boldsymbol{\alpha}) - (-1)^{[\boldsymbol{\beta}\circ\boldsymbol{\alpha}]}(\boldsymbol{\beta} \circ \boldsymbol{\alpha}) \circ \boldsymbol{d} \\ &= (\boldsymbol{d} \circ \boldsymbol{\beta}) \circ \boldsymbol{\alpha} - (-1)^{[\boldsymbol{\beta}]}(\boldsymbol{\beta} \circ \boldsymbol{d}) \circ \boldsymbol{\alpha} = \boldsymbol{\alpha}^*(\boldsymbol{D}(\boldsymbol{\beta}))\,. \end{aligned}$$

The correspondence

$$\mathbf{Hom}^\bullet_{\Bbbk}(\alpha, \boldsymbol{W}) : \big(\mathbf{Hom}^\bullet_{\Bbbk}(\boldsymbol{V}, \boldsymbol{W}), \boldsymbol{D}\big) \to \big(\mathbf{Hom}^\bullet_{\Bbbk}(\boldsymbol{V'}, \boldsymbol{W}), \boldsymbol{D}\big)\,,$$

which associates $(\boldsymbol{V}, \boldsymbol{d}) \rightsquigarrow \mathbf{Hom}^\bullet_{\Bbbk}(\boldsymbol{V}, \boldsymbol{W})$ and $\boldsymbol{\alpha} \rightsquigarrow \mathbf{Hom}^\bullet_{\Bbbk}(\alpha, \boldsymbol{W}) := \boldsymbol{\alpha}^*$, is then easily seen to be an additive functor:

$$\mathbf{Hom}^\bullet_{\Bbbk}(\_, \boldsymbol{W}) : \mathrm{DGV}(\Bbbk) \rightsquigarrow \mathrm{DGV}(\Bbbk)\,, \tag{2.9}$$

which is contravariant and exact.

2.1.11 *The Duality Functor*

This is the functor

$$(\_)^\vee := \mathbf{Hom}^\bullet_{\Bbbk}(\_, \Bbbk[0]) : \mathrm{DGV}(\Bbbk) \rightsquigarrow \mathrm{DGV}(\Bbbk)\,.$$

The *dual complex associated with a complex* $(\boldsymbol{V}, \boldsymbol{d})$ is then defined as

$$(\boldsymbol{V}, \boldsymbol{d})^{\vee} := (V^{\vee}, \boldsymbol{D}), \tag{2.10}$$

where $(\boldsymbol{V}^{\vee})^m := (\boldsymbol{V}^{-m})^{\vee}$ and $D_m = (-1)^{m+1}\, {}^t d_{-(m+1)}$.[8]

Remark 2.1.11.1 Take care that the natural embedding of graded vector spaces

$$\phi : \boldsymbol{V} \to \boldsymbol{V}^{\vee\vee}, \quad \phi(v)(\boldsymbol{\lambda}) := \boldsymbol{\lambda}(v),$$

gives only an embedding of complexes $(\boldsymbol{V}, -\boldsymbol{d}) \subseteq (\boldsymbol{V}, \boldsymbol{d})^{\vee\vee}$, where the sign of the differential has changed !

Indeed, applying the definition (2.7), we have

$$\begin{aligned} \boldsymbol{D}\big(\phi(v)\big)(\boldsymbol{\lambda}) = \; -(-1)^{[v]}\phi(v)(\boldsymbol{D}(\boldsymbol{\lambda})) &= (-1)^{1+[v]}(-1)^{1+[\boldsymbol{\lambda}]}\phi(v)(\boldsymbol{\lambda}\circ \boldsymbol{d}) \\ &= (-1)^{[v]+[\boldsymbol{\lambda}]}\boldsymbol{\lambda}(\boldsymbol{d}v) = \phi(-\boldsymbol{d}v)(\boldsymbol{\lambda}), \end{aligned}$$

since $[\boldsymbol{\lambda}] = [v] + 1$ is the only nonzero case to check.

The canonical isomorphism

$$\epsilon : (\boldsymbol{V}, \boldsymbol{d}) \to (\boldsymbol{V}, -\boldsymbol{d}), \quad \epsilon_m = (-1)^m \mathrm{id}_{V^m} \tag{2.11}$$

is then necessary to obtain an embedding $\phi \circ \boldsymbol{\epsilon} : (\boldsymbol{V}, \boldsymbol{d}) \hookrightarrow (\boldsymbol{V}, \boldsymbol{d})^{\vee\vee}$ which does not alter differentials.

Proposition 2.1.11.2

1. *There exists a canonical isomorphism between the cohomology of the dual and the dual of the cohomology, i.e.*

$$\boldsymbol{h}\big((\boldsymbol{V}, \boldsymbol{d})^{\vee}\big) \xrightarrow[\simeq]{} \big(\boldsymbol{h}(\boldsymbol{V}, \boldsymbol{d})\big)^{\vee}.$$

2. *A morphism of complexes* $\boldsymbol{\alpha} : (\boldsymbol{V}, \boldsymbol{d}) \to (\boldsymbol{V}', \boldsymbol{d}')$ *is a quasi-isomorphism if and only if so is* $\boldsymbol{\alpha}^{\vee}$.

Proof (1) Immediate since the duality functor is additive and that Vec($\mathbb{k}$) is a split category (*cf*. Exercise 2.1.6.1). (2) Results from (1) which entails the equivalences $\big(\boldsymbol{\alpha}$ is q-iso$\big) \Leftrightarrow \big(\boldsymbol{h}(\boldsymbol{\alpha})$ is iso$\big) \Leftrightarrow \big(\boldsymbol{h}(\boldsymbol{\alpha})^{\vee}$ is iso$\big) \Leftrightarrow \big(\boldsymbol{\alpha}^{\vee}$ is q-iso$\big)$. □

[8] Here, we denote by ${}^t(d_{-(m+1)})$ the adjoint map to $d_{-m-1} : V^{-m-1} \to V^{-m}$, hence the map $D_m := {}^t d_{-(m+1)} : (V^{\vee})^m \to (V^{\vee})^{m+1}$.

2.2 Categories of Manifolds

2.2.1 *Manifolds*

The names *manifold*, *map of manifolds* and *function* will be shortcuts respectively for *real Hausdorff paracompact differentiable manifold*,[9] for *differentiable map of class* C^∞, in short *smooth maps*, and for maps with values in $\mathbb{R}$. A manifold is said to be *equidimensional* if its connected components have the same dimension, in which case 'd_M' denotes this common dimension. Expressions such as 'M *is of dimension* d', presupposes that M is equidimensional.

2.2.2 *The Category of Manifolds*

By Man (resp. $\mathrm{Man}^{\mathrm{or}}$) we denote the *category of manifolds (resp. oriented) and (smooth) maps*. Over Man one has the *de Rham* contravariant functor

$$\Omega(\_) : \mathrm{Man} \rightsquigarrow \mathrm{DGV}(\mathbb{R}) , \quad M \rightsquigarrow \Omega(M, d_M) \tag{2.12}$$

and the *de Rham cohomology* contravariant functor

$$H(\_) := \mathbf{h}(\Omega(\_)) : \mathrm{Man} \rightsquigarrow \mathrm{Mod}^{\mathbb{N}}(\mathbb{R}) , \quad M \rightsquigarrow H(M) . \tag{2.13}$$

2.2.3 *The Category of Manifolds and Proper Maps*

By $\mathrm{Man}_{\mathrm{pr}}$ (resp. $\mathrm{Man}^{\mathrm{or}}_{\mathrm{pr}}$) we denote the subcategory of Man (resp. $\mathrm{Man}^{\mathrm{or}}$) of all manifolds and all **proper** maps.[10] (, p. 332).

Over $\mathrm{Man}_{\mathrm{pr}}$ one has, in addition to the functors (2.12) and (2.13), the *compactly supported de Rham* contravariant functor:

$$\Omega_{\mathrm{c}}(\_) : \mathrm{Man}_{\mathrm{pr}} \rightsquigarrow \mathrm{DGV}(\mathbb{R}) , \quad M \rightsquigarrow \Omega_{\mathrm{c}}(M, d_M) \tag{2.14}$$

---

[9]A connected manifold M which is paracompact, is automatically countable at infinity, second countable, and perfectly normal, which means that two disjoint closed subspaces F_1 and F_2 admit disjoint neighborhoods, $V_i \supseteq F_i$, and that there exist continuous functions $f : M \to \mathbb{R}$ verifying $f|_{V_1} = 1$ and $f|_{V_2} = 0$.

[10]A continuous map between locally compact spaces $f : M \to N$ is said to be *proper* if $f^{-1}(F)$ is compact for every compact subspace $F \subseteq N$, or, equivalently, if f is closed with compact fibers (exercise !) (See also 3.1.9.3–(3).)

and the *compactly supported de Rham cohomology* contravariant functor

$$H_c(\_) : \mathrm{Man}_{\mathrm{pr}} \rightsquigarrow \mathrm{Mod}^{\mathbb{N}}(\mathbb{R}) , \quad M \rightsquigarrow H_c(M, d_M) . \tag{2.15}$$

The inclusion $\Omega_c(\_) \subseteq \Omega(\_)$ induces a morphism of functors $H_c(\_) \to H(\_)$.

2.2.4 *G*-Manifolds

Let G denote a Lie group. A manifold endowed with a smooth (left) action of G is called *a G-manifold*. A map $f : M \to N$ between G-manifolds is said to be *G-equivariant* if it is compatible with the action of G, i.e. if $f(g(m)) = g(f(m))$, for all $g \in G$ and $m \in M$. The G-manifolds and the G-equivariant maps constitute *the category of G-manifolds* which we denote by G-Man. The categories $G\text{-}\mathrm{Man}^{\mathrm{or}}$, $G\text{-}\mathrm{Man}_{\mathrm{pr}}$ and $G\text{-}\mathrm{Man}^{\mathrm{or}}_{\mathrm{pr}}$ are the equivariant versions of the categories $\mathrm{Man}^{\mathrm{or}}$, $\mathrm{Man}_{\mathrm{pr}}$ and $\mathrm{Man}^{\mathrm{or}}_{\mathrm{pr}}$.

2.3 Orientation and Integration[11]

2.3.1 Orientability

A manifold M of dimension d_M is said to be *orientable* if it admits an atlas $\mathcal{A}$ all of whose transition maps T locally preserve the canonical orientation of $\mathbb{R}^{d_M}$, in other terms, are such that the *Jacobians* $J(T)$ are positive functions. Such an atlas $\mathcal{A}$ is called an *oriented atlas of M*.

Proposition 2.3.1.1

1. *A manifold M of dimension d_M is orientable if and only if there exists a nowhere vanishing differential form of highest degree $\omega \in \Omega^{d_M}(M)$.*
2. *A simply connected manifold is orientable.*

Proof

(1) If $\mathcal{A} = \{\varphi_\mathfrak{a} : U_\mathfrak{a} \to \mathbb{R}^{d_M}\}_{\mathfrak{a}\in\mathfrak{A}}$ is an oriented atlas of M, then, for any partition of unity $\Phi := \{\phi_\mathfrak{a} : U_\mathfrak{a} \to \mathbb{R}_{\geq 0}\}_{\mathfrak{a}\in\mathfrak{A}}$, the d_M-form $\omega(\mathcal{A}, \Phi)$

$$\omega(\mathcal{A}, \Phi) := \sum\nolimits_{\mathfrak{a}} \phi_\mathfrak{a}\, \varphi_\mathfrak{a}^*(dx_1 \wedge \cdots \wedge x_{d_M}) , \tag{2.16}$$

[11] See also Bott-Tu [18] §3 Orientation and Integration, p. 27.

is nowhere vanishing. Indeed, the pushforward of $\omega(\mathcal{A}, \Phi)$ through the chart $(\varphi_{\mathfrak{a}'} : U_{\mathfrak{a}'} \to \mathbb{R}^{d_M})$ is the differential form

$$\varphi_{\mathfrak{a}'*}(\omega(\mathcal{A}, \Phi)) = \sum\nolimits_{\mathfrak{a}} T^*_{\mathfrak{a},\mathfrak{a}'}(\phi_{\mathfrak{a}}\, dx_1 \wedge \cdots \wedge x_{d_M}) \tag{2.17}$$

where $T_{\mathfrak{a},\mathfrak{a}'} : \varphi_{\mathfrak{a}'}(U_{\mathfrak{a}'\mathfrak{a}}) \to \varphi_{\mathfrak{a}}(U_{\mathfrak{a}\mathfrak{a}'})$ is the transition map. But then

$$T^*_{\mathfrak{a},\mathfrak{a}'}(dx_1 \wedge \cdots \wedge x_{d_M}) = g_{\mathfrak{a},\mathfrak{a}'}\, dx_1 \wedge \cdots \wedge x_{d_M}$$

with $g_{\mathfrak{a},\mathfrak{a}'}$ a strictly positive function on $\varphi_{\mathfrak{a}'}(U_{\mathfrak{a}'\mathfrak{a}}) \subseteq \mathbb{R}^{d_M}$ and the sum (2.17) is nowhere vanishing on $U_{\mathfrak{a}'}$.

Conversely, for any connected open subset $U \subseteq M$ and any diffeomorphism $\varphi : U \to \varphi(U) \subseteq \mathbb{R}^{d_M}$ there exists a uniquely defined function $f : \varphi(U) \to \mathbb{R}$ such that $\varphi_*(\omega) = f_\omega\, dx_1 \wedge \cdots \wedge dx_{d_M}$. When ω is nowhere vanishing, we have either $f_\omega > 0$ or $f_\omega < 0$. We then say that φ is *ω-oriented* if $f_\omega > 0$. It is clear that if $\varphi := (\varphi_1, \varphi_2, \ldots, \varphi_{d_M})$ is not ω-oriented, then, by switching the first two coordinates, the diffeomorphism $\tilde{\varphi} := (\varphi_2, \varphi_1, \ldots, \varphi_{d_M})$ is ω-oriented. As a consequence, the charts $(\varphi_{\mathfrak{a}}, U_{\mathfrak{a}})$ in any atlas $\mathcal{A}$ of M can be easily modified to make all the charts ω-oriented, in which case the atlas becomes oriented.

(2) Let $TM \twoheadrightarrow M$ denote the tangent vector bundle of M. Statement (1) says that M is orientable if and only if the *determinant bundle* $\det TM := \Lambda^{d_M} TM$ admits a nowhere vanishing section. But since $\det TM$ is a line bundle, the bundle $\det TM \setminus \{0\}$ retracts to an $\mathbb{S}^0$-bundle above M, which is trivial if M is simply connected, after a well-known result in the theory of coverings. □

2.3.1.1 Orientations

Two atlases $\mathcal{A}_1, \mathcal{A}_2$ are said *to define the same orientation*, and we write $\mathcal{A}_1 \sim \mathcal{A}_2$, if the atlas $\mathcal{A}_1 \uplus \mathcal{A}_2$, defined by the collection of all charts of $\mathcal{A}_1$ and $\mathcal{A}_2$, is also an oriented atlas. The relation '$\sim$' is an equivalence relation on the set of all oriented atlases.

Let M be connected. If $\omega_1, \omega_2 \in \Omega^{d_M}(M)$ are nowhere vanishing, we have $\omega_1 = f_{12}\,\omega_2$ for a unique nowhere vanishing function $f_{12} : M \to \mathbb{R}$, which will therefore satisfy either $f_{12} > 0$ or $f_{12} < 0$. We say that ω_1 and ω_2 *define the same orientation*, and we write $\omega_1 \sim \omega_2$, if $f_{12} > 0$. The relation '$\sim$' is an equivalence relation on the set of nowhere vanishing differential forms of highest degree. It is thus clear that, in the proof of proposition 2.3.1.1, the correspondence $\mathcal{A} \rightsquigarrow \omega(\mathcal{A}, \Phi)$ verifies $\mathcal{A}_1 \sim \mathcal{A}_2$ if and only if $\omega(\mathcal{A}_1, \Phi_1) \sim \omega(\mathcal{A}_2, \Phi_2)$. As a consequence, if M is connected and orientable, there are exactly two classes of oriented atlases for M.

To *choose an orientation of M* then means to choose one these classes, or, equivalently, to choose a nowhere vanishing d_M-form on M. We denote such a choice by $[M]$ and we say that $(M, [M])$ *is oriented by* $[M]$.

When M is not connected, it is orientable if each of its connected components is orientable. In that case, *to choose an orientation for M* means to choose an orientation for each of its connected components.

2.3.2 Integration

Let $(M, [M])$ be an oriented manifold of dimension d_M.

Let $\mathcal{A} = \{\varphi_{\mathfrak{a}} : U_{\mathfrak{a}} \to \mathbb{R}^{d_M}\}_{\mathfrak{a}\in\mathfrak{A}}$ be an oriented atlas corresponding to the orientation $[M]$ and let $\Phi := \{\phi_{\mathfrak{a}} : U_{\mathfrak{a}} \to \mathbb{R}_{\geq 0}\}_{\mathfrak{a}\in\mathfrak{A}}$ a partition of unity.

Lemma 2.3.2.1 *Given $\beta \in \Omega_{\mathrm{c}}^{d_M}(M)$, there is a unique $f \in \Omega_{\mathrm{c}}^{0}(M)$ such that $\beta = f\,\omega(\mathcal{A}, \Phi)$* (cf. *(2.16)). Then, the sum*

$$\int_{[M]} \beta := \sum_{\mathfrak{a}} \int_{\mathbb{R}^{d_M}} (\varphi_{\mathfrak{a}}^{-1})^*(\phi_{\mathfrak{a}}\,\beta) = \sum_{\mathfrak{a}} \int_{\mathbb{R}^{d_M}} (f_{\mathfrak{a}} \circ \varphi_{\mathfrak{a}}^{-1})\, dx_1 \wedge \cdots \wedge x_{d_M}\,, \tag{2.18}$$

where $f_{\mathfrak{a}} := \phi_{\mathfrak{a}} f$ has compact support in the chart $(\varphi_{\mathfrak{a}} : U_{\mathfrak{a}} \to \mathbb{R}^n) \in \mathcal{A}$, is a finite sum independent of the choice of $\mathcal{A}$ and Φ.

In particular, if $|\beta| \subseteq U_{\mathfrak{a}}$, then

$$\int_{[M]} \beta = \int_{\mathbb{R}^{d_M}} (\varphi_{\mathfrak{a}}^{-1})^*(\beta)\,.$$

Proof We need only consider the case where $\mathcal{A}$ is the *full* atlas $\mathcal{A} := \{\varphi_{\mathfrak{a}} : U_{\mathfrak{a}} \to \varphi_{\mathfrak{a}}(U_{\mathfrak{a}}) \subseteq \mathbb{R}^{d_M}\}_{\mathfrak{a}\in\mathfrak{A}}$ corresponding to $[M]$.[12]

We begin proving that if $|\beta| \subseteq U_{\mathfrak{a}_0}$ for some $\mathfrak{a}_0 \in \mathfrak{A}$, then

$$\int_{\mathbb{R}^{d_M}} (\varphi_{\mathfrak{a}_0}^{-1})^*(\beta) = \sum_{\mathfrak{a}} \int_{\mathbb{R}^{d_M}} (\varphi_{\mathfrak{a}}^{-1})^*(\phi_{\mathfrak{a}}\,\beta)\,, \tag{2.19}$$

for any partition of unity $\Phi := \{\phi_{\mathfrak{a}} : U_{\mathfrak{a}} \to \mathbb{R}_{\geq 0}\}$. Indeed, in that case, for all $\mathfrak{a} \in \mathfrak{A}$, we have $|\phi_{\mathfrak{a}}\,\beta| \subseteq U_{\mathfrak{a}_0\mathfrak{a}}$, in which case

$$\int_{\mathbb{R}^{d_M}} (\varphi_{\mathfrak{a}_0}^{-1})^*(\phi_{\mathfrak{a}}\,\beta) = \int_{\mathbb{R}^{d_M}} (\varphi_{\mathfrak{a}}^{-1})^*(\phi_{\mathfrak{a}}\,\beta)$$

by change of variables since the transition map map $T_{\mathfrak{a}\mathfrak{a}_0}$ preserves the orientation. The right-hand term in (2.19) can then be rewritten as:

$$\begin{aligned}\sum_{\mathfrak{a}} \int_{\mathbb{R}^{d_M}} (\varphi_{\mathfrak{a}}^{-1})^*(\phi_{\mathfrak{a}}\,\beta) &= \sum_{\mathfrak{a}} \int_{\mathbb{R}^{d_M}} (\varphi_{\mathfrak{a}_0}^{-1})^*(\phi_{\mathfrak{a}}\,\beta) \\ &= \int_{\mathbb{R}^{d_M}} (\varphi_{\mathfrak{a}_0}^{-1})^*\big(\textstyle\sum_{\mathfrak{a}} \phi_{\mathfrak{a}}\,\beta\big) = \int_{\mathbb{R}^{d_M}} (\varphi_{\mathfrak{a}_0}^{-1})^*(\beta)\end{aligned}$$

proving the claim for $|\beta| \subseteq U_{\mathfrak{a}_0}$.

[12]See also Bott-Tu [18] Proposition 3.3, p. 30.

We can now approach the central question of the lemma. Let $\beta \in \Omega_c(M)$ be arbitrary. If $\Phi' := \{\phi'_\mathfrak{a} : U_\mathfrak{a} \to \mathbb{R}_{\geq 0}\}_{\mathfrak{a}\in\mathfrak{A}}$ is a second partition of unity, then

$$\sum_{\mathfrak{a}} \int_{\mathbb{R}^{d_M}} (\varphi_\mathfrak{a}^{-1})^*(\phi_\mathfrak{a}\,\beta) = \sum_{\mathfrak{a},\mathfrak{a}'} \int_{\mathbb{R}^{d_M}} (\varphi_\mathfrak{a}^{-1})^*(\phi_\mathfrak{a}\,\phi'_{\mathfrak{a}'}\,\beta)$$
$$=_1 \sum_{\mathfrak{a},\mathfrak{a}'} \int_{\mathbb{R}^{d_M}} (\varphi_{\mathfrak{a}'}^{-1})^*(\phi_\mathfrak{a}\,\phi'_{\mathfrak{a}'}\,\beta) = \sum_{\mathfrak{a}'} \int_{\mathbb{R}^{d_M}} (\varphi_{\mathfrak{a}'}^{-1})^*(\phi'_{\mathfrak{a}'}\,\beta)$$

where $(=_1)$ is justified by the fact that $|\phi_\mathfrak{a}\,\phi'_{\mathfrak{a}'}\,\beta| \subseteq U_{\mathfrak{a}\mathfrak{a}'}$. □

The sum (2.18) is called the *integral of* β *relative to* $[M]$, and will be simply denoted by $\int_M \beta$ when the orientation $[M]$ is understood.

Proposition 2.3.2.2 *Let M be an oriented manifold of dimension d_M. Extend the integration map $\int_M : \Omega_c^{d_M}(M) \to \mathbb{R}$ by linearity to the whole complex $\Omega_c(M)$ by setting $\int_M \beta := 0$ for all $\beta \in \Omega_c^i(M)$ and all $i < d_M$. The resulting map*

$$\int_M : \Omega_c(M)[d_M] \to \mathbb{R}[0]\,, \tag{2.20}$$

is then a morphism of complexes, which is invariant under the action of orientation preserving diffeomorphisms of M.

Proof Let $n := d_M$. We need only prove that

$$\int_M d\,\Omega_c^{n-1}(M) = 0\,.$$

Let $\mathcal{A} = \{\varphi_\mathfrak{a} : U_\mathfrak{a} \to \mathbb{R}^n\}$ be an oriented atlas of M corresponding to the orientation of M, and let $\Phi := \{\phi_\mathfrak{a} : U_\mathfrak{a} \to \mathbb{R}_{\geq 0}\}$ be a partition of unity.

For all $\beta \in \Omega_c^{n-1}(M)$, we have $\beta = \sum_\mathfrak{a} \phi_\mathfrak{a}\beta$, in which case

$$d\beta = \sum\nolimits_\mathfrak{a} d(\phi_\mathfrak{a}\beta)\,,$$

where the support of $d(\phi_\mathfrak{a}\beta)$ is compact and contained in $U_\mathfrak{a}$.

By the definition of the integration operation and Lemma 2.3.2.1, we then have

$$\int_M d\beta = \sum_\mathfrak{a} \int_{\mathbb{R}^n} d\beta_\mathfrak{a}\,,$$

with $\beta_\mathfrak{a} := (\varphi_\mathfrak{a}^{-1})^*(\phi_\mathfrak{a}\beta) \in \Omega_c^{n-1}(\mathbb{R}^n)$. Hence, to prove that $\int_M d\beta = 0$, we can restrict ourselves to $M := \mathbb{R}^n$. There, we have a system of global coordinates $\bar{x} := \{x_1, \ldots, x_n\}$ and $\beta \in \Omega_c^{n-1}(\mathbb{R}^n)$ is a linear combination over $\Omega_c^0(M)$ of the $(n-1)$-forms $dx_{i_1} \wedge \cdots \wedge dx_{i_{n-1}}$ so that to verify the equality $\int_{\mathbb{R}^n} d\beta = 0$, we need only check it for $\beta := f(\bar{x})\,dx_2 \wedge \cdots \wedge dx_n$ with $f \in \Omega^0_{\mathrm{c}}(M)$, in which case $d\beta = (\partial_{x_1} f)\,(\bar{x})\,dx_1 \wedge dx_2 \wedge \cdots \wedge dx_n$. But then,

$$\int_{\mathbb{R}^n} d\beta = \int_{\mathbb{R}} \cdots \int_{\mathbb{R}} \Big(\int_{\mathbb{R}} (\partial_{x_1} f)(x_1, x_2, \ldots, x_n)\,dx_1 \Big) dx_2 \ldots dx_n = 0\,,$$

by Fubini's theorem, and because, for all fixed $x_2, \ldots, x_n \in \mathbb{R}$,

$$\int_{\mathbb{R}} (\partial_{x_1} f)(x_1, x_2, \ldots, x_n)\,dx_1 = \lim_{t \mapsto +\infty} \big(f(t, x_2, \ldots, x_n) - f(-t, x_2, \ldots, x_n) \big) = 0$$

since f has compact support. □

2.4 Poincaré Duality

2.4.1 Poincaré Pairing

Let M be an oriented manifold of dimension d_M not necessarily connected. The composition of the exterior product

$$\wedge : \Omega^{d_M - i}(M) \times \Omega^i_{\mathrm{c}}(M) \to \Omega^{d_M}_{\mathrm{c}}(M)\,, \quad (\alpha, \beta) \mapsto \alpha \wedge \beta\,, \tag{2.21}$$

with integration $\int_M : \Omega^{d_M}_{\mathrm{c}}(M) \to \mathbb{R}$, gives rise to a pairing (2.1.2)

$$\langle \cdot, \cdot \rangle_M : \Omega^{d_M - i}(M) \times \Omega^i_{\mathrm{c}}(M) \to \mathbb{R}\,, \quad (\alpha, \beta) \mapsto \int_M \alpha \wedge \beta\,, \tag{2.22}$$

inducing the *Poincaré pairing in cohomology*

$$\langle \cdot, \cdot \rangle_M : H^{d_M - i}(M) \times H^i_{\mathrm{c}}(M) \to \mathbb{R}\,, \quad ([\alpha], [\beta]) \mapsto \int_M [\alpha] \cup [\beta]\,. \tag{2.23}$$

Proposition 2.4.1.1 (and Definitions) *The Poincaré pairing (2.22) is nondegenerate and the map*

$$\boxed{\mathbb{D}_M : \Omega(M)[d_M] \to \Omega_{\mathrm{c}}(M)^\vee} \tag{2.24}$$

defined by

$$\mathbb{D}_M(\alpha) := \Big(\beta \mapsto \int_M \alpha \wedge \beta \Big)\,, \tag{2.25}$$

is an injective morphism of complexes. It will be called the left Poincaré adjunction associated *with the pairing (2.22). It induces in cohomology the map*

$$\boxed{D_M : H(M)[d_M] \to H_c(M)^\vee} \tag{2.26}$$

which will be called the left Poincaré adjunction in cohomology

Proof Following 2.1.7–(S-3) and 2.1.11, we have

$$\begin{aligned} \mathbb{D}_M\big((-1)^{d_M} d\,\alpha\big)(\beta) &= \int_M (-1)^{d_M} d\,\alpha \wedge \beta \\ &= \int_M (-1)^{d_M} d(\alpha \wedge \beta) + (-1)^{d_M+[\alpha]+1} \int_M \alpha \wedge d\,\beta \\ &= (-1)^{[\beta]}\; \mathbb{D}_M(\alpha)\big(d\,\beta\big) \\ &= (-1)^{[\beta]}(-1)^{1+d_M+[\alpha]}(\boldsymbol{D}\mathbb{D}_M(\alpha))(\beta) \\ &= (\boldsymbol{D}\mathbb{D}_M(\alpha))(\beta)\,, \end{aligned} \tag{2.27}$$

where $\int_M d(\alpha \wedge \beta) = 0$, after Proposition 2.3.2.2, which proves that the left Poincaré adjunction (2.24) is a morphism of complexes.

Injectivity of $\mathbb{D}_M$ Let $\alpha \in \Omega^i(M)$ and assume $\alpha(x) \neq 0$ for some $x \in M$. We can then choose a chart $M \supseteq U_\mathfrak{a} \xrightarrow[\sim]{\varphi_\mathfrak{a}} W_\mathfrak{a} \subseteq \mathbb{R}^m$ $(m := d_M)$ such that

$$\alpha|_{U_\mathfrak{a}} = \varphi_\mathfrak{a}^*\big(f_\mathfrak{a}\, dx_1 \wedge \cdots \wedge dx_i + \cdots\big)\,,$$

for some $f_\mathfrak{a} \in \Omega^0(W_\mathfrak{a})$ such that $f_\mathfrak{a} > 0$.[13] In that case, for any $g_\mathfrak{a} \in \Omega_c(W_\mathfrak{a})$, we can consider the $(m-i)$-differential form

$$\beta := \varphi_\mathfrak{a}^*\big(g_\mathfrak{a}\; dx_{i+1} \wedge \cdots \wedge dx_m\big) \in \Omega_c^{m-i}(M)\,,$$

so that

$$\alpha \wedge \beta = \varphi_\mathfrak{a}^*\big(f_\mathfrak{a}\, g_\mathfrak{a}\; dx_1 \wedge \cdots \wedge dx_m\big) \in \Omega_c^m(M)\,.$$

In this way, if in addition $g_\mathfrak{a} \geq 0$ and $g_\mathfrak{a} \neq 0$, then

$$\big\langle \alpha, \beta \big\rangle_M := \int_{\mathbb{R}^m} f_\mathfrak{a}\, g_\mathfrak{a}\; dx_1 \wedge \cdots \wedge dx_m > 0\,.$$

[13] In fact, it can easily be seen that one can always assume $f_\mathfrak{a} = 1$.

The same argument obviously works for α with compact support. The Poincaré pairing (2.22) is therefore nondegenerate and $I\!D_M$ is injective. □

Exercise 2.4.1.2 Show that the Poincaré pairing (2.22) is perfect if and only if M is a finite set. Show that if $d_M > 0$ then none of the adjunctions induced by the pairing $\langle \cdot, \cdot \rangle_M$ is bijective. (🕯, p. 332)

Theorem 2.4.1.3 (Poincaré Duality) *Let M be an oriented manifold of dimension d_M not necessarily connected. We assume neither* $\dim_{\mathbb{R}} H(M) < +\infty$ *nor* $\dim_{\mathbb{R}} H_c(M) < +\infty$.

1. *The left Poincaré adjunction $I\!D_M : \Omega(M)[d_M] \hookrightarrow \Omega_c(M)^\vee$ is an injective quasi-isomorphism. It induces the* Poincaré duality isomorphism
$$D_M : H(M)[d_M] \underset{\simeq}{\longrightarrow} H_c(M)^\vee . \tag{2.28}$$
2. *The Poincaré pairing in cohomology*
$$\langle \cdot, \cdot \rangle_M : H(M) \times H_c(M) \to \mathbb{R} \tag{2.29}$$
is nondegenerate, *and is* perfect *(2.1.2) if and only if* $\dim(H(M)) < +\infty$.

Proof (1) Let $\mathcal{O}(M)$ denote the category of open subspaces of M where morphisms are inclusion maps.[14] Consider the functors

- $\Omega(\_) : \mathcal{O}(M) \rightsquigarrow \mathrm{DGV}(\mathbb{R})$, the de Rham complex;
- $\Omega_c(\_) : \mathcal{O}(M) \rightsquigarrow \mathrm{DGV}(\mathbb{R})$, the compactly supported de Rham complex;
- $I\!D(\_) : \Omega(\_)[d_M] \hookrightarrow \Omega_c(\_)^\vee$, the left Poincaré adjunction;
- $D(\_) : H(\_)[d_M] \hookrightarrow H_c(\_)^\vee$, the left Poincaré adjunction in cohomology.

Denote by

- $\mathcal{O}_{\mathrm{PD}}(M) \subseteq \mathcal{O}(M)$, subcategory on which $D(\_)$ is an isomorphism.

Lemma 1 *If $\mathcal{I} := \{U_i\}_{i \in \mathcal{I}}$ is an increasing family of open subspaces in $\mathcal{O}_{\mathrm{PD}}$, its union $\bigcup \mathcal{I} = \bigcup_i U_i$ also belongs to $\mathcal{O}_{PD}$.*

[14]The proof is close to that given in Bott-Tu [18] §5 The Mayer-Vietoris and Poincaré Duality on an Orientable Manifold, pp. 42–.

Proof of Lemma 1 Consider the following diagram where $U_j \subseteq U_i \in \mathcal{J}$

$$\begin{array}{ccc}
H(\cup\mathcal{J})[d_M] & \xrightarrow{\quad D(\cup\mathcal{J})\quad} & H_{\mathrm{c}}(\cup\mathcal{J})^{\vee} \\
\xi\, \Big\downarrow\, \text{q.i.} & & \text{q.i.}\, \Big\downarrow\, \xi' \\
\varprojlim_{\mathcal{J}} H(U)[d_M] & \xrightarrow[\text{q.i.}]{\varprojlim_{\mathcal{J}} D(U)} & \varprojlim_{\mathcal{J}} H_{\mathrm{c}}(U)^{\vee} \\
\Big\downarrow & & \Big\downarrow \\
H(U_i)[d_M] & \xrightarrow[\simeq]{D(U_j)} & H_{\mathrm{c}}(U_i)^{\vee} \\
\Big\downarrow & & \Big\downarrow \\
H(U_j)[d_M] & \xrightarrow[\simeq]{D(U_i)} & H_{\mathrm{c}}(U_j)^{\vee}
\end{array} \tag{2.30}$$

and where the vertical arrows are the natural restriction maps.

The morphism $\varprojlim_{\mathcal{J}} D(U)$ is an isomorphism since the third and fourth lines are isomorphisms by hypothesis. The vertical morphisms ξ and ξ' are also isomorphisms. Indeed, it is well-known that the natural maps

$$\begin{aligned}
\varinjlim_{\mathcal{J}} S_*(U;\mathbb{R}) &\to S_*(\cup\mathcal{J};\mathbb{R}) \\
\varinjlim_{\mathcal{J}} \Omega^*_{\mathrm{c}}(U) &\to \Omega^*_{\mathrm{c}}(\cup\mathcal{J})
\end{aligned} \tag{2.31}$$

(the first denotes the complex of singular chains) are isomorphisms of complexes. Applying the duality functor to (2.31), we get the two isomorphisms

$$\begin{aligned}
\varprojlim_{\mathcal{J}} H_*(\cup\mathcal{J};\mathbb{R})^{\vee} &\to H_*(U;\mathbb{R})^{\vee} \\
\xi' : \varprojlim_{\mathcal{J}} H^*_{\mathrm{c}}(\cup\mathcal{J})^{\vee} &\to H^*_{\mathrm{c}}(U)^{\vee}
\end{aligned} \tag{2.32}$$

since $(\_)^{\vee}$ is exact and transforms inductive limits into projective limits. The first line in (2.32) concerns duals of singular homology which, thanks to de Rham theorem, coincide with corresponding de Rham cohomology, so that ξ is also an isomorphism. The morphism $D(\cup\mathcal{J}) = \xi'^{-1} \circ \varprojlim_{\mathcal{J}} D(U) \circ \xi$ is therefore an isomorphism. ⊟

Lemma 2 *If* $U, V \in \mathcal{O}_{\mathrm{PD}}$, *then* $(U \cup V) \in \mathcal{O}_{\mathrm{PD}}$ *if and only if* $(U \cap V) \in \mathcal{O}_{\mathrm{PD}}$.

Proof of Lemma 2 Consider the two familiar Mayer-Vietoris short sequences

$$\begin{cases}
0 \to \Omega(U \cup V) \xrightarrow{\ p\ } \Omega(U) \oplus \Omega(V) \xrightarrow{\ q\ } \Omega(U \cap V) \to 0 \\
0 \to \Omega_{\mathrm{c}}(U \cap V) \xrightarrow{\ p'\ } \Omega_{\mathrm{c}}(U) \oplus \Omega(V) \xrightarrow{\ q'\ } \Omega_{\mathrm{c}}(U \cup V) \to 0
\end{cases} \tag{2.33}$$

where

$$\begin{cases} p(\omega) := (\omega|_U, \omega|_V) & \text{and} \quad q(\omega, \varpi) := \omega|_{U\cap V} - \varpi|_{U\cap V} \\ p'(\omega) := (\omega, -\omega) & \text{and} \quad q'(\omega, \varpi) := \omega + \varpi \end{cases} \tag{2.34}$$

While it is easy to check that both sequences are left exact, the surjectivity of q and q' needs justification, which we can do thanks to the existence of a smooth partition of unity in $U \cup V$ subordinate to the cover $\{U, V\}$. We will therefore consider two smooth positive functions $\phi_U, \phi_V : U \cup V \to \mathbb{R}$ satisfying the support conditions $|\phi_U| \subseteq U$ and $|\phi_V| \subseteq V$ and such that $\phi_U + \phi_V = 1$.

In that case,

- for $\omega \in \Omega(U \cap V)$, the differential form $\phi_V\,\omega$ extends by zero to U, and $\phi_U\,\omega$ to V. Hence, $q(\phi_V\,\omega, -\phi_U\,\omega) = \omega$, and the surjectivity of q.
- for all $\omega \in \Omega_c(U \cap V)$ we have $\phi_U\,\omega) \in \Omega_c(U)$ and $\phi_V\,\omega \in \Omega_c(V)$. Hence, $q'(\phi_U\,\omega, \phi_V\,\omega) = \omega$, and the surjectivity of q'.

These justifications are useful to describe the connecting morphisms c and c' in the following long exact sequences of cohomology associated with (2.33):

$$\begin{array}{l} \xrightarrow[{[1]}]{} H^\ell(U \cup V) \longrightarrow H^\ell(U) \oplus H^\ell(V) \longrightarrow H^\ell(U \cap V) \xrightarrow[{[1]}]{c} \\ \xleftarrow[{[1]}]{} H_c^{n-\ell}(U \cup V) \longleftarrow H_c^{n-\ell}(U) \oplus H_c^{n-\ell}(V) \longleftarrow H_c^{n-\ell}(U \cap V) \xleftarrow[{[1]}]{c'} \end{array} \tag{2.35}$$

Indeed, a simple computation shows that we have:

$$c(\alpha) = d\phi_V \wedge \alpha = -d\phi_U \wedge \alpha \quad \text{and} \quad c'(\beta) = d\phi_U \wedge \beta = -d\phi_V \wedge \beta\,. \tag{2.36}$$

If we now connect the sequences (2.35) with the adjunction morphisms, we get

$$\begin{array}{ccccccc} \xrightarrow[{[1]}]{} H(U \cup V)[d_M] & \longrightarrow & H(U)[d_M] \oplus H(V)[d_M] & \longrightarrow & H(U \cap V)[d_M] \xrightarrow[{[1]}]{c} \\ \downarrow {\scriptstyle D(U\cup V)} & (\mathrm{I}) & \downarrow {\scriptstyle D(U)} \quad \downarrow {\scriptstyle D(V)} & (\mathrm{II}) & \downarrow {\scriptstyle D(U\cap V)} \\ \xrightarrow[{[1]}]{} H_c(U \cup V)^\vee & \longrightarrow & H_c(U)^\vee \oplus H_c(V)^\vee & \longrightarrow & H_c(U \cap V)^\vee \xrightarrow[{[1]}]{c'^\vee} \end{array} \tag{2.37}$$

where the choices in (2.34) immediately give the commutativity of the sub-diagrams (I) and (II). For the commutativity of the connecting sub-diagram

$$\begin{array}{ccc} H^{\ell}(U\cap V)[d_M] & \xrightarrow[{[1]}]{c} & H^{\ell+1}(U\cup V)[d_M] \\ {\scriptstyle D(U\cap V)}\downarrow & & \downarrow{\scriptstyle D(U\cup V)} \\ H_{\mathrm{c}}^{d_M-\ell}(U\cap V)^{\vee} & \xrightarrow[{[1]}]{c'^{\vee}} & H_{\mathrm{c}}^{d_M-(\ell+1)}(U\cup V) \end{array}$$

the equalities (2.36) lead to

$$D(U\cup V)(c(\alpha))(\beta) = \int_M d\,\phi_V \wedge \alpha \wedge \beta$$

$$(D(U\cap V)(\alpha)(c'(\beta)) = \int_M -\alpha \wedge d\,\phi_V \wedge \beta = (-1)^{\ell+1} D(U\cup V)(c(\alpha))(\beta)\,,$$

which suggest modifying c_ℓ in $(-1)^{\ell+1}c$ if we want (2.37) to be a morphism of long exact sequences, which we do.

Applying the Five Lemma then finishes the proof. ⊟

The difficulty in using Lemma 2, is that while the open subspaces U and V can be very simple, the intersection $U\cap V$ can be problematic, with no reason why if U and V belong to $\mathcal{O}_{\mathrm{PD}}$, the same should be true for $U\cap V$. It is at this point that the idea of Leray of *good covers* comes on stage.

A *good cover of M* is an open cover $\mathcal{U} = \{U_i\}$ of M such that all finite intersections $U_{i_1}\cap\ldots\cap U_{i_k}$ are either vacuous or homeomorphic to $\mathbb{R}^{d_M}$.[15,16] Using good covers in connection with Poincaré duality is quite simple. Let $\mathcal{U} = \{B_1, B_2, \ldots\}$ be a good cover of M and define for $n\in\mathbb{N}$:

$$U_n := B_1\cup B_2\cup\cdots\cup B_n\,.$$

[15]The existence of good covers on manifolds (also called *simple covers or Leray covers*) is proved by a Riemannian geometry argument, see Bott-Tu [18] §5 Theorem 5.1, p. 42. Given any open cover $\mathcal{V}$ of M, there always exists a good cover $\mathcal{U}$ subordinate to $\mathcal{V}$. Since M is countable to infinity, the cover $\mathcal{U}$ can always be chosen to be locally finite (hence countable).

[16]As reported by Christian Houzel, in A Short History in Kashiwara-Schapira [61], p. 7, in a conversation with Henri Cartan and André Weil in 1945, Leray explained his idea of good covers, following which Weil proved later the de Rham theorems and Poincaré duality in the modern approach used today, see Weil [96]: *Sur les théorèmes de de Rham*, p. 17, and *Lettre à Henri Cartan*, p. 45. See also the interesting historical review in Bott's introduction to Bott-Tu [18], especially p. 7.

Then

- Each B_i belongs to $\mathcal{O}_{\text{PD}}$. Indeed, since $B_i \simeq \mathbb{R}^{d_M}$, we are lead to verify that the Poincaré pairing

$$\langle \cdot, \cdot \rangle_{\mathbb{R}^n} : H(\mathbb{R}^{d_M}) \times H_c(\mathbb{R}^{d_M}) \to \mathbb{R}$$

  is perfect.
  This results from Poincaré lemmas, both for arbitrary and compact supports, which show that cohomologies are concentrated respectively in degrees 0 and d_M. We then simply need to check that $\int_M : H_c(\mathbb{R}^{d_M}) \to \mathbb{R}$ is an isomorphism of vector spaces which is obvious.[17]
- For all n, the open subspace U_n belongs to $\mathcal{O}_{PD}$. Indeed, by induction on n, we can assume $U_{n-1} \in \mathcal{O}_{\text{DP}}$. Then, by Lemma 2, $U_n \in \mathcal{O}_{PD}$ if and only if $(U_{n-1} \cap B_n) \in \mathcal{O}_{PD}$. But here, the intersection is gentle since

$$(U_{n-1} \cap B_n) = (B_1 \cap B_n) \cup \cdots \cup (B_{n-1} \cap B_n) \tag{2.38}$$

  is a good cover with fewer that n terms. We can therefore apply the inductive hypothesis and conclude that $U_n = (U_{n-1} \cap B_n) \in \mathcal{O}_{PD}$.
- $M \in \mathcal{O}_{\text{PD}}$. Indeed, M is the union of the increasing family of open subspaces $\{U_n\} \subseteq \mathcal{O}_{\text{PD}}$ and we apply Lemma 1, which ends the proof of (1).

(2) For each i, the morphism $D^\vee_{M,i} : H_c^{d_M-i}(M)^{\vee\vee} \to H^i(M)^\vee$ is bijective and, composed with the canonical embedding $\epsilon_i : H_c^{d_M-i} \hookrightarrow (H_c^{d_M-i})^{\vee\vee}$ (2.11), gives the right adjunction $\rho_{\langle\cdot,\cdot\rangle} : H_c(M)[d_M] \to H(M)^\vee$. The finite dimensionality condition is then equivalent to the bijectivity of ϵ_i, hence of $\rho_{\langle\cdot,\cdot\rangle}$. □

2.4.2 *The Fundamental Class of an Oriented Manifold*

Let M be an oriented manifold of dimension d_M not necessarily connected. An immediate corollary of the Poincaré duality theorem 2.4.1.3 is that the left Poincaré adjunction gives a canonical isomorphism

$$D_M : H^0(M) \underset{\simeq}{\to} H_c^{d_M}(M)^\vee ,$$

where the image of $1 \in H^0(M)$, the integral operator $\int_{[M]}$ defined in 2.3, is a nonzero linear form over $H_c^{d_M}(C)$ for each connected component $C \in \Pi_0(M)$.

[17]See Bott-Tu [18]. Corollary 4.1.1 (Poincaré Lemma), p. 35; Corollary 4.7.1 (Poincaré Lemma for Compact Supports), p. 39.

When M is connected, we have $1 = \dim H^0(M) = \dim H_c^{d_M}(M)$, and there exists a unique cohomological class $\zeta_{[M]} \in H_c^{d_M}(M)$ such that

$$\int_{[M]} \zeta_{[M]} = 1\,.$$

This class is called *the fundamental class of the oriented manifold* $(M, [M])$. It will be denoted simply by ζ_M when the orientation $[M]$ is understood.

Exercise 2.4.2.1 Let M be an orientable manifold of dimension d_M, not necessary connected. (🕮, p. 333)

1. Show that to choose an orientation of M is equivalent to choosing a nonzero class in $H_c^{d_M}(C)$ for each connected component C of M.
2. Show that if M is oriented and $|\Pi_0(M)| < +\infty$, then $D_M(1) = \sum_{C\in\Pi_0(M)} \zeta_C$.
3. Can statement (2) be true when $|\Pi_0(M)|$ is not finite ?

2.5 Poincaré Adjunctions

2.5.1 Poincaré Adjoint Pairs

Let M and N be oriented manifolds of dimensions d_M and d_N respectively. A pair (L, R) of morphisms of complexes

$$L : \Omega(N) \to \Omega(M)[L] \quad \text{and} \quad R : \Omega_c(M) \to \Omega_c(N)[R]$$

is called a *Poincaré adjoint pair for* (M, N), if one has

$$[L] - [R] = d_M - d_N \quad \text{and} \quad \int_M L(\alpha) \wedge \beta = \int_N \alpha \wedge R(\beta) \tag{2.39}$$

for all $\alpha \in \Omega(N)$ and $\beta \in \Omega_c(M)$.

Analogously, in the cohomological framework, a pair (L, R) of graded morphisms

$$L : H(N) \to H(M)[L] \quad \text{and} \quad R : H_c(M) \to H_c(N)[R]$$

is called a *Poincaré adjoint pair for* (M, N), if one has

$$[L] - [R] = d_M - d_N \quad \text{and} \quad \int_M L([\alpha]) \cup [\beta] = \int_N [\alpha] \cup R([\beta]) \tag{2.40}$$

for all $[\alpha] \in H(N)$ and $[\beta] \in H_c(M)$.

Proposition 2.5.1.1 (and more Definitions)

1. *If* (L, R_1) *and* (L, R_2) *are Poincaré adjoint pairs, then* $R_1 = R_2$*, and we say that* $R := R_1$ *is the* Poincaré right adjoint of L.
2. *If* (L_1, R) *and* (L_2, R) *are Poincaré adjoint pairs, then* $L_1 = L_2$*, and we say that* $L := L_1$ *is the* Poincaré left adjoint of R.
3. *If* (L_1, R_1) *and* (L_2, R_2) *are Poincaré adjoint pairs respectively for* (M, N) *and* (N, L)*, then* $(L_1 \circ L_2, R_2 \circ R_1)$ *is a Poincaré adjoint pair for* (M, L).
4. *Define the* right Poincaré adjunction *(*cf. *Sect. 2.6.1)*

$$\mathbb{D}'_M : \Omega_c(M)[d_M] \to \Omega(M)^\vee ,$$

by

$$\mathbb{D}'_M(\beta) = \Big(\alpha \mapsto \int_M \alpha \wedge \beta\Big) .$$

If (L, R) *is a Poincaré adjoint pair, then*

$$\mathbb{D}_M \circ L = R^\vee \circ \mathbb{D}_N \quad \textit{and} \quad \mathbb{D}'_N \circ R = L^\vee \circ \mathbb{D}'_M ,$$

i.e. the following diagrams are commutative

$$\begin{array}{ccccccc} \Omega(M)[d_M] & \xrightarrow[\text{q.i.}]{\mathbb{D}_M} & \Omega_c(M)^\vee & \quad & \Omega_c(M)[d_M] & \xrightarrow{\mathbb{D}'_M} & \Omega(M)^\vee \\ {\scriptstyle L}\uparrow & & \uparrow{\scriptstyle R^\vee} & & {\scriptstyle R}\downarrow & & \downarrow{\scriptstyle L^\vee} \\ \Omega(N)[d_N] & \xrightarrow[\text{q.i.}]{\mathbb{D}_N} & \Omega_c(N)^\vee & & \Omega_c(N)[d_N] & \xrightarrow{\mathbb{D}'_N} & \Omega(N)^\vee \end{array}$$

5. *If* (L, R) *is a Poincaré adjoint pair of morphisms of complexes, then* $(H(L), H_c(R))$ *is an adjoint pair in cohomology, so that we have*

$$D_M \circ H(L) = H_c(R)^\vee \circ D_N , \qquad D'_N \circ H_c(R) = H(L)^\vee \circ D'_M .$$

In particular, $H(L)$ and $H_c(R)$ are adjoint operators via Poincaré duality.
6. *If* (L, R) *is a Poincaré adjoint pair and L (resp. R) is a quasi-isomorphism, then R (resp. L) is a quasi-isomorphism too.*

Proof

(1) If R_1 and R_2 are adjoints of the same L, we have,

$$\big\langle \alpha, R_1(\beta) - R_2(\beta) \big\rangle_N = 0 , \quad (\forall \alpha \in \Omega(N))\ (\forall \beta \in \Omega_c(M)) ,$$

in which case $R_1(\beta) - R_2(\beta) = 0$ since Poincaré pairing is nondegenerate on differential forms (*cf*. Proposition 2.4.1.1).

(2) Same proof as (1). If L_1 and L_2 are adjoints of the same R, we have,

$$\big\langle L_1(\alpha) - L_2(\alpha), \beta \big\rangle_M = 0, \quad (\forall \alpha \in \Omega(N))\ (\forall \beta \in \Omega_c(M)),$$

in which case $L_1(\alpha) - L_2(\alpha) = 0$ since Poincaré pairing is nondegenerate.

(3) Follows from the equalities:

$$\big\langle L_1 \circ L_2(\_), (\_) \big\rangle_M = \big\langle L_2(\_), R_1(\_) \big\rangle_N = \big\langle (\_), R_2 \circ R_1(\_) \big\rangle_L$$

(4) Straightforward verifications:

$$ID(L(\alpha))(\beta) = \big\langle L(\alpha), \beta \big\rangle_M = \big\langle \alpha, R(\beta) \big\rangle_N = R^\vee(ID(\alpha))(\beta)$$

$$ID'(R(\beta))(\alpha) = \big\langle \alpha, R(\beta) \big\rangle_N = \big\langle L(\alpha), \beta \big\rangle_M = L^\vee(ID'(\beta))(\alpha)$$

(5) Left to the reader.

(6) After the commutative diagram

$$\begin{array}{ccc} H(M)[d_M] & \xrightarrow[\simeq]{D_M} & H_c(M)^\vee \\ {\scriptstyle H(L)}\uparrow & & \uparrow{\scriptstyle H(R)^\vee} \\ H(N)[d_N] & \xrightarrow[\simeq]{D_N} & H_c(N)^\vee \end{array}$$

$$\big(H(L) \text{ is an iso}\big) \Leftrightarrow \big(H(R)^\vee \text{ is an iso}\big) \Leftrightarrow \big(H(R) \text{ is an iso}\big)$$

since the $\mathbb{R}$-duality functor is exact. □

Comments 2.5.1.2 We shall see that, given $f : M \to N$, the pullback morphism $f^* : \Omega(N) \to \Omega(M)$ may or may not admit a right Poincaré adjoint at the level of complexes, although such adjoint will always exist in derived category, hence in cohomology.

In cohomology, this right adjoint is the *Gysin morphism (for arbitrary maps)* $f_! : H_c(M) \to H_c(N)$. The pair $(H(f^*), f_!)$ is then a Poincaré adjoint pair in cohomology.

When f is a proper map, the pullback $f^* : \Omega_c(N) \to \Omega_c(M)$ is well-defined and one may look for a left Poincaré adjoint to f^*, i.e. one may look for some morphism of complexes $L : \Omega(M)[d_M] \to \Omega(N)[d_N]$ such that

$$\int_N L(\alpha) \wedge \beta = \int_M \alpha \wedge f^*(\beta).$$

As in the previous case, such L may or may not exist for complexes, but it always will in derived category, hence in cohomology.

In cohomology, this left adjoint is *the Gysin morphism for proper maps* $f_* : H(M)[d_M] \to H(N)[d_N]$. The pair $(f_*, H_c(f^*))$ is a Poincaré adjoint pair in cohomology.

2.5.2 Manifolds and Maps of Finite de Rham Type

Definition 2.5.2.1 A manifold M is said to be *of finite (de Rham) type*, if its de Rham cohomology $H(M)$ is finite dimensional.

A map between manifolds $f : M \to N$ is said to be *of finite (de Rham) type* if N is the union of a countable ascending chain $\mathscr{U} := \{U_0 \subseteq U_1 \subseteq \cdots\}$ of open subspaces of finite type such that each subspace $f^{-1}(U_m) \subseteq M$ is of finite type.

Proposition 2.5.2.2 *If M is a manifold, orientable or not, which admits a finite good cover (e.g. M compact), then* $\dim H(M) < +\infty$ *and* $\dim H_c(M) < +\infty$.

Proof By induction on the cardinality of a good cover $\{B_1, \dots, B_n\}$ of M, the open subspaces $U := B_1 \cup \cdots \cup B_{n-1}$ and $U \cap B_n = (B_1 \cap B_n) \cup \cdots \cup (B_{n-1} \cap B_n)$ verify the statement. Use of Mayer-Vietoris long exact sequence sequences for (U, B_n):

$$\xrightarrow[{[1]}]{} H(M) \longrightarrow H(U) \oplus (B_n) \longrightarrow H(U \cap B_n) \xrightarrow[{[1]}]{}$$

$$\xrightarrow[{[1]}]{} H_c(U \cap B_n) \longrightarrow H_c(U) \oplus H_c(B_n) \longrightarrow H_c(M) \xrightarrow[{[1]}]{}$$

then ends the proof. □

Remark 2.5.2.3 An *oriented* manifold is of finite type if and only if its Poincaré pairing in cohomology is perfect (2.4.1.3–(2)), in which case the compactly supported cohomology is also finite dimensional.

Exercises 2.5.2.4

1. Show that a locally trivial fibration $f : M \to N$ with finite type fiber (3.1), is a finite type map. (♝, p. 334)
2. Let M be a finite type manifold. Show that $\dim(H_c(M)) < +\infty$ if and only if the *orientation manifold* $\tilde{M}$ of M is also of finite type. (♝, p. 334)

2.5.3 Ascending Chain Property

Although general manifolds need not be of finite type, they are always the inductive limit of such. More precisely, any manifold M is the union of an ascending chain $\{U_0 \subseteq U_1 \subseteq \cdots\}$ of open subsets of finite type of M.

This weaker finiteness property, sufficient for our needs, results from the existence of countable good covers (fn. ([15]), p. 28). Indeed, if $\mathscr{V} := \{V_0, V_1, \dots\}$ is such a cover of M, the open subsets $U_n := V_0 \cup \cdots V_n$ then verify $\dim H(U_n) < +\infty$ (as well as $\dim H_c(U_n) < +\infty$).

When a manifold is endowed with the action of a Lie group G, we will also need each U_n to be G-stable.

Proposition 2.5.3.1 *Let G be a compact Lie Group. A G-manifold M is the union of a countable ascending chain $\mathscr{U} := \{U_0 \subseteq U_1 \subseteq \cdots\}$ of G-stable open subsets of M of finite type (as well as* $\dim H_c(U_n) < +\infty$*).*

The following sections summarize certain facts needed in the proof of this proposition, which we postpone to 2.5.6.

2.5.4 Existence of Proper Invariant Functions

The aim of this section is to show, for G compact, that there always exist positive proper G-invariant functions on a G-manifold (*cf*. (fn. ([10]), p. 18)).

Recall that, by our convention in 2.2.1, manifolds are paracompact spaces.

Fix a countable, locally finite cover $\mathscr{U} := \{U_n\}_{n\in\mathbb{N}}$ of M, where each U_n is a *relatively compact* open subset of M. Next, fix a smooth partition of unity $\{\varphi_n\}_{n\in\mathbb{N}}$ subordinate to $\mathscr{U}$. This means in particular that for each $n \in \mathbb{N}$, the equality $\varphi_n(x) = 0$ holds whenever $x \notin U_n$. Then one has, for every $N \in \mathbb{N}$,

$$1 = \sum\nolimits_{n>N} \varphi_n(x)\,, \quad \forall x \notin U_0 \cup \cdots \cup U_N\,. \tag{2.41}$$

Now, for every $x \in M$, the infinite sum

$$\phi(x) := \sum\nolimits_{n\in\mathbb{N}} n \cdot \varphi_n(x)$$

is finite and smooth on M, as it is a locally finite sum of smooth functions.

Lemma 2.5.4.1 *The function $\phi : M \to \mathbb{R}_{>0}$ is proper and differentiable.*

Proof By property (2.41) one has, for all $x \notin U_0 \cup \cdots \cup U_N$,

$$\phi(x) \geq \sum\nolimits_{n>N} n \cdot \phi_n(x) > N\Big(\sum\nolimits_{n>N} \phi_n(x)\Big) = N\,. \tag{2.42}$$

Now, to see that Φ is proper, note that if $F \subseteq \mathbb{R}$ is compact, then $F \subseteq [-N, N]$ for some $N \in \mathbb{N}$ and $\phi^{-1}(F) \subseteq U_0 \cup \cdots \cup U_N$ by (2.42). But the closure $\overline{U_0 \cup \cdots \cup U_N}$ is a compact subset of M because each $\overline{U}_i$ is assumed compact. As a closed subset of a compact set, $\phi^{-1}(F)$ is also compact. □

As a corollary of the previous lemma, we can now prove the existence of positive proper invariant functions.

Proposition 2.5.4.2 *A manifold M endowed with a smooth action of a compact Lie group G admits proper G-invariant positive functions $\Phi : M \to \mathbb{R}_{>0}$.*

Proof Let $\phi : M \to \mathbb{R}_{>0}$ denote a proper positive function (see 2.5.4.1), and set:

$$\Phi(x) := \int_G \phi(g\cdot x)\, dg\,,$$

where dg is a G-invariant differential form of top degree on the compact Lie group G, such that $1 = \int_G dg$.[18] The correspondence $x \mapsto \Phi(x)$ is clearly a well-defined nonnegative unbounded G-invariant function of M into $\mathbb{R}$. Now, for each $N \in \mathbb{N}$, the set $M_N := G\cdot\phi^{-1}([-N, N])$ is compact and G-stable, and if $y \notin M_N$, then $\phi(g\cdot y) > N$ for all $g \in G$, so that

$$\Phi(y) = \int_G \phi(g\cdot y)\, dg > N\,. \tag{2.43}$$

The properness of Φ then follows as for Lemma 2.5.4.1, i.e. if F is a compact subset of $\mathbb{R}$, then $F \subseteq [-N, N]$ for somc $N \in \mathbb{N}$, and $\Phi^{-1}(F) \subseteq M_N$ by (2.43). The subspace $\Phi^{-1}(F)$ is then compact since it is closed in the compact set M_N. □

2.5.5 Manifolds with Boundary

The following elementary finiteness property of compact manifolds with boundary will be needed in the proof of Proposition 2.5.3.1.

Proposition 2.5.5.1 *Let U be the* interior *of a compact manifold with boundary. Then,* $\dim H(U) < +\infty$ *and* $\dim H_c(U) < +\infty$.

Proof Let N be a compact manifold with boundary ∂N. Denote by $U := N \smallsetminus \partial N$ its *interior*. Gluing N with itself along its boundary, one gets the '*double*' $\overline{\overline{N}} := N \sqcup_{\partial N} N$, which is a compact manifold without boundary. We know from (2.5.2.2) that $H(\overline{\overline{N}})$ and $H(\partial N)$ are finite dimensional.

The finiteness of $\dim H_c(U)$ then results from the exactness of the long sequence of compactly supported cohomology associated with the closed embedding $\partial N \subseteq \overline{\overline{N}}$ (see 3.7.1–(1)):

$$\cdots \longrightarrow H^i_c(U) \oplus H^i_c(U) \longrightarrow H^i(\overline{\overline{N}}) \longrightarrow H^i(\partial N) \longrightarrow \cdots.$$

[18] See Tu [91] §13.2 Integration Over a Compact Connected Lie Group, p. 105.

Next, let N_ϵ be an open *collar neighborhood* of each copy of N within $\overline{\overline{N}}$. The open subspace $(\partial N)_\epsilon := N_\epsilon \cap N_\epsilon \subseteq \overline{\overline{N}}$ is homotopy-equivalent to ∂N, so that its cohomology is finite dimensional. The finiteness of $H_c(N_\epsilon)$ then results from the Mayer-Vietoris sequence associated with the pair (N_ϵ, N_ϵ):

$$\cdots \longrightarrow H^i(\overline{\overline{N}}) \longrightarrow H^i(N_\epsilon) \oplus H^i(N_\epsilon) \longrightarrow H^i((\partial N)_\epsilon) \longrightarrow \cdots .$$

We can then conclude, as before, that $\dim H(N_\epsilon) < +\infty$, and, since N_ϵ is homotopy-equivalent to U, that $\dim H(U) < +\infty$ too. □

2.5.6 Proof of Proposition 2.5.3.1

The connected components of a manifold M are always open and closed submanifolds of M. In particular, if $M = \coprod_{i\in\mathfrak{I}} C_i$ denotes the decomposition of M in connected components, then the indexing set $\mathfrak{I}$ is finite or countable, and the restriction of a proper function $\Phi : M \to \mathbb{R}$ to each C_i remains proper.

If all the connected components of M are compact, we may index them by natural numbers $C_0, C_1, \ldots$ and define $U_n := C_0 \cup C_1 \cup \cdots \cup C_n$. Each U_n is then open in M and is also a compact manifold, hence it is of finite type. The ascending chain $\{U_0 \subseteq U_1 \subseteq \cdots\}$ meets the requirements of Proposition 2.5.3.1.

If M contains a noncompact connected component C, fix any proper positive G-invariant function $\Phi : M \to \mathbb{R}$, which is possible due to 2.5.4.2, and note that $\Phi(C)$ is necessarily unbounded, since otherwise $C \subseteq \Phi^{-1}([0, T])$ for some $T \in \mathbb{R}$, and C would be compact as Φ is proper over C. Moreover, there exists $N \in \mathbb{N}$ such that $\Phi(M) \supseteq \Phi(C) \supseteq (]N, +\infty[)$, since $\Phi(C)$ is unbounded *and* connected. Now, by Sard's theorem, the interior of the set of *critical* values of $\Phi : M \to \mathbb{R}$ is empty so that there exists an unbounded increasing sequence of positive real numbers $\{N < t_0 < \cdots < t_n < \cdots\}_{n\in\mathbb{N}}$ which are *regular* values of Φ. (see Fig. 2.1)

Each subset $M_n := \Phi^{-1}(t_n)$ is then a submanifold of codimension 1 in M and, moreover, it is compact and G-stable since Φ is proper and G-invariant. Similarly, the sets $U_n := \Phi^{-1}(]-\infty, t_n[)$ and $W_n := \Phi^{-1}(]t_n, +\infty[)$, clearly nonempty, are G-stable open subsets of M.

We then easily check that $\overline{U}_n = U_n \sqcup M_n$ and $\overline{W}_n = M_n \sqcup W_n$ are in fact G-manifolds with boundary M_n embedded in M. Furthermore, $\overline{U}_n$ is compact as we have $\overline{U}_n := \Phi^{-1}(]-\infty, t_n]) = \Phi^{-1}([0, t_n])$ since Φ is positive.

By Proposition 2.5.5.1, the G-stable open subspace U_n verifies $\dim H(U) < +\infty$ and $\dim H_c(U) < +\infty$. Hence, the increasing chain $\{U_0 \subseteq U_1 \subseteq \cdots\}$ meets the requirements of Proposition 2.5.3.1. □

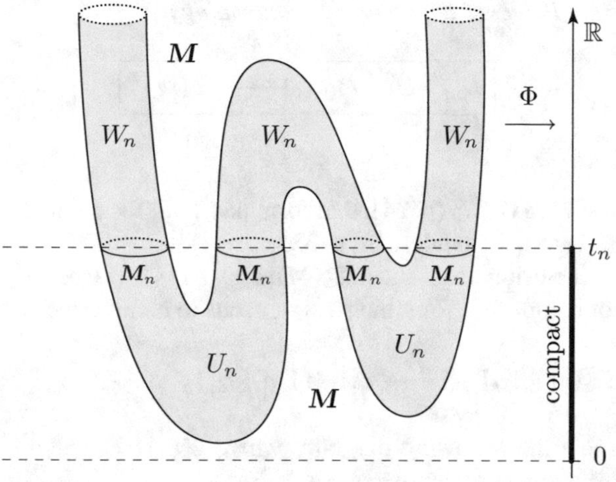

Fig. 2.1 Proof's figure

2.6 The Gysin Functor

2.6.1 The Right Poincaré Adjunction Map

In Sect. 2.4, we introduced the left Poincaré adjunction,

$$\mathbb{D}_M : \Omega(M)[d_M] \underset{\text{q.i.}}{\hookrightarrow} \Omega_c(M)^\vee \,. \tag{2.44}$$

By duality, this map yields $\mathbb{D}_M^\vee : \Omega_c(M)^{\vee\vee} \to \Omega(M)[d_M]^\vee$ which is also a quasi-isomorphism and, composed with the embedding $\Omega_c(M) \subseteq \Omega_c(M)^{\vee\vee}$, gives rise to the injection and quasi-injection (2.4.1.1, 2.1.6, 2.1.11.1)

$$\underbrace{\big(\Omega_c(M)[d_M], d\big) \hookrightarrow \big(\Omega_c(M)^{\vee\vee}[d_M], -\boldsymbol{D}\big) \xrightarrow[\text{q.i.}]{\mathbb{D}^\vee} \big(\Omega(M)^\vee, -\boldsymbol{D}\big)}_{\mathbb{D}'_M}$$

The resulting morphism of complexes is *the right Poincaré adjunction*:

$$\boxed{\mathbb{D}'_M : \big(\Omega_c(M)[d_M], d\big) \hookrightarrow \big(\Omega(M)^\vee, -\boldsymbol{D}\big)} \tag{2.45}$$

It is given by (*cf.* 2.5.1.1)

$$\mathbb{D}'_M(\beta) := \Big(\alpha \mapsto \int_M \alpha \wedge \beta\Big)\,, \tag{2.46}$$

inducing *the right Poincaré adjunction in cohomology*

$$\boxed{D'_M : H_c(M)[d_M] \hookrightarrow H(M)^\vee} \tag{2.47}$$

Exercises 2.6.1.1

1. Check, as for $I\!D$ in (2.27) (p. 24), that formula (2.46) for $I\!D'$ defines a morphism of differential graded modules. (p. 335)
2. The natural inclusion map $\iota : (\Omega_c(M)[d_M], d) \subseteq (\Omega(M), d)[d_M]$ is not a morphism of complexes, for which a sign needs to be introduced, for example,

$$(\Omega_c(M)[d_M], d) \xrightarrow{\epsilon} (\Omega(M), d)[d_M] \quad \beta \mapsto (-1)^{[\beta]d_M}\beta\,.$$

Show then that the following diagram, where $\Xi(\lambda) := (-1)^{|\lambda|+d_M}\lambda \circ \iota$, is a commutative diagram of complexes (p. 335)

$$\begin{array}{ccc}
(\Omega_c(M)[d_M], d) & \xrightarrow{\epsilon} & (\Omega(M), d)[d_M] \\
{\scriptstyle I\!D'_M}\downarrow & & \downarrow{\scriptstyle I\!D_M} \\
(\Omega(M), -\mathbf{D})^\vee & \xrightarrow{\Xi} & (\Omega_c(M)^\vee, \mathbf{D})\,.
\end{array}$$

Proposition 2.6.1.2 *Let M be an oriented manifold. The right Poincaré adjunction*

$$I\!D'_M : \big(\Omega_c(M)[d_M], d\big) \hookrightarrow \big(\Omega(M)^\vee, -\mathbf{D}\big)$$

is an injection and a quasi-injection.

Furthermore, it is a quasi-isomorphism if and only if M is of finite type.

Proof Same as 2.4.1.3–(2). □

2.6.2 The Gysin Morphism

The last statement shows that for oriented manifolds of finite type, the compactly supported cohomology canonically coincides with the dual of closed support cohomology so that if N is such kind of manifold, then the diagram

$$\begin{array}{ccc}
H_c(M)[d_M] & \xhookrightarrow{D'_M} & H(M)^\vee \\
{\scriptstyle f_!}\downarrow & \oplus & \downarrow{\scriptstyle H(f^*)^\vee} \\
H_c(N)[d_N] & \underset{\simeq}{\xhookrightarrow{D'_N}} & H(N)^\vee
\end{array} \tag{2.48}$$

can be commutatively closed in a unique way by a morphism of graded spaces

$$\boxed{f_! : H_c(M)[d_M] \to H_c(N)[d_N]} \tag{2.49}$$

It follows that the correspondence which assigns $M \rightsquigarrow M_! := H_c(M)[d_M]$ and $f \rightsquigarrow f_!$, is covariant and functorial.

When the manifold N in (2.48) is not of finite type, D'_N is still an injection but it is no longer surjective so that it is not obvious that the diagram can be closed. Statement (2) in the next theorem establishes that this is in fact the case. It is therefore always possible to define the morphism $f_! : M_! \to N_!$, which we call *the Gysin morphism for compact supports associated with f*. The resulting correspondence

$$(\_)_! : \mathrm{Man}^{\mathrm{or}} \rightsquigarrow \mathrm{GV}(\mathbb{R}) \qquad \begin{cases} M \rightsquigarrow M_! := H_c(M)[d_M] \\ f \rightsquigarrow f_! \end{cases}$$

is thus a well-defined covariant functor on the whole category $\mathrm{Man}^{\mathrm{or}}$, called *the Gysin functor*.[19]

Theorem 2.6.2.1 (and Definitions)

1. *Let M be oriented and endow its open subsets with induced orientations. For any inclusion of open subsets $j : V \subseteq W$, denote by $j_! : \Omega_c(V) \to \Omega_c(W)$ the map that assigns to $\beta \in \Omega_c(V)$ its* extension by zero *to W, also called* the pushforward of β. *Then, the following diagrams*

$$\begin{array}{ccc} \Omega_c(V)[d_M] & \xhookrightarrow{\mathbb{D}'_V} & \Omega(V)^\vee \\ {\scriptstyle j_*}\downarrow & & \downarrow{\scriptstyle (j^*)^\vee} \\ \Omega_c(W)[d_M] & \xhookrightarrow{\mathbb{D}'_W} & \Omega(W)^\vee \end{array} \qquad \begin{array}{ccc} H_c(V)[d_M] & \xhookrightarrow{D'_V} & H(V)^\vee \\ {\scriptstyle H_c(j_*)}\downarrow & & \downarrow{\scriptstyle H(j^*)^\vee} \\ H_c(W)[d_M] & \xhookrightarrow{D'_W} & H(W)^\vee \end{array}$$

are commutative, i.e. $(j^*, j_!)$ is a Poincaré adjoint pair (cf. *Proposition 2.5.1.1*).

2. *For any map $f : M \to N$ between oriented manifolds, we have the diagram*

$$\begin{array}{ccc} H_c(M)[d_M] & \xhookrightarrow{D'_M} & H(M)^\vee \\ {\scriptstyle f_!}\downarrow & & \downarrow{\scriptstyle H(f^*)^\vee} \\ H_c(N)[d_N] & \xhookrightarrow{D'_N} & H(N)^\vee \end{array} \tag{2.50}$$

[19] See Sect. 8.1 for a justification of the notation.

where $H(f^*)^\vee\big(\mathrm{Im}(D'_M)\big) \subseteq \mathrm{Im}(D'_N)$, *so that there exists a unique morphism of graded spaces*

$$f_! : H_c(M)[d_M] \longrightarrow H_c(N)[d_N] \tag{2.51}$$

called the Gysin morphism for compact supports associated with f, *such that the diagram (2.50) is commutative, i.e.* $(H(f^*), f_!)$ *is a Poincaré adjoint pair in cohomology, which means that, for any* $[\alpha] \in H(N)$ *and* $[\beta] \in H_c(M)$, *the equation in* X,

$$\int_M f^*([\alpha]) \cup [\beta] = \int_N [\alpha] \cup X\,, \tag{2.52}$$

admits a unique solution in $H_c(N)$, *namely* $X = f_![\beta]$.
Furthermore, $f_!$ *in (2.51) is a morphism of* $H(N)$*-modules, i.e. the equality, called the* projection formula,

$$\boxed{f_!\big(f^*([\alpha]) \cup [\beta]\big) = [\alpha] \cup f_!([\beta])} \tag{2.53}$$

holds for all $[\alpha] \in H(N)$ *and* $[\beta] \in H_c(M)$.

3. *The correspondence*

$$(\_)_! : \mathrm{Man}^{\mathrm{or}} \rightsquigarrow \mathrm{GV}(\mathbb{R}) \quad \text{with} \quad \begin{cases} M \rightsquigarrow M_! := H_c(M)[d_M] \\ f \rightsquigarrow f_! \end{cases}$$

is a covariant functor, called the Gysin functor.

4. *If* M *and* N *are oriented of finite type, then* $f^* : H(N) \to H(M)$ *is an isomorphism if and only if the Gysin morphism* $f_! : H_c(M)[d_M] \to H_c(N)[d_N]$ *is also an isomorphism.*

Proof

(1) The commutativity results from the equality

$$\int_V \alpha|_V \wedge \beta = \int_W \alpha \wedge j_!\beta$$

for $\alpha \in \Omega(W)$ and $\beta \in \Omega_c(V)$, which is clear since the support of $\alpha \wedge j_!\beta$ is contained in V.

(2) We need to verify that, given $[\beta] \in H_c(M)$, there exists $[\beta'] \in H_c(N)$ such that the linear form

$$[\alpha] \in H(N) \mapsto \int_M f^*[\alpha] \cup [\beta]$$

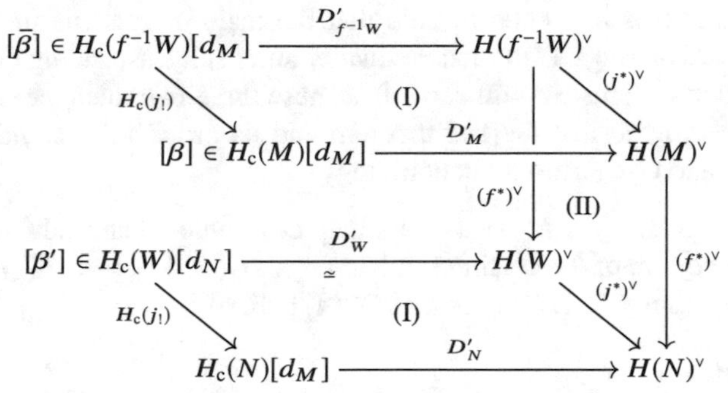

Fig. 2.2 Diagram (**D**)

coincides with

$$[\alpha] \in H(N) \mapsto \int_N [\alpha] \cup [\beta']$$

(3, 4) Thanks to Proposition 2.5.3.1, there exists an open subset $W \in N$ of finite type such that $f^{-1}W$ contains the support of β, denoted $\overline{\beta} := \beta|_{f^{-1}W}$.

We then have the commutative diagram (**D**) (Fig. 2.2) where the subdiagrams (I) are commutative after (2) and the commutativity of (II) is simply the functoriality of pullback morphisms.

Following the arrows, we see that

$$\begin{aligned}(f^*)^\vee \circ D'_M([\beta]) &= (i^*)^\vee \circ (f^*)^\vee \circ D'_{f^{-1}W}([\overline{\beta}]) \\ &= (j^*)^\vee \circ D'_W([\beta']) = D'_N \circ H_c(j_!)([\beta'])\end{aligned}$$

where $[\beta'] \in H_c(W)[d_N]$ verifies

$$D'_W([\beta']) = (f^*)^\vee \circ D'_{f^{-1}W}([\overline{\beta}])$$

which is possible since D'_W is **surjective** as W is of finite type !

The statement about Eq. (2.52) is clear and formally implies the projection formula since

$$\begin{aligned}\int_N [\omega] \cup f_!\big(f^*[\alpha] \cup [\beta]\big) &= \int_M f^*[\omega] \cup f^*[\alpha] \cup [\beta] \\ &= \int_M f^*([\omega] \cup [\alpha]) \cup [\beta] = \int_N [\omega] \cup [\alpha] \cup f_![\beta] .\end{aligned}$$

Finally, (3) is trivial since D' is bijective over its image, and (4) is clear. □

Remark 2.6.2.2 It is important to note that the main ingredients in the proof are (1) the Poincaré pairings, (2) Poincaré duality and (3) the ascending chain property (2.5.3). In later sections we will show that these three ingredients exist also in the equivariant setting so that the last theorem and its proof will extend *verbatim* to G-manifolds and G-equivariant cohomology.

Exercise 2.6.2.3 Let $f : M \to N$ be a map of oriented manifolds. Show that the left Poincaré adjoint of the Gysin morphism $f_! : H_c(M)[d_M] \to H_c(N)[d_N]$ is the *pullback morphism* $f^* : H(N) \to H(M)$. (☝, p. 336)

2.6.3 The Image of D'_M

The next proposition will be used when extending the Gysin functor to the equivariant context. It gives a description of the image of D'_M in terms of ascending chains of open finite type subspaces of M, which was the main reason for proving that such covers always exist (see 2.5.3.1).

Proposition 2.6.3.1 *Let $\mathscr{U}$ be a filtrant open cover*[20] *of a manifold M.*

1. *Let $j : V \subseteq W$ denote an inclusion of open subsets of M.*
   The extension by zero *morphism $j_! : \Omega_c(V) \subseteq \Omega_c(W)$, that assigns to $\beta \in \Omega_c(V)$ the differential form $j_!(\beta) \in \Omega_c(W)$, equal to β over V and 0 otherwise, is a morphism of complexes inducing, in cohomology, the morphism of graded spaces*

$$H_c(j_!) : H_c(V) \to H_c(W)$$

   We also have the morphism of complexes $j^* : \Omega(W) \to \Omega(V)$ that restricts a differential form of W to V, and the corresponding morphism of graded spaces

$$H(j^*) : H(W) \to H(V)\,.$$

   These constructions, applied to the elements of $\mathscr{U}$, give rise to the inductive systems $\{\Omega_c(U)\}_{U\in\mathscr{U}}$ and $\{H_c(U)\}_{U\in\mathscr{U}}$, and to the projective systems $\{\Omega(U)\}_{U\in\mathscr{U}}$ and $\{H(U)\}_{U\in\mathscr{U}}$, whence the canonical maps

$$\nu : \varinjlim_{U\in\mathscr{U}} \Omega_c(U) \to \Omega_c(M) \quad \textit{and} \quad H(\nu) : \varinjlim_{U\in\mathscr{U}} H_c(U) \to H_c(M)\,,$$

$$\mu : \Omega(M) \to \varprojlim_{U\in\mathscr{U}} \Omega(U) \quad \textit{and} \quad H(\mu) : H(M) \to \varprojlim_{U\in\mathscr{U}} H(U)\,.$$

   All these maps are bijective.

---

[20]We recall that $\mathscr{U} = \{U_i\}_{i\in\mathfrak{I}}$ is said to be *filtrant* whenever for all $U_1, U_2 \in \mathscr{U}$ there exists $U_3 \in \mathscr{U}$ such that $(U_1 \cup U_2) \subseteq U_3$.

2. *Suppose M is oriented, then the map*

$$\boxed{\begin{array}{rcl} \mathbb{D}'_{\mathscr{U}} : (\Omega_c(M), d)[d_M] & \longrightarrow & \varinjlim_{U\in\mathscr{U}} (\Omega(U)^\vee, -\mathbf{D}) \\ \beta & \mapsto & \left(\alpha \longmapsto \int_M \alpha\wedge\beta\right) \end{array}} \tag{2.54}$$

is a well-defined morphism of complexes inducing in cohomology the map

$$D'_{\mathscr{U}} : H_c(M)[d_M] \to \varinjlim_{U\in\mathscr{U}} H(U)^\vee$$

3. *Suppose further that each $U \in \mathscr{U}$ is of finite type. Then $\mathbb{D}'_{\mathscr{U}}$ is a quasi-isomorphism, and one has*

$$\mathrm{Im}(D'_M) = \varinjlim_{U\in\mathscr{U}} H(U)^\vee \subseteq H(M)^\vee \tag{2.55}$$

Moreover, the adjunction $D'^\vee_{\mathscr{U}}$ canonically identifies with D_M; more precisely, the following diagram is commutative:

$$\begin{array}{ccc} \varprojlim_{U\in\mathscr{U}} H(U)[d_M] = (\varinjlim_{U\in\mathscr{U}} H(U)^\vee)^\vee[d_M] & \xrightarrow{D'^\vee_{\mathscr{U}}} & H_c(M)^\vee \\ \uparrow \simeq & & \| \\ H(M)[d_M] & \xrightarrow{D_M} & H_c(M)^\vee \end{array}$$

Proof

(1) The map $\nu : \varinjlim_{U\in\mathscr{U}} \Omega^*_c(U) \to \Omega^*_c(M)$ is injective since it is the limit of a filtrant inductive system of injective maps. The image of ν is the union of $\Omega^*_c(U)$ for the same reason. Now, if $\omega \in \Omega^*_c(M)$, then its support, being compact, is contained in some $U \in \mathscr{U}$ so that ω is the pushforward of $\omega|_U \in \Omega^*_c(U)$. This justifies the equality $\Omega^*_c(M) = \bigcup_{U\in\mathscr{U}} \Omega^*_c(U)$ and proves that ν is surjective. Standard arguments on the homology of filtrant inductive systems of complexes prove that $H(\nu)$ is bijective.

The map $\mu : \Omega(M) \to \varprojlim_{U\in\mathscr{U}} \Omega(U)$ is injective since a differential form is null if and only if it is locally so. To see that μ also surjective, let $\{\alpha_U \in \Omega(U)\}_{U\in\mathscr{U}}$ be a given projective system of differential forms, then note that for any $x \in M$, the element $\tilde{\alpha}(x) := \alpha_U(x)$ is independent of the choice of $U \ni x$. Indeed, for $x \in U_1 \in \mathscr{U}$ and $x \in U_2 \in \mathscr{U}$, we can choose $U_3 \in \mathscr{U}$ such that $U_1 \cup U_2 \subseteq U_3$, since $\mathscr{U}$ is filtrant, in which case $\alpha_{U_1}(x) = \alpha_{U_3}(x) = \alpha_{U_2}(x)$. The differentiability of $\tilde{\alpha}$ is obvious since this is a local property. Finally, for all $U \in \mathscr{U}$ we have $\tilde{\alpha}|_U = \alpha_U$ by construction, which ends the proof of the surjectivity of μ.

We need now only justify that $H(\mu)$ is bijective. This is immediate when M is orientable, since $H(\mu)$ is then just the Poincaré dual of $H_c(\nu)$ which has already been shown to be bijective. Otherwise, when M is not orientable, we lift $\mathscr{U}$ to the orientation manifold $\tilde{M}$ associated with M through the canonical $\mathbb{Z}/2\mathbb{Z}$-covering $p : \tilde{M} \twoheadrightarrow M$, setting therefore $\tilde{\mathscr{U}} := \{\tilde{U} := p^{-1}(U) \mid U \in \mathscr{U}\}$. As $\tilde{M}$ is orientable, the map $H(\tilde{M}) \to \varprojlim_{U\in\mathscr{U}} H(\tilde{U})$ is now bijective, and because this map is also compatible with the reversing-orientation action of $\mathbb{Z}/2\mathbb{Z}$, it induces a bijection between invariants subspaces $H(\tilde{M})^{\mathbb{Z}/2\mathbb{Z}} \xrightarrow{\simeq} \varprojlim_{U\in\mathscr{U}} H(\tilde{U})^{\mathbb{Z}/2\mathbb{Z}}$, and one concludes since $H(U) = H(\tilde{U})^{\mathbb{Z}/2\mathbb{Z}}$.

(2) Endow each $U \in \mathscr{U}$ with the orientation induced by M. Taking the inductive limit of the maps $I\!D'_U : H_c(U)[d_M] \to H(U)^\vee$ and applying (1) one sees immediately that $I\!D_{\mathscr{U}} = \varinjlim_{U\in\mathscr{U}} I\!D'_U$.

(3) By 2.6.1.2 the maps $I\!D'_U : H_c(U)[d_M] \to H(U)^\vee$ are quasi-isomorphisms for each $U \in \mathscr{U}$, hence $I\!D_{\mathscr{U}} = \varinjlim_{U\in\mathscr{U}} I\!D'_U$ is also a quasi-isomorphism since $\mathscr{U}$ is filtrant. The rest of the statement is then clear by duality. □

2.7 The Gysin Functor for Proper Maps

In this section, the Gysin morphism for compact supports

$$f_! : H_c(M)[d_M] \to H_c(N)[d_N]$$

are extended to arbitrary closed supports

$$f_* : H(M)[d_M] \to H(N)[d_N]$$

when $f : M \to N$ is a **proper** map. We will see that this case is much simpler than the general one as it results immediately from Poincaré duality.

When $f : M \to N$ is proper, the pullback $f^* : \Omega(N) \to \Omega(M)$ respects compact supports inducing a morphism of complexes $f^* : \Omega_c(N) \to \Omega_c(M)$, which gives rise to the *covariant* functor from $\mathrm{Man}_{\mathrm{pr}}$ to $\mathrm{Vec}(\mathbb{R})$

$$M \rightsquigarrow H_c(M)^\vee, \qquad f \rightsquigarrow H_c(f^*)^\vee.$$

When M is oriented, the right Poincaré adjunction $I\!D'_M$ (Sect. 2.6.1) can be extended from $\Omega_c(M)$ to $\Omega(M)$ by

$$I\!D'_M(\alpha) = \Big(\beta \mapsto \int_M \beta \wedge \alpha\Big), \quad \forall \alpha \in \Omega(M), \quad \forall \beta \in \Omega_c(M),$$

so that the diagram of morphisms of complexes

$$\begin{array}{ccc}
\Omega(M)[d_M] & \xhookrightarrow[\text{q.i.}]{I\!D'_M} & \Omega_c(M)^\vee \\
{\scriptstyle\subseteq}\big\uparrow & & \big\uparrow \\
\Omega_c(M)[d_M] & \xhookrightarrow{I\!D'_M} & \Omega(M)^\vee
\end{array}$$

is commutative with its upper row a *quasi-isomorphism*, since it is simply the right Poincaré adjunction map $I\!D_M$ up to a sign ± 1 related to the anticommutativity of the wedge product $\beta \wedge \alpha = (-1)^{[\alpha][\beta]} \alpha \wedge \beta$.

Definition 2.7.1 If $f : M \to N$ is a proper map between oriented manifolds, the *Gysin morphism associated with* f is the map

$$\boxed{f_* : H(M)[d_M] \to H(N)[d_N]} \tag{2.56}$$

making commutative the diagram

$$\begin{array}{ccc}
H(M)[d_M] & \xrightarrow[\simeq]{D'_M} & H_c(M)^\vee \\
{\scriptstyle f_*}\big\downarrow & & \big\downarrow {\scriptstyle H_c(f^*)^\vee} \\
H(N)[d_N] & \xrightarrow[\simeq]{D'_N} & H_c(N)^\vee
\end{array}$$

Theorem 2.7.2 (and more Definitions)

1. *Let $f : M \to N$ be a proper map between oriented manifolds. Then, for $\beta \in H_c(N)$ and $\alpha \in H(M)$, the equation in X,*

$$\int_M f^*([\beta]) \cup [\alpha] = \int_N [\beta] \cup X\,, \tag{2.57}$$

admits a unique solution in $H(N)$, namely $X = f_*[\alpha]$. Furthermore,

(a) *f_* is a morphism of $H_c(N)$-modules, i.e. the following equality, called the* projection formula for proper maps,

$$\boxed{f_*\big(f^*([\beta]) \cup [\alpha]\big) = [\beta] \cup f_*[\alpha]} \tag{2.58}$$

holds for all $[\beta] \in H_c(N)$, $[\alpha] \in H(M)$.

(b) *The pullback $f^* : H_c(N) \to H_c(M)$ is an isomorphism if and only if the Gysin morphism $f_* : H(M)[d_M] \to H(N)[d_N]$ is an isomorphism.*

2. *The following correspondence is a covariant functor:*

$$(\_)_* : \mathrm{Man}^{\mathrm{or}}_{\mathrm{pr}} \rightsquigarrow \mathrm{GV}(\mathbb{R}) \qquad \textit{with} \qquad \begin{cases} M \rightsquigarrow M_* := H(M)[d_M] \\ f \rightsquigarrow f_* \end{cases}$$

We refer to it as the Gysin functor for proper maps

3. *The natural map* $\phi(\_) : H_c(\_)[d\_] \to H(\_)[d\_]$ *(Sect. 2.2.3) is a homomorphism of Gysin functors* $(\_)_! \to (\_)_*$ *on the category* $\mathrm{Man}^{\mathrm{or}}_{\mathrm{pr}}$*, i.e. the following diagrams are natural and commutative.*

$$\begin{array}{ccc} H_c(M)[d_M] & \xrightarrow{\phi(M)} & H(M)[d_M] \\ {\scriptstyle f_!}\downarrow & & \downarrow{\scriptstyle f_*} \\ H_c(N)[d_N] & \xrightarrow{\phi(N)} & H(N)[d_N] \end{array} \tag{2.59}$$

Proof (1,2) Same as 2.6.2.1. (3) Immediate after definitions. □

2.8 Constructions of Gysin Morphisms

We summarize the steps in the construction of the Gysin morphisms.

2.8.1 *The Proper Case*

Let $f : M \to N$ be a **proper** map of oriented manifolds. To $\alpha \in \Omega(M)$ we assign the linear form on $\Omega_c(N)$ defined by $I\!D'_f(\alpha) : \beta \mapsto \int_M f^*\beta \wedge \alpha$. In this way we obtain the diagram

$$\begin{array}{ccc} \Omega(M)[d_M] & \overset{f_*}{\dashrightarrow} & \Omega(N)[d_N] \\ & {\scriptstyle I\!D'_f}\searrow \quad \oplus & \downarrow{\scriptstyle I\!D'_N \text{ (quasi-iso)}} \\ & & \Omega_c(N)^\vee \end{array}$$

which may be closed in cohomology, since $I\!D'_N$ is a quasi-isomorphism. Note that the closing arrow f_*, the Gysin morphism for proper maps, in general exists *only* at the cohomology level.

2.8.2 The General Case

Let $f : M \to N$ be a map of oriented manifolds. To $\beta \in \Omega_c(M)$ we assign the linear form on $\Omega(N)$ defined by $\mathbb{D}'_f(\beta) : \alpha \mapsto \int_M f^*\alpha \wedge \beta$. In this way we obtain the diagram

$$\begin{array}{ccc} \Omega_c(M)[d_M] & \overset{f_!}{\cdots\cdots\rightarrow} & \Omega_c(N)[d_N \\ & \underset{\mathbb{D}'_f}{\searrow} \quad \oplus & \downarrow \mathbb{D}'_N \\ & & \Omega(N)^\vee \end{array} \quad \left(\begin{array}{c}\text{quasi-iso, when } N\\ \text{is of finite type}\end{array}\right) \tag{2.60}$$

which may be closed in cohomology (as in the proper case), when N **is of finite type**, since then $\mathbb{D}'_N$ is a quasi-isomorphism (2.6.1.2).

When N is not of finite type, we fix a filtrant cover $\mathscr{U}$ of N made up of open finite type subspaces of N (see 2.5.3.1), and replace $\mathbb{D}'_N$ with $\mathbb{D}'_{\mathscr{U}}$. In this way, we get (see 2.6.3.1–(2,3)), the following diagram:

$$\begin{array}{ccccc} \Omega_c(M)[d_M] & \overset{f_!}{\cdots\cdots\rightarrow} & \Omega_c(N)[d_N] & \overset{=}{\cdots\cdots} & \Omega_c(N)[d_N] \\ & \underset{\mathbb{D}'_{f,\mathscr{U}}}{\searrow} \quad \oplus & \downarrow \mathbb{D}'_{\mathscr{U}} \text{ (quasi-iso)} & & \vdots\, \mathbb{D}'_N \\ & & \varinjlim_{U \in \mathscr{U}} \Omega(U)^\vee & \underset{\subseteq}{\cdots\cdots\rightarrow} & \Omega(N)^\vee \end{array}$$

where $\mathbb{D}'_{f,\mathscr{U}}$ is defined as follows. For $\beta \in \Omega_c(M)$ denote by $|\beta|$ its support and by $\mathscr{U}_\beta \subseteq \mathscr{U}$ the system consisting of $U \in \mathscr{U}$ s.t. $|\beta| \subseteq f^{-1}U$.

One has a natural map $\varinjlim_{\mathscr{U}_\beta} \Omega(U)^\vee \to \varinjlim_{\mathscr{U}} \Omega(U)^\vee$ (which in fact is bijective). Now, for every $U \in \mathscr{U}_\beta$ the linear map $\left(\int_M f^*(\_) \wedge \beta\right) : \Omega(U) \to \mathbb{R}$, is well-defined and compatible with restriction, so that it defines an element of $\varinjlim_{\mathscr{U}_\beta} \Omega(U)^\vee$, and then of $\varinjlim_{\mathscr{U}} \Omega(U)^\vee$. This element is $\mathbb{D}'_{f,\mathscr{U}}(\beta)$ by definition.

The closing arrow $f_!$, *the Gysin morphism associated with a general map f*, is then defined in cohomology as the composition $D'^{\,-1}_{\mathscr{U}} \circ H(\mathbb{D}_{f,\mathscr{U}})$.

Remark 2.8.2.1 In all cases, the Gysin morphism appears as the composition of a morphism in the category $C(\mathrm{Vec}(\mathbb{R}))$ with the '*inverse*' of a quasi-isomorphism. While this is generally impossible in $C(\mathrm{Vec}(\mathbb{R}))$, it is possible in the *derived category of complexes* $\mathcal{D}(\mathrm{Vec}(\mathbb{R}))$, since its fundamental property is that: *a morphism in derived category is an isomorphism if and only if it induces an isomorphism in cohomology* (*cf*. Sects. A.1.6 and A.1.6.3). Gysin morphisms are hence naturally defined in derived categories.

Chapter 3
Poincaré Duality Relative to a Base Space

In this chapter we extend the concepts of Orientability, Differential Form with Compact Support, Integration and Poincaré Adjunctions from manifolds to fiber bundles with the aim of extending the definition of Poincaré Duality to make duality be *relative* to a base space. This is an important step towards the *equivariant cohomology of a G-space X*, as this is the name given to the cohomology of the total space of the fiber bundle $X_G := \mathbb{E}G \times_G X \twoheadrightarrow \mathbb{B}G$ where $\mathbb{B}G$ is a classifying space for the Lie group G (*cf*. Sect. 4.7).

3.1 Fiber Bundles

A map $\pi : E \to B$ between topological spaces is a *locally trivial fibration of fiber F*, if there is an open cover $\mathcal{U} := \{U_i\}_{i \in \mathfrak{I}}$ of B such that for each $i \in \mathfrak{I}$ there exists a homeomorphism $\Phi_i : \pi^{-1}(U_i) \to U_i \times F$ such that $p \circ \Phi_i = \pi$,

$$\begin{array}{ccc} \pi^{-1}(U_i) & \xrightarrow[\sim]{\Phi_i} & U_i \times F \\ {\scriptstyle \pi}\downarrow & \oplus & \downarrow{\scriptstyle p} \\ U_i & = & U_i \end{array} \qquad \text{where} \quad p(x, y) = x.$$

3.1.1 *Terminology*

F-1. The subspaces U_i are said to be *trivializing* for π. The trivialization Φ_i will also be denoted $(\pi(\_), \varphi_i(\_)) : \pi^{-1}(U_i) \to U_i \times F$, with $\varphi_i : \pi^{-1}(U_i) \to F$.

A. Arabia, *Equivariant Poincaré Duality on G-Manifolds*, Lecture Notes in Mathematics 2288, https://doi.org/10.1007/978-3-030-70440-7_3

F-2. For $b \in B$, we denote by $F_b := \pi^{-1}(b)$ the *fiber of* π *at* b, which is, after the definition, a closed subspace homeomorphic to F.
F-3. The quadruple (E, B, π, F) is called a *fiber bundle* of *total space* E, *base space* B, *projection map* π and *fiber space* F,
F-4. A *morphism of fiber bundles* from (E, B, π, F) to (E', B', π', F'), is a pair of maps $f : E' \to E$ and $\overline{f} : B' \to B$ such that

$$\pi \circ f = \overline{f} \circ \pi', \qquad \begin{array}{ccc} E' & \xrightarrow{f} & E \\ {\scriptstyle \pi'}\downarrow & \oplus & \downarrow{\scriptstyle \pi} \\ B' & \xrightarrow{\overline{f}} & B \end{array} \tag{3.1}$$

The map f therefore transforms the fiber $F'_{b'}$ into the fiber F_b, with $b := \overline{f}(b')$.
F-5. The definition of fiber bundles *of manifolds* and their morphisms are the same replacing spaces by manifolds and maps by differentiable maps.

Exercise 3.1.1.1 Let (E, B, π, F) be a fiber bundle of manifolds. Given a connected component E' of E, let $B' := \pi(E')$ and denote by $\pi' : E' \to B'$ the restriction of π to E'. Show that B' is a connected component of B and that π' is a locally trivial fibration of fiber F', union of connected components of F of dimension $d_{E'} - d_{B'}$. Hence, (E', B', π', F') is a fiber bundle of equidimensional manifolds. The inclusion maps $E' \subseteq E$, $B' \subseteq B$ are open and define a morphism of fiber bundles from (E', B', π', F') to (E, B, π, F). (♝, p. 336)

3.1.2 *The Categories* $\mathbf{Top}_B$ *and* $\mathbf{Man}_B$

The definition of a fibre bundle is the mathematical formulation of the familiar idea of a space E which is obtained by glueing together a family $\{F_b\}_{b\in B}$ of copies of a given fiber space F, parametrized by the elements of a base space B. When the fibers F_b are no longer the same, we have the more general concept of *a space* X *above* B, which, while in essence a simple map $\pi : X \to B$, may be better understood as the glueing of the family of fibers $\{F_b := \pi^{-1}(b)\}_{b\in B}$.

Given topological space B, we denote by Top_B the category whose objects are *spaces above* B, i.e. topological spaces X together with a continuous map $\pi : X \to B$. A morphism in Top_B from $\pi' : X' \to B$ to $\pi : X \to B$ is then a continuous map $f : X' \to X$ such that

$$\pi \circ f = \pi', \qquad \begin{array}{ccc} X' & \xrightarrow{f} & X \\ & {\scriptstyle \pi'}\searrow \quad \swarrow{\scriptstyle \pi} & \\ & B & \end{array}$$

A morphism thus appears as a continuous glueing of a family of continuous maps between fibers $\{f_b : \pi'^{-1}(b) \to \pi^{-1}(b)\}_{b\in B}$.

When data consist of differentiable manifolds and differentiable maps, we define in the same way the category Man_B of *manifolds above B*.

What makes interesting these categories are the constraints imposed by the presence of common base space B. Not only does it restrict the sets of morphisms between spaces, but, most importantly for us, the target of the cohomology functor becomes the category of $H(B)$-modules (which is a key point in equivariant cohomology).

3.1.3 The Relative Point of View

The idea of fixing a base space B turned out to be a very fruitful heuristic in the hands of Grothendieck who named it *the relative point of view*, as opposed to the *absolute point of view* where $B := \{\bullet\}$ and which, therefore, do not impose any constraint to morphisms. A fundamental transformation in this heuristic, which corresponds to *changing of point of view*, is the *base change functor*

$$B' \times_B (\_) : \mathrm{Top}_B \rightsquigarrow \mathrm{Top}_{B'}\,,$$

based in the operation of *fiber product*, which we now review.

3.1.4 Fiber Product

Given two morphisms $\pi_i : X_i \to Z$ with equal target in a category $\mathcal{C}$, a *fiber product*[1] of (π_1, π_2) is an object in $\mathcal{C}$, denoted by $X_1 \times_{(\pi_1,\pi_2)} X_2$, together with two morphisms $p_i : X_1 \times_{(\pi_1,\pi_2)} X_2 \to X_i$ such that $\pi_1 \circ p_1 = \pi_2 \circ p_2$

$$\begin{array}{ccccc} & & X_1 & & \\ & {\scriptstyle p_1}\nearrow & & \searrow{\scriptstyle \pi_1} & \\ X_1 \times_{(\pi_1,\pi_2)} X_2 & & \oplus & & Z\,, \\ & {\scriptstyle p_2}\searrow & & \nearrow{\scriptstyle \pi_2} & \\ & & X_2 & & \end{array}$$

[1] Also called *fibre product*, *fibered product* or *Cartesian square*.

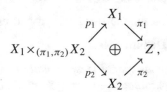

representing the functor $\mathrm{Mor}_{\mathcal{C}}(_, (\pi_1, \pi_2)) : \mathcal{C} \to \mathrm{Set}$, which associates with $W \in \mathcal{C}$, the set of morphisms $w_i : W \to X_i$ such that $\pi_1 \circ w_1 = \pi_2 \circ w_2$.

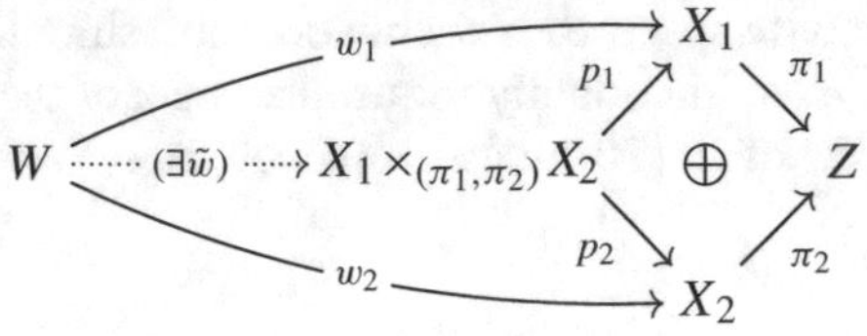

(3.2)

Recall that what is meant by this is that the correspondence

$$\mathrm{Mor}_{\mathcal{C}}(_, X_1 \times_{(\pi_1,\pi_2)} X_2) \to \mathrm{Mor}_{\mathcal{C}}(_, (\pi_1, \pi_2)), \quad \tilde{w} \mapsto (p_1 \circ \tilde{w}, p_2 \circ \tilde{w}),$$

is an isomorphism of functors.

The existence and uniqueness of $\tilde{w} : W \to X_1 \times_{(\pi_1,\pi_2)} X_2$ verifying $p_i \circ \tilde{w} = w_i$ is usually referred to as *the universal property of the fiber product*. It implies that fiber products, when they exist, are unique up to canonical isomorphism.

The category $\mathcal{C}$ is said to *have fiber products* if, for every pair of morphisms $\pi_i : X_i \to Z$, the fiber product $X_1 \times_{(\pi_1,\pi_2)} X_2$ exists.

Notice that if Z is a final object in $\mathcal{C}$, for example a singleton in a category of spaces, then the natural map $X_1 \times_Z X_2 \to X_1 \times X_2$ is an isomorphism. The concept of fiber product thus naturally extends that of product.

Convention The fiber product $X_1 \times_{(\pi_1,\pi_2)} X_2$ will be denoted simply by $X_1 \times_Z X_2$ when the morphisms π_i are understood and no confusion can arise.

Exercise 3.1.4.1 Let $\mathcal{C}$ be a category with fiber products. (☝, p. 336)

1. For $\pi \in \mathrm{Mor}_{\mathcal{C}}(X, W)$, let $X \times_W W := X \times_{(\pi,\mathrm{id}_W)} W$ and show that the morphism $(\mathrm{id}_X, \pi) : X \to X \times_W W$ is an isomorphism.
2. Define a natural isomorphism $X_1 \times_Z X_2 \to (X_1 \times_W W) \times_Z X_2$.

The following proposition summarizes basic properties of the fiber product.

Proposition 3.1.4.2 (And Definitions)

1. *The category of topological spaces (resp. locally compact) has fiber products. More precisely, given continuous maps $\pi_i : X_i \to Z$, a fiber product of (π_1, π_2) is given by the closed subspace of $X_1 \times X_2$:*

$$X_1 \times_Z X_2 := \{(x_1, x_2) \in X_1 \times X_2 \mid \pi_1(x_1) = \pi_2(x_2)\} \tag{3.3}$$

together with the maps

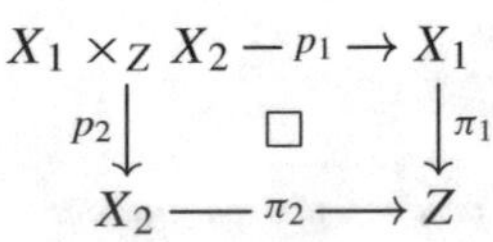

$$\begin{aligned} p_i : X_1 \times_Z X_2 &\to X_i \\ (x_1, x_2) &\mapsto x_i \end{aligned} \qquad \begin{array}{ccc} X_1 \times_Z X_2 & \xrightarrow{p_1} & X_1 \\ {\scriptstyle p_2}\downarrow & \square & \downarrow{\scriptstyle \pi_1} \\ X_2 & \xrightarrow{\pi_2} & Z \end{array} \tag{3.4}$$

Furthermore, the restriction to fibers:

$$p_1 : p_2^{-1}(x_2) \to \pi_1^{-1}(\pi_2(x_2)) \quad and \quad p_2 : p_1^{-1}(x_1) \to \pi_2^{-1}(\pi_1(x_1))$$

are homeomorphisms for all $x_i \in X_i$.

Terminology *A commutative diagram of spaces is said to be* Cartesian *if it is isomorphic to the diagram of a fiber product (3.4). A square box is then usually drawn inside the diagram indicate this.*

2. *The commutative diagram* (I) *of continuous maps is Cartesian if and only if it is* locally Cartesian *relative to Z, i.e. for every $z \in Z$ there exists an open neighborhood $V \subseteq Z$ such that the diagram* (II)*, with $U := f_1^{-1}(\pi_1^{-1}(V)) = f_2^{-1}(\pi_2^{-1}(V))$, is Cartesian.*

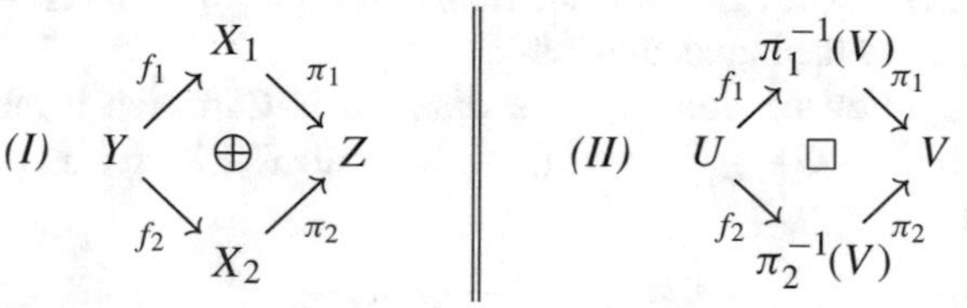

$$(I) \quad \begin{array}{ccc} & X_1 & \\ f_1 \nearrow & & \searrow \pi_1 \\ Y & \oplus & Z \\ f_2 \searrow & & \nearrow \pi_2 \\ & X_2 & \end{array} \qquad\qquad (II) \quad \begin{array}{ccc} & \pi_1^{-1}(V) & \\ f_1 \nearrow & & \searrow \pi_1 \\ U & \square & V \\ f_2 \searrow & & \nearrow \pi_2 \\ & \pi_2^{-1}(V) & \end{array} \qquad (3.5)$$

3. *The fiber product $X_1 \times_{(\pi_1,\pi_2)} X_2$ together with the map $\tilde{\pi} := \pi \circ p_1 = \pi \circ p_2$, is a product of π_1 and π_2 in Top_B.*
   Furthermore, if (E_1, B, π_1, F_1) and (E_2, B, π_2, F_2) are fiber bundles, then $\tilde{\pi} : E_1 \times_B E_2 \to B$ is a fiber bundle $(E_1 \times_B E_2, B, \tilde{\pi}, F_1 \times F_2)$.

Proof

(1) By the universal property of products, a pair of maps $w_i : W \to X_i$ factors as the composition of $(w_1, w_2) : W \to X_1 \times X_2$ and the canonical projections $p_i : X_1 \times X_2 \to X_i$. If moreover the maps w_i verify $\pi_1 \circ w_1 = \pi_2 \circ w_2$, then the image of (w_1, w_2) is the closed subspace $X_1 \times_{(\pi_1,\pi_2)} X_2$ (locally compact when the X_i's are so) and the first part of (1) is proved. The converse is obvious.

   The statement about the restrictions of the p_i's to fibers follows by a straightforward verification, after the definition (3.3).

(2) If (I) is a fiber product, then every pair of maps $(w_1, w_2) : W \to \pi_i^{-1}(V)$ such that $\pi_1 \circ w_1 = \pi_2 \circ w_2$, uniquely factors through Y, hence through U, proving that (II) in (3.5) is Cartesian.

   Conversely, by the universal property of fiber product, if (I) is commutative, we have a canonical continuous map $\xi : Y \to X_1 \times_{(\pi_1,\pi_2)} X_2$. This map is bijective since we are assuming (I) locally Cartesian and that, after (1), the underlying set of a fiber product of topological spaces is the fiber product of the underlying sets. The map ξ is then a homeomorphism since its restrictions to the open subspaces $U := f_1^{-1}(\pi_1^{-1}(V))$, which cover Y, are homeomorphisms.

(3) In the category Top_B, to give a morphism from $p_W : W \to B$ to the pair $\pi_i : X_i \to B$, means to give a pair of maps $w_i : W \to X_i$ such that $\pi_1 \circ w_i = p_W$,

hence, to give the map $(w_1, w_2) : W \to X_1 \times_B X_2$, after (3.2), which is a morphism in Top_B since $\tilde{\pi} \circ (w_1, w_2) = p_W$.

For the second part of (3), we note that, given an open subspace $U \subseteq B$ and setting $U_i := \pi_i^{-1}(U)$, the natural map $U_1 \times_U U_2 \to X_1 \times_B X_2$ is, after the definition (3.3), an open embedding. Therefore, if U is trivializing for both fiber bundles (E_i, B, π_i, F_i), we have $(\pi_i : U_i \to B) \sim (p_1 : U \times F_i \to B)$ in Top_B, whence

$$(U \times F_1) \times_U (U \times F_2) \sim U_1 \times_B U_2 \hookrightarrow X_1 \times_B X_2 \,,$$

but, $(U \times F_1) \times_U (U \times F_2) \sim U \times (F_1 \times F_2)$ (*cf*. Exercise 3.1.4.1). □

Exercise 3.1.4.3 Let $H \subseteq G$ be a group inclusion. Given a G-equivariant continuous map $f : X \to Y$ between G-spaces, denote by $\nu_{H,G,X} : X/H \to X/G$ and $\nu_{H,G,Y} : Y/H \to Y/G$ the orbit maps and let $f_H : X/H \to Y/H$ and $f_G : X/G \to Y/G$ be the induced maps.

Show that the following commutative diagram is Cartesian if and only if the restrictions $f_H : G \cdot \overline{x} \to G \cdot f_H(\overline{x})$ are bijective for all $\overline{x} \in X/H$, (🕮, p. 337)

$$\begin{array}{ccc} X/H & \xrightarrow{\nu_{H,G,X}} & X/G \\ {\scriptstyle f_H}\downarrow & \oplus & \downarrow{\scriptstyle f_G} \\ Y/H & \xrightarrow{\nu_{H,G,Y}} & Y/G\,. \end{array}$$

3.1.5 The Base Change Functor

Given a continuous map $\bar{h} : B' \to B$, consider the correspondence

$$\bar{h}^{-1}(\_) : \mathrm{Top}_B \rightsquigarrow \mathrm{Top}_{B'}$$

which associates with $(\pi : X \to B)$, the map $\left(\bar{h}^{-1}(\pi) : X \times_B B' \to B'\right)$ in the Cartesian diagram

$$\begin{array}{ccc} X \times_B B' & \xrightarrow{h} & X \\ {\scriptstyle \bar{h}^{-1}(\pi)}\downarrow & \square & \downarrow{\scriptstyle \pi} \\ B' & \xrightarrow{\bar{h}} & B \end{array} \quad \text{where} \quad \begin{cases} h(x, y) := x\,, \\ \bar{h}^{-1}(\pi)(x, y) := y\,, \end{cases}$$

and which associates with $\begin{array}{ccc} X' & \xrightarrow{f} & X \\ & {\scriptstyle \pi'}\searrow \quad \swarrow{\scriptstyle \pi} & \\ & B & \end{array} \in \mathrm{Mor}_{\mathrm{Top}_B}(\pi', \pi)$, the morphism

$$\begin{array}{ccc} X' \times_B B' & \xrightarrow{(f,\mathrm{id})} & X \times_B B' \\ & {\scriptstyle \bar{h}^{-1}(\pi')}\searrow \quad \swarrow{\scriptstyle \bar{h}^{-1}(\pi)} & \\ & B' & \end{array} \in \mathrm{Mor}_{\mathrm{Top}_{B'}}(\bar{h}^{-1}(\pi'), \bar{h}^{-1}(\pi))\,.$$

Proposition 3.1.5.1

1. *The correspondence $\bar{h}^{-1}(\_) : \mathrm{Top}_B \rightsquigarrow \mathrm{Top}_{B'}$ is a covariant functor.* *Furthermore, if $(\pi : X \to B) \in \mathrm{Top}_B$ is a locally trivial fibration of fiber F, then $h^{-1}(\pi)$ is a locally trivial fibration of fiber F.*
2. *Given a continuous map $\bar{h}' : B'' \to B'$, we have an isomorphism of functors*

$$\bar{h}'^{-1}(\_) \circ \bar{h}^{-1}(\_) \to (\bar{h} \circ \bar{h}')^{-1}(\_)\,.$$

Terminology The functor $\bar{h}^{-1}(\_) : \mathrm{Top}_B \rightsquigarrow \mathrm{Top}_{B'}$ is called the *pullback*, or the *base change*, functor induced by h.

Proof

(1) The fact that $\bar{h}^{-1}(\_)$ is a covariant functor, easily follows from the fact that, after 3.1.4.2-(3), base change results by composing two functorial operations, first the product in Top_B by $(h : B' \to B)$, which already gives the covariant endofunctor $(\_) \times_B B' : \mathrm{Top}_B \rightsquigarrow \mathrm{Top}_B$, and second the factorization of the projection map $\pi_{(\_)} \circ p_1$ through $p_2 : (\_) \times_B B' \to B'$:

$$\begin{array}{ccc} & & (\_) \times_B B' \\ & {\scriptstyle p_2}\swarrow & \downarrow{\scriptstyle \pi_{(\_)} \circ p_1} \\ B' & \xrightarrow{\bar{h}} & B \end{array}$$

To see that $\bar{h}^{-1}(\_)$ preserves locally trivial fibrations of given fiber F, we proceed as for 3.1.4.2-(3). If $U \subseteq B$ is an open trivializing subspace for $\pi : X \to B$, then $(\pi : \pi^{-1}(U) \to B)$ is isomorphic to $(p_1 : U \times F \to B)$ in Top_B, and we have the open embedding

$$(U \times F) \times_U \bar{h}^{-1}(U) \sim p^{-1}(U) \times_B \bar{h}^{-1}(U) \hookrightarrow X \times_B B'\,,$$

where $(U \times F) \times_U \bar{h}^{-1}(U) \simeq F \times \bar{h}^{-1}(U)$ (*cf*. Exercise 3.1.4.1).

(2) Concatenating the Cartesian diagrams generated by $\bar{h}^{-1}(\_)$ and $\bar{h}'^{-1}(\_)$:

$$\begin{array}{ccccc} X \times_B B' \times_{B'} B'' & \xrightarrow{h'} & X \times_B B' & \xrightarrow{h} & X \\ {\scriptstyle \bar{h}'^{-1}(\bar{h}^{-1}(\pi))}\downarrow & \square & {\scriptstyle \bar{h}^{-1}(\pi)}\downarrow & \square & \downarrow{\scriptstyle \pi} \\ B'' & \xrightarrow{\bar{h}'} & B' & \xrightarrow{\bar{h}} & B \end{array}$$

we immediately get a commutative diagram

$$\begin{array}{ccc} X \times_B B' \times_{B'} B'' & \longrightarrow & X \times_B B'' \\ {\scriptstyle \bar{h}'^{-1}(\bar{h}^{-1}(\pi))}\downarrow & \oplus & \downarrow{\scriptstyle (\bar{h}\circ\bar{h}')^{-1}(\pi)} \\ B'' & = & B'' \end{array}$$

which is natural with respect to $(\pi : X \to B)$ and where the horizontal arrow is a homeomorphism (*cf*. Exercise 3.1.4.1). □

3.1.6 Fiber Products of Fiber Bundles of Manifolds

Fiber products may not always exist in some categories of topological spaces, for example in the category of manifolds (exercise ! (𝕴, p. 337)). But they do exist when one of the maps $p_i : X_i \to Z$ is a locally trivial fibration.

Proposition 3.1.6.1

1. *Let (E, B, π, F) be a fiber bundle of manifolds and let B' be a manifold. For every map $\overline{f} : B' \to B$, the fiber product $E' := E \times_B B'$ exists in the category* Man *of manifolds and coincides as a topological space with the fiber product for locally compact spaces (3.3).*
   - *The map $\pi' : E' \to B'$, defined as $\pi'(x_1, x_2) := x_2$, is a locally trivial fibration of manifolds with fiber F.*
   - *The diagram*

$$\pi \circ f = \overline{f} \circ \pi', \qquad \begin{array}{ccc} E' & \xrightarrow{f} & E \\ {\scriptstyle \pi'}\downarrow & \square & \downarrow{\scriptstyle \pi} \\ B' & \xrightarrow{\overline{f}} & B \end{array} \tag{3.6}$$

   where $f : E' \to E$ denotes the map $f(x_1, x_2) := x_1$, is a Cartesian diagram.

   Terminology *The fiber bundle (E', B', π', F) is called* the pullback bundle of (E, B, π, F) by $\overline{f}$ *and is denoted by $\overline{f}^{-1}(E, B, \pi, F)$, or simply $\overline{f}^{-1}(E)$.*

2. *A commutative diagram of differentiable maps*

$$\begin{array}{ccc} E' & \xrightarrow{f} & E \\ {\scriptstyle \pi'}\downarrow & \oplus & \downarrow{\scriptstyle \pi} \\ B' & \xrightarrow{\overline{f}} & B \end{array}$$

   where π is a locally trivial fibration, is a Cartesian diagram if and only if it is locally Cartesian relative to B (cf. *Proposition 3.1.4.2–(2)).*

3. *Given two fiber bundles of manifolds* (E_i, B, π_i, F_i) *with base* B, *the product* $(E_1 \times_B E_2, B, \tilde{\pi}, F_1 \times F_2)$ *exists in the category* Man_B *of manifolds above* B *(see 3.1.8). It coincides as a topological space with the product in* Top_B.

Proof

(1) The differential structure of the topological space $E' := B' \times_B E$ (3.3), will be given by an atlas of differentiable manifolds. Let

$$\mathfrak{E} := \{\Phi_\mathrm{i} := (\pi, \varphi_\mathrm{i}) : \pi^{-1}(U_\mathrm{i}) \simeq U_\mathrm{i} \times F\}_{\mathrm{i} \in \mathfrak{I}}, \tag{3.7}$$

be an open cover of trivializations for the fiber bundle of manifolds $\pi : E \to B$ (see Exercise 3.1.9.3-(4)). The transition maps of $\mathfrak{E}$ are the diffeomorphisms

$$T_{\mathrm{ii}'} : U_{\mathrm{ii}'} \times F \to U_{\mathrm{ii}'} \times F, \quad T_{\mathrm{ii}'}(x, y) = (x, \varphi_{\mathrm{ii}'}(x, y)), \tag{3.8}$$

where $\varphi_{\mathrm{ii}'}(x, \_) : F \to F$ is a diffeomorphism depending on $x \in U_{\mathrm{i},\mathrm{i}'}$ differentiably, which is defined by the constraint

$$\varphi_{\mathrm{ii}'}(x, \varphi_\mathrm{i}(y)) = \varphi_{\mathrm{i}'}(y), \quad \forall y \in \pi^{-1}(U_{\mathrm{ii}'}). \tag{3.9}$$

Let $\mathfrak{B}' := \{V_\mathrm{i}\}_{\mathrm{i} \in \mathfrak{I}}$ be the open cover of B', where $V_\mathrm{i} := \overline{f}^{-1}(U_\mathrm{i})$. We have

$$\pi'^{-1}(V_\mathrm{i}) = V_\mathrm{i} \times_{U_\mathrm{i}} \pi^{-1}(U_\mathrm{i}),$$

by the universal property of fiber products for *topological spaces*.

Now, a trivialization $(\pi, \varphi_\mathrm{i}) : \pi^{-1}(U_\mathrm{i}) \simeq U_\mathrm{i} \times F$ in $\mathfrak{E}$, immediately gives us a trivialization $(\pi', \varphi'_\mathrm{i}) : \pi'^{-1}(V_\mathrm{i}) \simeq V_\mathrm{i} \times F$, by setting (*cf*. Exercise 3.1.4.1)

$$\begin{gathered} \pi'^{-1}(V_\mathrm{j}) = V_\mathrm{i} \times_{U_\mathrm{i}} \pi^{-1}(U_\mathrm{i}) \xrightarrow[\simeq]{} V_\mathrm{i} \times_{U_\mathrm{i}} (U_\mathrm{i} \times F) = V_\mathrm{i} \times F \\ (\pi', \varphi'_\mathrm{i})([x, y]) := (x, \varphi_\mathrm{i}(y)) \end{gathered} \tag{3.10}$$

where $[x, y]$ denotes an element in the fiber product $V_\mathrm{j} \times_{U_\mathrm{i}} \pi^{-1}(U_\mathrm{i})$.

The family of trivializations $\mathfrak{E}' := \{(\pi', \varphi'_\mathrm{i}) : \pi'^{-1}(V_\mathrm{i}) \to V_\mathrm{i} \times F\}_{\mathrm{i} \in \mathfrak{I}}$ is then an open cover of the space E' with transition homeomorphisms, the maps

$$T_{\mathrm{ii}'}(x', y) = (x', \varphi_{\mathrm{ii}'}(\overline{f}(x'), y)), \quad \forall x' \in V_{\mathrm{ii}'}, \tag{3.11}$$

which are clearly étale, hence diffeomorphic. The cover $\mathfrak{E}'$ therefore defines a differentiable structure on the topological fiber product $E' := B' \times_B E$.

The map f is differentiable since it appears through the trivializations $(\pi', \varphi'_{\mathrm{i}})$ and $(\pi, \varphi_{\mathrm{i}})$ as the differentiable map $(\overline{f}, \mathrm{id}_F)$

$$\begin{array}{ccc} \pi'^{-1}(V_{\mathrm{i}}) & \xrightarrow{f} & \pi^{-1}(U_{\mathrm{i}}) \\ {\scriptstyle (\pi',\varphi'_{\mathrm{i}})}\downarrow\simeq & & \simeq\downarrow{\scriptstyle (\pi,\varphi_{\mathrm{i}})} \\ V_{\mathrm{i}} \times F & \xrightarrow{(\overline{f},\mathrm{id}_F)} & U_{\mathrm{i}} \times F \end{array}$$

In the same way, for any manifold M and maps $g : M \to E$ and $\overline{g}' : M \to B'$, such that $\pi \circ g = \overline{g}' \circ \overline{f}$, the induced continuous map $\tilde{g} : M \to E' := B' \times_B E$, appears through the trivialization $(\pi', \varphi'_{\mathrm{i}})$ as the differentiable map

$$m \mapsto (\overline{g}'(m), \varphi_{\mathrm{i}}(g(m)))\,.$$

The manifold $E' := B' \times_B E$ with $f : E' \to E$ and $\pi' : E' \to B'$ is therefore a fiber product for $\pi : E \to B$ and $\overline{f} : B' \to B$, in the category of manifolds.

(2) Same proof as 3.1.4.2-(2).

(3) As in the proof of 3.1.5.1-(1), the fact that the product in Man_B is obtained by the base change $(\_) \times_B E' : \mathrm{Man}_B \rightsquigarrow \mathrm{Man}_{E'}$, which we proved it exists, followed by the extension of the base through the (differentiable) map $\pi' : E' \to B$, immediately shows that the product exists in Man_B.

It only remains to show that $\tilde{\pi} : E \times_B E' \to B$ is a locally trivial fibration of manifolds of fiber $F_1 \times F_2$. This follows as in 3.1.4.2-(3), by considering an atlas $\mathscr{U} := \{U_{\mathrm{j}}\}_{\mathrm{j}\in\mathfrak{J}}$ of B which is simultaneously trivializing for both fiber bundles (E_i, B, π_i, F_i) with respective trivializations, for $i = 1, 2$,

$$(\pi_i, \varphi_{i,\mathrm{j}}) : \pi_i^{-1}(U_{\mathrm{j}}) \to U_{\mathrm{j}} \times F_i\,.$$

We then have the trivialization of $\tilde{\pi}^{-1}(U_{\mathrm{j}})$ (*cf*. Exercise 3.1.4.1)

$$\pi_1^{-1}(U_{\mathrm{j}}) \times_B \pi_2^{-1}(U_{\mathrm{j}}) \to (U_{\mathrm{j}} \times F_1) \times_{U_{\mathrm{j}}} (U_{\mathrm{j}} \times F_1) = U_{\mathrm{j}} \times F_1 \times F_2$$

given by the map

$$\Phi_{\mathrm{j}} := (\tilde{\pi}, (\varphi_{1,\mathrm{j}} \circ p_1, \varphi_{2,\mathrm{j}} \circ p_2)) : \tilde{\pi}^{-1}(U_{\mathrm{j}}) \to U_{\mathrm{j}} \times F_1 \times F_2\,.$$

The transition map from $\Phi_{\mathrm{j}'}$ to Φ_{j},

$$T_{\mathrm{jj}'} : U_{\mathrm{jj}'} \times (F_1 \times F_2) \to U_{\mathrm{jj}'} \times (F_1 \times F_2)\,, \tag{3.12}$$

is therefore

$$T_{\mathrm{jj}'}(x, (y_1, y_2)) = (x, (\varphi_{1,\mathrm{jj}'}(x, y_1), \varphi_{2,\mathrm{jj}'}(x, y_2)))\,, \tag{3.13}$$

which is étale, hence diffeomorphic, since the maps $\varphi_{i,\mathrm{jj}'}$ are so (*cf*. 3.8). □

3.1.7 Orientable Fiber Bundles

A connected fiber bundle of manifolds (E, B, π, F) is said to be *orientable* if there exists $\omega_\pi \in \Omega^{d_F}(E)$ such that, for all $b \in B$, the restriction

$$\omega_\pi|_{F_b} \in \Omega^{d_F}(F_b), \tag{3.14}$$

is nowhere vanishing. A necessary, but not sufficient, condition for a fiber bundle to be orientable is that its fiber F be orientable (*cf*. Proposition 2.3.1.1). A non-connected fiber bundle is said to be orientable if its connected components are so.

Proposition 3.1.7.1 (and Definition) *A fiber bundle of manifolds (E, B, π, F), with orientable fiber F, is an orientable fiber bundle if and only if there exists an open cover $\mathscr{U} := \{U_\mathrm{i}\}_{\mathrm{i}\in\mathfrak{I}}$ of B by trivializing subspaces, and a trivialization $\Phi_\mathrm{i} := (\pi, \varphi_\mathrm{i}) : \pi^{-1}(U_\mathrm{i}) \to U_\mathrm{i} \times F$, for each $\mathrm{i} \in \mathfrak{I}$, such that in the transition maps from $\Phi_{\mathrm{i}'}$ to Φ_i (see (3.9))*

$$T_{\mathrm{i}\mathrm{i}'} : U_{\mathrm{i}\mathrm{i}'} \times F \to U_{\mathrm{i}\mathrm{i}'} \times F\,, \quad T_{\mathrm{i}\mathrm{i}'}(x, y) = (x, \varphi_{\mathrm{i}\mathrm{i}'}(x, y))\,,$$

the maps $\varphi_{\mathrm{i}\mathrm{i}'}(x, \_) : F \to F$ preserve the orientation, for all $x \in U_{\mathrm{i}\mathrm{i}'}$.

In that case, the family of trivializations $\{(\pi, \varphi_\mathrm{i}) : \pi^{-1}(U_\mathrm{i}) \to U_\mathrm{i} \times F\}_{\mathrm{i}\in\mathfrak{I}}$ is said to be oriented *(cf. 2.3.1).*

A nowhere vanishing differential form $\omega_F \in \Omega^{d_F}(F)$ then determines the differential form $\omega_\pi \in \Omega^{d_F}(E)$ in (3.14) by taking a partition of unity $\{\phi_\mathrm{i}\}_{\mathrm{i}\in\mathfrak{I}}$ subordinate to $\mathscr{U}$ and setting

$$\omega_\pi := \sum\nolimits_\mathrm{i} \phi_\mathrm{i}\, \varphi_\mathrm{i}^*(\omega_F) \in \Omega^m(E)\,.$$

The restriction to fibers $\omega_\pi|_{F_b} \in \Omega^{d_F}(F_b)$ are then nowhere vanishing.

Proof We assume, after Exercise 3.1.1.1, that E, B and F are equidimensional respectively of dimensions n, b and m.

Assume the orientability condition (3.14). Fix an orientation $[F]$ for F. Let $\mathscr{U} := \{U_\mathrm{i}\}_{\mathrm{i}\in\mathfrak{I}}$ be a cover of B by *connected* trivializing open subspaces.

Fix $\mathrm{i} \in \mathfrak{I}$, and fix a trivialization $(\pi, \varphi_\mathrm{i}) : \pi^{-1}(U_\mathrm{i}) \to U_\mathrm{i} \times F$, and denote its inverse by $\Psi_\mathrm{i} : U_\mathrm{i} \times F \to \pi^{-1}(U_\mathrm{i})$.

For each connected component F_c of F, we have the family of embeddings indexed by $x \in U_\mathrm{i}$

$$F_c \xrightarrow{\iota_{c,x}} U_\mathrm{i} \times F \xrightarrow{\Psi_\mathrm{i}} \pi^{-1}(U_\mathrm{i})\,,$$

where $\iota_{c,x}(y) = (x, y)$. Since U_i is connected, there exists $\epsilon_c = \pm 1$ such that

$$[F_c] = \epsilon_c\,[\iota_{c,x}^*\, \Psi_\mathrm{i}^*(\omega_\pi)]\,, \quad \forall x \in U_\mathrm{i}\,.$$

When $\epsilon_c = -1$, the maps $\iota_{c,x} : F_c \to \{x\} \times F_c$ inverse orientation for all $x \in U_{\mathfrak{i}}$, and we modify $(\pi, \varphi_{\mathfrak{i}})$ by composing $\varphi_{\mathfrak{i}} : \pi^{-1}(U_{\mathfrak{i}}) \to F = \coprod_c F_c$ with ι_{c,x_0}, for any arbitrary choice of $x_0 \in U_{\mathfrak{i}}$.

We proceed in the same way for each connected component F_c and we denote by $(\pi, \varphi'_{\mathfrak{i}}) : \pi^{-1}(U_{\mathfrak{i}}) \to U_{\mathfrak{i}} \times F$ the resulting modified trivialization, with inverse denoted by $\Psi'_{\mathfrak{i}}$. Then, by construction,

$$[F] = [\iota_x^* \Psi_{\mathfrak{i}}'^*(\omega_\pi)]\,, \quad \forall x \in U_{\mathfrak{i}}\,,$$

where $\iota_x : F \to U_{\mathfrak{i}} \times F$ is now simply $\iota_x(y) = (x, y)$.

We have thus a family of trivializations $\{(\pi, \varphi'_{\mathfrak{i}}) : \pi^{-1}(U_{\mathfrak{i}}) \to U_{\mathfrak{i}} \times F\}_{\mathfrak{i} \in \mathfrak{I}}$ whose transition maps respect well the orientations of fibers as announced.

Conversely, let $\omega_F \in \Omega^m(F)$ nowhere vanishing and such that $[\omega_F] = [F]$. Then, given an open cover $\mathscr{U} := \{U_{\mathfrak{i}}\}_{\mathfrak{i} \in \mathfrak{I}}$ of B by trivializing subspaces, and for each $\mathfrak{i} \in \mathfrak{I}$, a trivialization $(\pi, \varphi_{\mathfrak{i}}) : \pi^{-1}(U_{\mathfrak{i}}) \to U_{\mathfrak{i}} \times F$, we define

$$\omega_{\mathfrak{i}} := \varphi_{\mathfrak{i}}^*(\omega_F) \in \Omega^m(\pi^{-1}(U_{\mathfrak{i}}))\,.$$

Now, if $\{\phi_{\mathfrak{i}}\}_{\mathfrak{i} \in \mathfrak{I}}$ is a partition of unity subordinate to $\mathscr{U}$, the sum

$$\omega_\pi := \sum\nolimits_{\mathfrak{i}} \phi_{\mathfrak{i}}\, \omega_{\mathfrak{i}} \in \Omega^m(E)$$

is well-defined, and is such that, for $x \in B$, we have

$$\omega_\pi|_{F_x} = \sum\nolimits_{\mathfrak{i}} \phi_{\mathfrak{i}}(x)\, \varphi_{\mathfrak{i}\mathfrak{i}'}(x, \_)^*(\omega_F)|_{F_x} \tag{3.15}$$

for any $\mathfrak{i}' \in \mathfrak{I}$ such that $x \in U_{\mathfrak{i}'}$. In particular, if, as assumed, the maps $\varphi_{\mathfrak{i}\mathfrak{i}'}(x, \_)$ preserve the orientation of fibers for $x \in U_{\mathfrak{i}\mathfrak{i}'}$ (hence for x such that $\phi_{\mathfrak{i}}(x) \neq 0$), then 3.15 is a nowhere vanishing differential form in $\Omega^m(F_x)$ as required. □

Corollary 3.1.7.2

1. *Let $\overline{f} : B' \to B$ be a map of manifolds. The pullback $\overline{f}^{-1}(E)$ of an orientable fiber bundle (E, B, π, F) (*cf. *Proposition 3.1.6.1-(1)) is orientable.*
2. *The product in* Man_B (cf. *Sect. 3.1.8) of orientable fiber bundles is orientable.*
3. *A fiber bundle of manifolds (E, B, π, F) is orientable in the following cases.*

   (a) *$E = B \times F$;* (b) *E and B are orientable;* (c) *B is simply connected.*

Proof

(1) In describing the manifold structure of the pullback $\overline{f}^{-1}(E)$ in the proof of 3.1.6.1-(1), we fixed an open cover of trivializations for $\pi : E \to B$,

$$\mathfrak{E} := \big\{(\pi, \varphi_{\mathfrak{i}}) : \pi^{-1}(U_{\mathfrak{i}}) \simeq U_{\mathfrak{i}} \times F\big\}_{\mathfrak{i} \in \mathfrak{I}}\,,$$

with transition maps $T_{\mathrm{ii}'} : U_{\mathrm{ii}'} \times F \to U_{\mathrm{ii}'} \times F$,

$$T_{\mathrm{ii}'}(x, y) = (x, \varphi_{\mathrm{ii}'}(x, y)), \quad \forall x \in U_{\mathrm{ii}'}.$$

We then defined in (3.10) an open cover of trivializations for $E' := \overline{f}^{-1}(E)$,

$$\mathfrak{E}' := \left\{(\pi', \varphi'_{\mathrm{i}}) : \pi'^{-1}(V_{\mathrm{i}}) \to V_{\mathrm{i}} \times F\right\}_{\mathrm{i}\in\mathfrak{I}},$$

where $V_{\mathrm{i}} := \overline{f}^{-1}(U_{\mathrm{i}})$, whose transition maps have the form

$$T_{\mathrm{ii}'}(x', y) = (x', \varphi_{\mathrm{ii}'}(\overline{f}(x'), y)), \quad \forall x' \in V_{\mathrm{ii}'},$$

where the $\varphi_{\mathrm{ii}'}$ are the same for $\mathfrak{E}$ and $\mathfrak{E}'$.

In particular, applying Proposition 3.1.7.1, the family $\mathfrak{E}$ is oriented if and only if the family $\mathfrak{E}'$ is so, which proves (1).

(2) The proof is the same as for (1) so that we will be sketchy. We use the trivializing cover for the product of fiber bundles already described in the proof of 3.1.6.1-(3), where the transition maps are (3.13):

$$T_{\mathrm{jj}'}(x, (y_1, y_2)) = (x, (\varphi_{1,\mathrm{jj}'}(x, y_1), \varphi_{2,\mathrm{jj}'}(x, y_2))),$$

where the $\varphi_{i,\mathrm{jj}'}$ are the same that appear in the transition maps of trivializing covers of the fiber bundles (E_i, B, π_i, F_i). Hence, again by Proposition 3.1.7.1, an oriented trivialization open cover for the product $\tilde{\pi} : E \times_B E' \to B$ exists.

(3) It suffices to assume B connected.

(3a) Is obvious and is also consequence of (1) for the constant map $\overline{h} : B \to \{\bullet\}$.

(3b) Let $\omega_E \in \Omega^n(E)$, $\omega_B \in \Omega^b(B)$ and $\omega_F \in \Omega^m(F)$ be nowhere vanishing. Then, for any trivializing $U \subseteq B$ and any trivialization $\Phi : \pi^{-1}(U) \simeq U \times F$, there is a unique choice of sign $\epsilon_U \in \{1, -1\}$ such that ω_E and $\Phi^*(\omega_B \wedge (\epsilon_U\, \omega_F))$ define the same orientation on $\pi^{-1}(U)$ (2.3.1.1).

The differential forms $\omega_{\Phi,U} := \Phi^*(1 \otimes \epsilon_U\, \omega_F)$ are then m-forms defining the same orientation on every fiber F_x for $x \in U$, independently of the trivialization Φ of $\pi^{-1}(U)$. Hence the orientability of the fiber bundle after Proposition 3.1.7.1.

(3c) The proof of (3b) shows that the orientations of the fibers F_b define a two-sheeted covering $\mathcal{E}$ of B. When E and B are both oriented, there is a canonical way to choose one of the two sheets over each trivializing open subspace $U \subseteq B$, entailing the triviality of $\mathcal{E}$. When we do not know whether E is orientable, but we do know that B is simply connected, we can still assert that B is orientable and that $\mathcal{E}$ is trivial (*cf*. Proposition 2.3.1.1-(2)). We can therefore still choose the signs ϵ_U in a compatible way, and doing so, endow E of an orientation, in which case statement (3b) applies. □

Comment 3.1.7.3 The definition of orientability 3.1.7 is adapted to fiber bundles of manifolds (E, B, π, F), but in many important cases, while the fiber F is a manifold, the base B is not. For example, in the homotopy quotients $\mathbb{E}G \times_G F \twoheadrightarrow \mathbb{B}G$ where $\mathbb{B}G$ is a CW-complex of infinite dimension.

In those cases, where B is at least locally contractible, an alternative definition of orientability can be based on the idea in the proof of 3.1.7.2-(3) of considering the covering $\mathcal{E}$ of B defined by the orientations of the connected component of the fibers F_b, as b runs over B. One then says that (E, B, π, F) *is orientable* if $\mathcal{E}$ is a trivial covering, in which case *to choose an orientation* means to choose one sheet of $\mathcal{E}$ for each connected component of E.

With this definition of orientability, Propositions 3.1.7.1 and 3.1.7.2-(1,2) remain true and 3.1.7.2-(3) is verified by the classifying space $\mathbb{B}G$.

Exercises 3.1.7.4

1. Show that in an orientable fiber bundle of manifolds (E, B, π, F), the orientability of E and B are equivalent properties. (♝, p. 338)
2. Give an example of a nonorientable fiber bundle of manifolds (E, B, π, F) where B and F are orientable. (♝, p. 339)

3.1.8 *The Categories* $\mathbf{Man}_B$, $\mathbf{Fib}_B$ *and* $\mathbf{Fib}_B^{\mathrm{or}}$

As for topological spaces in 3.1.2, given a manifold B, we define the category of *manifolds above* B, denoted by Man_B. Its objects are the differentiable maps of manifolds $\pi : M \to B$, and the morphisms from $\pi : M' \to B$ to $\pi : M \to B$ are the differentiable maps $f : M' \to M$ such that $\pi \circ f = \pi'$.

We denote by $\mathrm{Fib}_B^{\mathrm{or}}$ and Fib_B the full subcategories of Man_B respectively of fiber bundles and oriented fiber bundles of base B.

Exercise 3.1.8.1 Check that by Propositions 3.1.6.1, 3.1.7.1 and 3.1.7.2, the analogue to Proposition 3.1.4.2 is verified, with the same proof, when replacing the category Top_B by the categories Man_B, Fib_B and $\mathrm{Fib}_B^{\mathrm{or}}$.

3.1.9 *Proper Subspaces of a Fiber Bundle*

We introduce the concept of *properness* in the category Top_B as the counterpart of *compactness* in the category Top. As we will see in 3.1.10, differential forms with *proper* supports on the total space of fiber bundles of manifolds constitute the right analogue to differential forms with *compact* supports on manifolds.

Definition 3.1.9.1 Let $\pi : X \to B$ be a continuous map. A subspace $P \subseteq X$ is said to be π*-proper*, or simply *proper* when π is understood, if $\pi : P \to Z$ is a proper map, i.e. if for every compact subspace $K \subseteq B$, the subspace $\pi^{-1}(K) \subseteq X$ is compact.

Given a commutative diagram of continuous maps

$$\begin{array}{ccc} X' & \xrightarrow{g} & X \\ {\scriptstyle \pi'}\downarrow & \oplus & \downarrow{\scriptstyle \pi} \\ B' & \xrightarrow{\bar{g}} & B \end{array} \tag{3.16}$$

we say that g^{-1} *preserves properness*, if for every π-proper subset $P \subseteq X$, the subspace $g^{-1}(P) \subseteq X'$ is π'-proper.

Proposition 3.1.9.2

1. *In a fiber product*

$$\begin{array}{ccc} X_1 \times_B X_2 & \xrightarrow{p_1} & X_1 \\ {\scriptstyle p_2}\downarrow & \square & \downarrow{\scriptstyle \pi_1} \\ X_2 & \xrightarrow{\pi_2} & B \end{array} \tag{3.17}$$

the inverse image p_1^{-1} preserves properness.
In particular, if π_1 is a proper map, so is p_2, and likewise exchanging $1 \leftrightarrow 2$.

2. *Given a commutative diagram*

$$\begin{array}{ccc} X' & \xrightarrow{g} & X \\ {\scriptstyle f'}\downarrow & \oplus & \downarrow{\scriptstyle f} \\ B' & \xrightarrow{\bar{g}} & B \end{array}$$

consider the factorization

$$\begin{array}{ccccc} & & \overset{g}{\frown\!\!\!\longrightarrow} & & \\ X' & \xrightarrow{\xi} & X \times_B B' & \xrightarrow{p_1} & X \\ {\scriptstyle f'}\downarrow & & {\scriptstyle p_2}\downarrow & \square & \downarrow{\scriptstyle f} \\ B' & = & B' & \xrightarrow{\bar{g}} & B \end{array}$$

where $\xi : X' \to X \times_B B'$ is the induced map from X' to the fiber product of $f : X \to B$ and $\bar{g} : B' \to B$, whence $g = p_1 \circ \xi$ and $p_2 \circ \xi = f'$. Then,

a. the map g^{-1} preserves properness if and only if ξ is a proper map;
b. if $\bar{g}$ is proper, then g^{-1} preserves properness if and only if g is proper.

Proof

(1) We must show that $p_1^{-1}(P) \cap p_2^{-1}(K)$ is compact for $K \subseteq X_2$ compact. By the definition of fiber product 3.1.4.2-(1), we have

$$p_1^{-1}(P) \cap p_2^{-1}(K) = \left(K \cap \pi_2^{-1}(\pi_1(P))\right) \times_B \left(\pi_1^{-1}(\pi_2(K)) \cap P\right), \qquad (3.18)$$

where $K \cap \pi_2^{-1}(\pi_1(P))$ is always compact.

When, in addition, $\pi_1 : P \to B$ is a proper map, then $\pi_1^{-1}(\pi_2(K)) \cap P$ is also compact, and the right-hand side of (3.18) is compact since it is closed in a product of compact spaces.

(2) *Lemma. Given a commutative diagram*

$$\begin{array}{ccc} X' & \xrightarrow{\ \xi\ } & W \\ & {\scriptstyle f'}\searrow & \downarrow{\scriptstyle q} \\ & & B', \end{array}$$

the map ξ is proper if and only if ξ^{-1} preserves properness.

Proof of Lemma For every $P \subseteq W$ and every compact $K \in B'$, we have

$$\xi^{-1}(P \cap q^{-1}(K)) = \xi^{-1}(P) \cap f'^{-1}(K). \qquad (3.19)$$

When $\xi : X' \to W$ and $q : P \to B'$ are proper, the left-hand side of (3.19) is compact, so that $\xi^{-1}(P)$ is f'-proper. Conversely, if $K \subseteq W$ is compact, it is q-proper and $\xi^{-1}(K)$ is f'-proper, so that $\xi^{-1}(K) \cap f'^{-1}(q(K))$ is compact. But, $\xi^{-1}(K) \cap f'^{-1}(q(K)) = \xi^{-1}(K)$. Hence ξ is proper. ⊟

(2a) Assume ξ proper. If $P \subseteq X$ is f-proper, then $p_1^{-1}(P)$ is p_2-proper, by (1), and $\xi^{-1}(p_1^{-1}(P)) = g^{-1}(P)$ is f'-proper after the preliminary result. Conversely, if $K \subseteq X \times_B B'$ is compact, then $\xi^{-1}(K)$ is compact too, since it is closed in $g^{-1}(p_1(K)) \cap f'^{-1}(p_2(K))$ which is compact because $p_1(K)$, being compact, is f-proper and that g^{-1} preserves properness.

(2b) If $\bar{g}$ is proper, then p_1 is proper, by (1). Hence, g is proper if and only if ξ is proper, and, by (2a), if and only if g^{-1} preserves properness. □

Exercises 3.1.9.3

1. In the definition of proper subspaces 3.1.9.1, show that if g is a proper map, then g^{-1} preserves properness, but that the converse is not true. (p. 339)
2. Show that in 3.1.9.2-(1), if π_1 is open, then p_2 is open, but that the same is not true for 'closed' in lieu of 'open'. (p. 339)
3. Call $\pi_1 : X_1 \to B$ *universally closed* if for all $\pi_2 : X_2 \to B$, the map $p_2 : X_1 \times_B X_2 \to X_2$ is closed. Show that in the category of locally compact spaces a map is proper if and only if it is universally closed. (p. 339)

4. In the definition of a locally trivial fibrations $\pi : E \to B$ of fiber F (3.1), a trivialization $\Phi : \pi^{-1}(U) \to U \times F$ is of the form $\Phi(\_) = (\pi(\_), \varphi(\_))$. Show that $(\pi(\_), \varphi(\_))$ is a trivialization if and only if the following diagram is Cartesian

$$\begin{array}{ccc} \pi^{-1}(U) & \xrightarrow{\varphi} & F \\ {\scriptstyle\pi}\downarrow & \oplus & \downarrow \\ U & \longrightarrow & \{\bullet\} \end{array}$$

or, equivalently, the restrictions of φ to fibers are homeomorphisms and that φ^{-1} preserves properness, or even, in the category of manifolds, if and only if the restrictions of φ to fibers are diffeomorphisms. (p. 340)

3.1.10 *Differential Forms with Proper Supports*

Given a map of manifolds $\pi : E \to B$, a differential form $\beta \in \Omega(E)$ is said to be of *proper* (or *compact vertical*) support, if $|\beta|$ is π-proper (3.1.9.1), i.e.

For every compact subspace $K \subseteq B$, the subspace $\pi^{-1}(K) \cap |\beta|$ is compact.

We denote by $\Omega_{\mathrm{cv}}(E) \subseteq \Omega(E)$ the subset of such differential forms.

Notice that if $B := \{\bullet\}$ then $\Omega_{\mathrm{cv}}(E) = \Omega_{\mathrm{c}}(E)$. The forthcoming propositions 3.1.10.2 and 3.2.1.3 show that the most relevant properties of $\Omega_{\mathrm{c}}(F)$ extend naturally to $\Omega_{\mathrm{cv}}(E)$.

Exercise 3.1.10.1 Given a map of manifolds $\pi : E \to B$, for $\phi \in \Omega^0(B)$ and $\beta \in \Omega(E)$, let $\phi\,\beta := (\phi \circ \pi)\beta$. Show that $\beta \in \Omega_{\mathrm{cv}}(E)$ if and only if $\phi\,\beta \in \Omega_{\mathrm{c}}(E)$ for all $\phi \in \Omega^0_{\mathrm{c}}(E)$. (p. 340)

Proposition 3.1.10.2 *Let (E, B, π, F) be a fiber bundle of manifolds.*

1. *The set $\Omega_{\mathrm{cv}}(E)$ is a differential graded ideal of $\Omega(E)$, in other words, $\Omega_{\mathrm{cv}}(E)$ is a subcomplex of $\Omega_{\mathrm{c}}(E)$ such that*

$$\Omega(E) \wedge \Omega_{\mathrm{cv}}(E) \subseteq \Omega_{\mathrm{cv}}(E)\,.$$

In particular, $\Omega_{\mathrm{cv}}(E)$ is a graded differential module over $\Omega(B)$ through the pullback morphism $\pi^* : \Omega(B) \to \Omega(E)$.

2. *If a morphism of fiber bundles,*

$$\begin{array}{ccc} E' & \xrightarrow{f} & E \\ {\scriptstyle\pi'}\downarrow & \oplus & \downarrow{\scriptstyle\pi} \\ B' & \xrightarrow{\overline{f}} & B \end{array}$$

is either Cartesian or is such that f is proper, then $f^*(\Omega_{cv}(E)) \subseteq \Omega_{cv}(E')$, thus inducing a diagram of pullback morphisms *of complexes*

$$\begin{array}{ccc} \Omega(B) & \xrightarrow{\overline{f}^*} & \Omega(B') \\ {\scriptstyle \pi^*}\downarrow & & \downarrow{\scriptstyle \pi'^*} \\ \Omega_{cv}(E) & \xrightarrow{f^*} & \Omega_{cv}(E') \\ \cap\!\!| & & \cap\!\!| \\ \Omega(E) & \xrightarrow{f^*} & \Omega(E') \end{array} \tag{3.20}$$

where, for all $\alpha \in \Omega(E)$ and $\beta \in \Omega_{cv}(E)$ (resp. $\beta \in \Omega(E)$), we have

$$f^*(\alpha \wedge \beta) = f^*(\alpha) \wedge f^*(\beta). \tag{3.21}$$

Proof

(1) Given $\beta_i \in \Omega_{cv}(E)$ and $\alpha \in \Omega(E)$, we have the following obvious relations on supports

$$|d\beta| \subseteq |\beta|, \quad |\beta_1 + t\beta_2| \subseteq |\beta_1| \cup |\beta_2|, \ \forall t \in \mathbb{R}, \qquad |\alpha \wedge \beta| \subseteq |\beta|,$$

which immediately imply that $\Omega_{cv}(E)$ is a differential graded ideal over $\Omega(E)$.

(2) The inclusion $f^*(\Omega_{cv}(E)) \subseteq \Omega_{cv}(E')$ results applying Propositions 3.1.9.2-(1,2b). The equality (3.21) is then obvious since $f^* : \Omega(E) \to \Omega(E')$ is a homomorphism of algebras. □

3.1.10.1 Poincaré Lemmas for $\Omega_{cv}(E)$ and $\Omega(E)$

We extend the classic Poincaré Lemmas from manifolds to fiber bundles.

Proposition 3.1.10.3

1. Given two homotopic morphisms of fiber bundles $(f_1, \overline{f}_1)$ and $(f_0, \overline{f}_0)$,

$$\begin{array}{ccc} E' & \xrightarrow{f_i} & E \\ {\scriptstyle \pi'}\downarrow & \oplus & \downarrow{\scriptstyle \pi} \\ B' & \xrightarrow{\overline{f}_i} & B \end{array}$$

either Cartesian or such that the f_i's are proper maps, the pullbacks

$$f_i^* : \Omega_{cv}(E) \to \Omega_{cv}(E') \quad \text{and} \quad f_i^* : \Omega(E) \to \Omega(E'),$$

are homotopic morphisms of complexes.

2. ***Poincaré Lemmas.*** *If (E, B, π, F) is the projection $\pi : \mathbb{R}^m \times F \twoheadrightarrow \mathbb{R}^m$, $\pi(x, y) := x$, then the pullbacks induced by $p_2 : \mathbb{R}^m \times F \twoheadrightarrow F$:*

$$p_2^* : \Omega_{\mathrm{c}}(F) \to \Omega_{\mathrm{cv}}(E) \quad \textit{and} \quad p_2^* : \Omega(F) \to \Omega(E)\,,$$

are homotopic morphisms of complexes.

Proof Let $(h, \bar{h})$ be a homotopy for the $(f_i, \overline{f}_i)$'s

$$\begin{array}{ccc} \mathbb{R} \times E' & \xrightarrow{\;h\;} & E \\ {\scriptstyle \pi'}\downarrow & \oplus & \downarrow{\scriptstyle \pi} \\ \mathbb{R} \times B' & \xrightarrow{\;\bar{h}\;} & B \end{array}$$

Denote by $t \in \mathbb{R}$ the variable parametrizing the homotopy. For every $\omega \in \Omega^i(E)$, we have a unique decomposition

$$h^*(\omega) = \alpha + dt \wedge \beta\,,$$

with $\alpha(t, y) \in \Omega^i(E')$ and $\beta(t, y) \in \Omega^{i-1}(E')$, for all $t \in \mathbb{R}$.

The map

$$\eta_i : \Omega^i(E) \to \Omega^{i-1}(E)\,, \quad \text{by} \quad \eta_i(\omega) := \int_0^1 \beta\, dt\,, \tag{3.22}$$

verifies

$$\eta \circ d + d \circ \eta = (f_1^* - f_2^*) : \Omega(E) \to \Omega(E')\,.$$

Indeed, we have $h^*(d\omega) = d(h^*(\omega)) = d_{E'}(\alpha) + dt \wedge \big(\partial_t \alpha - d_{E'}(\beta)\big)$, so that

$$\begin{cases} \eta(d_{E'}(\omega)) = \displaystyle\int_0^1 \big(\partial_t \alpha - d_{E'}(\beta)\big)\, dt \\ d_{E'}(\eta(\omega)) = d_{E'}\Big(\displaystyle\int_0^1 \beta\, dt\Big) = \displaystyle\int_0^1 d_{E'}(\beta)\, dt \end{cases}$$

Whence,

$$\begin{aligned} (\eta \circ d + d \circ \eta)(\omega)(y) = \int_0^1 \partial_t \alpha\, dt &= \alpha(1, y) - \alpha(0, y) \\ &= h^*(\omega)(t, y)|_{t=1} - h^*(\omega)(t, y)|_{t=0}\,, \end{aligned}$$

and the homotopy-equivalence of $f_i^* : \Omega(E) \to \Omega(E')$ is proved.

When $\omega \in \Omega_{\mathrm{cv}}(E)$, we have $\beta \in \Omega_{\mathrm{cv}}(\mathbb{R} \times E')$ by 3.1.10.2-(2), in which case, for every $\phi \in \Omega_{\mathrm{c}}^0(B')$, the support $|\phi\, \beta|$ is compact in $\mathbb{R} \times E'$. Hence, there exists

$K \subseteq E'$ such that $|\phi\,\beta| \subseteq \mathbb{R} \times K$, and $|\phi \int_0^1 \beta\, dt| \subseteq K$. We can therefore conclude that $\int_0^1 \beta dt \in \Omega_{\mathrm{cv}}(E')$ (Exercise 3.1.10.1). The same formula (3.22) thus gives a map $\eta : \Omega_{\mathrm{cv}}(E) \to \Omega_{\mathrm{cv}}(E)[-1]$, and the same arguments show that it gives a homotopy for the morphisms $f_i^* : \Omega_{\mathrm{cv}}(E) \to \Omega_{\mathrm{cv}}(E')$.
(2) The family of maps $\overline{f}_t : \mathbb{R}^m \to \mathbb{R}^m$, $f_t(\mathbf{v}) := t\,\mathbf{v}$ for $t \in \mathbb{R}$, induces a family of homotopic morphisms $(f_t, \overline{f}_t) : \mathbb{R}^m \times F \to \mathbb{R}^m \times F$. The induced morphism of complexes $f_1^* = \mathrm{id}$ is then homotopic to f_0^* by (1).

On the other hand, since $(f_0, \overline{f}_0)$ factors as the composition:

$$\begin{array}{ccccc}
\mathbb{R}^m \times F & \xrightarrow{p_2} & F & \xrightarrow{\iota_0} & \mathbb{R}^m \times F \\
\downarrow{\scriptstyle \pi} & & \downarrow{\scriptstyle \pi} & & \downarrow{\scriptstyle \pi} \\
\mathbb{R}^m & \xrightarrow{c_0} & \{0\} & \xrightarrow{\overline{\iota_0}} & \mathbb{R}^m
\end{array}
\qquad (f_0 = \iota_0 \circ p_2)$$

the pullbacks $p_2^* : \Omega(F) \to \Omega(\mathbb{R}^m \times F)$ and $p_2^* : \Omega_{\mathrm{c}}(F) \to \Omega_{\mathrm{cv}}(\mathbb{R}^m \times F)$ admit as homotopy left inverse the morphism ι_0^*. But this same ι_0^* is clearly also a right inverse. Hence the proof of Poincaré Lemmas. □

3.1.10.2 Sheafification of $\Omega_{\mathrm{cv}}(E)$ and $\Omega(E)$ on the Base Space B

For every open inclusion $\iota_U : U \subseteq B$, denote by $E_U := (\pi^{-1}(U), U, \pi, F)$ the pullback $\iota_U^{-1}(E, B, \pi, F)$. Then, thanks to the functoriality of the base change functors (3.1.5.1), the correspondences which associate with an open subspace $U \subseteq B$, the complexes

$$U \rightsquigarrow \Omega_{\mathrm{cv}}(E_U) \quad \text{and} \quad U \rightsquigarrow \Omega(E_U)\,,$$

and which associate with the inclusion maps $\overline{\iota_{VU}} : V \subseteq U$, the pullbacks

$$\iota_{VU}^* : \Omega_{\mathrm{cv}}(E_U) \to \Omega_{\mathrm{cv}}(E_V) \quad \text{and} \quad \iota_{VU}^* : \Omega(E_U) \to \Omega(E_V)\,,$$

define presheaves of $\underline{\underline{\Omega}}_B$-differential graded modules, respectively denote by

$$\boxed{\pi_!\, \underline{\underline{\Omega}}_E \ \text{ and } \ \pi_*\, \underline{\underline{\Omega}}_E} \tag{3.23}$$

Proposition 3.1.10.4

1. *The presheaves $\pi_!\, \underline{\underline{\Omega}}_E$ and $\pi_*\, \underline{\underline{\Omega}}_E$ are sheaves.*
2. *The graded sheaves of cohomology they define:*

$$\mathcal{H}(\pi_!) := \mathcal{H}(\pi_!\, \underline{\underline{\Omega}}_E) \quad \textit{and} \quad \mathcal{H}(\pi_*) := \mathcal{H}(\pi_*\, \underline{\underline{\Omega}}_E)\,, \tag{3.24}$$

are locally trivial sheaves of fibers respectively $H_{\mathrm{c}}(F)$ and $H(F)$.

Furthermore, if E is connected, then the sheaf $\mathcal{H}^{d_F}(\pi_!)$ admits nowhere vanishing global sections if and only if (E, B, π, F) is an orientable fiber bundle. In particular, a fiber bundle (E, B, π, F) with connected fiber F is orientable if and only if the sheaf $\mathcal{H}^{d_F}(\pi_!)$ is isomorphic to the trivial sheaf $\underline{\underline{\mathbb{R}}}_B$.

Proof

(1) The correspondence $U \rightsquigarrow \Omega(\pi^{-1}(U))$ is the direct image presheaf $\pi_*\underline{\underline{\Omega}}_E$ which is well-known to be a sheaf. For the correspondence $U \rightsquigarrow \Omega_{\rm cv}(\pi^{-1}(U))$ we need to justify the sheaf condition, i.e. that for every open cover $U = \bigcup_{\mathfrak{i}\in\mathfrak{I}} U_\mathfrak{i}$, the sequence

$$0 \to \Omega_{\rm cv}(\pi^{-1}(U)) \to \prod\nolimits_{\mathfrak{i}\in\mathfrak{I}} \Omega_{\rm cv}(\pi^{-1}(U_\mathfrak{i})) \to \prod\nolimits_{\mathfrak{i},\mathfrak{i}'\in\mathfrak{I}} \Omega_{\rm cv}(\pi^{-1}(U_{\mathfrak{i}\mathfrak{i}'}))\,, \tag{3.25}$$

is well-defined and exact.

For this, we need to justify two facts about a continuous map $p : X \to U$.

- *First. If f is proper, then the restriction $p : p^{-1}(V) \to V$ is proper for every open subspace $V \subseteq U$*, which is clear after the definition of proper maps. Then, taking for X the supports of differential forms, this first fact shows that (3.25) is well-defined.
- *Second. For every open cover $U = \bigcup_{\mathfrak{i}\in\mathfrak{I}} U_\mathfrak{i}$, if $p : p^{-1}(U_\mathfrak{i}) \to U_\mathfrak{i}$ is proper for all $\mathfrak{i} \in \mathfrak{I}$, then $f : X \to U$ is proper.* For this, we need only recall that, given $K \subseteq U$ compact, there exists a finite decomposition $K = \bigcup_{j\in\mathfrak{I}} K_\mathfrak{j}$, where $K_\mathfrak{j}$ is compact and contained in some $U_\mathfrak{i}$. But then, $p^{-1}(K) = \bigcup_{j\in\mathfrak{I}} p^{-1}(K_\mathfrak{j})$, where $p^{-1}(K_\mathfrak{j})$ is compact in $p^{-1}(U_\mathfrak{i})$, hence in X.

Now, as $\pi_!\,\underline{\underline{\Omega}}_E$ is contained in the sheaf $\pi_*\,\underline{\underline{\Omega}}_E$, to prove that (3.25) is exact, we need only check exactness at its central term. Given a family of differential forms $\{\beta_\mathfrak{i} \in \Omega_{\rm cv}(\pi^{-1}(U_\mathfrak{i}))\}_{\mathfrak{i}\in\mathfrak{I}}$ such that $\beta_\mathfrak{i}|_{U_{\mathfrak{i}\mathfrak{i}'}} = \beta_{\mathfrak{i}'}|_{U_{\mathfrak{i}\mathfrak{i}'}}$, for all $\mathfrak{i}, \mathfrak{i}' \in \mathfrak{I}$, we already know that there exists a unique $\beta \in \Omega(\pi^{-1}(U))$ such that $\beta|_{U_\mathfrak{i}} = \beta_\mathfrak{i}$, and, again, taking for X supports of differential forms, the second fact shows that β has proper support.

(2) Local triviality in (3.24) is immediate by Poincaré Lemmas 3.1.10.3-(2).
By definition, a connected fiber bundle is orientable if and only if there exists $\omega_\pi \in \Omega^{d_F}(E)$ such that $\omega_\pi\,|_{F_b}$ is nowhere vanishing for all $b \in B$. This form ω_π therefore defines a global section of $\mathcal{H}^{d_F}(\pi_!)$ since it determines a fundamental class $\zeta_{C_b} \in H_c^{d_M}(C_b)$ on each connected component C_b of the fiber F_b (*cf.* Sect. 2.4.2). For the converse, one notes that a nowhere vanishing global section σ of $\mathcal{H}^{d_F}(\pi_!)$, although it determines a fundamental class ζ_{C_b} for at least one component C_b of F_b, it could vanish on the others. However, this cannot occur since we assumed E connected. Indeed, in that case any component C_b' of F_b can be joined to C_b by some path γ in E, and it is clear, by the local triviality

of π, that if we follow the section σ along the path $\pi \circ \gamma$, the fundamental class ζ_{C_b} moves on a fundamental class of C'_b, hence proving that the section σ determines an orientation of the fiber bundle. □

Corollary 3.1.10.5 *Given two manifolds B and F, the following inclusions are quasi-isomorphisms :*

$$\underline{\underline{\Omega}}_B \otimes \Omega_c(F) \subseteq \pi_! \underline{\underline{\Omega}}_{B\times F} \quad \text{and} \quad \underline{\underline{\Omega}}_B \otimes \Omega(F) \subseteq \pi_* \underline{\underline{\Omega}}_{B\times F}\,,$$

$$\Omega(B) \otimes \Omega_c(F) \subseteq \Omega_{cv}(B \times F) \quad \text{and} \quad \Omega(B) \otimes \Omega(F) \subseteq \Omega(B \times F)\,.$$

Proof Recall that given two topological spaces X and Y, and sheaves of vector spaces $\mathscr{F} \in \mathrm{Sh}(X; \Bbbk)$ and $\mathscr{G} \in \mathrm{Sh}(Y; \Bbbk)$, one defines the *external tensor product* $\mathscr{F} \boxtimes \mathscr{G} \in \mathrm{Sh}(X \times Y; \Bbbk)$ as the sheaf associated with the presheaf

$$U \times V \rightsquigarrow \mathscr{F}(U) \otimes_{\Bbbk} \mathscr{G}(V)\,.$$

In the case of the product $B \times F$, we have a natural inclusion of complexes of sheaves $\underline{\underline{\Omega}}_B \boxtimes \underline{\underline{\Omega}}_F \subseteq \underline{\underline{\Omega}}_{B\times F}$ giving rise to the inclusions of complexes

$$\underline{\underline{\Omega}}_B \otimes \Omega_c(F) \subseteq \pi_! \underline{\underline{\Omega}}_{B\times F} \quad \text{and} \quad \underline{\underline{\Omega}}_B \otimes \Omega(F) \subseteq \pi_* \underline{\underline{\Omega}}_{B\times F}\,, \tag{3.26}$$

inducing the morphisms in the cohomology sheaves:

$$\underline{\underline{\mathbb{R}}}_B \otimes H^p_c(F) \to \mathcal{H}^p(\pi_!) \quad \text{and} \quad \underline{\underline{\mathbb{R}}}_B \otimes H^p(F) \to \mathcal{H}^p(\pi_*) \tag{3.27}$$

which are isomorphisms since this is so at stalks level after 3.1.10.4-(2).

The terms of the complexes (3.26), being $\underline{\underline{\Omega}}^0_B$-modules, are $\Gamma(B; \_)$-acyclic, so that the natural map $\Gamma(B; \_) \to I\!R\,\Gamma(B; \_)$ induces quasi-isomorphisms when applied to the complexes (3.26) (Theorem B.6.3.4-(3)).[2] The morphisms

$$\begin{cases} I\!R\,\Gamma(B; \underline{\underline{\Omega}}_B \otimes \Omega_c(F)) \to I\!R\,\Gamma(B; \pi_! \underline{\underline{\Omega}}_{B\times F}) \\ I\!R\,\Gamma(B; \underline{\underline{\Omega}}_B \otimes \Omega(F)) \to I\!R\,\Gamma(B; \pi_* \underline{\underline{\Omega}}_{B\times F}) \end{cases} \tag{3.28}$$

[2]See also Bredon [20], ch. II-§9 *Φ-soft and Φ-fine sheaves, p. 65*, in particular examples 9.4 and 9.17 which state that the sheaf of rings $\underline{\underline{\Omega}}^0(M)$ of differentiable functions on a manifold M is Φ-soft for any paracompactifying family of supports Φ. This implies that every $\underline{\underline{\Omega}}^0(M)$-module is Φ-soft, hence acyclic for the functor $\Gamma_\Phi(M; \_)$ of global sections with supports in Φ, which includes the functor $\Gamma(M; \_)$ and $\Gamma_c(M; \_)$ (*cf*. Sect. B.6.2). Other references for these questions are Godement [46] ch. II-§3.7. *Faisceaux fins, p. 156*, and Iversen [58] ch. III-§2. *Soft sheaves.* p. 149.

then generate, by Godement's simultaneous resolutions (*cf*. proof of 3.2.2.1-(2)), spectral sequences which, in the second page, give rise to the natural maps

$$\begin{cases} (\mathbb{E}_2^{p,q}) & \mathbb{R}^p\,\Gamma(B;\underline{\underline{\mathbb{R}}}_B\otimes H_c^q(F)))\to\mathbb{R}^p\,\Gamma(B;\mathcal{H}^q(\pi_!)) \\ (\mathbb{E}_2^{p,q}) & \mathbb{R}^p\,\Gamma(B;\underline{\underline{\mathbb{R}}}_B\otimes H^q(F)))\to\mathbb{R}^p\,\Gamma(B;\mathcal{H}^q(\pi_*)) \end{cases}$$

where one recognizes $\mathbb{R}^p\,\Gamma(B;\,\_)$ acting on the isomorphisms (3.27). The spectral sequences $(\mathbb{E}_r)$ are therefore isomorphic for $r\geq 2$ and the morphisms (3.28) are quasi-isomorphisms, which proves the corollary. □

3.2 Integration Along Fibers on Fiber Bundles

In 2.3 we defined, for an oriented manifold M of dimension d_M, the integration map

$$\int_M : \Omega_c(M)[d_M]\to\mathbb{R}[0]\,,$$

which is both a morphism of complexes (Proposition 2.3.2.2), and the main ingredient of another morphism of complexes: the left Poincaré adjunction map (Proposition 2.4.1.1)

$$\mathbb{D}_M : \Omega(M)[d_M]\longrightarrow \mathrm{Hom}_{\mathbb{R}}(\Omega_c(M),\mathbb{R})\,, \tag{3.29}$$

which is the basis of the Poincaré Duality theorem.

These facts extend to any oriented fiber bundle of manifolds (E, B, π, M) with equidimensional fiber.[3] For this, we will define an operation of integration *along the fiber M*

$$\int_M : \Omega_{cv}(E)[d_M]\to\Omega(B)\,,$$

and establish it is a morphism of $\Omega(B)$-differential graded modules. We will then extend the left Poincaré adjunction map (3.29) to a morphism of complexes

$$\mathbb{D}_{B,M} : \Omega(E)[d_M]\longrightarrow \mathrm{Hom}_{\Omega(B)}(\Omega_{cv}(E),\Omega(B))\,,$$

which, in turn, will be the basis of the Poincaré Duality theorem *relative to B*.

But before dealing with the general case, we take a closer look at the special case of *trivial Euclidean fiber bundles*.

[3]Most of the time we shall assume the total space to be connected only for the purpose of guaranteeing equidimensionality of fibers (*cf*. Exercise 3.1.1.1).

3.2.1 The Case of Trivial Euclidean Bundles

By this terminology we mean a fiber bundle $(U \times V, U, p_1, V)$, with $U \subseteq \mathbb{R}^b$ and $V \subseteq \mathbb{R}^m$ open subspaces, and where $p_1 : U \times V \to U$ is the map $(x, y) \mapsto x$.

What makes trivial Euclidean fiber bundles particularly handy is that, in them, we have the global systems of coordinates given by the Euclidean coordinates, in U the set $\bar{x} := \{x_1, \dots, x_b\}$, and in V the set $\bar{y} := \{y_1, \dots, y_m\}$.

For $I \subseteq [\![1, b]\!]$, let $d_I := dx_{i_1} \wedge \cdots \wedge dx_{i_r}$, if $I := \{i_1 < \cdots < i_r\}$ is nonempty, and let $d_\emptyset = 1$. Proceed likewise for $J \subseteq [\![1, m]\!]$. Then, a differential form $\omega \in \Omega(U \times V)$ can be written in a unique way as a sum

$$\omega = \sum\nolimits_{I,J} f_{IJ}(\bar{x}, \bar{y})\, d_I \wedge d_J\,, \tag{3.30}$$

with $f_{IJ} \in \Omega^0(U \times V)$. In particular,

$$\omega \in \Omega_{\mathrm{cv}}(U \times V) \leftrightarrow f_{IJ} \in \Omega^0_{\mathrm{cv}}(U \times V)\,, \quad \forall I \subseteq [\![1, b]\!]\,,\ J \subseteq [\![1, m]\!]\,.$$

We define

$$\int_V : \Omega_{\mathrm{cv}}(U \times V) \to \Omega(U)\,, \tag{3.31}$$

by the following conditions:

- $\int_V$ is $\mathbb{R}$-linear;
- if $J \neq [\![1, m]\!]$, then $\int_V f_{IJ}(\bar{x}, \bar{y})\, d_I \wedge d_J = 0$;
- if $J = [\![1, m]\!]$, then $\int_V f_{IJ}(\bar{x}, \bar{y})\, d_I \wedge d_J = (-1)^m \Big(\int_V f_{IJ}(\bar{x}, \bar{y})\, d_{[\![1,m]\!]} \Big)\, d_I\,.$

Proposition 3.2.1.1 (And Definitions)

1. Let $U \subseteq \mathbb{R}^b$ and $V \subseteq \mathbb{R}^m$ be open. The map

$$\int_V : \Omega_{\mathrm{cv}}(U \times V)[m] \to \Omega(U)\,, \tag{3.32}$$

is a morphism of left $\Omega(U)$-differential graded modules.
Furthermore, for any map $\overline{f} : U' \to U$, where $U' \subseteq \mathbb{R}^{b'}$ is open, we have a Cartesian diagram

$$\begin{array}{ccc} U' \times V & \xrightarrow{\ f\ } & U \times V \\ {\scriptstyle p_1}\downarrow & \square & \downarrow{\scriptstyle p_1} \\ U' & \xrightarrow{\ \overline{f}\ } & U \end{array} \qquad f(x, y) = (\overline{f}(x), y)\,, \tag{3.33}$$

and a corresponding commutative diagram of morphisms of complexes

$$\begin{array}{ccc} \Omega_{\mathrm{cv}}(U \times V)[m] & \xrightarrow{\;f^*\;} & \Omega_{\mathrm{cv}}(U' \times V)[m] \\ {\scriptstyle \int_V}\big\downarrow & & \big\downarrow{\scriptstyle \int_V} \\ \Omega(U) & \xrightarrow{\;\;\overline{f}^*\;\;} & \Omega(U')\,. \end{array}$$

2. *The map*

$$I\!D_{U,V} : \Omega(U \times V)[m] \hookrightarrow \mathrm{Hom}_{\Omega(U)}\big(\Omega_{\mathrm{cv}}(U \times V), \Omega(U)\big)\,, \tag{3.34}$$

defined by

$$I\!D_{U,V}(\alpha) := \Big(\beta \mapsto \int_V \alpha \wedge \beta\Big)\,, \tag{3.35}$$

is an injective *morphism of left* $\Omega(U)$*-differential graded modules, called* the left Poincaré adjunction map relative to U.

Proof

(1) The map $\int_V$ is, by definition, a morphism of left $\Lambda(\mathbb{R}^b)$-graded modules, so that to prove that it is a morphism of left $\Omega(U)$-graded modules, it remains only to verify that $\int_V$ commutes with the action of $g \in p^*(\Omega^0(U))$, which is also clear since these functions are independent of the integration variables.

For the compatibility with differentials, we have, on the one hand

$$\begin{aligned} d\int_V f(\bar{x},\bar{y})\,d_I \wedge d_J &= (-1)^{|I|+|J|}\, d\Big(\int_V f(\bar{x},\bar{y})\,d_J\Big)\,d_I \\ &= (-1)^{|I|+|J|}\sum\nolimits_{i=1}^{b} \frac{\partial}{\partial x_i}\Big(\int_V f(\bar{x},\bar{y})\,d_J\Big)\,dx_i \wedge d_I \\ &=_1 (-1)^{|I|+|J|}\sum\nolimits_{i=1}^{b}\Big(\int_V \frac{\partial f}{\partial x_i}(\bar{x},\bar{y})\,d_J\Big)\,dx_i \wedge d_I \end{aligned} \tag{3.36}$$

where $(=_1)$ is justified since, for $\bar{x}$ fixed, the support of $f(\bar{x}\ \_)$ is compact.

And we have, on the other hand,

$$\begin{aligned} \int_V d(f(\bar{x},\bar{y}))\,d_I \wedge d_J &= \int_V \Big(\sum\nolimits_{i=1}^{b} \frac{\partial f}{\partial x_i}(\bar{x},\bar{y})\,dx_i \wedge d_I \wedge d_J \\ &\quad + \sum\nolimits_{j=1}^{m} \frac{\partial f}{\partial y_j}(\bar{x},\bar{y})\,dy_j \wedge d_I \wedge d_J\Big) \\ &=_2 (-1)^{1+|I|+|J|}\sum\nolimits_{i=1}^{b}\int_V \Big(\frac{\partial f}{\partial x_i}(\bar{x},\bar{y})\,d_J\Big)\,dx_i \wedge d_I \end{aligned} \tag{3.37}$$

where $(=_2)$ is due to the vanishing of the last term in the previous line, in the only case where it could not vanish, i.e. for $J = [\![1, m]\!] \smallsetminus \{j\}$. Indeed, in that case, since $f(\bar{x}, \_)$ has compact support in V, we can write, for all fixed $\bar{x}$,

$$\int_V \frac{\partial f}{\partial y_j}(\bar{x}, \bar{y})\, dy_j \wedge d_I \wedge d_J = (-1)^{|I|+|J|} \int_{\mathbb{R}^{n-1}} \Big(\int_{\mathbb{R}} \frac{\partial f}{\partial y_j}(\bar{x}, \bar{y})\, dy_j \Big)\, d_J \wedge d_I \,,$$

by Fubini's theorem. But then

$$\int_{\mathbb{R}} \frac{\partial f}{\partial y_j}(\bar{x}, \bar{y})\, dy_j = f(\bar{x}, \bar{y})\Big|_{y_j=+\infty} - f(\bar{x}, \bar{y})\Big|_{y_j=-\infty} = 0$$

since, again, $f \in \Omega^0_{\mathrm{cv}}(U \times V)$.

Putting together (3.36) and (3.37), we get

$$d \circ \int_V = (-1)^m \int_V \circ\, d \,.$$

The statement in (1) concerning the Cartesian diagram (3.33) is now quite straightforward and results from elementary checks, left to the reader.

(2) The compatibility of $I\!D_{U,V}$ with differentials results almost as for Proposition 2.4.1.1, except that now we have also to consider the differential in $\Omega(U)$. We have

$$\begin{aligned} I\!D_{U,V}\big((-1)^m d\,\alpha\big)(\beta) &= \int_V (-1)^m d(\alpha \wedge \beta) + (-1)^{m+[\alpha]+1} \int_V \alpha \wedge d\,\beta \\ &=_1 d\big(I\!D_{U,V}(\alpha)(\beta)\big) - (-1)^{[\beta]-1}\, I\!D_{U,V}(\alpha)\big(d\,\beta\big) = (D I\!D_{U,V}(\alpha))(\beta)\,, \end{aligned}$$

where, for the first term after $(=_1)$, we applied (1).

– Injectivity of $I\!D_{U,V}$ results from the description (3.30) of $\alpha \in \Omega(U \times V)$:

$$\alpha = \sum\nolimits_{I,J} f_{IJ}(\bar{x}, \bar{y})\, d_I \wedge d_J \,,$$

where $f_{IJ} \in \Omega^0(U \times V)$. Indeed, for every $I_0 \subseteq [\![1, b]\!]$ and $J_0 \subseteq [\![1, m]\!]$, denote $I_0' := [\![1, b]\!] \smallsetminus I_0$ and $J_0' := [\![1, m]\!] \smallsetminus J_0$. Then, for $g \in \Omega^0_{\mathrm{c}}(U \times V)$, we have

$$\alpha \wedge g(\bar{y})\big(d_{I_0'} \wedge d_{J_0'}\big) = \pm f_{I_0 J_0}(\bar{x}, \bar{y})\, g(\bar{x}, \bar{y})\, d_{[\![1,b]\!]} \wedge d_{[\![1,m]\!]} \,,$$

and

$$I\!D_{U,V}(\alpha)\big(g(\bar{x}, \bar{y})\, d_{[\![1,b]\!]} \wedge d_{[\![1,m]\!]}\big) = \pm \Big(\int_V f_{I_0 J_0}(\bar{x}, \bar{y})\, g(\bar{x}, \bar{y}) \Big) d_{[\![1,b]\!]} \,. \tag{3.38}$$

But then, if $f_{I_0 J_0} \neq 0$, we can choose g with compact support small enough to have both: $f_{I_0 J_0}\, g \neq 0$ and $f_{I_0 J_0}\, g \geq 0$. In which case the right-hand side of (3.38) is clearly nonzero, showing that $\mathbb{D}_{U,V}(\alpha) \neq 0$, as announced. □

3.2.1.1 The Case of General Fiber Bundles

We can now define for any connected and oriented fiber bundle of manifolds (E, B, π, M) (*cf*. fn. ([3]), p. 71), the operation of *integration along fibers*

$$\int_M : \Omega_{\mathrm{cv}}(E)[d_M] \to \Omega(B)\,.$$

For this, consider an atlas $\mathscr{U} := \{U_\mathrm{i}\}_{\mathrm{i}\in\mathfrak{I}}$ for B, where the U_i's are trivializing, and a corresponding partition of unity $\{\phi_\mathrm{i}\}_{\mathrm{i}\in\mathfrak{I}}$. Then, following Proposition 3.1.7.1, fix an *oriented* family of trivializations

$$\{(\pi, \varphi_\mathrm{i}) : \pi^{-1}(U_\mathrm{i}) \to U_\mathrm{i} \times M\}_{\mathrm{i}\in\mathfrak{I}}\,.$$

Let $\mathscr{V} := \{V_\mathrm{j}\}_{\mathrm{j}\in\mathfrak{J}}$ be an atlas for M, let $\{\phi_\mathrm{j}\}_{\mathrm{j}\in\mathfrak{J}}$ be corresponding partitions of unity, and define

$$\phi_{\mathrm{ij}}(y) := \phi_\mathrm{i}\big(\pi(y)\big)\, \phi_\mathrm{j}\big(\varphi_\mathrm{i}(y)\big)\,.$$

For each $\mathrm{i} \in \mathfrak{I}$, the family $\Phi_\mathrm{i} := \{\phi_{\mathrm{ij}}\}_{\mathrm{j}\in\mathfrak{J}}$ is a partition of the nonnegative function $\phi_\mathrm{i} \circ \pi : \pi^{-1}(U_\mathrm{i}) \to \mathbb{R}_{\geq 0}$, and the family $\Phi := \{\phi_{\mathrm{ij}}\}_{(\mathrm{i},\mathrm{j})\in\mathfrak{I}\times\mathfrak{J}}$ is a partition of unity in E, subordinate to the *Euclidean cover*

$$\mathscr{W} := \{W_{\mathrm{ij}} := \varphi_\mathrm{i}^{-1}(V_\mathrm{j})\}_{(\mathrm{i},\mathrm{j})\in\mathfrak{I}\times\mathfrak{J}}\,. \tag{3.39}$$

A key point to notice is that if $\beta \in \Omega_\mathrm{c}(E)$, then for each $\mathrm{i} \in \mathfrak{I}$, the sum $\sum_\mathrm{j} \phi_{\mathrm{ij}}\beta$ is a *finite* sum of nonzero differential forms in $\Omega_{\mathrm{cv}}(\pi^{-1}(U_\mathrm{i}))$, and this, because the differential form $(\phi_\mathrm{i} \circ \pi)\,\beta$ is of compact support in $\pi^{-1}(U_\mathrm{i})$ and that the family $\Phi_\mathrm{i} := \{\phi_{\mathrm{ij}}\}_{\mathrm{j}\in\mathfrak{J}}$ is a locally finite partition of $\phi_\mathrm{i} \circ \pi$.

As a consequence, if for each $(\mathrm{i}, \mathrm{j}) \in \mathfrak{I} \times \mathfrak{J}$, we denote by

$$\int_{V_\mathrm{j}} : \Omega_{\mathrm{cv}}(W_{\mathrm{ij}}) \to \Omega(U_\mathrm{i})$$

the integration along the fibers of the trivial Euclidean fiber bundle $(W_{\mathrm{ij}}, U_\mathrm{i}, \pi, V_\mathrm{j})$, then the sum

$$\int_M \beta := \sum\nolimits_{\mathrm{ij}} \int_{V_\mathrm{j}} \phi_{\mathrm{ij}}\, \beta \tag{3.40}$$

is a well-defined differential form in $\Omega(B)$.

Lemma 3.2.1.2 *The sum (3.40) is independent of the choices of atlases* $\mathscr{U} := \{U_\mathrm{i}\}_{\mathrm{i}\in\mathfrak{I}}$, $\mathscr{V} := \{V_\mathrm{j}\}_{\mathrm{j}\in\mathfrak{J}}$ *and partitions of unity* $\{\phi_\mathrm{i}\}_{\mathrm{i}\in\mathfrak{I}}$ *and* $\{\phi_\mathrm{j}\}_{\mathrm{j}\in\mathfrak{J}}$. *In particular, if* $\beta \in \Omega_{\mathrm{cv}}(E)$ *is such that* $|\beta| \subseteq \Omega_{\mathrm{cv}}(W_{\mathrm{ij}})$, *then, for all* $\alpha \in \Omega(B)$,

$$\int_M \pi^*(\alpha) \wedge \beta = \alpha \wedge \int_{V_\mathrm{j}} \beta\,.$$

Proof The proof of the independence upon the choices of atlases is the same as for the Lemma 2.3.2.1. The displayed equality results from:

$$\begin{aligned}\int_M \pi^*(\alpha) \wedge \beta &= \sum\nolimits_\mathrm{i} \int_M \pi^*(\phi_\mathrm{i}\,\alpha) \wedge \beta = \sum\nolimits_\mathrm{i} \int_{V_\mathrm{j}} \pi^*(\phi_\mathrm{i}\,\alpha) \wedge \beta \\ &= \Big(\sum\nolimits_\mathrm{i} \phi_\mathrm{i}\,\alpha\Big) \wedge \int_{V_\mathrm{j}} \beta = \alpha \wedge \int_{V_\mathrm{j}} \beta\,,\end{aligned}$$

after Proposition 3.2.1.1. □

We have now enough tools to extend Proposition 3.2.1.1 to the general case.

Proposition 3.2.1.3 (and Definitions) *Let* (E, B, π, M) *be a connected, oriented fiber bundle of manifolds.*

1. *The map defined by (3.40)*

$$\int_M : \Omega_{\mathrm{cv}}(E)[d_M] \to \Omega(B)\,, \tag{3.41}$$

is a morphism of left $\Omega(B)$*-differential graded modules. In particular, for all* $\alpha \in \Omega(B)$ *and* $\beta \in \Omega_{\mathrm{cv}}(E)$, *we have*

$$\int_M \pi^*(\alpha) \wedge \beta = \alpha \wedge \int_M \beta\,. \tag{3.42}$$

Furthermore, for every Cartesian diagram

$$\begin{array}{ccc} E' & \xrightarrow{f} & E \\ {\scriptstyle \pi'}\downarrow & \square & \downarrow{\scriptstyle \pi} \\ B' & \xrightarrow{\overline{f}} & B \end{array}$$

the fiber bundle E' *is orientable, and if* f *preserves orientations, then we have a commutative diagram of morphisms of complexes*

$$\begin{array}{ccc} \Omega_{\mathrm{cv}}(E)[d_M] & \xrightarrow{f^*} & \Omega_{\mathrm{cv}}(E')[d_M] \\ {\scriptstyle \int_M}\downarrow & \oplus & \downarrow{\scriptstyle \int_M} \\ \Omega(B) & \xrightarrow{\overline{f}^*} & \Omega(B')\,. \end{array} \tag{3.43}$$

2. *The map*

$$\mathbb{D}_{B,M} : \Omega(E)[d_M] \hookrightarrow \mathbf{Hom}^{\bullet}_{\Omega(B)}\left(\Omega_{\mathrm{cv}}(E), \Omega(B)\right), \tag{3.44}$$

defined by

$$\mathbb{D}_{B,M}(\alpha) := \left(\beta \mapsto \int_M \alpha \wedge \beta\right), \tag{3.45}$$

is an injective *morphism of left $\Omega(B)$-differential graded modules, called* the left Poincaré adjunction map relative to B.

Proof All the statements result by restriction to the subspaces of an Euclidean cover $\mathscr{W} := \{W_{\mathrm{ij}}\}_{(\mathrm{i},\mathrm{j})\in\mathfrak{I}\times\mathfrak{J}}$ of E together with a corresponding partition of unity $\Phi := \{\phi_{\mathrm{ij}}\}_{(\mathrm{i},\mathrm{j})\in\mathfrak{I}\times\mathfrak{J}}$, as described in 3.2.1.1 -(3.39).

For example, to show that

$$\int_M \pi^*(\alpha) \wedge (\beta + d\beta') = \alpha \wedge \left(\int_M \beta + (-1)^{d_M} d \int_M \beta'\right)$$

we apply formula (3.40) and Lemma 3.2.1.2 to reduce to the case of trivial Euclidean fiber bundles already settled by Proposition 3.2.1.1. We then get

$$\begin{aligned}
\int_M \pi^*(\alpha) \wedge (\beta + d\beta') &= \sum_{\mathrm{ij}} \int_{V_{\mathrm{j}}} \pi^*(\alpha) \wedge (\phi_{\mathrm{ij}}\, \beta + d\, \phi_{\mathrm{ij}}\, \beta') \\
&= \sum_{\mathrm{ij}} \alpha \wedge \left(\int_{V_{\mathrm{j}}} \phi_{\mathrm{ij}}\, \beta + \int_{V_{\mathrm{j}}} d\, \phi_{\mathrm{ij}}\, \beta'\right) \\
&= \alpha \wedge \sum_{\mathrm{ij}} \left(\int_{V_{\mathrm{j}}} \phi_{\mathrm{ij}}\, \beta + (-1)^{d_M} d \int_{V_{\mathrm{j}}} \phi_{\mathrm{ij}}\, \beta'\right) \\
&= \alpha \wedge \left(\int_M \beta + (-1)^{d_M} d \int_M \beta'\right)
\end{aligned}$$

We check the commutativity of (3.43) in the same way by choosing an Euclidean cover $\mathscr{W}' := \{W'_{\mathrm{i}'\mathrm{j}'}\}_{(\mathrm{i}',\mathrm{j}')\in\mathfrak{I}'\times\mathfrak{J}'}$ of E' which is a refinement of the inverse image $f^{-1}\mathscr{W} := \{f^{-1}(W_{\mathrm{ij}})\}_{(\mathrm{i},\mathrm{j})\in\mathfrak{I}\times\mathfrak{J}}$. Details are left to the reader.
(2) Same as for the trivial Euclidean case 3.2.1.1-(2) □

Comment 3.2.1.4 It is worth noting that the injectivity of the Poincaré adjunction relative to B in 3.2.1.3-(2) establishes the nondegeneracy of what should be called *the Poincaré pairing relative to* B, i.e. the $\Omega(B)$-bilinear map:

$$\langle \cdot, \cdot \rangle_{B,M} : \Omega(E) \times \Omega_{\mathrm{cv}}(E) \to \Omega(B), \quad (\alpha, \beta) \mapsto \int_M \alpha \wedge \beta, \tag{3.46}$$

thus extending the nondegeneracy property of the (absolute) case $B := \{\bullet\}$, established in Proposition 2.4.1.1.

3.2.2 Sheafification of Integration Along Fibers

Given a connected, oriented fiber bundle (E, B, π, M), the compatibility of $\int_M$ with open base changes $U \subseteq B$ established in 3.2.1.3-(1), gives rise to the sheafification of the integration along fibers, i.e. to a morphism of $\underline{\underline{\Omega}}_B$-differential graded modules:

$$\int_M : \pi_! \, \underline{\underline{\Omega}}_E[d_M] \to \underline{\underline{\Omega}}_B \,, \tag{3.47}$$

Proposition 3.2.2.1 *Let (E, B, π, M) be a connected, oriented fiber bundle.*

1. *The morphism (3.47) induces a morphism in the cohomology sheaves*
$$\int_M : \mathcal{H}^{q+d_M}(\pi_!) \to \mathcal{H}^q(\underline{\underline{\Omega}}_B) = \underline{\underline{\mathbb{R}}}_B[0]^q \tag{3.48}$$
which is zero for $q \neq 0$. For $q = 0$ it is always surjective, and is an isomorphism if, in addition, M is connected (cf. *Proposition 3.1.10.4 -(2)).*
Furthermore, for $M = \mathbb{R}^m$ the morphism (3.48) is an isomorphism for all q.
2. *The morphism (3.47) can be filtered so a to get a morphism of convergent spectral sequences*

$$\begin{array}{ccc}
I\!E(\pi_!)_2^{p,q} := H^p\big(B; \mathcal{H}^{q+d_M}(\pi_!)\big) & \Longrightarrow & H_{\mathrm{cv}}^{p+q+d_M}(E) \\
{\scriptstyle H^p(B,\int_M)}\big\downarrow & & \big\downarrow{\scriptstyle \int_M} \\
I\!E_2^{p,q} := H^p(B; \underline{\underline{\mathbb{R}}}_B[0]^q) & \Longrightarrow & H^{p+q}(B)
\end{array}$$

where the left vertical arrow is induced by the sheaf morphism (3.48).

Proof

(1) The important fact here is that $\int_M$ is a morphism of sheaves, which, as already mentioned, is consequence of the compatibility of integration along fibers and base change. The statements in (1) are then local in nature, and we have commutative diagrams

$$\begin{array}{ccc}
\Omega(U) \otimes \Omega_c(M)[d_M] & \xrightarrow[\subseteq]{} & \Omega_{\mathrm{cv}}(\pi^{-1}(U))[d_M] \\
{\scriptstyle \mathrm{id} \otimes \int_M}\big\downarrow & \oplus & \big\downarrow{\scriptstyle \int_M} \\
\Omega(U) \otimes \mathbb{R} & = & \Omega(U)
\end{array}$$

inducing in cohomology

$$\begin{array}{ccc}
H(U) \otimes H_c(M)[d_M] & \xrightarrow[\simeq]{} & \Gamma(U; \mathcal{H}(\pi_!)[d_M]) \\
{\scriptstyle \mathrm{id} \otimes \int_M}\big\downarrow & \oplus & \big\downarrow{\scriptstyle \int_M} \\
H(U) \otimes \mathbb{R} & = & \Gamma(U; \underline{\underline{\mathbb{R}}}_B[0])
\end{array} \tag{3.49}$$

$$
\begin{array}{ccccccccc}
 & & 0 & & 0 & & 0 & & \\
 & & \downarrow & & \downarrow & & \downarrow & & \\
0 & \longrightarrow & \pi_! \,\underline{\underline{\Omega}}^0_E & \longrightarrow & \pi_! \,\underline{\underline{\Omega}}^1_E & \longrightarrow & \pi_! \,\underline{\underline{\Omega}}^2_E & \longrightarrow & \\
 & & \downarrow & & \downarrow & & \downarrow & & \\
0 & \longrightarrow & \mathscr{I}^0_0 & \longrightarrow & \mathscr{I}^1_0 & \longrightarrow & \mathscr{I}^2_0 & \longrightarrow & \\
 & & \downarrow & & \downarrow & & \downarrow & & \\
0 & \longrightarrow & \mathscr{I}^0_1 & \longrightarrow & \mathscr{I}^1_1 & \longrightarrow & \mathscr{I}^2_1 & \longrightarrow & \\
 & & \downarrow & & \downarrow & & \downarrow & &
\end{array}
\tag{3.50}
$$

Fig. 3.1 Godement's simultaneous flasque resolution of $\pi_!\underline{\underline{\Omega}}_E$

where $U \subseteq B$ is any π-trivializing *contractible* open subspace and where the horizontal '$\subseteq$' arrow in the left-hand diagram is the quasi-isomorphism 3.1.10.5.

In degree 0, we have $\Gamma(U; \mathcal{H}(\pi_!))[d_M]^0 = H_c^{d_M}(M)$ and $\Gamma(U; \underline{\mathbb{R}}_B[0])^0 = \mathbb{R}$, and the right-hand side vertical arrow is the integration $\int_M : H_c^{d_M}(M) \to \mathbb{R}$, which is always surjective and is bijective if M is connected.

When $M = \mathbb{R}^m$, we know already that $\int_M H_c(M)[d_M] = \mathbb{R}[0]$ (Poincaré Duality), which implies that the morphisms in (3.49) are all isomorphisms.

(2) We give a fairly detailed proof of this statement as it will serve as a model for future constructions of spectral sequences.

To study the action in cohomology of the morphism of complexes of sheaves

$$\int_M : \Gamma(B; \pi_! \,\underline{\underline{\Omega}}_E[d_M]) \to \Gamma(B; \underline{\underline{\Omega}}_B),$$

we recall that since the sheaves are $\underline{\underline{\Omega}}^0(B)$-modules, they are $\Gamma(B; \_)$-acyclic (B.6.3.4-(3)) (*cf*. fn. (2), p. 70) and that we can work in the derived category $\mathcal{D}^+(B; \mathbb{R})$ of bounded below complexes of sheaves on B, hence replacing the functor $\Gamma(B; \_)$ by its right derived functor $I\!R\,\Gamma(B; \_)$ (*cf*. Sect. A.2.3) and the complexes of sheaves $\pi_! \,\underline{\underline{\Omega}}_E$ and $\underline{\underline{\Omega}}(B)$ by quasi-isomorphic complexes. For example, the simultaneous Godement resolutions, the bicomplex (3.50), where each column is a flasque resolution of the sheaf at its top, and where, if we apply the cohomology functor $\mathcal{H}(\_)$ to the rows, then the i'th column $0 \to \pi_! \,\underline{\underline{\Omega}}^i_E \to \mathscr{I}^i_0 \to \mathscr{I}^i_1 \to \cdots$ becomes the complex of sheaves $0 \to \mathcal{H}^i(\pi_!) \to \mathcal{H}^i(\mathscr{I}^*_0) \to \mathcal{H}^i(\mathscr{I}^*_1) \to \cdots$ which is again a flasque resolution. It is this particularity that explains the terminology *simultaneous* resolution for the bicomplex (3.50).

Likewise, denote by $\underline{\underline{\Omega}}^*_B \to \mathscr{I}^*_\star$ be the Godement's resolution of $\underline{\underline{\Omega}}^*_B$.

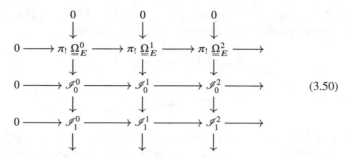

Godement's flasque resolution is functorial and exact from the category $C^*(B)$ of complexes of sheaves on B to the category $C^{\star,*}_{\mathrm{fl}}(B)$ of bicomplexes of flasque sheaves on B.[4] Applied to the morphism of complexes $\int_M : \pi_! \underline{\underline{\Omega}}_E[d_M] \to \underline{\underline{\Omega}}_B$, gives the commutative diagram of morphisms of complexes

$$\begin{array}{ccc} \underline{\underline{\Omega}}_E[d_M] & \xrightarrow{\ \int_M\ } & \underline{\underline{\Omega}}_B \\ \downarrow & & \downarrow \\ \mathbf{Tot}_\oplus \mathcal{I}_\star^* & \xrightarrow{\ \int_M\ } & \mathbf{Tot}_\oplus \mathcal{J}_\star^* \end{array} \tag{3.51}$$

where $\mathbf{Tot}_\oplus \mathcal{K}_\star^*$ is the total complex associated with a bicomplex $\mathcal{K}_\star^*$ (*cf*. Sect. 5.4.2), and where the arrow in the second row denotes the induced morphism by the Godement's resolution functor. Furthermore, the vertical arrows are quasi-isomorphisms, which is a standard property of the $\mathbf{Tot}_\oplus$ functor (*cf*. Proposition 5.4.3.1).

As a consequence, if we apply the functor $\mathbb{R}\Gamma(B; \_)$ to (3.51), we get a commutative diagram

$$\begin{array}{ccc} \mathbb{R}\Gamma(B; \underline{\underline{\Omega}}_E[d_M]) & \xrightarrow{\ \int_M\ } & \mathbb{R}\Gamma(B; \underline{\underline{\Omega}}_B) \\ \downarrow & & \downarrow \\ \mathbb{R}\Gamma(B; \mathbf{Tot}_\oplus \mathcal{I}_\star^*) & \xrightarrow{\ \int_M\ } & \mathbb{R}\Gamma(B; \mathbf{Tot}_\oplus \mathcal{J}_\star^*) \end{array}$$

where the vertical arrows are still quasi-isomorphisms. The study of the induced morphism in cohomology by the top row can therefore be accomplished in the bottom row. There, the morphism $\int_M$ comes from a morphism of bicomplexes (3.50) and, as such, it respects the row filtration (clearly regular), thus inducing a morphism of convergent spectral sequences whose $\mathbb{E}_1^p$ terms are respectively

$$\mathbb{E}(\pi_!)_1^{p,q} := \mathbb{R}\Gamma(B; \mathcal{H}^q(\mathcal{I}_p^*)) \quad \text{and} \quad \mathbb{E}_1^{p,q} := \mathbb{R}\Gamma(B; \mathcal{H}^q(\mathcal{J}_p^*)) .$$

But, for q fixed, the $\star$-complex $\mathcal{H}^q(\mathcal{I}_\star^*)$ is a flasque resolution of $\mathcal{H}^q(\pi_!)$, and the $\star$-complex $\mathcal{H}^q(\mathcal{J}_\star^*)$ is a flasque resolution of $\mathcal{H}^q(\underline{\underline{\Omega}}_B)$, i.e. of $\mathbb{R}_B[0]^q$. Furthermore, by exactness of Godement's resolution, the morphism $\int_M : \mathbb{E}(\pi_!)_1 \to \mathbb{E}_1$ is the one induced by the morphism $\int_M : \mathcal{H}^{q+d_M}(\pi_!) \to \mathcal{H}^q(\underline{\underline{\Omega}}_B)$ of (1).

Therefore, in the second page of spectral sequences, we get the morphism

$$\begin{array}{ccc} \mathbb{E}(\pi_!)_2^{p,q} & \xrightarrow{\ \int_M\ } & \mathbb{E}_2^{p,q} \\ \| & & \| \\ H^p(B; \mathcal{H}^q(\pi_!)) & \xrightarrow{\ H^p(B; \int_M)\ } & H^p(B; \mathbb{R}_B[0]^q) \end{array}$$

which ends the proof of (2). □

[4]The existence of *simultaneous resolutions* is easy to establish thanks to the exactness of Godement's *flasque resolution*. See Godement [46] §4.3 Resolution canonique d'un faisceau, p. 167, or Bredon [20] §II.2 The canonical resolution and sheaf cohomology, p. 36.

3.2.3 Thom Class of an Oriented Vector Bundle

Let $(M, [M])$ be a connected, oriented manifold of dimension d_M. In 2.4.2 we recalled the definition of the fundamental class of M as the unique cohomological class $\zeta_M \in H_c^{d_M}(M)$ verifying the equality $\int_M \zeta_M = 1$. The analogue for an oriented fiber bundle of manifolds (E, B, π, M) would be a cohomological class $\zeta_\pi \in H_{cv}^{d_M}(E)$ verifying $\int_M \zeta_M = 1$. But, although it is easy to construct differential forms $\zeta_\pi \in \Omega_{cv}^{d_M}(E)$ verifying $\int_M \zeta_\pi = 1$ (exercise ! (♝, p. 340)), it may not always be possible to have a closed form (see counterexample 3.7.2.1).

A particular case where such a cohomology class $\zeta_\pi \in H_{cv}^{d_M}(E)$ does exist is when the fiber M is a vector space, for example if (E, B, π, M) is an oriented *vector bundle*.[5]

Proposition 3.2.3.1 (Thom Isomorphism) *Let (E, B, π, M) be an oriented fiber bundle such that $H_c(M)[d_M] = \mathbb{R}[0]$. Then, the morphism*

$$\int_M : H_{cv}(E)[d_M] \to H(B)\,, \tag{3.52}$$

is an isomorphism of $H(B)$-module. There exists therefore a unique cohomology class $\Phi_\pi \in H_{cv}^{d_M}(E)$ such that,

$$\int_M \pi^*([\alpha]) \cup \Phi_\pi = [\alpha]\,, \quad \forall [\alpha] \in H(B)\,. \tag{3.53}$$

The class Φ_π is called the Thom class *of the fiber bundle (E, B, π, M).*

The inverse map of 3.52 is the isomorphism

$$\pi^*(\_) \cup \Phi_\pi : H(B) \to H_{cv}(E)[d_M]\,, \tag{3.54}$$

which is called the Thom isomorphism.

Proof Proposition 3.2.2.1-(2) says us that (3.52) is the abutment of a morphism of spectral sequences which, in the second page, is the morphism

$$H^p(B; \textstyle\int_m) : \mathbb{E}_2^{p,q}(\pi_!) := H^p(B; \mathcal{H}^q(\pi_!)) \to \mathbb{E}_2^{p,q} := H^p(B; \underline{\mathbb{R}}_B[0]^q)$$

induced by the morphism $\int_M : \mathcal{H}(\pi_!)[d_M] \to \underline{\mathbb{R}}_B[0]$, which is an isomorphism since M has the cohomology of a vector space (3.2.2.1-(1)). We can then state that (3.52) is an isomorphism.

[5]Recall that a *vector bundle* is a fiber bundle (E, B, π, V), where the fiber V is a vector space, and which is defined by a trivializing cover $\{\Phi_i : \pi^{-1}(U_i) \to U_i \times V\}_{i\in\mathfrak{I}}$ such that the transition maps are linear isomorphisms on the fibers. This enables us to give an intrinsic definition of a structure of vector spaces of the fibers of E, and, in particular, to define the zero section $\sigma : B \to E, x \mapsto 0_V$, which is an obvious homotopy equivalence.

(Note that $\mathbb{E}_2(\pi_!)$ and $\mathbb{E}_2$ are concentrated in $q = 0$, so that the spectral sequences $(\mathbb{E}_r(\pi_!), d_r)$ and $(\mathbb{E}_r, d_r)$ already degenerate for $r \geq 2$.) □

3.2.3.1 On Tubular Neighborhoods

Giving a closed embedding of manifolds $\iota : N \hookrightarrow M$, a *tubular neighborhood* of N consists of a vector bundle $(E, N, \pi, \mathbb{R}^{d_M - d_N})$ and an open map $\tau : E \to M$ which is a diffeomorphism onto its image $U_\tau := \mathrm{im}(\tau)$ and is such that the restriction of τ to the zero section $\sigma(N)$ is the embedding $\iota : N \hookrightarrow M$.

$$\begin{array}{ccccc} E & \xrightarrow[\simeq]{\ \tau\ } & U_\tau & \xrightarrow[\subseteq]{\ j_\tau\ } & M \\ {\scriptstyle \pi}\big\downarrow\big\uparrow{\scriptstyle \sigma} & & {\scriptstyle \subseteq}\big\uparrow{\scriptstyle \iota} & & \\ N & = & N & & \end{array}$$

In that case, we have an induced projection map $\pi : U_\tau \to N$ and we can define the compact vertical cohomology $H_{\mathrm{cv}}(U_\tau)$ of U_τ.

The map τ is an *open tube in M around* N, and the manifold M is said to *admit tubular neighborhoods* if for every closed embedding $N \hookrightarrow M$ and every neighborhood $U \supseteq N$ there exists a an open tube τ in U around N. Note that, under our assumptions in 2.2.1, manifolds admit tubular neighborhoods.[6]

When the conditions of orientability of the fiber bundle $(U_\tau, N, \pi, \mathbb{R}^{d_M - d_N})$ are also satisfied, for example M and N are orientable (3.1.7.2), we can apply Proposition 3.2.3.1 and state that the morphisms of graded spaces

$$\iota^* : H(U_\tau) \to H(N) \quad \text{and} \quad (\_) \cup \Phi_\tau : H(N) \to H_{\mathrm{cv}}(U_\tau)[d_M - d_N],$$

where Φ_τ is a second notation for the Thom class Φ_π, are isomorphisms.

Now, since the differential forms in $\Omega_{\mathrm{cv}}(U_\tau)$ have compact vertical support, they can be extended by zero to the whole M. Denote by $j_\tau : U_\tau \subseteq M$ the inclusion map, and by $j_{\tau,!} : \Omega_{\mathrm{cv}}(U_\tau) \to \Omega(M)$ the extension by zero map, which is clearly a morphism of complexes. It therefore induces a morphism in cohomology $j_{\tau,!} : H_{\mathrm{cv}}(U_\tau) \to H(M)$, allowing us to extend Thom isomorphism (3.54) to the graded morphism

$$(\_) \cup \Phi_\tau : H(N) \to H(M)[d_M - d_N]. \tag{3.55}$$

This morphism depends a priori on the choice of tube τ, since the inclusion $U_{\tau_{12}} \subseteq U_{\tau_1}$ does not necessarily respect fibers. But, we will show as a consequence of

---

[6]For N compact, see Spivak [83] ch. 9, Tubular Neighborhoods, Theorem. 20, p. 346. For N general, see Lang [72] ch. IV, §5. Existence Of Tubular Neighborhoods, Theorem 5.1, p. 110.

Poincaré duality relative to the base of the tube, the manifold N, that the morphism (3.55) it is indeed intrinsic (3.6.6.1).

3.3 Poincaré Duality for Fiber Bundles

3.3.1 *Sheafification of the Poincaré Adjunction*

In Proposition 3.2.1.3-(2), we introduced, for a given connected, oriented fiber bundle of manifolds (E, B, π, M), the left Poincaré adjunction relative to B, which is the injective morphism of Ω_B-differential graded modules

$$\mathbb{D}_{B,M} : \Omega(E)[d_M] \hookrightarrow \mathbf{Hom}^{\bullet}_{\Omega(B)}\left(\Omega_{\mathrm{cv}}(E), \Omega(B)\right), \tag{3.56}$$

defined by $\mathbb{D}_{B,M}(\alpha) := \left(\beta \mapsto \int_M \alpha \wedge \beta\right)$.

We will proceed to sheafify this construction as we did for $\int_M$ in 3.2.2, but this time relatively to the fibration $\pi : E \twoheadrightarrow B$.

The compatibility of $\int_M$ with base changes (3.2.1.3-(1)), implies that, given open subspaces $W \subseteq V \subseteq U \subseteq B$, and differential forms $\alpha \in \Gamma(U; \pi_* \underline{\underline{\Omega}}_E)$ and $\beta \in \Gamma(V; \pi_! \underline{\underline{\Omega}}_E)$, we have $\left(\int_M \alpha|_V \wedge \beta\right)|_W = \int_M \alpha|_W \wedge \beta|_W$. In other terms, the following diagram where the vertical arrows are the restriction maps, is commutative.

$$\begin{array}{ccc} \Gamma(V; \pi_! \underline{\underline{\Omega}}_E) & \xrightarrow{\ \mathbb{D}_{V,M}(\alpha|_V)\ } & \Gamma(V; \underline{\underline{\Omega}}_B) \\ {\scriptstyle (-)|^U_V}\downarrow & \oplus & \downarrow{\scriptstyle (-)|^U_V} \\ \Gamma(W; \pi_! \underline{\underline{\Omega}}_E) & \xrightarrow{\ \mathbb{D}_{W,M}(\alpha|_W)\ } & \Gamma(W; \underline{\underline{\Omega}}_B) \end{array} \tag{3.57}$$

In this way, $\alpha \in \Gamma(U, \pi_* \Omega_E)$ determines a *morphism of sheaves*

$$\mathbb{D}_{U,M}(\alpha) \in \mathrm{Hom}^{[\alpha]}\left(\pi_! \underline{\underline{\Omega}}_E|_U, \underline{\underline{\Omega}}_B|_U\right) = \Gamma\left(U; \underline{Hom}^{[\alpha]}(\pi_! \underline{\underline{\Omega}}_E, \underline{\underline{\Omega}}_B)\right)$$

almost tautologically compatible with the action of $\underline{\underline{\Omega}}_B$ and the restriction maps $\Gamma(U, \pi_* \Omega_E) \to \Gamma(U', \pi_* \Omega_E)$, for $U' \subseteq U$. Hence, the morphism of sheaves

$$\boxed{\underline{\underline{\mathbb{D}}}_{B,M} : \pi_* \underline{\underline{\Omega}}_E \hookrightarrow \underline{\underline{\boldsymbol{Hom}}}^{\bullet}_{\underline{\underline{\Omega}}_B}(\pi_! \underline{\underline{\Omega}}_E, \underline{\underline{\Omega}}_B)} \tag{3.58}$$

which is the announced sheafification of the left Poincaré adjunction (3.56).

Notice that the morphism $\underline{\underline{\mathbb{D}}}_{B,M}$ is injective since the horizontal arrows in the diagram (3.57) are nonzero are as long as the corresponding restriction of α do not vanish, and this since the Poincaré adjunction (3.56) is injective.

3.3.2 Deriving the Sheafified Poincaré Adjunction Functors

Until now, the fact that $\underline{\underline{\Omega}}_B$ (resp. $\Omega(B)$) is a differential graded algebra (dga) has played only a marginal role, but since the introduction of the Poincaré adjunction relative to B (3.58), it appears in the functor $\mathbf{Hom}^\bullet_{\underline{\underline{\Omega}}_B}(\_,\_)$, which obliges us to work in the category of $(\underline{\underline{\Omega}}_B, d)$-differential graded modules (dgm).

An $(\underline{\underline{\Omega}}_B, d)$-dgm is an $\underline{\underline{\Omega}}_B$-graded module $\mathcal{M} := (\mathcal{M}, d)$ equipped with a differential d of degree 1 and verifying the well-known compatibility relation between differentials $d(\alpha\, m) = (d\alpha)m + (-1)^{[\alpha]}\alpha\, dm$, for all $\alpha \in \underline{\underline{\Omega}}_B$ homogeneous and all $m \in \mathcal{M}$. A morphism of $(\underline{\underline{\Omega}}_B, d)$-dgm's $\alpha : (\mathcal{M}, d) \to (\mathcal{N}, d)$ is a morphism of graded $\underline{\underline{\Omega}}_B$-modules compatible with differentials.

Denote by $\mathrm{DGM}(\underline{\underline{\Omega}}_B, d)$ (resp. $\mathrm{DGM}(\Omega(B), d)$) the category of $(\underline{\underline{\Omega}}_B, d)$-dgm's. We have the following commutative diagram of categories and functors

$$\begin{array}{ccccccc} C\,\mathrm{GM}(\underline{\underline{\Omega}}_B) & \xrightarrow{\mathbf{Tot}} & \mathrm{DGM}(\underline{\underline{\Omega}}_B, d) & \xrightarrow{\Gamma(B;\_)} & \mathrm{DGM}(\Omega(B), d) & \xleftarrow{\mathbf{Tot}} & C\,\mathrm{GM}(\Omega(B)) \\ \downarrow & \text{(I)} & \downarrow & \text{(II)} & \downarrow & \text{(III)} & \downarrow \\ C\,\mathrm{GM}(\underline{\underline{\mathbb{R}}}_B) & \xrightarrow{\mathbf{Tot}} & \mathrm{DGM}(\underline{\underline{\mathbb{R}}}_B, 0) & \xrightarrow{\Gamma(B;\_)} & \mathrm{DGM}(\mathbb{R}, 0) & \xleftarrow{\mathbf{Tot}} & C\,\mathrm{GM}(\mathbb{R}) \end{array} \tag{3.59}$$

where **Tot** is the *total complex* functor, and $\Gamma(B;\_)$ the global section functor. The vertical arrows are the restriction functors induced by the augmentation morphism of dga's $\epsilon : \underline{\underline{\mathbb{R}}}_B \to \underline{\underline{\Omega}}_B$ and $\epsilon : \mathbb{R} \to \Omega(B)$. Notice that since $\underline{\underline{\mathbb{R}}}_B$ is concentrated in degree 0, the category $\mathrm{DGM}(\underline{\underline{\mathbb{R}}}_B, 0)$ (resp. $\mathrm{DGM}(\mathbb{R}, 0)$) is simply the category of complexes of $\underline{\underline{\mathbb{R}}}_B$-modules $C\,\mathrm{Mod}(\underline{\underline{\mathbb{R}}}_X)$ (resp. $C\,\mathrm{Mod}(\mathbb{R})$), which we also denoted $\mathrm{Sh}(B; \mathbb{R})$ (resp. $C\,\mathrm{Vec}(\mathbb{R})$).

The bifunctor $\mathbf{Hom}^\bullet(\_,\_)$ is well-defined in each of the categories appearing in diagram (3.59). The definitions are compatible through corresponding functors, and, more importantly, they extend to derived categories using techniques adapted to categories $\mathrm{DGM}(A, d)$, where (A, d) is a general differential graded algebra, for example as they are developed in Stacks Project [90] §24.26, and that we briefly recall in A.3. The commutativity of subdiagrams (I) and (III) is discussed in sections 5.3–5.4.6, and we establish that of (II) in Appendix B.9.

We denote $\mathcal{K}\mathrm{DGM}(\underline{\underline{\Omega}}_B, d)$ and $\mathcal{D}\mathrm{DGM}(\underline{\underline{\Omega}}_B, d)$ the corresponding *homotopy category* and *derived category* of $\mathrm{DGM}(\underline{\underline{\Omega}}_B, d)$ (and likewise for $\mathrm{DGM}(\Omega(B), d)$).

In the category of sheaves, we also have the sheafified bifunctor $\underline{\underline{\boldsymbol{Hom}}}^\bullet_{\underline{\underline{\Omega}}_B}(\_,\_)$, internal bifunctor of $\mathrm{DGM}(\underline{\underline{\Omega}}_B, d)$ with derived bifunctor

$$R\,\underline{\underline{\boldsymbol{Hom}}}^\bullet_{(\underline{\underline{\Omega}}_B,d)}(\_,\_) : \mathcal{D}\mathrm{DGM}(\underline{\underline{\Omega}}_B, d) \times \mathcal{D}\mathrm{DGM}(\underline{\underline{\Omega}}_B, d) \rightsquigarrow \mathcal{D}\mathrm{DGM}(\underline{\underline{\Omega}}_B, d)$$

where the subscript in the notation '$R\,\underline{\underline{\boldsymbol{Hom}}}^\bullet_{(\underline{\underline{\Omega}}_B,d)}$' recalls that we are deriving in $\mathcal{D}\mathrm{DGM}(\underline{\underline{\Omega}}_B, d)$ and not in $\mathcal{D}\mathrm{GM}(\underline{\underline{\Omega}}_B)$.

The following facts are established in Appendix B (Corollary B.9.1.3).

Proposition 3.3.2.1

1. *For all* $\mathcal{M}, \mathcal{N} \in \mathrm{DGM}(\underline{\underline{\Omega}}_B)$, *the natural map*

$$I\!R\,\underline{\underline{\mathbf{Hom}}}^\bullet_{(\underline{\underline{\Omega}}_B,d)}(\mathcal{M},\mathcal{N}) \to I\!R\,\underline{\underline{\mathbf{Hom}}}^\bullet_{\underline{\underline{\mathbb{R}}}_B}(\mathcal{M},\mathcal{N})\,.$$

is an isomorphism in the derived category $\mathcal{D}(B;\mathbb{R})$. *In particular,*

$$I\!R\,\underline{\underline{\mathbf{Hom}}}^\bullet_{(\underline{\underline{\Omega}}_B,d)}(\pi_!\underline{\underline{\Omega}}_E,\underline{\underline{\Omega}}_B) \simeq I\!R\,\underline{\underline{\mathbf{Hom}}}^\bullet_{\underline{\underline{\mathbb{R}}}_B}(\pi_!\underline{\underline{\Omega}}_E,\underline{\underline{\Omega}}_B) \simeq I\!R\,\underline{\underline{\mathbf{Hom}}}^\bullet_{\underline{\underline{\mathbb{R}}}_B}(\pi_!\underline{\underline{\Omega}}_E,\underline{\underline{\mathbb{R}}}_B)\,.$$

2. *The derived functor of the restriction functor* $\mathrm{DGM}(\underline{\underline{\Omega}}_B,d) \rightsquigarrow \mathrm{DGM}(\underline{\underline{\mathbb{R}}}_B,0)$ *induced by the augmentation morphism of dga's* $\epsilon : \underline{\underline{\mathbb{R}}}_B \to \underline{\underline{\Omega}}_B$, *is an equivalence of derived categories*

$$\mathcal{D}\,\mathrm{DGM}(\underline{\underline{\Omega}}_B,d) \simeq \mathcal{D}\,\mathrm{DGM}(\underline{\underline{\mathbb{R}}}_B,0) =: \mathcal{D}(B;\mathbb{R})\,,$$

with inverse, the functor induced by the functor

$$\underline{\underline{\Omega}}_B \otimes_{\underline{\underline{\mathbb{R}}}_B} (\_) : \mathrm{Sh}(B;\mathbb{R}) \rightsquigarrow \mathrm{DGM}(\underline{\underline{\Omega}}_B)\,.$$

3.3.3 The Poincaré Duality Theorem for Fiber Bundles

The following theorem extends the Poincaré Duality Theorem for oriented manifolds 2.4.1.3, to oriented fiber bundles.

Theorem 3.3.3.1 (Poincaré Duality for Fiber Bundles) *Let* (E, B, π, M) *be an oriented fiber bundle with fiber* M *of dimension* d_M.

1. *The following morphisms induced by the left Poincaré adjunction (3.58),*

$$\begin{array}{ccc} \pi_*\underline{\underline{\Omega}}_E[d_M] & \xrightarrow[\text{q.i.}]{I\!D_{B,M}} & I\!R\,\underline{\underline{\mathbf{Hom}}}^\bullet_{(\underline{\underline{\Omega}}_B,d)}\big(\pi_!\underline{\underline{\Omega}}_E,\underline{\underline{\Omega}}_B\big) \\ & \searrow^{\text{q.i.}}_{\textit{Grothendieck-Verdier Duality}} & \downarrow \text{q.i.} \\ & & I\!R\,\underline{\underline{\mathbf{Hom}}}^\bullet_{\underline{\underline{\mathbb{R}}}_B}\big(\pi_!\underline{\underline{\Omega}}_E,\underline{\underline{\mathbb{R}}}_B\big) \end{array} \tag{3.60}$$

are isomorphisms in $\mathcal{D}(B;\mathbb{R})$.

2. *Applying* $I\!R\,\Gamma(B;\_)$ *to (3.60), we get quasi-isomorphisms of complexes*

$$\begin{array}{ccc} \Omega(E)[d_M] & \xrightarrow[\text{q.i.}]{I\!R\,\Gamma(B;\,\underline{\underline{I\!D}}_{B,M})} & I\!R\,\mathbf{Hom}^\bullet_{(\Omega(B),d)}\big(\Omega_{\mathrm{cv}}(E),\Omega(B)\big) \\ & \searrow^{\text{q.i.}} & \downarrow \text{q.i.} \\ & & I\!R\,\mathbf{Hom}^\bullet\big(\pi_!\underline{\underline{\Omega}}_E,\underline{\underline{\mathbb{R}}}_B\big) \end{array} \tag{3.61}$$

3. *The sheaf-cohomology functor $\mathcal{H}(\_)$ applied to (3.60), gives an isomorphism of local systems*

$$\mathcal{H}(I\!D_{B,M}) : \mathcal{H}(\pi_*)[d_M] \underset{\simeq}{\longrightarrow} \underline{\underline{\mathbf{Hom}}}^{\bullet}_{\underline{\underline{\mathbb{R}}}_B}\big(\mathcal{H}(\pi_!), \underline{\underline{\mathbb{R}}}_B\big).$$

4. *If M is of finite type, the previous statements remain true if we swap terms $\pi_!\underline{\underline{\Omega}}_E \leftrightarrow \pi_*\underline{\underline{\Omega}}_E$ and $\Omega_{\mathrm{cv}}(E) \leftrightarrow \Omega(E)$. The* right *Poincaré adjunction:*

$$\pi_!\underline{\underline{\Omega}}_E[d_M] \xrightarrow[\text{q.i.}]{\underline{\underline{I\!D}}'_{B,M}} I\!R\,\underline{\underline{\mathbf{Hom}}}^{\bullet}_{(\underline{\underline{\Omega}}_B,d)}\big(\pi_*\underline{\underline{\Omega}}_E, \underline{\underline{\Omega}}_B\big) \tag{3.62}$$

is therefore a quasi-isomorphism in $\mathcal{D}(B;\mathbb{R})$ too.

Proof

(1) The vertical arrow is 3.3.2.1-(1). For the morphism $\underline{\underline{I\!D}}_{B,M}$, since it concerns sheaves, we need only show that it is locally an isomorphism. We can then confine ourselves to only consider trivializing open subspaces $U \subseteq B$, and take advantage of the fact that in that case, the inclusions $\underline{\underline{\Omega}}_U \otimes \Omega_{\mathrm{c}}(M) \subseteq \pi_!\,\underline{\underline{\Omega}}_{U\times M}$ and $\underline{\underline{\Omega}}_B \otimes \Omega(M) \subseteq \pi_*\,\underline{\underline{\Omega}}_{B\times M}$, are quasi-isomorphisms after Corollary 3.1.10.5. The restrictions of the sheafified left Poincaré adjunction to $\pi^{-1}(U)$ then reads, in $\mathcal{D}(U;\mathbb{R})$, as the isomorphism $\mathrm{id}\otimes I\!D_M : \underline{\underline{\Omega}}_B \otimes \Omega(M)[d_M] \to \underline{\underline{\Omega}}_B \otimes \Omega_{\mathrm{c}}(M)^{\vee}$, where $D_M : H(M)[d_M] \to H_{\mathrm{c}}(M)^{\vee}$ is the Poincaré duality isomorphism 2.4.1.3.

Hence, the commutative diagram

$$\begin{array}{ccc}
\underline{\underline{\Omega}}_U \otimes \Omega(M)[d_M] & \xrightarrow{\underline{\underline{I\!D}}_{U,M}} & I\!R\,\underline{\underline{\mathbf{Hom}}}^{\bullet}_{(\underline{\underline{\Omega}}_U,d)}\big(\underline{\underline{\Omega}}_U \otimes \Omega_{\mathrm{c}}(M), \underline{\underline{\Omega}}_U\big) \\
 & \searrow^{\simeq}_{\mathrm{id}\otimes D_M} & \| \\
 & & \underline{\underline{\Omega}}_U \otimes \Omega_{\mathrm{c}}(M)^{\vee}
\end{array}$$

which ends the proof of (1).

(2) Follows from (1), since by Proposition B.9.1.1-(1), we have

$$I\!R\,\mathbf{Hom}^{\bullet}_{(\Omega(B),d)}(\_,\_) = I\!R\,\Gamma\big(B, I\!R\,\underline{\underline{\mathbf{Hom}}}^{\bullet}_{(\underline{\underline{\Omega}}(B),d)}(\_,\_)\big).$$

(3) Applying $\mathcal{H}(\_)$ to $\pi_*\underline{\underline{\Omega}}_E \to I\!R\,\underline{\underline{\mathbf{Hom}}}^{\bullet}\big(\pi_!\underline{\underline{\Omega}}_E, \underline{\underline{\mathbb{R}}}_B\big)$ in (3.60), we get

$$\begin{aligned}
\mathcal{H}(\pi_*) \simeq \mathcal{H}\big(\underline{\underline{\mathbf{Hom}}}^{\bullet}(\pi_!\underline{\underline{\Omega}}_E, \mathcal{I}_\star)\big) \Leftarrow_1 I\!E_2^{p,q} &:= \mathbf{h}^p\,\underline{\underline{\mathbf{Hom}}}^{\bullet}(\mathcal{H}^q(\pi_!), \mathcal{I}_\star) \\
&\simeq_2 \underline{\underline{\mathbf{Hom}}}^{\bullet}(\mathcal{H}(\pi_!), \underline{\underline{\mathbb{R}}}_B),
\end{aligned}$$

where $\mathcal{I}_\star$ is an injective resolution $\underline{\underline{\mathbb{R}}}_B$, $(\Leftarrow_1)$ is the spectral sequence given by the (regular) $\star$-filtration, and $(\simeq_2)$ since $\mathcal{H}^q(\pi_!)$ is locally free (3.1.10.4-(2)), which implies that $\underline{\underline{\boldsymbol{Hom}}}^\bullet(\mathcal{H}^q(\pi_!), \_)$ is an exact functor, in which case the spectral sequence $\mathbb{E}_r^{p,q}$ is concentrated in $p = 0$, hence degenerated, for all $r \geq 2$.

(4) Should be obvious. The proof is exactly the same as for the previous statements, so details are left to the reader. □

Comments 3.3.3.2

1. Although clear enough, it is still worth emphasizing that, for $B = \{\bullet\}$, we recover Poincaré Duality for a manifold. The reader may also notice, that changing the point of view from a point $\{\bullet\}$ to a general base space B, entails changing the category of oriented manifolds $\mathrm{Man}^{\mathrm{or}}$ by the category of oriented fiber bundles $\mathrm{Fib}_B^{\mathrm{or}}$ (3.1.8), changing the category of complexes $C\,\mathrm{Vec}(\mathbb{R}) = \mathrm{DGM}(\Omega(\{\bullet\}), d)$ by the category of differential graded modules $\mathrm{DGM}(\Omega(B), d)$, and, therefore, changing the target category $\mathrm{GM}(\mathbb{R})$ of the cohomology functor by $\mathrm{GM}(H(B))$ (*cf*. Sect. 3.1.3).
2. The attentive reader has noticed the reference to the Grothendieck -Verdier duality in diagram (3.60), which is fully justified since we give its original formulation there (*cf*. Chap. 8). In fact, we can say that, in a sense, this chapter is a detailed example of the Grothendieck-Verdier Duality formalism in the simplified framework of fiber bundles of manifolds.

3.3.4 Poincaré Duality for Fiber Bundles and Base Change

The Poincaré Duality Theorem for fiber bundles is well behaved relative to the base change (3.1.5). More precisely, given a map $\bar{h} : B' \to B$ and a fiber bundle $E := (E, B, \pi, M)$, let $\bar{h}^{-1}(E) := (E', B', \pi', M)$ (*cf*. Corollary 3.1.7.2-(1)). By base change $\bar{h}^{-1}(\_) : \mathrm{Fib}_B \rightsquigarrow \mathrm{Fib}_{B'}$ (3.1.5), we have the Cartesian diagram

$$\begin{array}{ccc} E' & \xrightarrow{\;h\;} & E \\ {\scriptstyle \pi'}\downarrow & \square & \downarrow{\scriptstyle \pi} \\ B' & \xrightarrow{\;\bar{h}\;} & B \end{array}$$

and we know, from 3.1.10.2-(2), that $\bar{h}^{-1}(E)$ is oriented if E is so, in which case integration along fibers is compatible with base change in the sense that the following diagram is commutative

$$\begin{array}{ccc} \Omega_{\mathrm{cv}}(E)[d_M] & \xrightarrow{\;h^*\;} & \Omega_{\mathrm{cv}}(E')[d_M] \\ {\scriptstyle \int_M}\downarrow & \oplus & \downarrow{\scriptstyle \int_M} \\ \Omega(B) & \xrightarrow{\;\bar{h}^*\;} & \Omega(B')\,. \end{array} \tag{3.63}$$

It is therefore natural to expect some kind of compatibility between Poincaré Duality relative to B and to B'. The following theorem addresses this question.

Theorem 3.3.4.1 *Let $E := (E, B, \pi, M)$ and $E' := (E', B', \pi', M)$ be two oriented fiber bundles of manifolds with same fiber M of dimension d_M, and let*

$$\begin{array}{ccc} E' & \xrightarrow{h} & E \\ {\scriptstyle \pi'}\downarrow & \Box & \downarrow{\scriptstyle \pi} \\ B' & \xrightarrow{\bar{h}} & B \end{array} \tag{3.64}$$

be a Cartesian diagram of fiber bundles. Then, there exists a canonical commutative diagram of isomorphisms in $\mathcal{D}(B';\mathbb{R})$

$$\begin{array}{ccc} \bar{h}^{-1}(\pi_* \underline{\underline{\Omega}}_E)[d_M] & \xrightarrow[\text{q.i.}]{\bar{h}^{-1}(I\!D_{B,M})} & \bar{h}^{-1}\big(I\!R\, \underline{\underline{\boldsymbol{Hom}}}^{\bullet}_{(\underline{\underline{\Omega}}_B,d)}(\pi_! \underline{\underline{\Omega}}_E, \underline{\underline{\Omega}}_B)\big) \\ \text{q.i.}\downarrow & \oplus & \downarrow\text{q.i.} \\ \pi'_* \underline{\underline{\Omega}}_{E'}[d_M] & \xrightarrow[\text{q.i.}]{I\!D_{B',M}} & I\!R\, \underline{\underline{\boldsymbol{Hom}}}^{\bullet}_{(\underline{\underline{\Omega}}_{B'},d)}(\pi_! \underline{\underline{\Omega}}_{E'}, \underline{\underline{\Omega}}_{B'}) \end{array} \tag{3.65}$$

and likewise, if M is of finite type, swapping $\pi_! \underline{\underline{\Omega}}_E \leftrightarrow \pi_* \underline{\underline{\Omega}}_E$.

Proof Since we are working in derived categories of sheaves, we can replace the complexes $\underline{\underline{\Omega}}$ by the constant sheaf $\underline{\underline{\mathbb{R}}}$, and even by the constant sheaf $\underline{\underline{\mathbb{k}}}$ for any field $\mathbb{k}$ since the theorem is still verified in this case. The diagram (3.65) is then a straightforward consequence of three technical facts which we first justify.[7]

Lemma A *There exists natural canonical quasi-isomorphisms:*

$$\begin{cases} \text{(i) } \bar{h}^{-1}(\pi_* \underline{\underline{\mathbb{k}}}_E) \xrightarrow[\text{q.i.}]{} \pi'_*(h^{-1}\underline{\underline{\mathbb{k}}}_E)\,, \\ \text{(ii) } \bar{h}^{-1}(\pi_! \underline{\underline{\mathbb{k}}}_E) \xrightarrow[\text{q.i.}]{} \pi'_!(h^{-1}\underline{\underline{\mathbb{k}}}_E)\,. \end{cases} \tag{3.66}$$

Proof of Lemma A To prove (i), we begin observing the well-known relations

$$\begin{aligned} \mathrm{Hom}\big(h^{-1}(\_), h^{-1}(\_)\big) &=_1 \mathrm{Hom}\big((\_), f_* \circ h^{-1}(\_)\big) \\ &\to_2 \mathrm{Hom}\big(\pi_*(\_), \pi_* \circ f_* \circ h^{-1}(\_)\big) \\ &=_3 \mathrm{Hom}\big(\pi_*(\_), \bar{h}_* \circ \pi'_* \circ h^{-1}(\_)\big) \\ &=_4 \mathrm{Hom}\big(\bar{h}^{-1} \circ \pi_*(\_), \pi'_* \circ h^{-1}(\_)\big) \end{aligned} \tag{3.67}$$

[7] See also the proof of Proposition 8.2.2.1-(2), p. 252.

where $(=_{1,4})$ are the usual adjunctions between the inverse and direct image functors in categories of sheaves, $(\to_2)$ is the map defined by direct image functor π_*, and $(=_3)$ results from the obvious equality $\pi_* \circ h_* = \bar{h}_* \circ \pi'_*$. The morphism $\mathrm{id}(\_) \in \mathrm{Hom}(h^{-1}(\_) \to h^{-1}(\_))$ therefore defines naturally a morphism for every complex of sheaves $\mathcal{F}^\bullet \in \mathcal{D}^+(E;\Bbbk)$:

$$\bar{h}^{-1} \circ I\!R\,\pi_*(\mathcal{F}^\bullet) \to I\!R\,\pi'_* \circ h^{-1}(\mathcal{F}^\bullet)\,. \tag{3.68}$$

Taking $\mathcal{F}^\bullet := \underline{\underline{\Bbbk}}_E$, the morphism (3.68) induces the morphism of cohomology sheaves

$$\bar{h}^{-1}(\mathcal{H}^i(\pi_*(\underline{\underline{\Bbbk}}_E))) \to \mathcal{H}^i(I\!R\,\pi'_*(h^{-1}(\underline{\underline{\Bbbk}}_E))) = \mathcal{H}^i(\pi'_*(\underline{\underline{\Bbbk}}_{E'}))\,,$$

which is clearly an isomorphism since the sheaves $\mathcal{H}^i(\pi_*\underline{\underline{\Bbbk}}_E)$ and $\mathcal{H}^i(\pi'_*(\underline{\underline{\Bbbk}}_{E'}))$ are both locally trivial with the same fiber $H(M)$, because the diagram of fiber bundles (3.64) is Cartesian (*cf*. Proposition 3.1.10.4-(2)).

To prove (ii) we proceed as in (i), replacing π_* with $\pi_!$. The equality $(=_3)$ in (3.67) is then a consequence of the equality $\pi_! \circ h_* = \bar{h}_* \circ \pi'_!$, which is easy to prove since (3.64) is the diagram Cartesian (*cf*. Proposition 3.1.9.2-(1,2b)). ⊟

Lemma B *Let $\mathcal{F}^*$ and $\mathcal{G}^\star$ be bounded below complexes of sheaves in* $\mathrm{Sh}(B;\Bbbk)$, *and assume that the cohomology sheaves $\mathcal{H}^q(\mathcal{F}^*)$ are locally trivial.*[8] *Then the natural morphism*

$$\text{(iii)}\quad \bar{h}^{-1}\big(I\!R\,\underline{\underline{\mathbf{Hom}}}^\bullet(\mathcal{F}^*,\mathcal{G}^\star)\big) \to I\!R\,\underline{\underline{\mathbf{Hom}}}^\bullet(\bar{h}^{-1}(\mathcal{F}^*),\bar{h}^{-1}(\mathcal{G}^\star)) \tag{3.69}$$

is a quasi-isomorphism.

Proof of Lemma B We have already seen in the proof of 3.3.3.1-(3) that under the current assumptions in $\mathcal{F}^*$ and $\mathcal{G}^\star$, we have a canonical quasi-isomorphism

$$\mathcal{H}(I\!R\,\underline{\underline{\mathbf{Hom}}}^\bullet(\mathcal{F}^*,\mathcal{G}^\star)) \simeq \underline{\underline{\mathbf{Hom}}}^\bullet(\mathcal{H}(\mathcal{F}^*),\mathcal{G}^\star)\,.$$

As a consequence, and given that $\mathcal{H}^q(\bar{h}^{-1}(\mathcal{F}^*)) = \bar{h}^{-1}(\mathcal{H}^q(\mathcal{F}^*))$ is also locally trivial, we can show that (3.69) is a quasi-isomorphism simply by showing that the natural morphism

$$\bar{h}^{-1}\big(\underline{\underline{\mathbf{Hom}}}^\bullet(\mathcal{H}(\mathcal{F}^*),\mathcal{G}^\star)\big) \to \underline{\underline{\mathbf{Hom}}}^\bullet\big(\bar{h}^{-1}(\mathcal{H}(\mathcal{F}^*)),\bar{h}^{-1}\mathcal{G}^\star\big)\,.$$

is an isomorphism, which is clear by looking at stalk level. ⊟

Taking $\mathcal{F}^* := \underline{\underline{\pi}}_!\Bbbk_E$ and $\mathcal{G}^\star := \underline{\underline{\Bbbk}}_B$ in lemmas A and B, and recalling the compatibility between integration along fibers and base change (3.63), establishes

[8] See Borel [16], Proposition 10.21, p. 171.

the existence of the commutative diagram of isomorphisms in $\mathcal{D}^+(B;\Bbbk)$

$$\begin{array}{ccc}
\bar{h}^{-1}(\pi_*\underline{\Bbbk}_E)[d_M] & \xrightarrow[\text{q.i.}]{\bar{h}^{-1}(\underline{I\!D}_{B,M})} & \bar{h}^{-1}\big(I\!R\,\underline{\mathcal{H}om}^\bullet(\pi_!\underline{\Bbbk}_E,\underline{\Bbbk}_B)\big) \\
{\scriptstyle\text{q.i.}}\downarrow & \oplus & \downarrow{\scriptstyle\text{q.i.}} \\
\pi'_*\underline{\Bbbk}_{E'}[d_M] & \xrightarrow[\text{q.i.}]{\underline{I\!D}_{B',M}} & I\!R\,\underline{\mathcal{H}om}^\bullet(\pi_!\underline{\Bbbk}_{E'},\underline{\Bbbk}_{B'})
\end{array}$$

hence the diagram (3.65). □

Comment 3.3.4.2 Statement (ii) in Lemma A, is a well-known result valid for any $\mathcal{F}^* \in \mathcal{D}^+(E)$ rather than the constant sheaf $\underline{\Bbbk}_E$ and for any Cartesian diagram (3.64), not necessarily of fiber bundles.[9] By contrast, care must be taken with (i) since it does not generally hold. Here, the hypothesis that E is a fiber bundle is critical.

3.4 Poincaré Duality Relative to a Formal Base Space

Although absolute and relative viewpoints emerge from the same formalism, an important difference arises when comparing the statements of the absolute (2.4.1.3) and the relative (3.3.3.1) Poincaré duality theorems, which is that the second states only quasi-isomorphisms of complexes, saying nothing about the induced morphisms in cohomology.

More precisely, in the relative case it is stated (for example) that the left Poincaré adjunction induces an isomorphism in cohomology

$$H(E)[d_M] \xrightarrow[\simeq]{} \boldsymbol{h}\, I\!R\,\mathbf{Hom}^\bullet_{(\Omega(B),d)}(\Omega_{\mathrm{cv}}(E),\Omega(B))\,,$$

and we know, besides, that there exists a natural map (see 5.4.7.2-(1))

$$\boldsymbol{h}\, I\!R\,\mathbf{Hom}^\bullet_{(\Omega(B),d)}(\Omega_{\mathrm{cv}}(E),\Omega(B)) \xrightarrow{\ \xi\ } \mathbf{Hom}^\bullet_{H(B)}(H_{\mathrm{cv}}(E),H(B))\,,$$

[9] See Kashiwara-Schapira [61], Prop. 2.5.11, p. 106, or Borel [16] Prop. 10.7, p. 159.

leading us to wonder if the composition $D_{B,M} := \xi \circ H(B; \mathbb{D}_{B,M'})$:

$$\begin{array}{ccc} H(E) & \xrightarrow[\simeq]{H(B;\mathbb{D}_{B,M'})} & \mathbf{h}\,\mathbb{R}\,\mathbf{Hom}^{\bullet}_{(\Omega(B),d)}((\Omega_{\mathrm{cv}}(E),d),(\Omega(B),d)) \\ & \searrow^{D_{B,M}}_{(\not\simeq)} & \big\downarrow \xi \ (\not\simeq) \\ & & \mathbf{Hom}^{\bullet}_{H(B)}(H_{\mathrm{cv}}(E), H(B))\,, \end{array} \tag{3.70}$$

is also an isomorphism. But it turns out that ξ is generally not an isomorphism, which means we cannot always expect the natural morphism of $H(B)$-modules

$$D_{B,M} : H(E) \xrightarrow[(\not\simeq)]{} \mathbf{Hom}^{\bullet}_{H(B)}(H_{\mathrm{cv}}(E), H(B)) \tag{3.71}$$

to be an isomorphism, contrary to the absolute case $B = \{\bullet\}$.

The following section recalls an important property of a base space B, particularly relevant in equivariant cohomology where B will be a classifying space $\mathbb{B}G$, which allows us a better understanding of (3.71).

3.4.1 Formality of Topological Spaces

Following Deligne-Griffiths-Morgan-Sullivan,[10] we recall the concept of *formality* of differential graded algebras and of topological spaces, in a way suited to our needs.

Definition 3.4.1.1

1. A dg-algebra (A, d) is said to be *formal* if there exists a zig-zag diagram

$$\begin{array}{ccccccccc} (A,d) & & (A_2,d) & & (A_4,d) & \cdots & (A_n,d) & & \\ & \searrow \ \swarrow & & \searrow \ \swarrow & & & & \searrow & \\ & (A_1,d) & & (A_3,d) & & & & & \mathbf{h}(A,d) \end{array}$$

   of quasi-isomorphic morphisms of dg-algebras, beginning at (A, d) and ending at its cohomology viewed as a dg-algebra with zero differential.
2. A manifold M is said to be *formal* if its de Rham complex $(\Omega(M), d)$ is a formal dg-algebra. More generally, a topological space X is said to be $\Bbbk$-*formal* if the complex $\Omega(X; \Bbbk)$ of its Alexander Spanier cochains with coefficients in $\Bbbk$ is a formal dg-algebra.

[10] DGMS [36], Section 4. Formality of Differential Algebras, p. 260.

Examples 3.4.1.2 Compact Kähler manifolds (e.g. smooth projective varieties), Lie groups, classifying spaces $\mathbb{B}G$ of Lie groups G are examples of $\mathbb{R}$-formal spaces (*cf*. Sect. 4.1.1.3 and especially Theorem 4.9.2.1-(2)).[11]

The following Proposition is proved in Appendix A as Corollary A.3.3.2.

Proposition 3.4.1.3 *Let (A, d) be a formal dg-algebra such that the cohomology algebra $H(A) := \mathbf{h}(A, d)$ is of finite homological dimension. Let (M, d) be an (A, d)-dg-module and denote by $H(M)$ the $H(A)$-graded module $\mathbf{h}(M, d)$.*

1. *There exists a convergent spectral sequence*

$$\begin{cases} \mathbb{E}_2^{p,q} := \mathbf{h}^p\, \mathbb{R}\,\mathbf{Hom}^\bullet_{H(A)}(H^q(M), H(A)) \\ \qquad\qquad \Downarrow \\ \mathbf{h}^{p+q}\, \mathbb{R}\,\mathbf{Hom}^\bullet_{(A,d)}((M, d), (A, d)) \end{cases}$$

2. *If* $\dim_{\text{proj}}(H(M)) \leq 1$, *then there exists an isomorphism in* $\mathcal{D}\mathrm{DGM}(H(A))$*:*

$$\mathbb{R}\,\mathbf{Hom}^\bullet_{(A,d)}((M, d), (A, d)) \simeq \mathbb{R}\,\mathbf{Hom}^\bullet_{H(A)}(H(M), H(A))\,.$$

and, the spectral sequence $(\mathbb{E}_r, d_r)$ in (1) degenerates for $r \geq 2$.

Returning to our preliminary discussion in 3.4, we can now state the following enhancement of the Poincaré duality of fiber bundles when the base manifold B is a connected and simply connected formal space.

Theorem 3.4.1.4 (Poincaré Duality for Fiber Bundles with Formal Base Space) *Let (E, B, π, M) be an oriented fiber bundle of manifolds where M is of dimension d_M and where B is a formal manifold such that $H(B)$ is of finite homological dimension. Then, the following holds:*

1. *The Poincaré adjunction (3.3.3.1-(1)) induces a convergent spectral sequence:*

$$\mathbb{E}_2^{p,q} := \mathbf{h}^p\, \mathbb{R}\,\mathbf{Hom}^\bullet_{H(B)}(H^q_{\text{cv}}(E), H(B)) \Rightarrow H^{p+q+d_M}(E)\,.$$

2. *If* $\dim_{\text{proj}}(H_{\text{cv}}(E)) \leq 1$ *as $H(B)$-gm, then the Poincaré adjunction induces an isomorphism in* $\mathcal{D}\mathrm{DGM}(H(B))$*:*

$$D_{B,M} : \Omega(E)[d_M] \simeq \mathbb{R}\,\mathbf{Hom}^\bullet_{H(B)}(H_{\text{cv}}(E), H(B))\,, \tag{3.72}$$

and the spectral sequence $(\mathbb{E}_r, d_r)$ in (1) degenerates, i.e. $d_r = 0$ for $r \geq 2$.

[11]From Sullivan [85] §12. Formal computation and Kaehler manifolds, p. 317.

Furthermore, it $H_{cv}(E)$ is a projective $H(B)$-gm, then isomorphism (3.72) is canonical and induces an isomorphism of $H(B)$-gm's

$$H(E)[d_M] \simeq \mathbf{Hom}^{\bullet}_{H(B)}\big(H_{cv}(E), H(B)\big)\,. \tag{3.73}$$

3. *If M is of finite type, the statements (1) and (2) remain true if we swap the terms $\Omega_G(M) \leftrightarrow \Omega_{G,c}(M)$ and $H(E) \leftrightarrow H_{cv}(E)$.*

Proof Apply Proposition 3.4.1.3 to Poincaré Duality of Fiber Bundles 3.3.3.1. □

3.4.2 Poincaré Duality Relative to Classifying Spaces

We explain how Equivariant Poincaré duality for G-manifolds, where G is a compact Lie group, can be seen as a particular case of Poincaré Duality relative to the classifying space $\mathbb{B}G$.

When writing Sect. 3.3 on Poincaré Duality for fiber bundles, we were aware that the same results can be stated in categories larger than that of fiber bundles of manifolds. The reason being that we essentially only need separateness, local contractibility, and paracompactness of topological spaces. This is why we included Appendix B which extends the considerations of Sects. 3.1 to 3.3 to fiber bundles of *mild* spaces. The only significant change to be done is to substitute the complexes $\underline{\underline{\Omega}}(\_)$ of sheaves of differential forms by the complexes $\underline{\underline{\Omega}}(\_;\Bbbk)$ of sheaves of Alexander-Spanier cochains with coefficients in the field $\Bbbk$.

It is therefore possible to state Poincaré Duality theorem for fiber bundles, as stated in 3.3.3.1 and 3.4.1.4, in the category $\mathrm{Fib}^{\mathrm{or}}_{\mathbb{B}G}$ of fiber bundles $(E, \mathbb{B}G, \pi, M)$, where M is an oriented manifold of dimension d_M (or more generally a mild Poincaré Duality space), and where the base space $\mathbb{B}G$ is a classifying space for a compact *connected* Lie group G. We will later see that $\mathbb{B}G$ (thoroughly discussed in 4.6), is not a manifold, but, rather, a CW-complex inductive limit of compact manifolds.

To fully understand the relevance of Poincaré duality for fiber bundles in equivariant cohomology, we need recall the *Borel construction* (*cf*. Sect. 4.7). This is a covariant functor $(\_)_G : G\text{-Man} \rightsquigarrow \mathrm{Fib}_{\mathbb{B}G}$ from the category G-Man of G-manifolds and G-equivariant maps, which associates with a G-manifold M, the fiber bundle $(M_G, \mathbb{B}G, \pi, M)$, where M_G denotes the homotopy quotient $M_G := \mathbb{E}G \times_G M$ and where $\pi : M_G \to \mathbb{B}G$ is the locally trivial fibration $[x, m] \mapsto [x]$ of fiber space M (*cf*. Definition 4.7.1.1). The *equivariant cohomology of M with coefficients in* $\Bbbk$, denoted by $H_G(M;\Bbbk)$, is then defined as the composition of the two functors

$$\begin{array}{ccccc}
 & & \overset{H_G(_;\Bbbk)}{\frown\!\!\!\frown\!\!\!\frown\!\!\!\frown\!\!\!\frown\!\!\!\frown\!\!\!\frown\!\!\!\frown\!\!\!\longrightarrow} & & \\
G\text{-Man} & \overset{(_)_G}{\rightsquigarrow} & \mathrm{Fib}_{\mathbb{B}G} & \overset{H(_;\Bbbk)}{\rightsquigarrow} & \mathrm{GM}(H(\mathbb{B}G;\Bbbk)) \\
M & \longmapsto & (M_G,\mathbb{B}G,\pi,M) & \longmapsto & H(M_G;\Bbbk)
\end{array}$$

When the G-manifold M is oriented, the fiber bundle $M_G := (M_G, \mathbb{B}G, \pi, M)$ is oriented (*cf*. Proposition 4.7.3.1) and Poincaré Duality relative to $\mathbb{B}G$ applies to M_G (3.4.1.4). But in addition, and this is fairly important, the classifying space $\mathbb{B}G$ is an $\mathbb{R}$-formal space (4.9.2.1-(2)) and since $H(\mathbb{B}G;\mathbb{R}) \simeq S(\mathfrak{g}^\vee)^G$ is a polynomial algebra, it has finite homological dimension, and fill the conditions for applying Theorem 3.4.1.4. The following immediate corollary of theorems 3.3.3.1 and 3.4.1.4 can then be stated.

Corollary 3.4.2.1 (Equivariant Poincaré Duality) *Let G be a compact Lie group, and let M be an orientable G-manifold of dimension d_M.*

Denote by

- *$(M_G, \mathbb{B}G, \pi, M)$, the associated Borel fiber bundle (4.7).*
- *$\Omega_G(M) := \Omega(M_G;\mathbb{R})$ and $\Omega_{G,\mathrm{c}}(M) := \Omega_{\mathrm{cv}}(M_G,\mathbb{R})$, the complexes of Alexander-Spanier cochains on M_G of respectively closed and proper supports and with coefficients in $\mathbb{R}$; and then $\Omega_G := \Omega_G(\{\bullet\}) = \Omega(\mathbb{B}G;\mathbb{R})$.*
- *$H_G(M) := \mathbf{h}(\Omega_G(M))$, $H_{G,\mathrm{c}}(M) := \mathbf{h}(\Omega_{G,\mathrm{c}}(M))$ and $H_G := H_G(\{\bullet\})$.*

1. *The left Poincaré adjunction isomorphism 3.3.3.1-(2) in $\mathcal{D}\mathrm{DGM}(\Omega_G)$:*

$$D_{B,M} : \Omega_G(M)[d_M] \simeq \mathbb{R}\,\mathbf{Hom}^\bullet_{(\Omega_G,d)}(\Omega_{G,\mathrm{c}}(M), \Omega_G)\,, \tag{3.74}$$

induces a convergent spectral sequence:

$$\mathbb{E}_2^{p,q} := \mathbf{h}^p\, \mathbb{R}\,\mathbf{Hom}^\bullet_{H_G}(H^q_{G,\mathrm{c}}(M), H_G) \Rightarrow H_G^{p+q+d_M}(M)\,.$$

2. *If $\dim_{\mathrm{proj}}(H_{G,\mathrm{c}}(M)) \le 1$ as H_G-graded module, then the Poincaré adjunction induces an isomorphism in $\mathcal{D}\mathrm{DGM}(H_G)$:*

$$D_{B,M} : \Omega_G(M)[d_M] \simeq \mathbb{R}\,\mathbf{Hom}^\bullet_{H_G}(H_{G,\mathrm{c}}(M), H_G)\,, \tag{3.75}$$

and the spectral sequence $(\mathbb{E}_r, d_r)$ in (1) degenerates, i.e. $d_r = 0$ for $r \ge 2$. Furthermore, if $H_{G,\mathrm{c}}(M)$ is a projective H_G-graded module, then (3.75) induces an isomorphism of H_G-graded modules

$$H_G(M)[d_M] \simeq \mathbf{Hom}^\bullet_{H_G}\big(H_{G,\mathrm{c}}(M), H_G\big)\,.$$

3. *If M is of finite type, the statements (1) and (2) remain true if we swap the terms $\Omega_G(M) \leftrightarrow \Omega_{G,\mathrm{c}}(M)$ and $H_G(M) \leftrightarrow H_{G,\mathrm{c}}(M)$.*

Sketch of Proof The only difference with respect to theorems 3.3.3.1 and 3.4.1.4 is that here we are working in the category of fiber bundles above the classifying space $\mathbb{B}G$, which is not a manifold but a CW-complex, and which obliges us to replace de Rham differential forms by, for example, Alexander-Spanier cochains. Theorem 3.3.3.1 is then valid over any field, and Theorem 3.4.1.4 over a field $\mathbb{k}$, such that $\mathbb{B}G$ is $\mathbb{k}$-formal, i.e. of characteristic not dividing the cardinality of the Weyl group $W(G)$ (*cf*. Sect. 8.4). (The reader will find detailed proof of this corollary by a somewhat different route in Chap. 8 devoted to Equivariant Poincaré Duality over arbitrary fields (*cf*. Theorem 8.4.1.3).) ⊡

Comment 3.4.2.2 When G is the circle group $\mathbb{S}^1$, the ring $H(\mathbb{B}G; \mathbb{R})$ is isomorphic to the polynomial algebra $\mathbb{R}[X]$ which is of homological dimension 1. In that case, the hypothesis $\dim_{\text{proj}}(H_{G,c}(M)) \leq 1$ in 3.4.2.1-(2) is automatically verified so can be omitted.

3.5 Gysin Morphisms for Fiber Bundles

3.5.1 Gysin Morphisms Relative to a Base Space

Although we will not develop this section as thoroughly as in the absolute case, where $B := \{\bullet\}$, we want to make clear that the definition of Gysin morphisms, which we base on the quasi-isomorphisms given by Poincaré adjunctions, can be repeated word for word in the relative case.

Given a proper morphism $f : (E', B, \pi', M') \to (E, B, \pi, M)$ in Fib_B^{or}

$$\begin{array}{ccc} E' & \xrightarrow{f} & E \\ & {\scriptstyle \pi'}\searrow \quad \swarrow{\scriptstyle \pi} & \\ & B & \end{array}$$

we know after 3.1.10.2-(2) that the pullback morphism $f^* : \Omega(E) \to \Omega(E')$ induces a morphism $f^* : \Omega_{\text{cv}}(E) \to \Omega_{\text{cv}}(E')$ in $\text{DGM}(\Omega(B), d)$. We can then apply the relative Poincaré Duality theorem 3.3.3.1-(2) and consider the diagram

$$\begin{array}{ccc} \Omega(E')[d_{M'}] & \xrightarrow[\text{q.i.}]{\Gamma(B; \underline{\mathbb{D}}_{B,M'})} & \mathbb{R}\mathbf{Hom}^\bullet_{(\Omega(B),d)}\big(\Omega_{\text{cv}}(E'), \Omega(B)\big) \\ {\scriptstyle f_*}\vdots & & \downarrow \qquad \uparrow{\scriptstyle f^*} \\ \Omega(E)[d_M] & \xrightarrow[\text{q.i.}]{\Gamma(B; \underline{\mathbb{D}}_{B,M})} & \mathbb{R}\mathbf{Hom}^\bullet_{(\Omega(B),d)}\big(\Omega_{\text{cv}}(E), \Omega(B)\big) \end{array} \tag{3.76}$$

inducing a canonical morphism in $\mathcal{D}\text{DGM}(\Omega(B), d)$

$$\boxed{f_* : \Omega(E')[d_{M'}] \to \Omega(E)[d_M]} \tag{3.77}$$

which we call *the Gysin morphism relative to B associated with* f.

In the same way, for every morphism $f : (E', B, \pi', M') \to (E, B, \pi, M)$ in Fib_B^{or} (proper or not), but where M is of finite type, the same formalism, exchanging $\Omega_{\text{cv}}(E) \leftrightarrow \Omega(E)$ and applying the relative Poincaré Duality theorem 3.3.3.1-(4), induces a morphism of $(\Omega(B), d)$-dgm's

$$\boxed{f_! : \Omega_{\mathrm{cv}}(E')[d_{M'}] \to \Omega_{\mathrm{cv}}(E)[d_M]} \tag{3.78}$$

which is canonical in the derived category $\mathcal{D}\,\mathrm{DGM}(\Omega(B), d)$ and which we call *the Gysin morphism relative to B for proper supports associated with f.*

Proposition 3.5.1.1 *Denote by*

- $\mathrm{Fib}^{\mathrm{or}}_{B,\mathrm{pr}}$ *the category of oriented fiber bundles over B and proper maps;*
- $\mathrm{Fib}^{\mathrm{or}}_{\mathrm{f.t.}B}$ *the category of oriented fiber bundles over B with finite type fiber spaces and arbitrary maps;*
- (E, B, π, M) *a fiber bundle of manifolds.*

1. *The correspondence:*

$$(\_)_* : \mathrm{Fib}^{\mathrm{or}}_{B,\mathrm{pr}} \rightsquigarrow \mathcal{D}(\mathrm{DGM}(\Omega(B), d)) \quad \textit{with} \quad \begin{cases} E \rightsquigarrow E_* := (\Omega(E), d)[d_M] \\ f \rightsquigarrow f_*, \end{cases}$$

   is a covariant functor. It is the Gysin functor relative to B for proper maps.

2. *The correspondence:*

$$(\_)_! : \mathrm{Fib}^{\mathrm{or}}_{\mathrm{f.t.}B} \rightsquigarrow \mathcal{D}(\mathrm{DGM}(\Omega(B), d)) \quad \textit{with} \quad \begin{cases} E \rightsquigarrow E_! := (\Omega_{\mathrm{cv}}(E), d)[d_M] \\ f \rightsquigarrow f_!, \end{cases}$$

   is a covariant functor. It is the Gysin functor relative to B for arbitrary maps.

Proof Straightforward, as in the absolute case (*cf*. Sect. 2.6.2), since the definitions of Gysin morphisms are based on Poincaré adjunctions which are functorial. □

Comment 3.5.1.2 Unlike the absolute case, it is not clear if the need of finiteness hypothesis can be waived in the definition of the Gysin morphism $f_!$. Indeed, although the restrictions $f_{U!} : \Omega_{\mathrm{cv}}(E'_U)[d_{M'}] \to \Omega_{\mathrm{cv}}(E_U)[d_M]$ exist above trivializing open subspaces $U \subseteq B$, whether M is of finite type or not, they are only defined in the derived categories $\mathcal{D}(U)$ and the question of glueing them together in a morphism of $\mathcal{D}(B)$ is uncertain.

3.5.2 Gysin Morphisms for Fiber Bundles and Base Change

In Theorem 3.3.4.1 we established that Poincaré Duality for fiber bundles commutes with base change, and this immediately gives the same property for Gysin morphisms.

Indeed, for every map $\bar{h} : B' \to B$, the base change functor

$$\bar{h}^{-1}(\_) : \mathrm{Fib}^{\mathrm{or}}_{B,\mathrm{pr}} \rightsquigarrow \mathrm{Fib}^{\mathrm{or}}_{B',\mathrm{pr}}$$

is well defined (*cf*. Corollary 3.1.7.2-(1)) and when applied to a proper morphism of fiber bundles $f : (E', B, \pi', M') \to (E, B, \pi, M)$, where M and M' are equidimensional, gives rise to the inverse image of the diagram of complexes of sheaves in $\mathcal{D}^+(B;\mathbb{R})$, corresponding to (3.76), which we used in defining the Gysin morphism f_* associated with the pullback $f^* : \Omega_{\rm cv}(E) \to \Omega_{\rm cv}(E')$, i.e. we have a commutative diagram in $\mathcal{D}^+(B';\mathbb{R})$

$$\begin{array}{ccc}
\bar{h}^{-1}\big(\pi'_*\underline{\underline{\Omega}}(E')\big)[d_{M'}] & \xrightarrow[\text{q.i.}]{\bar{h}^{-1}(\mathbb{D}_{B,M'})} & \bar{h}^{-1}\big(\mathbb{R}\,\mathbf{Hom}^\bullet_{(\underline{\underline{\Omega}}(B),d)}\big(\pi'_!\underline{\underline{\Omega}}(E'),\underline{\underline{\Omega}}(B)\big)\big) \\
{\scriptstyle \bar{h}^{-1}(f_*)}\downarrow & & \uparrow{\scriptstyle f^*} \\
\bar{h}^{-1}\big(\pi_*\underline{\underline{\Omega}}(E)\big)[d_M] & \xrightarrow[\text{q.i.}]{\bar{h}^{-1}(\mathbb{D}_{B,M})} & \bar{h}^{-1}\big(\mathbb{R}\,\mathbf{Hom}^\bullet_{(\underline{\underline{\Omega}}(B),d)}\big(\pi_!\underline{\underline{\Omega}}(E),\underline{\underline{\Omega}}(B)\big)\big)
\end{array}$$

which coincides, thanks to Theorem 3.3.4.1, with the diagram defining the Gysin morphism $\bar{h}^{-1}(f)_*$ associated with $\bar{h}^{-1}(f)^* : \Omega_{\rm cv}(\bar{h}^{-1}(E)) \to \Omega_{\rm cv}(\bar{h}^{-1}(E'))$.

We have thus justified the following Proposition.

Proposition 3.5.2.1 *Gysin morphisms are compatible with base change.*

More precisely, given a map $\bar{h} : B' \to B$ *and a morphism of oriented fiber bundles* $f : (E', B, \pi', M') \to (E, B, \pi, M)$, *where* M *and* M' *are equidimensional, we have the base change diagram*

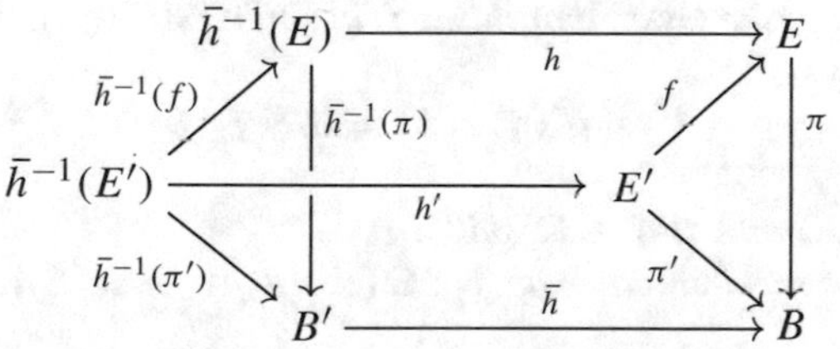

where all the parallelogram subdiagrams are Cartesian, and the following diagrams combining pullback and Gysin morphisms are commutative

$$\begin{array}{ccc}
\Omega(E) & \xrightarrow[\text{(if } f \text{ is proper)}]{f_*} & \Omega(E') \\
{\scriptstyle h^*}\downarrow & \oplus & \downarrow{\scriptstyle h'^*} \\
\Omega(\bar{h}^{-1}(E)) & \xrightarrow[\bar{h}^{-1}(f)_*]{} & \Omega(\bar{h}^{-1}(E'))
\end{array}
\qquad
\begin{array}{ccc}
\Omega_{\rm cv}(E) & \xrightarrow{f_!} & \Omega_{\rm cv}(E') \\
{\scriptstyle h^*}\downarrow & \oplus & \downarrow{\scriptstyle h'^*} \\
\Omega_{\rm cv}(\bar{h}^{-1}(E)) & \xrightarrow[\bar{h}^{-1}(f)_!]{} & \Omega_{\rm cv}(\bar{h}^{-1}(E'))
\end{array}$$

3.6 Examples of Gysin Morphisms

In this section, fiber bundles are implicitly assumed to be connected fiber bundles of manifolds with finite type fibers.

3.6.1 Adjointness of Gysin Morphism

The discussion on Poincaré adjoint pairs 2.5 is also meaningful in the relative case since it only depends on the nondegeneracy of the Poincaré adjunction at the level of cochain complexes (*cf*. Comment 3.2.1.4). This property gives a useful way to discover explicit definitions of Gysin morphisms.

Proposition 3.6.1.1 *Let $f : (E', B, \pi', M') \to (E, B, \pi, M)$ be a morphism of oriented fiber bundles above B.*

1. *A set-theoretic map $f_\natural : \Omega_{\mathrm{cv}}(E')[d_{M'}] \to \Omega_{\mathrm{cv}}(E)[d_M]$ verifying,*

$$\int_{M'} f^*(\alpha) \wedge \beta = \int_M \alpha \wedge f_\natural(\beta)\,, \tag{3.79}$$

   for all $\alpha \in \Omega(E)$ and $\beta \in \Omega_{\mathrm{cv}}(E')$, is automatically a morphism of $\Omega(B)$-dg-modules inducing the Gysin morphism $f_!$ in $\mathcal{D}(\mathrm{DGM}(\Omega(B), d))$. Furthermore,

$$f_\natural\big(f^*(\alpha) \wedge \beta\big) = \alpha \wedge f_\natural(\beta)\,, \tag{3.80}$$

   for all $\alpha \in \Omega(E)$ and all $\beta \in \Omega_{\mathrm{cv}}(E')$.
2. *If f is proper, a set-theoretic map $f_\natural : \Omega(E')[d_{M'}] \to \Omega(E)[d_M]$ verifying,*

$$\int_{M'} f^*(\beta) \wedge \alpha = \int_M \beta \wedge f_\natural(\alpha)\,, \tag{3.81}$$

   for all $\alpha \in \Omega(E')$ and $\beta \in \Omega_{\mathrm{cv}}(E)$, is automatically a morphism of $\Omega(B)$-dg-modules inducing the Gysin morphism f_* in $\mathcal{D}(\mathrm{DGM}(\Omega(B), d))$. Furthermore,

$$f_\natural\big(f^*(\beta) \wedge \alpha\big) = \beta \wedge f_\natural(\alpha)\,, \tag{3.82}$$

   for all $\alpha \in \Omega(E')$ and all $\beta \in \Omega_{\mathrm{cv}}(E)$.

Proof Consequence of the injectivity of the relative Poincaré adjunctions at cochain level 3.2.1.3-(2), which implies the nondegeneracy of the relative Poincaré pairing (3.2.1.4). For example, the fact that $f_\natural$ commutes with differential results from the obvious equalities

$$\begin{aligned}(-1)^{[\alpha]}\int_M \alpha\wedge f_\natural(d\beta) &= \int_{M'} f^*(\alpha)\wedge d\beta\\ &= d\int_{M'} f^*(\alpha)\wedge\beta - \int_{M'} d\, f^*(\alpha)\wedge\beta\\ &= d\int_M \alpha\wedge f_\natural(\beta) - \int_M d\alpha\wedge f_\natural(\beta)\\ &= (-1)^{[\alpha]}\int_M \alpha\wedge d\, f_\natural(\beta)\,,\end{aligned}$$

which imply that $f_\natural\circ d = d\circ f_\natural$. □

Comment 3.6.1.2 Note that while adjointness completely determine the Gysin morphisms at cochain level, the same is not always true in cohomology, in other words, the formulas (3.80) and (3.82) do not define Gysin morphisms by themselves, for which we would need to have Poincaré duality isomorphisms

$$\begin{cases} H(E)[d_M]\simeq \mathbf{Hom}^\bullet_{H(B)}\big(H_{\mathrm{cv}}(E), H(B)\big)\\ H_{\mathrm{cv}}(E)[d_M]\simeq \mathbf{Hom}^\bullet_{H(B)}\big(H(E), H(B)\big)\,.\end{cases}$$

And this actually happens when B is a formal space, that $H(B)$ is of finite homological dimension, and that $H_{\mathrm{cv}}(E)$ is a projective $H(B)$-module (3.4.1.4-(2)). For example, when $B=\{\bullet\}$ (2.6.2.1-(2)). The pairs $(f^*, f_!)$ and (f^*, f_*) are then *Poincaré adjoint pairs in cohomology* (*cf*. Proposition 2.5.1.1).

3.6.2 *Constant Map and Locally Trivial Fibrations*

Let (E, B, π, M) be and oriented fiber bundle. The relative version of the constant map $c_M : M\to\{\bullet\}$ is the morphism of fiber bundles

$$\begin{array}{ccc} E & \xrightarrow{\ \pi\ } & B\\ & {\scriptstyle\pi}\searrow\quad\swarrow{\scriptstyle\mathrm{id}_B} & \\ & B & \end{array}$$

and formula (3.79), with $\alpha\in\Omega(B)$ and $\beta\in\Omega_{\mathrm{cv}}(E)$, gives the equality:

$$\alpha\wedge\pi_\natural(\beta) = \int_{\{\bullet\}}\alpha\wedge\pi_\natural(\beta) = \int_M \pi^*(\alpha)\wedge\beta = \alpha\wedge\int_M\beta\,.$$

by 3.2.1.3-(1). Hence, the identification of Gysin morphism and integration

$$\pi_! = \textstyle\int_M : \Omega_{\mathrm{cv}}(E)[d_M]\to\Omega(B)\,.$$

The pair $(\pi^*, \int_M)$ is therefore a Poincaré adjoint pair in $\mathcal{D}\mathrm{DGM}(\Omega(B), d)$.

3.6.3 Open Embedding

Let (E, B, π, M) be and oriented fiber bundle, and let $j : U \subseteq E$ be an open embedding of fiber bundles above B. The formula (3.80), with $\alpha \in \Omega(E)$ and $\beta \in \Omega_{cv}(U)$, then gives, in $\Omega(B)$, the equality:

$$\int_M j^*\alpha \wedge \beta = \int_M \alpha \wedge j_! \beta , \tag{3.83}$$

where $j_! \beta \in \Omega_c(M)$ is the *extension by zero* of β.

The relative Gysin morphism in cohomology

$$j_! : \Omega_{cv}(U)[d_U] \to \Omega_{cv}(M)[d_M]$$

naturally extends the *pushforward* morphism $j_! : \Omega_c(U) \to \Omega_c(E)$ (2.6.3.1-(1)). The pair $(j^*, j_!)$ is a Poincaré adjoint pair in $\mathcal{D}\mathrm{DGM}(\Omega(B), d)$.

3.6.4 Proper Base Change

Let (E, B, π, M) and (E', B', π, M) be oriented fiber bundles, where M is equidimensional, let $g : B \to B'$ be a proper map and let

$$\begin{array}{ccc} E' & \xrightarrow{g} & E \\ \pi\downarrow & \square & \downarrow\pi \\ B' & \xrightarrow{g} & B \end{array}$$

be a Cartesian diagram (with the obvious abuse of notation).

The statements in following proposition are part of the so-called *(proper) base change theorems*.

Proposition 3.6.4.1 *If $g : B' \to B$ is proper, then $g : E' \to E$ is proper, and*

$$\begin{cases} \text{(i)}\ g^* \circ \pi_! = \pi_! \circ g^* : \Omega_{cv}(E)[d_M] \to \Omega(B') , \\ \text{(ii)}\ g^* \circ \pi_! = \pi_! \circ g^* : \Omega_c(E)[d_M] \to \Omega(B') , \\ \text{(iii)}\ \pi^* \circ g_* = g_* \circ \pi^* : \Omega(B') \to \Omega(E)[d_M] . \end{cases}$$

Proof The map $g : E' \to E$ is proper, since a compact subspace $K \subseteq E$ can always be expressed a finite union of compact subspaces K_i, where $K_i \subseteq \pi^{-1}(U_i)$ for some open trivialization $(\pi(\_), \varphi_i(\_)) : \pi^{-1}(U_i) \to U_i \times M$ (*cf*. Sect. 3.1-(F-1)). It then suffices to prove that $g^{-1}(K_i)$ is compact. For this, we replace K_i by the product of compact spaces $X \times Y$ with $X = \pi(K_i)$ and $Y := \varphi_i(K_i)$. But then $g^{-1}(X \times Y) = g^{-1}(X) \times Y$ is compact since $g : B' \to B$ is proper (cf. Proposition 3.1.6.1).

Notice that formulas (i,ii,iii) now make sense since the two g's are proper maps.

Identity (i) is the particular case of Proposition 3.5.2.1 where $f := (\pi : E \to B)$, and we know by 3.6.2 that $\pi_!$ is given by integration along fibers $\int_M$. Notice also that the equality of differential forms

$$g^*\left(\int_M \omega\right) = \int_M g^*(\omega), \quad \forall \omega \in \Omega(E),$$

which results by a local check in B',[12] then fully justifies (i). Statement (ii) results from the obvious equality $\int_M(\Omega_c(E)) \subseteq \Omega_c(B)$, and (ii) and (iii) are equivalent by adjointness.   □

3.6.5   Zero Section of a Vector Bundle

Let $V := (V, B, \pi, \mathbb{R}^n)$ be an oriented vector bundle. A notable difference of V compared to the case of a general fiber bundle $E := (E, B, \pi, M)$ is the existence of a distinguished section of π, the *zero section map* $\sigma : B \to V$, which is a homotopy equivalence.

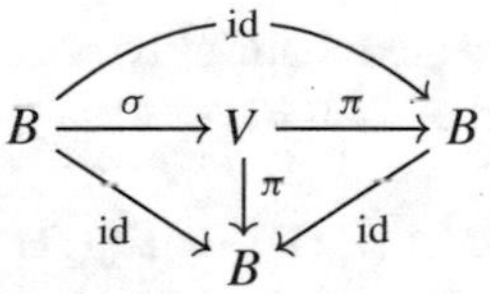

We therefore have the Gysin morphisms $\sigma_!$ and $\pi_!$ relative to B, which are inverse of each other (since their composition is the identity), and that both are adjoint to homotopy equivalences, namely the pullbacks σ^* and π^*,

$$\Omega(B) \xleftarrow[\text{q.i.}]{\sigma^*} \Omega(V) \xleftarrow[\text{q.i.}]{\pi^*} \Omega(B)$$
$$\underbrace{\qquad\qquad\qquad\qquad}_{\text{id}}$$

$$\Omega(B) \xrightarrow[\text{q.i.}]{\sigma_!} \Omega_{cv}(V)[n] \xrightarrow[\text{q.i.}]{\pi_!} \Omega(B)$$
$$\underbrace{\qquad\qquad\qquad\qquad}_{\text{id}}$$

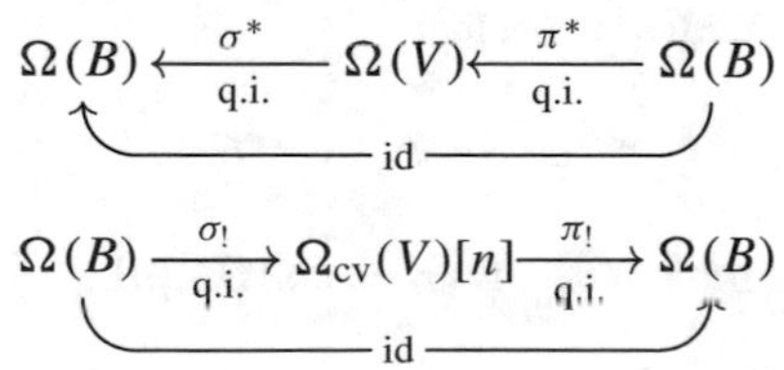

(3.84)

---

[12]Bott-Tu [18] **I** §6 pp. 61–63.

where, an adjoint $\sigma_\natural$ of σ^* at the level of cochains, has to verify

$$\int_{\mathbb{R}^n} \alpha \wedge \sigma_\natural(\beta) = \sigma^*(\alpha) \wedge \beta = \sigma^*(\alpha \wedge \pi^*\beta) \wedge \mathbf{1} = \int_{\mathbb{R}^n} \alpha \wedge \pi^*(\beta) \wedge \sigma_\natural(1) .$$

Therefore, by uniqueness of the adjoint, we have $\sigma_\natural(\beta) = \pi^*(\beta) \wedge \sigma_\natural(\mathbf{1})$, where $\sigma_\natural(\mathbf{1}) \in \Omega^n_{\mathrm{cv}}(V)$ is a cocycle such that

$$\pi({}_!\sigma_\natural(\mathbf{1})) = \int_{\mathbb{R}^n} \sigma_\natural(1) = 1 .$$

We then set, after 3.2.3.1,

$$\sigma_\natural(1) := \Phi_\pi ,$$

where $\Phi_\pi \in \Omega^n_{\mathrm{cv}}(V)$ is a cocycle representing the Thom class Φ_π of $(V, B, \pi, \mathbb{R}^n)$.
We have thus proved the following proposition.

Proposition 3.6.5.1 *The Gysin morphism* $\sigma_! : \Omega(B) \to \Omega_{\mathrm{cv}}(V)[n]$ *relative to* B *is represented by the morphism of* $\Omega(B)$*-dgm's*

$$\sigma_!(\_) := \pi^*(\_) \wedge \Phi_\pi ,$$

which induces the Thom isomorphism in cohomology (3.2.3.1-(3.54)).

When B and V are also oriented manifolds, we can consider the absolute point of view and seek for an adjoint $\sigma_\natural : \Omega(B) \to \Omega(E)[n]$ of the restriction morphism

$$\sigma^* : \Omega_{\mathrm{c}}(V) \to \Omega_{\mathrm{c}}(B) ,$$

which is generally not a quasi-isomorphism.[13] We then write

$$\int_V \sigma_\natural(\alpha) \wedge \beta = \int_B \alpha \wedge \sigma^*(\beta) = \int_B 1 \wedge \sigma^*(\pi^*(\alpha) \wedge \beta) = \int_V \sigma_\natural(1) \wedge \pi^*(\alpha) \wedge \beta ,$$

and conclude, as in the relative case, that

$$\sigma_*(\alpha) := \Phi_\pi \wedge \pi^*(\alpha) ,$$

represents the Gysin morphism $\sigma_* : \Omega(B) \to \Omega(V)[n]$.

[13]For example, if B is compact and $d_B < d_V$, then $H^0_{\mathrm{c}}(B) \neq 0$ and $H^0_{\mathrm{c}}(V) = 0$.

Proposition 3.6.5.2 *Let (π, V, B) and (π, V', B') be oriented vector bundles with zero section maps $\sigma : B \to V$ and $\sigma : B' \to V'$. Let*

$$\begin{array}{ccc} V' & \xrightarrow{g} & V \\ {\scriptstyle\pi}\downarrow & \square & \downarrow{\scriptstyle\pi} \\ B' & \xrightarrow{g} & B \end{array}$$

be a Cartesian diagram with $g : B' \to B$ proper. Then $g : V' \to V$ is proper and the following equalities hold

$$\begin{cases} \text{(i) } g^* \circ \sigma_! \ = \sigma_! \circ g^* \ : \Omega_c(B) \to \Omega_c(V')\,, \\ \text{(ii) } g^* \circ \sigma_! \ = \sigma_! \circ g^* \ : \Omega(B) \to \Omega_{cv}(V')\,, \\ \text{(iii) } \sigma^* \circ g_* = g_* \circ \sigma^* : \Omega(V') \to \Omega(B)\,. \end{cases}$$

Proof Corollary of 3.6.4.1 since $\sigma_!$ is the inverse of $\pi_!$. □

3.6.6 Closed Embedding

Given a closed embedding $\iota : N \subseteq M$ of connected oriented manifolds, we seek for a map $\iota_\natural : \Omega(N)[d_N] \to \Omega(M)[d_M]$ adjoint to the restriction morphism

$$\iota^* : \Omega_c(M) \to \Omega_c(N)\,.$$

For that, fix a tubular neighborhood $U_\tau \supseteq N$ of N in M (3.2.3.1), and decompose $\iota : N \subseteq M$ as the composition of the zero section $\sigma : N \to U_\tau$ and the open inclusion $j_\tau : U_\tau \to M$. For $\beta \in \Omega_c(M)$ and $\alpha \in \Omega(N)$, we then have:

$$\int_N \alpha \wedge \iota^*(\beta) = \int_N \alpha \wedge \iota^*(\phi_\epsilon\, \beta)$$

where $\phi_\epsilon \in \Omega^0_{cv}(U_\tau)$ is any function verifying $\phi = 1$ in a neighborhood of the support of a Thom form Φ_τ of the vector bundle $\pi : U_\tau \to N$ (*cf*. Sect. 3.2.3.1).

In that case, $\phi_\epsilon\, \beta \in \Omega_{cv}(U_\tau)$ and $\iota^*(\phi_\epsilon\, \beta) = \sigma^*(\phi_\epsilon\, \beta)$. We can then write

$$\int_N \alpha \wedge \iota^*(\phi_\epsilon\, \beta) = \int_{U_\tau} \Phi_\tau \wedge \pi^*(\alpha) \wedge (\phi_\epsilon\, \beta) = \int_{U_\tau} \big(\Phi_\tau \wedge \pi^*(\alpha)\big) \wedge \beta\,,$$

which proves the following proposition.

Proposition 3.6.6.1 *The map*

$$j_{\tau!} \circ \big(\Phi_\tau \wedge \pi^*(\_)\big) : \Omega(N)[d_N] \to \Omega(M)[d_M]\,,$$

is Poincaré adjoint to the restriction morphism $\iota^* : \Omega_c(M) \to \Omega_c(N)$ *and represents the Gysin morphism* $\iota_* : \Omega(N)[d_N] \to \Omega(M)[d_M]$*, which is independent of the choice of the tubular neighborhood* U_τ*, in the derived category* $\mathcal{D}\mathrm{Vec}(\mathbb{R})$.

3.7 Applications

We now give two important applications of the existence of Gysin morphisms in the form of exercises whose solutions are to be founded in Helpful Hints. We encourage the reader to solve them without looking at the solutions.

3.7.1 Gysin Long Exact Sequence

Let $i : F \subseteq M$ be a closed embedding of oriented manifolds. Assume F compact.[14] Put $U := M \smallsetminus F$ and $j : U \subseteq M$ the open injection map.

1. (a) Let $\mathscr{F}$ denote the set of open neighborhood of F. Restriction morphisms $R_V^W : \Omega(W) \to \Omega(V)$ for all $W \supseteq V \supseteq F$, give rise to a filtrant inductive system $\{R_V^W \mid W \supseteq V \text{ in } \mathscr{F}\}$ and a canonical morphism of complexes $R_{\mathscr{F}}^M : \Omega(M) \to \varinjlim_{V \in \mathscr{F}} \Omega(V)$. Show that the short sequence

$$0 \to \Omega_c(U) \xrightarrow{j_!} \Omega_c(M) \xrightarrow{R_{\mathscr{F}}^M} \varinjlim_{\mathscr{F}} \Omega(V) \to 0$$

where $j_!$ is the extension by zero, (3.6.3), is exact. (, p. 341)

(b) The restrictions $R_F^V : \Omega(V) \to \Omega(F)$ for $V \supseteq F$, define a morphism of the inductive system $\{R_V^W \mid W \supseteq V \text{ in } \mathscr{F}\}$ to $\Omega(F)$. Denote by $R_F^{\mathscr{F}} := \varinjlim_{\mathscr{F}} R_F^V$. Show that the morphism

$$R_F^{\mathscr{F}} : \varinjlim_{\mathscr{F}} \Omega(V) \to \Omega(F)\,,$$

is a quasi-isomorphism.

Hint. Use the Tubular Neighborhood Theorem (cf. *question 1d*). (, p. 342)

(c) Derive from the previous questions the *the long exact sequence of compactly supported cohomology* (, p. 342)

$$\cdots \to H_c^k(U) \xrightarrow{j_!} H_c^k(M) \xrightarrow{i^*} H_c^k(F) \xrightarrow{c_k} H_c^{k+1}(U) \to \cdots \tag{3.85}$$

---

[14]Although not necessary, compactness of N simplifies the formulation of the exercise.

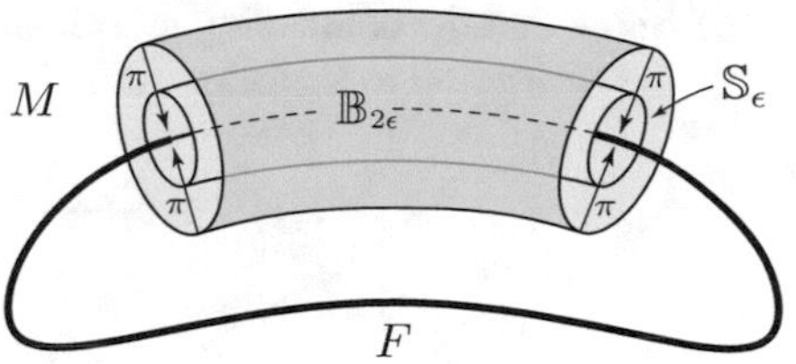

Fig. 3.2 Tubular neighborhood

(d) Endow M with a Riemannian metric $d : M \times M \to \mathbb{R}$. For $\epsilon \in \mathbb{R}$, denote

$$\mathbb{B}_\epsilon(F) := \{m \in M \mid d(m, F) < \epsilon\}.$$

Since F is compact, the Tubular neighborhood theorem states that for ϵ small enough, $\mathbb{B}_{2\epsilon}(F)$ is a fiber bundle with fiber $\mathbb{R}^{d_M - d_F}$ over F via the geodesic projection $\pi : \mathbb{B}_{2\epsilon}(F) \to F$. By restriction, denote

$$\pi' : \mathbb{S}_\epsilon(F) \to F \tag{3.86}$$

the fiber bundle with fiber the sphere $\mathbb{S}^{d_M - d_F - 1} \subseteq \mathbb{R}^{d_M - d_F}$ (see Fig. 3.2).

Let $\sigma : \mathbb{S}_\epsilon \hookrightarrow \mathbb{B}_{2\epsilon} \setminus F$ and $j_\epsilon : \mathbb{B}_{2\epsilon} \setminus F \hookrightarrow U$ denote the inclusion maps, and consider the diagram

$$\begin{array}{ccccc}
\mathbb{S}_\epsilon & \xrightarrow{\sigma} & \mathbb{B}_{2\epsilon} \setminus F & \xrightarrow[\subset]{j_\epsilon} & U \\
{\scriptstyle \pi'}\downarrow & {\scriptstyle \pi}\swarrow & & & \downarrow{\scriptstyle j}\,\subset \\
F & & \xrightarrow[\subset]{i} & & M
\end{array} \tag{3.87}$$

Show that in the long exact sequence (3.85), the connecting morphism $c : H(F) \to H_c(U)[1]$, is the following composition of morphisms

$$H(F) \xrightarrow{\pi^*} H_c(\mathbb{S}_\epsilon) \xrightarrow{\sigma_![-d_{\mathbb{S}_\epsilon}]} H_c(\mathbb{B}_{2\epsilon} \setminus F)[1] \xrightarrow{j_{\epsilon!}[1]} H_c(U)[1]$$

$$\underbrace{\hspace{10cm}}_{c}$$

where $\sigma_!$ and $j_{\epsilon!}$ denote the Gysin morphism respectively associated with σ and j_ϵ. (✶, p. 342)

2. (a) Dualizing and shifting the long exact sequence of compactly supported cohomology (3.85), justify the exactness of the *Gysin long exact sequence*

$$\xrightarrow{\delta[-1]} H(F)[d_F - d_M] \xrightarrow{i_*[-d_M]} H(M) \xrightarrow{j^*} H(U) \xrightarrow{\delta} \tag{3.88}$$

where $i : F \to N$ and $j : U \to N$ are the canonical injections and δ is adjoint to the shift of the connecting morphism c in (3.85). (✶, p. 343)

(b) Show that the connecting morphism $\delta : H(U) \to H(F)[-(d_M - d_F - 1)]$ is the restriction to $\mathbb{S}_\epsilon$ followed by integration along fibers of π' (p. 344)

$$H(U) \ni \alpha \mapsto \delta(\alpha) = \int_{\mathbb{S}_\epsilon} \alpha|_{\mathbb{S}_\epsilon} \in H(F)\,.$$

3.7.2 Lefschetz Fixed Point Theorem

Let M be a compact connected oriented manifold. Denote by $\delta : M \to M \times M$ the diagonal embedding $x \mapsto (x, x)$ and let $\Delta_M := \mathrm{Im}(\delta)$. Given $f : M \to M$, denote $\mathrm{Gr}(f) : M \to M \times M$ *the graph map* $x \mapsto (f(x), x)$.

The *Lefschetz class of* f is by definition

$$L(f) := \mathrm{Gr}(f)^*(\delta_*(1)) \in H^{d_M}(M)\,,$$

and the *Lefschetz number of* f is the number $\Lambda_f := \int_M L(f)\,.$

1. Explain the following equalities (p. 344)

$$\begin{aligned}\Lambda_f &:= \int_M \mathrm{Gr}(f)^*(\delta_*(1)) \qquad (3.89)\\ &= \int_{M\times M} \delta_*(1) \cup \mathrm{Gr}(f)_*(1) = (-1)^{d_M} \int_M \delta^*(\mathrm{Gr}(f)_*(1))\,.\end{aligned}$$

2. Assuming that f has no fixed points, show that the Gysin morphism

$$\mathrm{Gr}(f)_* : H(M)[d_M] \to H(M \times M)[2d_M]$$

factors through $H_c(M \times M \smallsetminus \Delta_M)$ and $\Lambda_f = 0$. (p. 344)

3. Let $\mathscr{B} := \{e_i\}_{i\in I}$ be a graded basis of $H(M)$ and let $\mathscr{B}' := \{e'_i\}_{i\in I}$ denote the Poincaré dual basis of $\mathscr{B}$, i.e. such that $e_i \cup e'_j = \delta_{i,j}\zeta_M$, where ζ_M denotes the fundamental class of M (2.4.2). Using the projection formula for diagonal map $\delta : M \to M \times M$ show that

$$\delta_*(1) = \sum\nolimits_{i\in I} (-1)^{\deg(e_i)}\, e_i \otimes e'_i\,,$$

Deduce the equality: $\int_M \delta_*(1)|_{\Delta_M} = \sum_{k\in\mathbb{N}} (-1)^k \dim\big(H^k(M)\big)$. (p. 345)

4. Combining (3.89) with the last result, show the *Lefschetz fixed point formula*

$$\Lambda_f = \sum\nolimits_{k\in\mathbb{N}} (-1)^k \mathrm{Tr}\big(f^* : H^k(M) \to H^k(M)\big)\,.$$

In particular, if this sum does not vanish, then f has fixed points ! (p. 345)
Notice that if $f := \mathrm{id}_M$, then Λ_f is the *Euler characteristic* χ_M of M.

Exercise 3.7.2.1 Let M be a compact connected oriented manifold. The notations are the same as in 3.7.2. Denote by $(M_\epsilon, \Delta_M, \pi, \mathbb{R}^{d_M})$ a tubular neighborhood of the diagonal Δ_M in $M \times M$, and define the *Euler class of the pair* $(\Delta_M, M \times M)$ by

$$\mathrm{Eu}(\Delta_M) := \delta_!(1)|_{\Delta_M} \in H^{d_M}_{\mathrm{cv}}(M_\epsilon)\,.$$

1. Show that if the *Euler characteristic* of M is nonzero, then $\mathrm{Eu}(\Delta_M) \neq 0$.
2. Let $M'_\epsilon := M_\epsilon \smallsetminus \Delta_M$, and let $\pi' : M'_\epsilon \to M$ be the restriction of π to M'_ϵ. Then, consider the diagram of open/closed embeddings

$$\begin{array}{ccccc} M'_\epsilon & \xrightarrow[\subset]{j} & M_\epsilon & \xleftarrow[\supset]{\delta} & M \\ & \searrow^{\pi'} & \downarrow{\scriptstyle\pi} & \swarrow_{\mathrm{id}} & \\ & & M & & \end{array} \tag{3.90}$$

   (a) Show that $\pi' : M'_\epsilon \to M$ is an oriented fiber bundle.
   (b) Describe the fiber F of π'.
   (c) Consider the long exact sequence of compact vertical cohomology associated with (3.90), and show that there exists a natural surjective map

$$H^{d_M-1}(M) \twoheadrightarrow H^{d_M}_{\mathrm{cv}}(M'_\epsilon)\,.$$

   (d) Give an example of manifold M such that $H^{d_M}_{\mathrm{cv}}(M'_\epsilon) = 0$. Conclude that although there are differential forms $\zeta_{\pi'} \in \Omega^{d_M}_{\mathrm{cv}}(M'_\epsilon)$ verifying $\int_F \zeta_{\pi'} = 1$, none of them can be a cocycle, thus providing the counterexample announced in 3.2.3. (☞, p. 345)

3.8 Conclusion

We have reached the end of the preliminaries on Poincaré duality and Gysin morphisms in the nonequivariant setting and using the de Rham model for cohomology, both in the absolute and in the relative case. As shown, the key ingredient in the approach is to define the Poincaré adjunction morphism in some derived category of differential graded modules.

In the next chapter, we will concentrate on G-manifolds and de Rham G-equivariant cohomology, and so recall the Cartan Model for equivariant cohomology. We will then introduce the Poincaré pairing and the corresponding adjunctions, in such a way that we can continue to apply the same approach. Chap. 5 is entirely devoted to this, while in Chap. 6, the G-equivariant Gysin functors will be defined following the procedures described in 2.8.

But, before going into those subjects, the next chapter begins with quick historical review of the origins of equivariant cohomology theory.

Chapter 4
Equivariant Background

4.1 Significant Dates in Equivariant Cohomology Theory

In this section, we recall some of the more important dates in the genesis of equivariant cohomology. We do not aim for completeness and do not address developments after the 1990s.

4.1.1 Cartan's ENS Seminar (1950)[1]

In lectures n° 19/20 of the *Séminaire Cartan* at the *École Normale Supérieure de Paris,*[2] delivered in May and June of 1950, Henri Cartan is interested in principal G-bundles $p : \mathscr{E} \to \mathscr{B}$, where G is a compact *and connected* (see p. 7) Lie group of Lie algebra $\mathfrak{g}$, and where $\mathscr{B}$ is a manifold. His focus was to settle an algebraic framework for studying the relationship between the cohomologies of $\mathscr{E}$, $\mathscr{B}$ and G, extending the approach of the Chern–Weil homomorphism $\mathrm{ch} : S(\mathfrak{g})^G \to H(\mathscr{B})$, and hence the construction of characteristic classes, well beyond the category of manifolds.

To construct Chern–Weil homomorphisms, we need only the *de Rham complex* $(\Omega(\mathscr{E}), d)$ equipped with *interior products* $\iota(X)$ and *Lie derivatives* $\theta(X)$ for all $X \in \mathfrak{g}$, and with an *infinitesimal connection* $f : \mathfrak{g}^\vee \to \Omega^1(\mathscr{E})$.

Since the operators $\iota(X)$ and $\theta(X)$ admit purely algebraic descriptions, the underlying manifold $\mathscr{E}$ can be disregarded, and with this in mind, Cartan introduces,

[1] See Chevalley's review [29], and also Chapter III of Tu [91], The Cartan Model, p. 141.

[2] Lecture 19 on May 15 [25], and lecture 20 in two sessions: May 23 and June 19 [26]. The contents of these lectures were published with some additions in [32] *Colloque de topologie (espaces fibrés)*, held in Brussels on June 5–8 1950, (1951).

A. Arabia, *Equivariant Poincaré Duality on G-Manifolds*, Lecture Notes in Mathematics 2288, https://doi.org/10.1007/978-3-030-70440-7_4

in his first lecture, the category DGA($\mathfrak{g}$) of $\mathfrak{g}$-*differential graded algebras* ($\mathfrak{g}$-dga), which are dga's equipped with operators $\iota(X)$ and $\theta(X)$ satisfying axioms which reflect the way they interact in the context of manifolds (*cf*. Sect. 4.2.3). In the same way, the concept of *infinitesimal connection* for principal G-bundles is extended to every $E \in$ DGA($\mathfrak{g}$) in what Cartan calls *algebraic connection*. This is a θ-equivariant linear map $f : \mathfrak{g}^\vee \to E^1$ verifying $\iota(X)f(\lambda) = \lambda(X)$, for all $X \in \mathfrak{g}$. The correspondence $\mathrm{Conn}_{\mathrm{alg}} : \mathrm{DGA}(\mathfrak{g}) \rightsquigarrow \mathrm{Set}$ associating with $E \in \mathrm{DGA}(\mathfrak{g})$ the set $\mathrm{Conn}_{\mathrm{alg}}(E)$ of its algebraic connections is a *representable* functor. It therefore characterizes a $\mathfrak{g}$-dga $(W(\mathfrak{g}), d)$: *the Weil algebra of* $\mathfrak{g}$, and an isomorphism of functors

$$\mathrm{Conn}_{\mathrm{alg}}(\_) \cong \mathrm{Mor}_{\mathrm{DGA}(\mathfrak{g})}(W(\mathfrak{g}), \_)\,. \tag{4.1}$$

As a graded algebra, $W(\mathfrak{g}) := \Lambda(\mathfrak{g}^\vee) \otimes S(\mathfrak{g}^\vee)$ where $\Lambda(\mathfrak{g}^\vee)$ and $S(\mathfrak{g}^\vee)$ are respectively the *exterior* and the *symmetric* algebras on $\mathfrak{g}$. The grading of $W(\mathfrak{g})$ is defined by setting $\mathfrak{g}^\vee$ in degree 1 in $\Lambda(\mathfrak{g}^\vee)$, and in degree 2 in $S(\mathfrak{g}^\vee)$.

As a differential algebra, Cartan shows that $(W(\mathfrak{g}), d)$ has the cohomology of a contractible G-space endowed with a canonical connection f_0 (corresponding to $\mathrm{id} : W(\mathfrak{g}) \to W(\mathfrak{g})$ by (4.1)). The full subcategory of $\mathfrak{g}$-dga's admitting algebraic connection is therefore a category of $W(\mathfrak{g})$-modules and the canonical connection f_0 appears as a *universal connection*.[3]

Taken together, this all already justifies Cartan saying (see top lines in Fig. 4.1, p. 113): *the algebra* $W(\mathfrak{g})$ *plays the role of a cochain algebra of a fiber bundle* $p : \mathbb{E}G \to \mathbb{B}G$ *which would be classifying for* <u>*all*</u> *principal* G*-bundles*. (See 4.9.)

Cartan's second lecture focuses on different ways of relating the cohomologies of $\mathscr{E}$, $\mathscr{B}$ and G. Of these, the one interesting us is the construction of a $\mathfrak{g}$-dga canonically quasi-isomorphic to $(\Omega(\mathscr{B}), d)$ which takes as its main ingredients the de Rham complex $(\Omega(\mathscr{E}), d)$ and the Weil algebra $(W(\mathfrak{g}), d)$.

This is the purpose of §5 in [28], where Cartan observes that for a principal G-bundle $p : \mathscr{E} \to \mathscr{B}$, the pullback morphism $p^* : \Omega(\mathscr{B}) \to \Omega(\mathscr{E})$ identifies $\Omega(\mathscr{B})$ by with the subcomplex $\Omega(\mathscr{E})^{\mathrm{bas}}$ of *basic elements* of $\Omega(\mathscr{E})$, which is the name Cartan gives to $\mathfrak{g}$-invariant and horizontal differential forms.[4] The point here is that, for G connected, the notion of *basic element* is again purely algebraic and can be extended to the whole category DGA($\mathfrak{g}$). The question then arises of how to algebraically determine E^{bas} for any given $E \in$ DGA($\mathfrak{g}$).

4.1.1.1 The Cartan-Weil Morphisms

For $E := \Omega(\mathscr{E})$, Cartan proposes the $\mathfrak{g}$-dga $W(\mathfrak{g}) \otimes \Omega(\mathscr{E})$ as algebraic '*de Rham like*' model for the cohomology of the space $\mathbb{E}G \times \mathscr{E}$, product of the CW-complex $\mathbb{E}G$ and the manifold $\mathscr{E}$. Then, if f is a connection for $p : \mathscr{E} \to \mathscr{B}$, the

[3] See Kumar [70], A remark on universal connections.

[4] A differential form in the total space $\mathscr{E}$ of a principal G-bundle is *horizontal* if it is killed by the interior products with G-fundamental fields (Sect. 4.4.2).

representability equivalence (4.1) associates with it the *Weil morphism* of $\mathfrak{g}$-dga's denoted (by abuse) $f : W(\mathfrak{g}) \to \Omega(\mathscr{E})$, which leads Cartan to consider the diagram of morphisms of $\mathfrak{g}$-dga's

$$\begin{array}{ccccccc} & & & & W(\mathfrak{g}) & \xrightarrow{f} & \Omega(\mathscr{E}) \\ & & & & \downarrow & (\mathrm{I}) & \| \\ \Omega(\mathscr{B}) & \xrightarrow{p^*} & \Omega(\mathscr{E}) & \xrightarrow{\mathfrak{i}} & W(\mathfrak{g}) \otimes \Omega(\mathscr{E}) & \xrightarrow{\mathfrak{f}} & \Omega(\mathscr{E}) \end{array} \tag{4.2}$$

where $\mathfrak{i}(\omega) := 1 \otimes \omega$ and $\mathfrak{f}(\alpha \otimes \omega) := f(\alpha) \wedge \omega$. By restriction to basic elements, the bottom row gives the sequence of morphisms of complexes

$$\Omega(\mathscr{B}) \underset{\simeq}{\xrightarrow{p^*}} \Omega(\mathscr{E})^{\mathrm{bas}} \xrightarrow{\bar{\mathfrak{i}}} (W(\mathfrak{g}) \otimes \Omega(\mathscr{E}))^{\mathrm{bas}} \xrightarrow{\bar{\mathfrak{f}}} \Omega(\mathscr{E})^{\mathrm{bas}}, \tag{4.3}$$

where $\bar{\mathfrak{f}} \circ \bar{\mathfrak{i}}$ is the identity. We call $\bar{\mathfrak{i}}$ and $\bar{\mathfrak{f}}$ *the Cartan-Weil morphisms*.

The morphism $\bar{\mathfrak{i}}$ can clearly be defined, in the same way, for any $E \in \mathrm{DGA}(\mathfrak{g})$, and likewise for $\bar{\mathfrak{f}}$ as long as E admits an algebraic connection. The main result concerning this sequence (4.3) is Cartan's Theorem 3, in §5 of [28]:[5,6,7]

Theorem 4.1.1.1 (Cartan) *If* $E \in \mathrm{DGA}(\mathfrak{g})$ *admits algebraic connections, then the Cartan-Weil morphism* $\bar{\mathfrak{i}} : E^{\mathrm{bas}} \to (W(\mathfrak{g}) \otimes E)^{\mathrm{bas}}$ *is a quasi-isomorphism. In particular, the Cartan-Weil morphism*

$$H(\bar{\mathfrak{f}}) : H\big((W(\mathfrak{g}) \otimes E)^{\mathrm{bas}}\big) \to H(E^{\mathrm{bas}})$$

is an isomorphism too, inverse of $H(\bar{\mathfrak{i}})$*, hence independent of the connection.*

Following immediately the theorem, Cartan more closely examines (in §6 [28]) the complex $(W(\mathfrak{g}) \otimes E)^{\mathrm{bas}}$ for any given $E := (E, d) \in \mathrm{DGA}(\mathfrak{g})$. He shows that there exists a natural isomorphism,[8] which we call *Cartan isomorphism*,

$$\boxed{(W(\mathfrak{g}) \otimes E)^{\mathrm{bas}} \xrightarrow{\simeq} ((S(\mathfrak{g}^\vee) \otimes E)^G, d_{\mathfrak{g}})} \tag{4.4}$$

---

[5]Cartan gave summary indications of the proof. Later Michel André in his Ph.D. thesis [6] directed by Claude Chevalley gave complete proofs of all the statements in Cartan's lectures. The theorem is also proved in Tu [91] Appendix A, Theorem A.2, p. 264.

[6]An important, but rarely mentioned fact about sequence (4.3), is that it remains meaningful for any $\mathfrak{g}$-submodule $K \subseteq E$ which is also a graded ideal of E, and that, in that case, the theorem remains true for K instead of E (*cf*. Appendix C). In particular, for any G-equivariant fiber bundle $\pi : E \to B$, the morphism $\Omega_{\mathrm{cv}}(E)^{\mathrm{bas}} \to (W(\mathfrak{g}) \otimes \Omega_{\mathrm{cv}}(E))^{\mathrm{bas}}$ is a quasi-isomorphism although there are generally no algebra homomorphisms $W(\mathfrak{g}) \to \Omega_{\mathrm{cv}}(E)$.

[7]The theorem corroborates the idea that *algebraic connections* on $\mathfrak{g}$-dg-algebras are the right analogue to *free actions* on G-manifolds.

[8]The isomorphism is completely justified in Cartan [28] §6 Transformation de l'algèbre différentielle $\overline{B}$, p. 63. We recall this proof in Sect. 4.3.2, p. 124. See also Tu [91], The Cartan Model in General §21-1,2, pp. 167–172.

where

$$\boxed{d_{\mathfrak{g}}(P \otimes \omega) = P \otimes d\omega + \sum\nolimits_i Pe^i \otimes \iota(e_i)\,\omega} \tag{4.5}$$

where d is the differential in E, $\{e_i\}$ is a basis of $\mathfrak{g}$ of dual basis $\{e^i\}$, and $\iota(e_i)$ is the interior product corresponding to $e_i \in \mathfrak{g}$ (*cf*. Sect. 4.3.2).

Another important by-product of diagram (4.2) results from the restriction of the subdiagram (I) to basic elements. We have $W(\mathfrak{g})^{\text{bas}} = S(\mathfrak{g}^\vee)^G$ and the restriction of $\bar{\mathfrak{f}}$ to this subalgebra is, by construction, the well-known Chern–Weil homomorphism. Hence, $\mathbf{ch} = f^{\text{bas}}$. We therefore have commutative diagrams

$$\begin{array}{c|c}
p : \mathscr{E} \to \mathscr{B} \text{ any principal } G\text{-bundle} & E \text{ any } \mathfrak{g}\text{-dga with connection} \\
\begin{array}{ccc}
S(\mathfrak{g}^\vee)^G & \xrightarrow{\ \text{ch}\ } & \Omega(\mathscr{B}) \\
\downarrow & & \| \\
(S(\mathfrak{g}^\vee) \otimes \Omega(\mathscr{E}))^G & \xrightarrow[\text{q.i.}]{\ \bar{\mathfrak{f}}\ } & \Omega(\mathscr{B})
\end{array} &
\begin{array}{ccc}
S(\mathfrak{g}^\vee)^G & \xrightarrow[f^{\text{bas}}]{\ \text{ch}\ } & E^{\text{bas}} \\
\downarrow & & \| \\
(S(\mathfrak{g}^\vee) \otimes E)^G & \xrightarrow[\text{q.i.}]{\ \bar{\mathfrak{f}}\ } & E^{\text{bas}}
\end{array}
\end{array}$$

which, besides proving that the usual Chern–Weil homomorphism in cohomology is independent of the infinitesimal connection, show that the abstract Cartan-Weil formalism is the right framework for dealing with connections and characteristic classes.

The following commutative diagram summarizes the Cartan-Weil approach.

$$\begin{array}{ccc}
\text{Mor}_{\mathfrak{g}\text{-dga}}(W(\mathfrak{g}), \Omega(\mathscr{E})) & \overset{\text{Weil}}{\underset{\text{Cartan}}{=\!=\!=}} & \left\{\text{algebraic connections on } \Omega(\mathscr{E})\right\} \\
\downarrow \text{cohomology of basic subcomplexes} & & \uparrow \\
\left\{\text{ch} : S(\mathfrak{g})^G \quad \Omega(\mathscr{B})\right\} & \overset{\text{Chern}}{\underset{\text{Weil}}{\longleftarrow}} & \left\{ \text{infinitesimal connections on } \Omega(\mathscr{E}) \right\} \\
\uparrow \text{cohomology} & & \uparrow\uparrow\uparrow\ (\dagger) \\
\text{Hot}(\mathscr{B}, \mathbb{B}G) & \overset{\text{Steenrod}}{=\!=\!=} & \left\{ \begin{array}{c} \text{equivalence classes of} \\ \text{principal } G\text{-bundles above } \mathscr{B} \end{array} \right\}
\end{array} \tag{4.6}$$

Here, the third row is the representability theorem of the set of equivalence classes of principal G-bundles above a manifold $\mathscr{B}$.[9] The second row represents the construction of characteristic classes through the Chern–Weil homomorphism, where the multiple arrow (†) recalls that for a given principal G-bundle there are many infinitesimal connections. The first row, the representability of the set

[9] See Steenrod [84], §19.3 Classification Theorem, p. 101 and the Historical Note in p. 105, where the Theorem for Lie groups is independently granted to Steenrod, Whitehead and jointly by Chern and Sun. See also Husemöller [56] §Preface to the First Edition p. xi.

```
Revenant à l'algèbre de Weil W(G), on voit qu'elle se comporte,
du point de vue homologique, comme une algèbre universelle pour
les espaces fibrés de groupe G, c'est-à-dire comme une algèbre
de cochaines d'un espace fibré qui serait classifiant pour tous
les espaces fibrés de groupe G, quelle que soit la dimension de
leur espace de base. L'algèbre IS(G) joue le rôle de l'algèbre de
cochaînes de l'espace de base d'un tel espace fibré universel, avec
la particularité que les éléments de IS(G) sont tous des cocycles...
```

Fig. 4.1 Excerpt from the last paragraph in Cartan's lecture [32]

of algebraic connections, and the corresponding upper subdiagram represent the contribution of Cartan's lectures to characteristic classes.

4.1.1.2 The Cartan Complex

It is worth emphasizing that for any G-manifold M, whether G acts freely or not, the de Rham complex $\Omega(M)$ is well equipped with a structure of $\mathfrak{g}$-dga. We can hence still consider the complex $(W(\mathfrak{g}) \otimes \Omega(G))^{\text{bas}}$ (4.4), which we denote

$$\boxed{\Omega_G(M) := ((S(\mathfrak{g}^\vee) \otimes \Omega(M))^G, d_\mathfrak{g})} \tag{4.7}$$

Although this was clear in Cartan's lectures, the subject was out of focus at the time, as research was mainly concentrated on principal G-bundles rather than on general G-manifolds, and still less on general topological G-spaces.

The complex $\Omega_G(M)$ (4.7) is commonly referred to as *the Cartan complex of* M, and also as the *Cartan Model for the cohomology of the* homotopy quotient $M_G := \mathbb{E}G \times_G M$, since it is indeed quasi-isomorphic to the complex of Alexander-Spanier cochains $\mathit{\Omega}(M_G; \mathbb{R})$ (*cf*. 4.8.3.1, 4.10.1.2).

4.1.1.3 Homotopy Quotients and Formality of the Classifying Space

We will see in Theorem 4.9.2.1 that the dg-algebras $(\Omega_G(M), d)$ and $(\mathit{\Omega}(M_G; \mathbb{R}), d_\mathfrak{g})$ can be joined by a zig-zag ($\leftrightsquigarrow$) of quasi-isomorphisms of dga's and, in the particular case where $M := \{\bullet\}$, we will then have $(\mathit{\Omega}(\{\bullet\}_G; \mathbb{R}), d) \leftrightsquigarrow (\Omega_G(\{\bullet\}), d_\mathfrak{g}) = (S(\mathfrak{g}^\vee)^G, 0)$, where $\{\bullet\}_G = \mathbb{B}G$. But here, the complex $(S(\mathfrak{g}^\vee)^G, 0)$ is canonically isomorphic to its cohomology, whence the $\mathbb{R}$-formality of $\mathbb{B}G$ (*cf*. Sect. 3.4.1).

This is a remarkable fact which has not been sufficiently highlighted in Cartan's lectures, although Cartan was clearly aware of it as he says in the very last lines of his first lecture (see bottom lines in Fig. 4.1): *The algebra* $S(\mathfrak{g}^\vee)^G$ *plays the role of the cochain algebra of the base space of a universal fiber bundle, with the particularity that the elements of* $S(\mathfrak{g}^\vee)^G$ *are all cocycles.*

> 3.9. REMARK. All our discussion will center around the space X_G, and the remarks 3.6 and 6.7 will be basic. Similar arguments have been used by Conner [33] when G is a circle, in rational cohomology. For an algebraic analogue when G is discrete, see Grothendieck [49] Chap. V. The space X_G and the embeding $F(X;G) \times B_G \subset X_G$ were also mentioned to the author by A. Shapiro. The proof of Smith's theorem 4.3 is also related to that of Borel [14].

Fig. 4.2 Excerpt from A. Borel [15], IV-§3 , p. 55 (1960), where Borel refers to the sources of the homotopy quotient

4.1.2 *Borel's IAS Seminar (1960)*

Some years later, in 1958–1959, Armand Borel, who had, since his arrival in Paris in 1949, been an active participant in the Cartan Seminar and in the Leray's courses at the *Collège de France*, held his *Seminar on transformation groups* at the Institute for Advanced Study in Princeton (Borel [15]). There, Borel drew attention to the advantages of considering for *any locally compact* G-space X, the orbit space of $\mathbb{E}G \times X$ under the diagonal action of G:

$$X_G := \mathbb{E}G \times_G X \, ,$$

as the homotopically best-suited substitute for the orbit space X/G, *whether* G *acts freely or not on* X *.*

The space X_G is the total space of the following two surjective maps

$$X_G := \mathbb{E}G \times_G X \;\begin{cases} \xrightarrow[\pi]{[X]} \mathbb{B}G \\ \xrightarrow[p]{[\mathbb{B}G_x]} X/G \, , \end{cases}$$

where the fibers are shown in brackets. More precisely

- $\pi : X_G \twoheadrightarrow \mathbb{B}G$, $\overline{(w,x)} \mapsto \overline{w}$, a locally trivial fibration of fiber space X, and
- $p : X_G \twoheadrightarrow X/G$, $\overline{(w,x)} \mapsto \overline{x}$, where the fibers are the classifying spaces $\mathbb{B}G_x$ of the different stability groups G_x for $x \in X$.

As Borel says in its introduction: X_G *allows us to tie together the cohomology groups of* X, X/G, *and the fixed point set* $F := X^G$, *with those of the classifying spaces of the stability groups and of* G.

4.1.2.1 The Borel Construction

The space X_G, which Borel called *twisted product*, is known today as *the homotopy quotient*, *the homotopy orbit space* and also *the Borel construction* (see Fig. 4.2).

Beyond its immediate aim, which was the homological study of the set of fixed points $F := X^G$, the seminar laid most of the foundations of what would later be known as *the equivariant cohomology of locally compact G-spaces*. Orbit types, slice theorems, spectral sequences and fixed points theorems, were already present, if not yet in their final form, at least at a level that would appeal to other mathematicians for further development.

The restriction map

$$H(X_G) \to H(F_G) \tag{4.8}$$

appears in almost every section of applications of the Borel Seminar, often with restrictive conditions to ensure it is an isomorphism. And in the case of the circle group action, $G := T^1$, while it is clear that Borel was aware that (4.8) is an isomorphism modulo $H(\mathbb{B}G)$-torsion, he never stated it in those terms.

4.1.3 Atiyah-Segal: Equivariant K-Theory (1968)

In 1968, the bases of *equivariant K-theory* are set out in the works of Atiyah-Segal and Segal [8, 82], where the following enhanced analogue to (4.8) appears for the first time.

Localization theorem ([8, 82]) [10] *Let G be a compact Lie group and let X be a locally compact G-space. The localized restriction map*

$$K_G(X)_{\mathfrak{p}} \to K_G(G.X^S)_{\mathfrak{p}}\,, \tag{4.9}$$

where $\mathfrak{p}$ is a prime ideal of $K_G(\bullet)$, S is minimal among the subgroups of G such that $\mathfrak{p}$ is the inverse image of a prime ideal of $K_S(\bullet)$ under the restriction map $K_G(\bullet) \to K_S(\bullet)$, and X^S is the set of S-fixed points in X, is an isomorphism.

4.1.4 Quillen: Equivariant Cohomology (1971)

As K-theory has a cohomological behavior, equivariant *cohomology* soon came to light. We owe it to Daniel Quillen [80] who, merging the ideas of Atiyah-Segal and Borel, defines[11] *the equivariant cohomology $H_G(X)$ of a G-space X with coefficients in a ring A*, as the ordinary cohomology of Borel's construction X_G, i.e.

$$H_G(X) := H(X_G\,;A)\,.$$

[10] Proved in Segal [82], Section §4. Localization, Proposition 4.1, p. 144.

[11] In the first page of Part I, page 549 of the journal.

Quillen proves the analogue to the localization theorem (4.9) for the case where G is an elementary p-group, and for the case where G is a torus T.

Theorem ([80]) *Assume either X is compact or paracompact with* $\dim_{\mathrm{h}}(X) < +\infty$ *and that the set of identity components of the isotropy groups of points of X is finite. Then the inclusion of X^T in X induces an isomorphism*

$$H_T(X)\big[\big(H_T(\bullet) - 0\big)^{-1}\big] \to H_T(XT)\big[\big(H_T(\bullet) - 0\big)^{-1}\big]. \tag{4.10}$$

4.1.5 Hsiang's Book (1975)

1975 saw the appearance of Wu-yi Hsiang's book '*Cohomology theory of topological transformation groups*' [55] in which the third chapter promptly introduces the reader to the foundations of equivariant cohomology for locally compact G-spaces. It includes a version of the localization theorem more in the vein of Atiyah-Segal that Hsiang calls the '*Borel-Atiyah-Segal localization theorem*'. It is stated as follows:

Theorem [12] *Assume either X is compact or paracompact with* $\dim_{\mathrm{h}}(X) < +\infty$ *and that the set of identity components of the isotropy groups of points of X is finite. For a multiplicative system $S \subseteq H_G(\bullet) := H(\mathbb{B}G)$, put*

$$X^S = \big\{x \in X \;\big|\; \text{no element of } S \text{ maps to zero in } H(\mathbb{B}G) \to H(\mathbb{B}G_x)\big\}.$$

Then, the localized restriction map

$$S^{-1}H_G(X) \to S^{-1}H_G(X^S),$$

is an isomorphism.

For example, if G is a torus T and $S = H_T(\bullet) \setminus \{0\}$, then $X^S = X^F$ and, again, the kernel and cokernel of the restriction map

$$H_T(X) \to H_T(X^T)$$

are torsion $H_T(\bullet)$-modules as stated in Quillen's (4.10).

[12] Hsiang [55] Chapter III. Equivariant Cohomology Theory. Theorem (III.1), p. 40.

4.1.6 *Atiyah-Bott and Berline-Vergne: Equivariant de Rham Cohomology (1980)*

The reader may have noticed that the Cartan complex, although introduced in 1950, was not mentioned in the previous paragraphs. This is because, at the time, Borel, Quillen, Hsiang, ... were mostly interested in applying equivariant cohomology with coefficients in fields of positive characteristic (which excludes differential forms) to find conditions for the existence of fixed points in locally compact G-spaces and to infer cohomological properties of the fixed point sets from the ambient space X (for example being a cohomological [Poincaré duality] manifold when X is so).

In the early 1980s the whole theory underwent an unexpected development when N. Berline and M. Vergne succeeded in proving the Duistermaat-Heckman formula on the pushforward of the Liouville measure on a symplectic manifold under the moment map (Duistermaat and Heckman [38]) by a new fixed point theorem for G-manifolds inspired by an old paper by R. Bott [17].

Let G be a compact Lie group and M a G-manifold. For $X \in \mathfrak{g} := \mathrm{Lie}(G)$, let X^* be the fundamental vector field on M generated by the infinitesimal action of X (*cf*. Sect. 4.4.2). Denote by $\iota(X)$ and $\mathcal{L}(X)$ respectively the interior product and the Lie derivative associated with X^*, acting on the dga $(\mathbb{C} \otimes \Omega(M), d)$. The vector X is called *nondegenerate* if, for $m \in M$, fixed by the one-parameter group $\exp(tX)$, the linear operator $L_m(X)$ on $T_m(M)$ induced by $\mathcal{L}(X)$, is invertible.

With these data, Berline-Vergne introduced the linear operator on $\mathbb{C} \otimes \Omega(M)$:

$$d_X := d - 2\pi\, i\, \iota(X^*)\,, \tag{4.11}$$

which verifies $d_X^2 = -2\pi i\, \mathcal{L}(X)$. Thereby, denoting by $\Omega(M)^X$ the sub-algebra of differential forms on M which are invariant under the action of $\exp(tX)$, the pair $(\Omega(M)^X, d_X)$ is a ($\mathbb{Z}/2\mathbb{Z}$-graded) differential algebra. Berline-Vergne denoted by $H_X(M)$ its cohomology, and proved the following fixed point theorem.

Theorem (Berline-Vergne [9–11]) *Let G be a compact Lie group and M an oriented compact G-manifold (of even dimension). Then, for all $X \in \mathfrak{g}$ nondegenerate and all $\mu \in H_X(M)$, we have:*

$$\int_M \mu = \sum\nolimits_{m \in M^X} \frac{\mu(m)}{\mathrm{Pf}(L_m(X))}$$

where $\mu(m)$ is the restriction of μ to the singleton $\{m\}$, M^X is the fixed point set (necessarily finite) of $\exp(tX)$ *and* $\mathrm{Pf}(L_m(X))$ *is the Pfaffian of $L_m(X)$.*

At about the same time the Atiyah-Bott paper [7] appeared. Motivated by the same Duistermaat-Heckman work, as well as a recent work of E. Witten [97], it introduced a de Rham model for the equivariant cohomology of manifolds and stated the corresponding localization theorems. In [7] (th. 4.13, p. 12), taking finite dimensional approximations of $\mathbb{E}G$, Atiyah-Bott gave a sketched a proof of the Equivariant de Rham Theorem which states that the cohomology of the Cartan complex $(\Omega_G(M), d_\mathfrak{g})$ is the ordinary cohomology of the topological space M_G (see 4.8.3.1). In this way the original, and oddly neglected, Cartan complex $(\Omega_G(M), d_\mathfrak{g})$ turned out to have been an excellent model for the equivariant cohomology of manifolds. The elements of $\Omega_G(M)$ have since become known as the *G-equivariant (de Rham) differential forms*.

In comparing the Berline-Vergne complex (4.11) to the Cartan complex (4.7), one is naturally led to consider the map $\mathrm{ev}_X : \big(\Omega_G(M), d_\mathfrak{g}\big) \to \big(\Omega(M)^X, d_X\big)$, $\mathrm{ev}_X(P \otimes \mu) := P(-2\pi i X)\mu$, which is a morphism of complexes. Denote the induced homomorphism in cohomology by

$$\mathrm{ev}_X : \mathbb{C} \otimes H_G(M) \to H_X(M)\,.$$

If $T \subseteq G$ is the torus topologically generated by $X \in \mathfrak{g}$, then $M^X = M^T$ and we have the commutative diagram of restrictions to fixed point sets:

$$\begin{array}{ccc}
\mathbb{C} \otimes H_T(M) & \xrightarrow{\ \mathrm{ev}_X\ } & H_X(M) \\
{\scriptstyle (\sim)}\big\downarrow & \oplus & \big\downarrow{\scriptstyle \simeq} \\
\mathbb{C} \otimes H_T(M^T) & \underset{\mathrm{ev}_X}{\twoheadrightarrow} & H_X(M^X)
\end{array} \tag{4.12}$$

where the left vertical arrow is an isomorphism modulo H_T-torsion after Quillen.

Now, the proof of the Berline-Vergne fixed-point theorem proves also that the right-hand vertical arrow in (4.12) is a *true* isomorphism.[13] As a consequence, the map $\mathrm{ev}_X : \mathbb{C} \otimes H_T(M) \to H_X(M)$ is surjective and the Berline-Vergne fixed point theorem could also be justified through the Atiyah-Bott's de Rham version of the localization theorem. Indeed, the equivariant integration map $\int_M$ gives rise to the commutative diagram

$$\begin{array}{ccc}
\mathbb{C} \otimes H_T(M) & \xrightarrow{\ \mathrm{ev}_X\ } & H_X(M) \\
{\scriptstyle \int_M}\big\downarrow & \oplus & \big\downarrow{\scriptstyle \int_M} \\
H_T(\bullet) & \xrightarrow[\ \mathrm{ev}_X\ ]{} & \mathbb{C}
\end{array}$$

[13]This results from the fact that, thanks to the Poincaré Lemma for Berline-Vergne cohomology stating that the pullback map $H_X(M) \to H_X(\mathbb{R} \times M)$ is an isomorphism, it is easy to check that we have a long exact sequence of Berline-Vergne cohomology:

$$\to H_{X,c}(M \smallsetminus M^X) \to H_X(M) \to H_X(M^X) \to$$

where $H_{X,c}(M \smallsetminus M^X) = 0$, after the original proof of the Berline-Vergne fixed point theorem.

with, in the bottom row, $ev_X(P) = P(-2\pi i X)$. Then, by the localization theorem for $H_T(M)$, we see that for all $\mu \in H_X(M)$ and every $\tilde{\mu} \in H_T(M)$ such that $ev_X(\tilde{\mu}) = \mu$, we have:

$$\int_M \mu = \Big(\int_M \tilde{\mu}\Big)(-2\pi i X) = \sum\nolimits_{m\in M^T} \frac{\mu(m)}{\mathrm{Eu}_T(m, M)(-2\pi i X)},$$

where $\mathrm{Eu}_T(m, M)$ is the equivariant Euler class of $m \in M$, as introduced by Atiyah-Bott in [7]-3.19.

The Berline-Vergne's and Atiyah-Bott's discoveries stimulated renewed interest in equivariant cohomology, in particular because of its applications to Lie group representation theory. We direct curious readers to the interesting bibliographic notes included at the end of the chapters of Berline-Getzler-Vergne's book [12].

What happened next goes well beyond the scope of this work. New routes of research appeared: *Equivariant Intersection Cohomology*, Joshua [60] and Brylinski [24]; *Equivariant Chow rings*, Brion [21]; *Equivariant Homology and Cohomology on G-CW-complexes*, Costenoble [34]; *Equivariant Derived Category*, Goresky et al. [47];[14] etc.

4.2 Category of $\mathfrak{g}$-Differential Graded Modules[15]

4.2.1 Field in Use

Lie groups and Lie algebras, vector spaces, tensor products and related material, are defined over the field of real numbers $\mathbb{R}$.

4.2.2 The Category of $\mathfrak{g}$-Modules

Let $\mathfrak{g}$ be a real *Lie algebra*. A *representation of* $\mathfrak{g}$, also called *a* $\mathfrak{g}$*-module*, is a real vector space V together with a Lie algebra homomorphism $\rho_V : \mathfrak{g} \to \mathrm{End}_{\mathbb{R}}(V)$. For simplicity, for all $Y \in \mathfrak{g}$ and $v \in V$, the notation '$Y \cdot v$' will replace '$\rho_V(Y)(v)$' when the representation ρ_V is understood.

Given $\mathfrak{g}$-modules V and W, a $\mathfrak{g}$*-module morphism from* V *to* W is a linear map $\lambda : V \to W$ s.t. $\lambda \circ \rho_V(Y) = \rho_W(Y) \circ \lambda$ for all $Y \in \mathfrak{g}$. We denote by $\mathrm{Hom}_{\mathfrak{g}}(V, W)$ the subspace of $\mathrm{Hom}_{\mathbb{R}}(V, W)$ of such maps.

---

[14] As the referee points out, there are several mistakes in [47] in particular for non abelian groups. See for instance Alekseev and Meinrenken [1, p. 481].

[15] Other references for this section and the next are Tu [91] §III. The Cartan Model p. 141, and Guillemin-Sternberg [50] §5. Cartan's formula p. 53.

The $\mathfrak{g}$-modules and their morphisms constitute the *category of* $\mathfrak{g}$*-modules*, denoted by $\mathrm{Mod}(\mathfrak{g})$.

Terminology A $\mathfrak{g}$-module V is said to be:

- a *trivial module*, or a *trivial representation of* $\mathfrak{g}$, if $\rho_V = 0$. The vector space $\mathbb{R}$ with the trivial representation is *'the' trivial representation of* $\mathfrak{g}$.
- *finite dimensional*, if $\dim_{\mathbb{R}}(V) < +\infty$.
- *simple or irreducible*, if it is nonzero and has no nontrivial submodules;
- *reducible* if it is a direct sum of two nonzero submodules;
- *semisimple* or *completely reducible*, if it is a direct sum of irreducible submodules.

Exercise 4.2.2.1 Let V be a $\mathfrak{g}$-module. Consider the following properties.

1. V is semisimple.
2. V is a sum of irreducible modules.
3. Every submodule or quotient module of V is semisimple.
4. If W is a submodule of V, then $V = W \oplus W'$ for some submodule W'.

Show that (1)$\Leftrightarrow$(2)$\Leftrightarrow$(3)$\Rightarrow$(4), and (4)$\Leftrightarrow$(1) if V is finite dimensional. (♝, p. 346)

Definition 4.2.2.2 An element v in a $\mathfrak{g}$-module V, is said to be $\mathfrak{g}$*-invariant* if $Y \cdot v = 0$, for all $Y \in \mathfrak{g}$. The set of such elements is a vector subspace of V which will be denoted by $V^{\mathfrak{g}}$.

Exercise 4.2.2.3 Let V be a $\mathfrak{g}$-module.

1. Show that for all $\varphi \in \mathrm{Hom}_{\mathfrak{g}}(V, W)$, $\varphi(V^{\mathfrak{g}}) \subseteq W^{\mathfrak{g}}$. Derive the fact that the correspondence $V \rightsquigarrow V^{\mathfrak{g}}$, $\varphi \rightsquigarrow \varphi|_{V^{\mathfrak{g}}}$ is functorial from $\mathrm{Mod}(\mathfrak{g})$ into $\mathrm{Vec}(\mathbb{R})$.
2. Endow $\mathbb{R}$ with the trivial action of $\mathfrak{g}$. Show that the map

$$\mathrm{Hom}_{\mathfrak{g}}(\mathbb{R}, V) \to V^{\mathfrak{g}}, \quad \varphi \mapsto \varphi(1),$$

is an isomorphism of functors $\mathrm{Hom}_{\mathfrak{g}}(\mathbb{R}, \_) \to (\_)^{\mathfrak{g}}$, and $(\_)^{\mathfrak{g}}$ is left exact.

4.2.3 $\mathfrak{g}$-*Differential Graded Modules*

A $\mathfrak{g}$*-dg-module* ($\mathfrak{g}$-dgm), or $\mathfrak{g}$*-complex*, is a quadruple (C, d, θ, ι), where:

- (C, d) is a complex in $\mathrm{DGV}(\mathbb{R})$ (*cf*. 2.1.6);
- $\theta : \mathfrak{g} \to \mathrm{Endgr}_{\mathrm{GV}(\mathbb{R})}(C)$ is a Lie algebra homomorphism, the $\mathfrak{g}$*-derivation*;
- $\iota : \mathfrak{g} \to \mathrm{Mor}_{\mathrm{GV}(\mathbb{R})}(C, C[-1])$ is a linear map, the $\mathfrak{g}$*-interior product*;[16]

[16]Recall that given $C, D \in \mathrm{GV}(\mathbb{R})$, we denote by $\mathrm{Mor}_{\mathrm{GV}(\mathbb{R})}(C, D)$ the group of graded homomorphisms of degree zero from C into D. The terms *derivation* and *interior product* derive from the fact that in the main example where (C, d) is the de Rham complex of a G-manifold, $\iota(X)$

such that, for all $X, Y \in \mathfrak{g}$

$$\begin{cases} \text{i) } \iota(X) \circ \iota(Y) + \iota(Y) \circ \iota(X) = 0\,, \\ \text{ii) } d \circ \iota(X) + \iota(X) \circ d = \theta(X)\,, \\ \text{iii) } \theta(X) \circ \iota(Y) - \iota(Y) \circ \theta(X) = \iota([X,Y])\,. \end{cases} \tag{4.13}$$

A *morphism of* $\mathfrak{g}$*-dgm's* $\alpha : (C, d, \theta, \iota) \to (D, d, \theta, \iota)$, is a morphism of dgm's $\alpha : (C, d) \to (D, d)$ commuting with derivations and interior products, i.e.

$$\alpha \circ \theta(X) = \theta(X) \circ \alpha \quad \text{and} \quad \alpha \circ \iota(X) = \iota(X) \circ \alpha\,, \quad \forall X \in \mathfrak{g}\,.$$

The $\mathfrak{g}$-complexes (C, d, θ, ι) and their morphisms constitute the *category of* $\mathfrak{g}$*-complexes* denoted by $\mathrm{DGM}(\mathfrak{g})$.

We denote by $\mathrm{DGM}^b(\mathfrak{g})$, $\mathrm{DGM}^+(\mathfrak{g})$ and $\mathrm{DGM}^-(\mathfrak{g})$ the full subcategories of $\mathrm{DGM}(\mathfrak{g})$ of respectively, bounded, bounded below and bounded above $\mathfrak{g}$-complexes (Sect. 2.1.4).

In the sequel, a $\mathfrak{g}$-complex (C, d, θ, ι) will also be denoted by (C, d) or simply C when the remaining data are understood.

4.2.4 $\mathfrak{g}$-Differential Graded Algebras

A $\mathfrak{g}$*-differential graded algebra* ($\mathfrak{g}$-dga), is a $\mathfrak{g}$-dgm (A, d, θ, ι), where (A, d) is a positively graded differential graded $\mathbb{R}$-algebra, which is graded commutative, and where, for all $X \in \mathfrak{g}$, the $\mathbb{R}$-linear operators $\theta(X)$ and $\iota(X)$ are graded derivations of degrees 0 an -1 respectively, i.e. such that for all homogeneous $a \in A^d$ and all $b \in A$, we have:

$$\begin{cases} \text{iv) } \iota(X)(a\,b) = \iota(X)(a)\,b + (-1)^d a\,\iota(X)(b)\,, \\ \text{v) } \theta(X)(a\,b) = \theta(X)(a)\,b + a\,\theta(X)(b)\,. \end{cases} \tag{4.14}$$

A *morphism of* $\mathfrak{g}$*-dga's*, is a morphism of $\mathfrak{g}$-dgm's $\alpha : (A, d, \theta, \iota) \to (A', d, \theta, \iota)$, which is also a morphism of dga's $\alpha : (A, d) \to (A', d)$, i.e. a morphism of graded $\mathbb{R}$-algebras commuting with differentials.

The $\mathfrak{g}$-dga's (A, d, θ, ι) and their morphisms constitute the *category of* $\mathfrak{g}$*-dga's* denoted by $\mathrm{DGA}(\mathfrak{g})$.

Remark 4.2.4.1 Condition (4.13)-(ii) gives the equality $d \circ \theta(\_) = \theta(\_) \circ d$, which implies that θ induces an action of $\mathfrak{g}$ on the cohomology of (C, d). However, by the same condition, the operator $\iota(X)$ is a homotopy for $\theta(X)$, and the action in question is trivial.

---

is the interior product with the fundamental vector field $\vec{X}$ (*cf.* 4.4.2) and $\theta(X)$ is the corresponding Lie derivative.

4.2.5 Split $\mathfrak{g}$-Complexes

Notation 4.2.5.1 Given an inclusion of $\mathfrak{g}$-modules $N \subseteq M$, we denote by '$N|M$' the fact that the natural map

$$\mathrm{Hom}_{\mathfrak{g}}(V, M) \to \mathrm{Hom}_{\mathfrak{g}}(V, M/N) \tag{4.15}$$

is **surjective** for every **finite** dimensional $\mathfrak{g}$-modules V.

Exercise 4.2.5.2 Show that the condition $N|M$ is equivalent to the fact that for every $\mathfrak{g}$-submodule $M' \subseteq M$ such that $N \subseteq M'$ is of finite codimension, there exists a $\mathfrak{g}$-submodule $H \subseteq M'$ such that $M' = H \oplus N$. (, p. 347)

Definition 4.2.5.3 For a $\mathfrak{g}$-complex (C, d), let $B^i := \mathrm{im}(d_{i-1})$ and $Z^i := \ker(d_i)$ respectively be *the* $\mathfrak{g}$*-submodules of* i*-coboundaries and* i*-cocycles of* (C, d). The $\mathfrak{g}$-complex (C, d) will be called $\mathfrak{g}$*-split* whenever we have

$$B^i | Z^i | C^i \,, \qquad \text{for all } i \in \mathbb{Z}.$$

Lemma 4.2.5.4 *Keeping the above notations, we have the following.*

1. *If $N|M$, the natural map $\dfrac{M^{\mathfrak{g}}}{N^{\mathfrak{g}}} \longrightarrow \left(\dfrac{M}{N}\right)^{\mathfrak{g}}$ is an isomorphism.*
2. *The condition $B^i|Z^i$ is equivalent both to the surjectivity of $(Z^i)^{\mathfrak{g}} \to (Z^i/B^i)$, and to the existence of a $\mathfrak{g}$-submodule H^i of Z^i such that $Z^i = B^i \oplus H^i$, in which case H^i is a trivial $\mathfrak{g}$-module isomorphic to Z^i/B^i.*
3. *A $\mathfrak{g}$-complex (C, d) such that each C^i is semisimple, is $\mathfrak{g}$-split.*

Proof

(1) After 4.2.2.3, the functor $(\_)^{\mathfrak{g}}$ is isomorphic to $\mathrm{Hom}_{\mathfrak{g}}(\mathbb{R}; \_)$ and the sequence $0 \to N^{\mathfrak{g}} \to M^{\mathfrak{g}} \to (M/N)^{\mathfrak{g}}$ is left exact. The split condition ensures it is also right exact.

(2) Recall that $\mathcal{H}^i := Z^i/B^i$ is a trivial $\mathfrak{g}$-module (see 4.2.4.1). Following (1), the split condition gives the surjection $(Z^i)^{\mathfrak{g}} \twoheadrightarrow (\mathcal{H}^i)^{\mathfrak{g}} = \mathcal{H}^i$. Conversely, we clearly have $\mathrm{Hom}_{\mathfrak{g}}(\mathcal{H}^i, \_) = \mathrm{Hom}_{\mathbb{R}}(\mathcal{H}^i, (\_)^{\mathfrak{g}})$ and, thereafter, the diagram

$$\begin{array}{ccc}
\mathrm{Hom}_{\mathfrak{g}}(\mathcal{H}^i, Z^i) & \longrightarrow & \mathrm{Hom}_{\mathfrak{g}}(\mathcal{H}^i, \mathcal{H}^i) \\
\| & \oplus & \| \\
\mathrm{Hom}_{\mathbb{R}}(\mathcal{H}^i, (Z^i)^{\mathfrak{g}}) & \longrightarrow & \mathrm{Hom}_{\mathbb{R}}(\mathcal{H}^i, \mathcal{H}^i)
\end{array}$$

where the surjectivity of the second row implies that of the first. In particular, there exists $\sigma \in \mathrm{Hom}_{\mathfrak{g}}(\mathcal{H}^i, Z^i)$ such that $\pi \circ \sigma = \mathrm{id}$ where $\pi : Z^i \twoheadrightarrow \mathcal{H}^i$ denotes the canonical projection. Setting $H^i := \mathrm{Im}(\sigma)$ completes the proof.

(3) Clear after Exercise 4.2.2.1. □

Proposition 4.2.5.5 *Let (C, d) be a $\mathfrak{g}$-split $\mathfrak{g}$-complex.*

1. *The inclusion $C^{\mathfrak{g}} \subseteq C$ is a quasi-isomorphism*
2. *If V is a finite dimensional* **semi-simple** *$\mathfrak{g}$-module, the inclusions*

$$V^{\mathfrak{g}} \otimes C \quad \supseteq \quad V^{\mathfrak{g}} \otimes C^{\mathfrak{g}} \quad \subseteq \quad (V \otimes C)^{\mathfrak{g}}$$

$$\mathbf{Hom}^\bullet_{\mathbb{R}}(V^{\mathfrak{g}}, C) \supseteq \mathbf{Hom}^\bullet_{\mathbb{R}}(V^{\mathfrak{g}}, C^{\mathfrak{g}}) \subseteq \mathbf{Hom}^\bullet_{\mathfrak{g}}(V, C)$$

are quasi-isomorphisms.

Proof

(1) Immediate from (4.2.5.4-(1)).
(2) Let us first show that if S is a simple $\mathfrak{g}$-module different from $\mathbb{R}$, then the complexes $(S \otimes C)^{\mathfrak{g}}$ and $\mathbf{Hom}^\bullet_{\mathfrak{g}}(S, C)$ are acyclic.[17]

We need treat only the $\mathbf{Hom}^\bullet$ case, since we have

$$\mathbf{Hom}^\bullet_{\mathfrak{g}}(S, C) = \mathbf{Hom}^\bullet_{\mathbb{R}}(S, C)^{\mathfrak{g}} = (S^\vee \otimes C)^{\mathfrak{g}} .$$

An i-cocycle of $\mathbf{Hom}^\bullet_{\mathfrak{g}}(S, C)$ is a $\mathfrak{g}$-module morphism $\lambda : S \to C^i$ such that $d \circ \lambda = 0$, i.e. such that $\mathrm{im}(\lambda) \subseteq Z^i$. But the composition of λ with the surjection $Z^i \twoheadrightarrow Z^i/B^i$ is null since $\mathfrak{g}$ acts trivially on cohomology, so that in fact $\mathrm{im}(\lambda) \subseteq B^i$. Now, thanks to the fact that $Z^i | C^i$, we can lift $\lambda : S \to B^i$ to $\mu : S \to C^{i-1}$ thereby proving that $\lambda = d \circ \mu$, i.e. that λ is a coboundary.

If V is a semisimple $\mathfrak{g}$-module, we decompose V as $V^{\mathfrak{g}} \oplus S$, where S is a direct sum of simple $\mathfrak{g}$-modules different from $\mathbb{R}$. Then

$$\mathbf{Hom}^\bullet_{\mathfrak{g}}(V, C) = \mathbf{Hom}^\bullet_{\mathfrak{g}}(V^{\mathfrak{g}}, C) \oplus \mathbf{Hom}^\bullet_{\mathfrak{g}}(S, C)$$

is quasi-isomorphic to $\mathbf{Hom}^\bullet_{\mathfrak{g}}(V^{\mathfrak{g}}, C)$ after the previous paragraph. But

$$\mathbf{Hom}^\bullet_{\mathfrak{g}}(V^{\mathfrak{g}}, C) = \mathbf{Hom}^\bullet_{\mathfrak{g}}(V^{\mathfrak{g}}, C^{\mathfrak{g}}) = \mathbf{Hom}^\bullet_{\mathbb{R}}(V^{\mathfrak{g}}, C^{\mathfrak{g}}),$$

so that $\mathbf{Hom}^\bullet_{\mathbb{R}}(V^{\mathfrak{g}}, C^{\mathfrak{g}}) \subseteq \mathbf{Hom}^\bullet_{\mathfrak{g}}(V, C)$ is clearly a quasi-isomorphism.

Finally, that the inclusion $\mathbf{Hom}^\bullet_{\mathbb{R}}(V^{\mathfrak{g}}, C^{\mathfrak{g}}) \subseteq \mathbf{Hom}^\bullet_{\mathbb{R}}(V^{\mathfrak{g}}, C)$ is a quasi-isomorphism results from (1) since $V^{\mathfrak{g}} \simeq \mathbb{R}^r$ and the inclusion in question becomes simply $\prod_{1 \leqslant i \leqslant r} C^{\mathfrak{g}} \subseteq \prod_{1 \leqslant i \leqslant r} C$. □

[17] A complex (C, d) is said to be *acyclic* if it is exact, i.e. if it has zero cohomology.

4.3 Equivariant Cohomology of $\mathfrak{g}$-Complexes

4.3.1 The $\mathfrak{g}$-dg-Algebra $S(\mathfrak{g}^\vee)$

We denote by $S(\mathfrak{g}^\vee)$ the graded $\mathbb{R}$-algebra of polynomial maps $P : \mathfrak{g} \to \mathbb{R}$ graded by twice the polynomial degree. Hence, $S^2(\mathfrak{g}^\vee) = \mathfrak{g}^\vee$ and $S^m(\mathfrak{g}^\vee) = 0$, for every odd integer m.

The $\mathfrak{g}$-dg-algebra $(S(\mathfrak{g}^\vee), d, \theta, \iota)$ is defined by setting $\theta : \mathfrak{g} \to \mathrm{Der}^0_{\mathbb{R}}(S(\mathfrak{g}^\vee))$ to be the Lie algebra homomorphism induced by coadjoint representation of $\mathfrak{g}$, i.e. define $(\theta(X)(\lambda))(Y) := \lambda([Y, X]) = \lambda(-\theta(X)(Y))$ for a linear form λ, and extend to the whole $S(\mathfrak{g}^\vee)$ by Leibniz rule, and by setting $\iota : \mathfrak{g} \to \mathrm{Der}^{-1}_{\mathbb{R}}(S(\mathfrak{g}^\vee))$ to be zero, which implies and the differential d is zero too.

4.3.2 Cartan Complexes

Given a $\mathfrak{g}$-dgm $C := (C, d, \theta, \iota)$, the tensor product $S(\mathfrak{g}^\vee) \otimes C$ is endowed with the action of $\mathfrak{g}$ which extends the operators $\theta(X)$ by the Leibniz rule, i.e. set

$$\theta(X)(P \otimes \omega) := \theta(X)(P) \otimes \omega + P \otimes \theta(X)(\omega)\,, \quad \forall X \in \mathfrak{g}\,, P \in S(\mathfrak{g}^\vee)\,, \omega \in C\,.$$

The map $\theta : \mathfrak{g} \to \mathrm{Endgr}_{\mathrm{GV}(\mathbb{R})}(S(\mathfrak{g}^\vee) \otimes C)$ is a representation of Lie algebra.

When, C is a $\mathfrak{g}$-dg-algebra, $S(\mathfrak{g}^\vee) \otimes C$ is endowed with the structure of graded algebra defined by $(a \otimes b) \cdot (a' \otimes b') = (aa') \otimes (bb')$ (since $S(\mathfrak{g}^\vee)$ is evenly graded) and the operators $\theta(X)$ are all $\mathbb{R}$-derivations.

In 4.1.1-(4.4), we recalled that Cartan gave an explicit isomorphism,[18]

$$(W(\mathfrak{g}) \otimes C)^{\mathrm{bas}} \xrightarrow[\simeq]{} (S(\mathfrak{g}^\vee) \otimes C)^{\mathfrak{g}}\,. \tag{4.16}$$

For this, Cartan uses the fact that $W^+(\mathfrak{g}) := \Lambda^+(\mathfrak{g}^\vee) \otimes S(\mathfrak{g}^\vee)$ is a graded ideal of $W(\mathfrak{g})$, stable by $\mathfrak{g}$-derivations (but not by $\mathfrak{g}$-interior products, nor by the differential in $W(\mathfrak{g})$), and that the quotient map, which we denote

$$\Xi : W(\mathfrak{g}) \otimes C \twoheadrightarrow \big(W(\mathfrak{g})/W^+(\mathfrak{g})\big) \otimes C = S(\mathfrak{g}^\vee) \otimes C\,, \tag{4.17}$$

is a θ-equivariant epimorphism of graded spaces. But, most importantly, Cartan explains that the restriction of Ξ to the submodule $(W(\mathfrak{g}) \otimes C)^{\mathrm{hor}}$ of *horizontal* ele-

[18] Cartan introduces the concept of *basic element* in [27, p. 20]. An element ω in a $\mathfrak{g}$-dgm $C := (C, d, \theta, \iota)$ is called *basic* if it is $\mathfrak{g}$-*invariant* and *horizontal*, i.e. respectively $\theta(X)(\omega) = 0$ and $\iota(X)(\omega) = 0$, for all $X \in \mathfrak{g}$. The subset of basic elements of C is a subcomplex of (C, d) denoted by C^{bas}.

ments (*cf*. fn. ([18]), p. 124), is a θ-equivariant isomorphism (exercise !) (♝, p. 347), which we denote:

$$\Xi' : (W(\mathfrak{g}) \otimes C)^{\mathrm{hor}} \xrightarrow[\simeq]{} S(\mathfrak{g}^\vee) \otimes C\,. \tag{4.18}$$

Cartan's isomorphism (4.16) is then the restriction of Ξ' to θ-invariants.

Then, as $(W(\mathfrak{g}) \otimes C)^{\mathrm{bas}}$ is a subcomplex of $(W(\mathfrak{g}) \otimes C, d)$, Cartan transfers the differential d to a differential $d_{\mathfrak{g}}$ on $(S(\mathfrak{g}^\vee) \otimes C)^{\mathfrak{g}}$, given by formula (4.5), which, interpreting $\omega \in S(\mathfrak{g}^\vee) \otimes C$ as a polynomial maps $\omega : \mathfrak{g} \mapsto C, Y \mapsto \omega(Y)$, reads as

$$(d_{\mathfrak{g}}(\omega))(Y) := d(\omega(Y)) + \iota(Y)(\omega(Y))\,, \quad \forall Y \in \mathfrak{g}\,. \tag{4.19}$$

Notice that the formula is meaningful on the whole tensor product $S(\mathfrak{g}^\vee) \otimes C$, where it defines an endomorphism of $S(\mathfrak{g}^\vee)$-graded module of degree $+1$ commuting with $\mathfrak{g}$-derivations, and, although the theory tells us that it is a differential on $(S(\mathfrak{g}^\vee) \otimes C)^{\mathfrak{g}}$, the fact is not obvious at first glance, especially if we take as starting point the data $\{S(\mathfrak{g}^\vee), C, d_{\mathfrak{g}}\}$. Below we give an independent justification for this fact.

Lemma 4.3.2.1 *On $S(\mathfrak{g}^\vee) \otimes C$, we have the equality of commutators*

$$[\theta(X), d_{\mathfrak{g}}] = [\theta(X), \iota(Y)] = 0\,.$$

In particular, $d_{\mathfrak{g}}\big((S(\mathfrak{g}^\vee) \otimes C)^{\mathfrak{g}}\big) \subseteq (S(\mathfrak{g}^\vee) \otimes C)^{\mathfrak{g}}$.

Proof The first equality $[\theta(X), d_{\mathfrak{g}}] = [\theta(X), \iota(Y)]$ comes from $[\theta(X), d] = 0$, which is clear.

For the second equality, fix, as usual, dual bases $\{e_i\}$ and $\{e^i\}$ of $\mathfrak{g}$. Then,

$$\begin{aligned}
\big[\theta(X), \iota(Y)\big] &= \Big[\theta(X), \sum\nolimits_i e^i \otimes \iota(e_i)\Big] \\
&= \sum\nolimits_i \Big(\big[\theta(X), e^i\big] \otimes \iota(e_i) + e^i \otimes \big[\theta(X), \iota(e_i)\big]\Big) \\
&= \sum\nolimits_i \Big(\theta(X)(e^i) \otimes \iota(e_i) + e^i \otimes \iota\big(\theta(X)(e_i)\big)\Big)\,,
\end{aligned} \tag{4.20}$$

and, since $\theta(X)(e_i) = \sum_j e^j\big(\theta(X)(e_i)\big)\, e_j = -\sum_j \theta(X)(e^j)(e_i)\, e_j$, we have

$$\sum\nolimits_i e^i \otimes \iota\big(\theta(X)(e_i)\big) = -\sum\nolimits_{i,j} \theta(X)(e^j)(e_i)\, e^i \otimes \iota(e_j) = -\sum\nolimits_j \theta(X)(e^j) \otimes \iota(e_j)$$

which, substituted in (4.20), gives the announced equality $[\theta(X), \iota(Y)] = 0$. □

Furthermore,

$$\begin{aligned}
d_{\mathfrak{g}}^2 &= (d + c(Y))^2 = d \circ c(Y) + c(Y) \circ d \\
&= \sum\nolimits_i e^i \otimes \big(d \circ \iota(e_i) + \iota(e_i) \circ d\big) = \sum\nolimits_i e^i \otimes \theta(e_i)\,,
\end{aligned}$$

so that, in $(S(\mathfrak{g}^\vee)\otimes C)^\mathfrak{g}$, we will have

$$d_\mathfrak{g}^2 = -\sum_i e^i\theta(e_i)\otimes \mathrm{id}\,, \tag{4.21}$$

since $(1\otimes\theta(X))(\omega) + (\theta(X)\otimes 1)(\omega) = 0$, for all $X\in\mathfrak{g}$ and all θ-invariant ω.

But in (4.21), the term $\Theta := \sum_i e^i\theta(e_i)$ is the null operator on $S(\mathfrak{g}^\vee)$. Indeed, as it acts as a derivation on $S(\mathfrak{g}^\vee)$, we need show only that it vanishes on every linear form $\lambda\in\mathfrak{g}^\vee$. This results from the straightforward computation

$$\Theta(\lambda)(e_1) = \Big(\sum_i e^i\theta(e_i)(\lambda)\Big)(e_1) = \sum_i e^i(e_1)\,\lambda([e_1,e_i]) = \lambda([e_1,e_1]) = 0\,.$$

where e_1 can be any element in $\mathfrak{g}\setminus\{0\}$. The vanishing of $\Theta(\lambda)$ then follows from the fact that it is a homogeneous polynomial function of degree $d>0$ ($d=2$).

As a consequence, the homomorphism $d_\mathfrak{g}$ is a differential on $(S(\mathfrak{g}^\vee)\otimes C)^\mathfrak{g}$. It is commonly referred to as *the Cartan differential.*

Definition 4.3.2.2 Given a $\mathfrak{g}$-dgm $C := (C,d,\theta,\iota)$, the *Cartan complex associated with* C is the complex $\big(C_\mathfrak{g}, d_\mathfrak{g}\big)$, where

$$C_\mathfrak{g} := (S(\mathfrak{g}^\vee)\otimes C)^\mathfrak{g} \quad\text{and}\quad d_\mathfrak{g}(\omega))(Y) := d(\omega(Y)) + \iota(Y)(\omega(Y)\,,\ \forall Y\in\mathfrak{g}\,.$$

Its cohomology will be called the $\mathfrak{g}$*-equivariant cohomology of* C, and will be denoted by

$$H_\mathfrak{g}(C) := \mathbf{h}\big(C_\mathfrak{g}, d_\mathfrak{g}\big)\,.$$

Notice that:

- $H_\mathfrak{g}(C)$ is a graded $H_\mathfrak{g}(\mathbb{R}) = S(\mathfrak{g}^\vee)^\mathfrak{g}$-module.
- After Cartan's theorem 4.1.1.1, if C admits an algebraic connection, then $H_\mathfrak{g}(C) = H(C^{\mathrm{bas}})$. In particular, $H_\mathfrak{g}(W(\mathfrak{g})) = S(\mathfrak{g}^\vee)^\mathfrak{g}$.
- If $A := (A,d,\theta,\iota)$ is a $\mathfrak{g}$-dg-algebra, then $A_\mathfrak{g}$ is an $S(\mathfrak{g}^\vee)^\mathfrak{g}$-dg-algebra and $H_\mathfrak{g}(A)$ is an $S(\mathfrak{g}^\vee)^\mathfrak{g}$-algebra.

4.3.3 Induced Morphisms on Cartan Complexes

A morphism of $\mathfrak{g}$-complexes $\alpha : (C,d,\theta,\iota) \to (D,d,\theta,\iota)$ induces a canonical $S(\mathfrak{g}^\vee)^\mathfrak{g}$-linear morphism of complexes $\alpha_\mathfrak{g} : C_\mathfrak{g} \to D_\mathfrak{g}$, by $\alpha_\mathfrak{g} = \mathrm{id}\otimes\alpha$. Furthermore, if α is a morphism of $\mathfrak{g}$-dg-algebras, then $\alpha_\mathfrak{g}$ is a morphism of dg-algebras.

Theorem 4.3.3.1 *With the above notations, we have the following.*

1. *The correspondence* $(C, d, \theta, \iota) \rightsquigarrow (C_{\mathfrak{g}}, d)$ *and* $\alpha \rightsquigarrow \alpha_{\mathfrak{g}}$ *is a covariant functor from* DGM($\mathfrak{g}$) *into* DGV($\mathbb{R}$).
2. *For every* $\mathfrak{g}$*-complex* $(C, d, \theta, \iota) \in \mathrm{DGM}^+(\mathfrak{g})$*, there exists a spectral sequence converging to* $H_{\mathfrak{g}}(C)$ *with*

$$\left(\mathbb{E}_0^{p,q} = \left(S^p(\mathfrak{g}^\vee) \otimes C^q\right)^{\mathfrak{g}}\,,\; d_0 = 1 \otimes d\right) \Rightarrow H_{\mathfrak{g}}^{p+q}(C)\,.$$

3. *Let* G *be a compact Lie group, let* $\mathfrak{g} := \mathrm{Lie}(G)$*, and let* C *and* D *be* $\mathfrak{g}$***-split*** $\mathfrak{g}$*-complexes (4.2.5.1) in* $\mathrm{DGM}^+(\mathfrak{g})$*. Then, the following statements hold.*
   a. *The* $(\mathbb{E}_2, d_2)$ *spectral sequence term in* (2) *is given by*

$$\left(\mathbb{E}_2^{p,q} = S^p(\mathfrak{g}^\vee)^{\mathfrak{g}} \otimes H^q(C)\,,\; d_2 = \textstyle\sum_i e^i \otimes \iota(e_i)\right) \Rightarrow H_{\mathfrak{g}}^{p+q}(C)\,.$$

   b. *If* $H^m(C) = 0$ *for all odd (or for all even)* m*, then*

$$H_{\mathfrak{g}}(C) = S(\mathfrak{g}^\vee)^{\mathfrak{g}} \otimes \mathbf{h}(C)\,.$$

   c. *If* $\alpha : C \to D$ *is a quasi-isomorphism, then* $\alpha_{\mathfrak{g}}$ *is a quasi-isomorphism.*
4. *Let* T *be a* **commutative** *compact Lie group and* $\mathfrak{g} := \mathrm{Lie}(G)$.
   a. *For every* $\mathfrak{g}$*-complex* (C, d, θ, ι)*, the subcomplex* $(C^{\mathfrak{g}}, d)$ *is stable under* θ *and* ι*, i.e.* $(C^{\mathfrak{g}}, d, \theta, \iota)$ *is a well-defined* $\mathfrak{g}$*-complex.*
   In the next statements C *and* D *are* $\mathfrak{g}$*-complexes in* $\mathrm{DGM}^+(\mathfrak{g})$.
   b. *The* $(\mathbb{E}_2, d_2)$ *spectral sequence term in* (2) *is given by*

$$\left(\mathbb{E}_2^{p,q} = S^p(\mathfrak{g}^\vee) \otimes H^q(C^{\mathfrak{g}})\,,\; d_2 = \textstyle\sum_i e^i \otimes \iota(e_i)\right) \Rightarrow H_{\mathfrak{g}}^{p+q}(C)\,.$$

   c. *If* $H^m(C^{\mathfrak{g}}) = 0$ *for all odd (or for all even)* m*, then*

$$H_{\mathfrak{g}}(C) = S(\mathfrak{g}^\vee) \otimes \mathbf{h}(C^{\mathfrak{g}})\,.$$

   d. $\alpha : C \to D$ *is a quasi-isomorphism, then* $\alpha_{\mathfrak{g}}$ *is a quasi-isomorphism.*

Proof

(1) Clear. (2) For $m \in \mathbb{Z}$, let $K_m = \left(S^{\geq m}(\mathfrak{g}^\vee) \otimes C\right)^{\mathfrak{g}}$. Each K_m is clearly a subcomplex of $(C_{\mathfrak{g}}, d_{\mathfrak{g}})$ and $\left(C_{\mathfrak{g}} = K_0 \supseteq K_1 \supseteq \cdots\right)$ is a *regular* decreasing filtration of $(C_{\mathfrak{g}}, d_{\mathfrak{g}})$ since C is bounded below (see [46] §4 pp. 76-) giving rise to the stated spectral sequence.

(3a) The assumption that G is compact ensures that each (finite dimensional) $\mathfrak{g}$-module $S^p(\mathfrak{g}^\vee)$ is semisimple. Proposition 4.2.5.5-(2) may be used, and $\left((S^p(\mathfrak{g}^\vee) \otimes C)^{\mathfrak{g}}, 1 \otimes d\right)$ is quasi-isomorphic to $\left(S^p(\mathfrak{g}^\vee)^{\mathfrak{g}} \otimes C, 1 \otimes d\right)$. Consequently $(\mathbb{E}_0, d_0)$ in (2) is quasi-isomorphic to $\left(S(\mathfrak{g}^\vee)^{\mathfrak{g}} \otimes C, 1 \otimes d\right)$

and $\mathbb{E}_1^{p,q} = S^p(\mathfrak{g}^\vee)^\mathfrak{g} \otimes H^q(C)$. But the differential $d_1 : \mathbb{E}_1^{p,q} \to \mathbb{E}_1^{p+1,q}$ is equal to zero since the $S(\mathfrak{g}^\vee)$ vanishes in odd degrees, therefore $\mathbb{E}_1 = \mathbb{E}_2$, which completes the proof of the claim.

(3b) Since the differential d_r is of total degree 1 and that $\mathbb{E}_r^{p,q} = 0$ if p or q is odd for all $r \geq 2$, then we have $d_r = 0$ for $r \geq 2$, and $\mathbb{E}_2 = \mathbb{E}_\infty$.

(3c) Follows immediately from (3-i) and 4.2.5.5-(1).

(4a) We must check that $\theta(Y)\iota(X)C^\mathfrak{g} = 0$ for all $X, Y \in \mathfrak{g}$, but, on $C^\mathfrak{g}$ we have $\theta(Y)\iota(X) = \theta(Y)\iota(X) + \iota(X)\theta(Y) = \iota([Y, X]) = \iota(0)$ since $\mathfrak{g}$ is abelian and from property (iii) of $\mathfrak{g}$-complexes (see Sects. 4.2.3–(4.13)).

(4b,c,d) Left to the reader. □

4.3.4 Split G-Complexes

Let G be a Lie group (not necessarily compact) with Lie algebra $\mathfrak{g}$. It's worth noting that the proof of 4.3.3.1-(3) uses the split condition 4.2.5.1 *only* on the finite dimensional vector spaces $S^p(\mathfrak{g}^\vee) \subseteq S(\mathfrak{g}^\vee)$ endowed with the structure of $\mathfrak{g}$-module obtained by differentiating their natural G-module structure. We are thus led to extend the split definition 4.2.5.1 to G-modules.

Definition 4.3.4.1 For any inclusion of G-modules $N \subseteq M$ we write '$N|M$' whenever the natural map

$$\mathrm{Hom}_G(V, M) \to \mathrm{Hom}_G(V, M/N) \tag{4.22}$$

is **surjective** for all **finite dimensional** G-modules V.

A complex of G-modules (C, d) is said to be *G-split* is $B^i|Z^i|C^i$, for all $i \in \mathbb{Z}$.

Exercise 4.3.4.2 Let G be a compact Lie group. Denote by G_0 the connected component of the identity element $e \in G$. Show that a complex of G-modules (C, d) is G-split if and only if it is G_0-split. (♝, p. 348)

The proof of the following proposition is then the same as for Proposition 4.2.5.5.

Proposition 4.3.4.3 *Let G be a compact Lie group, and let (C, d) be a G-split complex of G-modules such that the action of G in $\mathbf{h}(C, d)$ is trivial. Then,*

1. *The inclusion $C^G \subseteq C$ is quasi-isomorphism.*
2. *If V is a **semisimple** finite dimensional G-module, the inclusions*

$$V^G \otimes C \supseteq V^G \otimes C^G \subseteq (V \otimes C)^G$$

$$\mathbf{Hom}^\bullet_\mathbb{R}(V^G, C) \supseteq \mathbf{Hom}^\bullet_\mathbb{R}(V^G, C^G) \subseteq \mathbf{Hom}^\bullet_G(V, C)$$

are quasi-isomorphisms.

4.4 Equivariant Differential Forms

4.4.1 *Fields in Use*

Manifolds, Lie groups and Lie algebras, vector spaces, linear maps, tensor products and related material will be defined over the field of real numbers $\mathbb{R}$.

4.4.2 *G-Fundamental Vector Fields*

Let G be a Lie group with Lie algebra $\mathfrak{g}$. On every G-manifold M, an element $Y \in \mathfrak{g}$ defines a vector field, called *the G-fundamental vector field on M associated with Y*, by the formula

$$\vec{Y}(m) := \frac{d}{dt}\big(t \mapsto \exp(tY) \cdot m\big)_{t=0}. \tag{4.23}$$

The vector field $\vec{Y}$ is G-invariant and the map $(\vec{\_}) : \mathfrak{g} \to \mathfrak{X}(M)^G$, where $\mathfrak{X}(M)$ denotes the Lie algebra of vector fields on M, is a Lie algebra homomorphism.

When $M := G$, the map $(\vec{\_}) : \mathfrak{g} \to \mathfrak{X}(G)^G$ is an isomorphism onto the set of left invariant vector fields. Moreover $\vec{\mathfrak{g}}(x) = T_x(G)$ for every $x \in G$, and, as a consequence, a differential form $\omega \in \Omega^i(G)$ is completely determined by the functions $\omega((\vec{\_}), \ldots, (\vec{\_})) : G \to \Lambda(\mathfrak{g}^\vee)$. Hence, a canonical identification

$$\Omega(G) = \Omega^0(G) \otimes \Lambda(\mathfrak{g}^\vee). \tag{4.24}$$

4.4.3 *Interior Products and Lie Derivatives*

– The *interior product with a vector field* $\xi \in \mathfrak{X}(M)$ is the map

$$\iota(\xi) : \Omega(M) \to \Omega(M)[-1],$$

defined by

$$\iota(\xi)(\omega)(\xi_1, \ldots, \xi_{d-1}) := \omega(\xi, \xi_1, \ldots, \xi_{d-1}), \quad \forall \omega \in \Omega^d(M),\ \xi_i \in \mathfrak{X}(M),$$

it is a *graded derivation* (aka *antiderivation*) of degree -1 , i.e.

$$\iota(\xi)(\omega_1 \wedge \omega_2) = \iota(\xi)(\omega_1) \wedge \omega_2) + (-1)^{d_1}\omega_1 \wedge \iota(Y)(\omega_2),$$

for all $\omega_1 \in \Omega^{d_1}(M)$ and $\omega_2 \in \Omega(M)$. The map

$$\iota : \mathfrak{g} \to \mathrm{Mor}_{\mathrm{GV}(\Bbbk)}(\Omega(M), \Omega(M)[-1]), \quad Y \mapsto \iota(\vec{Y}), \tag{4.25}$$

verifies conditions (i)-(4.13) and (iv)-(4.14) for $\mathfrak{g}$-complexes.

– The *Lie derivative with respect to* $\xi \in \mathfrak{X}(M)$, is the map

$$\mathcal{L}(\xi) : \Omega(M) \to \Omega(M),$$

defined by $\mathcal{L}(\xi) := d \circ \iota(\xi) + \iota(\xi) \circ d$, it is a *graded derivation* of degree 0, i.e.

$$\mathcal{L}(\xi)(\omega_1 \wedge \omega_2) = \mathcal{L}(\xi)(\omega_1) \wedge \omega_2) + \omega_1 \wedge \mathcal{L}(Y)(\omega_2),$$

for all $\omega_1 \in \Omega^{d_1}(M)$ and $\omega_2 \in \Omega(M)$. The map

$$\theta : \mathfrak{g} \to \mathrm{Endgr}_{\mathrm{GV}(\Bbbk)}(\Omega(M)), \quad Y \mapsto \mathcal{L}(\vec{Y}), \tag{4.26}$$

is a Lie algebra representation by $\mathbb{R}$-*derivations* and verifies conditions (ii,iii)-(4.13) (p. 121) and (v)-(4.14) (p. 121) for $\mathfrak{g}$-complexes.

The operators $\theta(Y)$ and $\iota(Y)$ stabilize $\Omega_c(M)$, and the quadruples

$$(\Omega(M), d, \theta, \iota) \quad \text{and} \quad (\Omega_c(M), d, \theta, \iota),$$

are $\mathfrak{g}$-complexes since all the conditions in Sect. 4.2.3 are satisfied.

4.4.4 Complexes of Equivariant Differential Forms

Let G be a compact Lie group. The *complex of G-equivariant differential forms, resp. compactly supported, of M*, modeled on the Cartan complex (4.3.2.2), is the complex:

$$\begin{aligned} \big(\Omega_G(M), d_G\big) &:= \big(\big(S(\mathfrak{g}^\vee) \otimes \Omega(M)\big)^G, d_{\mathfrak{g}}\big), \\ \textit{resp. } \big(\Omega_{G,c}(M), d_G\big) &:= \big(\big(S(\mathfrak{g}^\vee) \otimes \Omega_c(M)\big)^G, d_{\mathfrak{g}}\big), \end{aligned} \tag{4.27}$$

with

$$d_G(\omega))(Y) := d(\omega(Y)) + \iota(Y)(\omega(Y), \quad \forall Y \in \mathfrak{g}. \tag{4.28}$$

Its cohomology, denoted by $H_G(M)$, resp. $H_{G,c}(M)$, is the *G-equivariant cohomology, resp. compactly supported, of M*.

Note that the complexes $\Omega_G(M)$, $\Omega_{G,\mathrm{c}}(M)$ and the cohomology vector spaces $H_G(M)$ and $H_{G,\mathrm{c}}(M)$ are $S(\mathfrak{g}^\vee)^G$-graded modules (*cf.* 5.1.2).

Convention 4.4.4.1 When $M := \{\bullet\}$, we shrink notations to

$$\boxed{\Omega_G := \Omega_G(\{\bullet\}) \text{ and } H_G := H_G(\{\bullet\})}$$

Notice that in that case $d_G = 0$, so that $\Omega_G = S(\mathfrak{g}^\vee)^G = H_G$.

Comment 4.4.4.2 The reader will have noticed that in (4.27), we do not assume G connected, and that we have replaced the $\mathfrak{g}$-invariants involved in the definition of Cartan complexes 4.3.2.2, by G-invariants. This is possible since the actions of G on $\mathfrak{g}$ and on M are differentiable, and that it is this differentiability that induces the action of the Lie algebra $\mathfrak{g}$ on the $\mathfrak{g}$-complexes $\Omega(M)_\mathfrak{g}$ and $\Omega_\mathrm{c}(M)_\mathfrak{g}$. As a consequence, when G is connected there is no difference between G-invariance and $\mathfrak{g}$-invariance and the results in Cartan's lectures apply. On the contrary, when G is not connected, we deviate from Cartan's lectures and we need to justify the relevance of our definition and its relationship with the connected case. This is the aim of the next section.

4.4.5 On the Connectedness of G

Let G be a Lie group and denote by G_0 the connected component of the identity element $e \in G$. We recall that G_0 is an open normal subgroup of G and that $W := G/G_0$ is a discrete group (finite if G is compact).

Let M be a G-manifold, and denote by $\Omega_?(M)$ either $\Omega(M)$ or $\Omega_\mathrm{c}(M)$. The group G acts on $\mathfrak{g} := \mathrm{Lie}(G_0)$ by the adjoint representation, and on M by diffeomorphisms, whence its action on $S(\mathfrak{g}^\vee) \otimes \Omega_?(M)$, and the equality:

$$(S(\mathfrak{g}^\vee) \otimes \Omega_?(M))^G = \big((S(\mathfrak{g}^\vee) \otimes \Omega_?(M))^\mathfrak{g}\big)^W . \tag{4.29}$$

Lemma 4.4.5.1

1. *The map $d_G : S(\mathfrak{g}^\vee) \otimes \Omega_?(M) \to S(\mathfrak{g}^\vee) \otimes \Omega_?(M)$, defined in 4.4.4-(4.28) by the formula*

$$(P \otimes \omega) \mapsto P \otimes d\omega + \sum\nolimits_i Pe^i \otimes \iota(e_i)\,\omega ,$$

is an $S(\mathfrak{g}^\vee)$-linear G-equivariant map.
2. *The differential in the Weil algebra $(W(\mathfrak{g}), d)$ is G-equivariant.*

Proof We denote by a dot '$\cdot$' the (left) actions of G.

For all $g \in G$, we have:

- For all $\lambda \in \mathfrak{g}^\vee$ and all $X \in \mathfrak{g}$:

$$\langle \lambda, X \rangle = \langle g \cdot \lambda, g \cdot X \rangle . \tag{4.30}$$

  Hence, if $\{e_i\}$ and $\{e^i\}$ are dual basis of $\mathfrak{g}$, then so are $\{g \cdot e_i\}$ and $\{g \cdot e^i\}$.
- For all $\omega \in \Omega_?(M)$, $X \in \mathfrak{g}$ and $x \in M$, we have:

$$\begin{aligned}(g \cdot \omega)(x)(\vec{X}(x), \dots) &= \omega(g^{-1} \cdot x)\big(T_x g^{-1}(\vec{X}(x)), \dots)\big) \\ &= \omega(g^{-1} \cdot x)\big(\overrightarrow{g^{-1} \cdot X}(g^{-1} \cdot x), \dots)\big) ,\end{aligned}$$

  whence

$$g \cdot \big(\iota(X)(\omega)\big) = \iota(g \cdot X)(g \cdot \omega) , \tag{4.31}$$

  and also

$$g \cdot \big(\theta(X)(\omega)\big) = \theta(g \cdot X)(g \cdot \omega) , \tag{4.32}$$

  since $\theta(X) = d \circ \iota(X) + \iota(X) \circ d$.

(1) For all $P \in S(\mathfrak{g}^\vee)$ and $\omega \in \Omega_?(M)$, we have:

$$\begin{aligned} d_G\big(g \cdot (P \otimes \omega)\big) &= d_G\big((g \cdot P) \otimes (g \cdot \omega)\big) \\ &= (g \cdot P) \otimes d(g \cdot \omega) + \sum\nolimits_i (g \cdot P)\, e^i \otimes \iota(e_i)(g \cdot \omega) \\ &=_1 (g \cdot P) \otimes \cdot (d\omega) + \sum\nolimits_i g \cdot \big(P\,(g^{-1} \cdot e^i)\big) \otimes \iota(g^{-1} \cdot e_i)(\omega) \\ &=_2 g \cdot \big(d_G(P \otimes \omega)\big) ,\end{aligned}$$

where, for $(=_1)$ we used (4.31), and for $(=_2)$ we used (4.30).

(2) Since the differential d of the Weil algebra $(W(\mathfrak{g}), d)$ is an $\mathbb{R}$-derivation, and since G acts by algebra isomorphisms, the G-equivariance of d results by simply checking this property on the elements $\lambda \otimes 1 \in \Lambda^1(\mathfrak{g}^\vee) \otimes 1$ and $1 \otimes \lambda \in 1 \otimes S^2(\mathfrak{g}^\vee)$.

After Cartan's description[19] of the differential d, we have, for all $g \in G$,

$$\begin{aligned} g \cdot d(\lambda \otimes 1) &= g \cdot \big(d_K(\lambda) \otimes 1 + 1 \otimes \lambda\big) \\ &= d_K(g \cdot \lambda) \otimes 1 + 1 \otimes \cdot \lambda \\ &= g \cdot d_K(\lambda) \otimes 1 + 1 \otimes \cdot \lambda = d\big(g \cdot (\lambda \otimes 1)\big) ,\end{aligned}$$

[19] See Cartan [27], §6 L'algèbre de Weil d'une algèbre de Lie, formulas (9–10), p. 23–24.

where d_K is the Koszul differential on $\lambda(\mathfrak{g}^\vee)$, well-known to be G-equivariant. We also have:

$$\begin{aligned} g\cdot d(1\otimes\lambda) &= g\cdot\Big(\sum_i e^i\otimes\theta(e_i)(\lambda)\Big)\\ &= \sum_i (g\cdot e^i)\otimes\theta(g\cdot e_i)(g\cdot\lambda) = d\big(1\otimes(g\cdot\lambda)\big), \end{aligned}$$

by (4.30) and (4.32). □

Thanks to the lemma, we see that, whether the group G is connected or not, the complexes

$$\boxed{\big(S(\mathfrak{g}^\vee)\otimes\Omega_?(M), d_G\big)^G = \big((S(\mathfrak{g}^\vee)\otimes\Omega_?(M), d_G)^{\mathfrak{g}}\big)^W} \tag{4.33}$$

are well defined.

This all immediately leads us to ask if this extension of the definition of *Cartan complexes* is compatible with other parts of Cartan's lectures (Sect. 4.1.1), among which, if, given a principal G-bundle (E, B, π, G), the Cartan-Weil morphisms $\bar{\mathfrak{i}}$, $\bar{\mathfrak{f}}$ (in (4.3), p. 111, and Appendix C) and the Cartan morphism Ξ' (in (4.18), p. 125):

$$\begin{array}{ccccc} \Omega_?(E)^{\mathrm{bas}} & \xrightarrow{\bar{\mathfrak{i}}} & (W(\mathfrak{g})\otimes\Omega_?(E))^{\mathrm{bas}} & \xrightarrow{\bar{\mathfrak{f}}} & \Omega_?(E)^{\mathrm{bas}} \\ & & \downarrow{\scriptstyle \Xi'} & & \\ & & (S(\mathfrak{g}^\vee)\otimes\Omega_?(E))^{\mathfrak{g}} & & \end{array} \tag{4.34}$$

are W-equivariant quasi-isomorphisms?

The answer is yes. Indeed, the morphisms $\bar{\mathfrak{i}}$ and Ξ' are W-equivariant, since $\mathfrak{i}:\Omega_?(E)\to W(\mathfrak{g})\otimes\Omega_?(E), \omega\mapsto 1\otimes\omega$, and $\Xi: W(\mathfrak{g})\otimes\Omega_?(E)\to S(\mathfrak{g}^\vee)\otimes\Omega_?(E)$ (*cf.* (4.18), p. 125) are G-equivariant. The same is true for $\bar{\mathfrak{f}}$ provided that we can choose the infinitesimal connection $f:\mathfrak{g}^\vee\to\Omega^1(E)$ to be G-invariant, and we know this is possible when G is compact (which is always the case in this book), since we can then G-average any infinitesimal connection.

The following proposition summarizes all these observations.

Proposition 4.4.5.2 *Let G be a compact Lie Group. Denote by G_0 the connected component of the identity element $e\in G$, and set $W := G/G_0$. Let*

$$(\_)^{\mathrm{Bas}} := \big((\_)^{\mathrm{bas}}\big)^W = (\_)^{\mathrm{hor}}\cap(\_)^G .$$

Then, for every principal G-bundle (E, B, π, G), the morphisms in (4.34) are W-equivariant and induce the following quasi-isomorphic morphisms of dg-algebras,

$$\Omega(B)\xrightarrow[\simeq]{\pi^*}\Omega(E)^{\mathrm{Bas}}\xrightarrow[\text{q.i.}]{\bar{\mathfrak{i}}'}(W(\mathfrak{g})\otimes\Omega(E))^{\mathrm{Bas}}\xrightarrow[\simeq]{\Xi''}\big(S(\mathfrak{g}^\vee)\otimes\Omega(E)\big)^G, \tag{4.35}$$

(resp. of dg-modules if we replace $\Omega(M)$ by $\Omega_c(M)$), and where $\bar{\imath}'$ and Ξ'' denote the restrictions of $\bar{\imath}$ and Ξ' to the subspaces $(\_)^{\mathrm{Bas}} \subseteq (\_)^{\mathrm{bas}}$. In particular, $H_{G_0}(\Omega_?(M))$ is a W-module, and we have canonical isomorphisms

$$\begin{cases} H(B) \simeq H_G(E) \simeq H_{G_0}(E)^W \\ H_c(B) \simeq H_{G,c}(E) \simeq H_{G_0,c}(E)^W . \end{cases} \tag{4.36}$$

Proof The fact that $\pi^* : \Omega(B) \to \Omega(E)^{\mathrm{Bas}}$ is an isomorphism is standard in differential geometry, and is true whether G is compact or not. The morphism Ξ'' is an isomorphism by elementary algebraic reasons. At the end, compactness of G is needed to justify the existence of a connection of principal G-bundle $f : \mathfrak{g}^\vee \to \Omega^1(E)$, hence to apply Cartan's theorem 4.1.1.1. We therefore get W-equivariant quasi-isomorphisms

$$\Omega(E)^{\mathrm{bas}} \xrightarrow[\text{q.i.}]{\bar{\imath}} (W(\mathfrak{g}) \otimes \Omega(E))^{\mathrm{bas}} \xrightarrow[\text{q.i.}]{\Xi'} \big(S(\mathfrak{g}^\vee) \otimes \Omega(E)\big)^{\mathfrak{g}} . \tag{4.37}$$

Compactness of G implies also that the group W is finite, hence that the restrictions $\bar{\imath}'$ and Ξ'' of the quasi-isomorphisms $\bar{\imath}$ and Ξ' in (4.37), to W-invariants remain quasi-isomorphic, proving (4.35) for $\Omega(M)$. The same arguments apply, *mutatis mutandis*, to $\Omega_c(M)$ in lieu of $\Omega(M)$, replacing Cartan's theorem by its enhancement in Appendix C. The isomorphisms (4.36) then easily follow. □

4.4.6 Splitness of Complexes of Equivariant Differential Forms

Theorem 4.4.6.1 ([20])

1. *Let G be a compact Lie group. For every G-manifold M, the complexes of G-modules $(\Omega(M), d)$ and $(\Omega_c(M), d)$ are G-split (4.3.4).*
2. *For all $m \in \mathbb{N}$, the inclusions*

$$S^{\geq m}(\mathfrak{g}^\vee)^G \otimes C \supseteq S^{\geq m}(\mathfrak{g}^\vee)^G \otimes C^G \subseteq (S^{\geq m}(\mathfrak{g}^\vee) \otimes C)^G ,$$

where C denotes $(\Omega(M), d)$ or $(\Omega_c(M), d)$, are quasi-isomorphisms if and only if G acts trivialy on $\mathbf{h}(C)$, for example if G is connected.[21]

[20] A different approach of this theorem can be found in Tu [91], Appendix C, p. 283.

[21] On the connectedness hypothesis see (4.4.6.3)–(1).

Proof

(1) After Exercise 4.3.4.2 we can assume G connected.

The *pushforward action* of G on $\Omega^i(M)$ is defined, for $g \in G$ and $\omega \in \Omega^i$, by

$$g_*(\omega) := (g^{-1})^*(\omega)\,,$$

where $(g^{-1})^*$ denotes the usual *pullback* of differential forms. The pushforward action is such that we have $(g_1 g_2)_* = g_{1*} \circ g_{2*}$ for all $g_i \in G$.

If V is a (smooth) finite dimensional representation of G over $\mathbb{C}$, we make the group G act on $\mathrm{Hom}(V, \Omega^i(M))$ by the formula

$$(g \cdot \lambda)(v) = g_*\big(\lambda(g^{-1}v)\big)\,, \quad \forall \lambda \in \mathrm{Hom}(V, \Omega^i(M))\,,$$

so that λ is a G-module morphism if and only if $g \cdot \lambda = \lambda$. We claim that there exists a '*symmetrization*' operator

$$\Sigma : \mathrm{Hom}(V, \Omega^i(M)) \to \mathrm{Hom}(V, \Omega^i(M))^G\,,$$

such that $\Sigma^2 = \Sigma$ and $\Sigma(\lambda) = \lambda$ if and only if λ is a G-module morphism.

Indeed, let λ be a linear map from V to $\Omega^i(M)$. For every i-tuple of vector fields $\{\chi_1, \dots, \chi_i\}$ over M and each $v \in V$, the real function

$$M \ni x \mapsto \Big(\int_G g_*\big(\lambda(g^{-1}v)\big)(x)\big(\chi_1(x), \dots, \chi_i(x)\big)\, dg\Big) \in \mathbb{R}\,,$$

where dg is a G-invariant form of top degree on G, such that $1 = \int_G dg$, is a smooth function, since V is finite dimensional, depending linearly on $v \in V$, and multilinearly and antisymmetrically on the χ_*'s. We therefore have an i-differential form which we denote by

$$\Sigma(\lambda)(v) := \int_G g_*\big(\lambda(g^{-1}v)\big)\, dg\,, \tag{4.38}$$

and whose fundamental properties are

- $\Sigma(d \circ \lambda) = d \circ \Sigma(\lambda)$;
- $\Sigma(\lambda) : V \to \Omega^i(M)$ is a G-module morphism;
- $\Sigma(\lambda) = \lambda$, if λ is already a G-module morphism.

We can now resume the proof that $Z^i(M)|\Omega^i(M)$. Given a G-module morphism $\mu \in \mathrm{Hom}_G(V, B^{i+1}(M))$, there always exists a linear map $\lambda : V \to \Omega^i(M)$ lifting μ, i.e. such that $\mu = d \circ \lambda$, but then we apply the symmetrization operator Σ and we get $\mu = \Sigma(\mu) = \Sigma(d \circ \lambda) = d \circ \Sigma(\lambda)$, which shows that the G-module morphism $\Sigma(\lambda)$ lifts μ.

For $Z^i_c(M)|\Omega^i_c(M)$, note that, since V is finite dimensional, the supports of the elements in $\lambda(V)$ are contained in one and the same compact subset $C \subseteq M$, but then the supports of the $g_*(\lambda(g^{-1}v))$ in (4.38) are contained in $G\cdot C$ which is obviously compact. Therefore, given $\lambda : V \to \Omega_c(M)$, we get a linear map $\Sigma(V) : V \to \Omega_c(M)$ which is a G-module morphism, and the preceding arguments apply to the compactly supported case.

To prove that $B^i(M)|Z^i(M)$, we need only, from 4.2.5.4-(2), show that every cocycle is cohomologous to a G-invariant cocycle. But before doing so, let us recall a general homotopy argument. Given a smooth map $\varphi : \mathbb{R} \times M \to N$, if $\omega \in \Omega^i(N)$ the pullback $\varphi^*\omega$ belongs to $\Omega^i(\mathbb{R} \times M)$, i.e. is a section of the exterior algebra bundle of the cotangent bundle $T^*(\mathbb{R} \times M)$ of $\mathbb{R} \times M$. Now, the canonical decomposition $T^*(\mathbb{R} \times M)$ as the direct sum of cotangent bundles $T^*(\mathbb{R}) \oplus T^*(M)$, leads to a canonical decomposition of the i-th exterior power of the cotangent bundle

$$\bigwedge^i T^*(\mathbb{R} \times M) = \mathbb{R} \otimes \bigwedge^i (T^*M) \oplus \left(T^*(\mathbb{R}) \otimes \bigwedge^{i-1}(T^*M)\right).$$

Consequently, the pullback $\varphi^*(\omega)$ canonically decomposes as

$$\varphi^*(\omega)(t,x) = \alpha(t,x) + dt \wedge \beta(t,x),$$

where α (resp. β) is a section of the vector bundle $\bigwedge^i T^*(M)$ (resp. $\bigwedge^{i-1} T^*(M)$) over the base space $\mathbb{R} \times M$.

When ω is in addition a cocycle, so is $\varphi^*(\omega)$ and, in view of the previous decomposition, this amounts to the following two conditions

$$d\alpha(t,x) = 0, \qquad \frac{\partial}{\partial t}\alpha(t,x) = d\beta(t,x),$$

where d is the coboundary in $\Omega(M)$ (t is then assumed constant). In particular, denoting by $\varphi_t : M \to N$ the map $x \mapsto \varphi(t,x)$, we get

$$\begin{aligned} \varphi_t^*(\omega) - \varphi_0^*(\omega) &= \alpha(t) - \alpha(0) \\ &= \int_0^t \frac{\partial}{\partial t}\alpha(t)\,dt = \int_0^t d\beta(t)\,dt = d\left(\int_0^t \beta(t)\,dt\right), \end{aligned} \tag{4.39}$$

and the cocycles $\varphi_t^*(\omega)$ are all cohomologous to $\varphi_0^*(\omega)$.

It is worth noting that this process gives a canonical element $\varpi(x) = \int_0^1 \beta(t,x)\,dt \in \Omega^{i-1}(M)$, depending on ω, such that $\varphi_1^*(\omega) - \varphi_0^*(\omega) = d\varpi$.

Under the hypothesis of the theorem, a first consequence of the previous observations, is that if $\omega \in Z^i(M)$, then $g^*\omega$ is cohomologous to ω for all $g \in G$. Indeed, since G we assumed connected, there is a smooth path $\gamma : \mathbb{R} \to G$ such that $\gamma(0) = e$ and $\gamma(1) = g$, and then taking $\varphi : \mathbb{R} \times M \to M$, $(t,x) \mapsto \gamma(t)\cdot x$, we conclude that $g^*\omega = \gamma_1^*(\omega) \sim \gamma_0^*(\omega) = \omega$.

More generally, given a diffeomorphism $\phi : \mathbb{R}^{d_G} \to G$ onto an open subset $U \subseteq G$, we define a smooth multiplicative action of $\mathbb{R}$ over U by setting $t \star g := \phi(t \cdot \phi^{-1}(g))$ for all $t \in \mathbb{R}$ and $g \in U$, and consider, for each $g \in U$, the map $\varphi_g : \mathbb{R} \times M \to M$, $\varphi_g(t,x) = (t \star g)x$.

Following that, if ω is a cocycle of $\Omega^i(M)$, we then have

$$g^*\omega - g_0^*\omega = d\Big(\int_0^1 \beta(t,g)\,dt\Big), \tag{4.40}$$

with $g_0 := \phi(0)$ and where $\beta(t,g)$ denotes a family of elements of $\Omega^{i-1}(M)$ depending smoothly on $(t,g) \in \mathbb{R} \times U$, i.e. for any $(i-1)$-tuple $(\chi_1,\dots,\chi_{i-1})$ of vector fields over M, the following map is smooth:

$$\mathbb{R} \times U \times M \ni (t,g,x) \mapsto \beta(t,g,x)(\chi_1(x),\dots,\chi_{i-1}(x)) \in \mathbb{R}\,.$$

We now come to a key point. If, in addition, we have a compactly supported function $\rho : U \to \mathbb{R}$, then, for any top degree form dg on G, we have

$$\begin{aligned}\int_G \rho(g)\,g^*\omega\,dg &= \int_G \rho(g)\,(g^*\omega - g_0^*\omega)\,dg + \Big(\int_G \rho(g)\,dg\Big)g_0^*\omega\\ &= d\Big(\int_G\int_0^1 \rho(g)\,\beta(t,g)\,dg\Big) + \Big(\int_G \rho(g)\,dg\Big)g_0^*\omega\end{aligned}$$

where $\int_G\int_0^1 \rho(g)\,\beta(t,g)\,dg$ is a **smooth** differential form over M. But, as we already show that $g_0^*\omega \sim \omega$, since G is connected, we may conclude that

$$\int_G \rho(g)\,g^*\omega\,dg \sim \Big(\int_G \rho(g)\,dg\Big)\,\omega,$$

which is satisfied by any compactly supported function $\rho : G \to \mathbb{R}$ whose support is contained in any open subset of M diffeomorphic to $\mathbb{R}^d_G$.

If we now make use of the fact that G is compact (which we haven't done so far), we can choose the measure dg to be G-invariant such that $\int_G dg = 1$, and we can fix a smooth partition of unity $\{\rho_i\}$ subordinate to a finite good cover of G (fn. ([15]), p. 28). Then, for any cocycle $\omega \in Z^i(M)$, we have

$$\begin{aligned}\Sigma(\omega) := \int_G g^*\omega\,dg &= \int_G \textstyle\sum_i \rho_i(g)\,g^*\omega\,dg = \sum_i \int_G \rho_i(g)\,g^*\omega\,dg\\ &\sim \Big(\sum_i \int_G \rho_i(g)\,dg\Big)\,\omega = \Big(\int_G \textstyle\sum_i \rho_i(g)\,dg\Big)\,\omega = \omega\,,\end{aligned} \tag{4.41}$$

and $\Sigma(\omega)$ is then obviously a G-**invariant** cocycle, thus completing the proof that $B^i(M)|Z^i(M)$ as G-modules.

If we denote by $|\_|$ the support of a differential form, we see in the preceding lines that for $t \in [0,1]$ and $g \in G$, we have

$$\begin{cases} |\beta(t)| \subseteq \gamma([0,1])\cdot|\omega| & \text{in (4.39)},\\ |\rho(g)\beta(t,g)| \subseteq ([0,1] \star |\rho|)|\omega| & \text{in (4.40)}.\end{cases}$$

Therefore, if $|\omega|$ is compact, the arguments show that $\Sigma(\omega) - \omega$ is in fact the differential of a compactly supported differential form, i.e. we have also proved that $B^i_c(M)|Z^i_c(M)$, which ends the proof of the fact that the complexes of G-modules $(\Omega(M), d)$ and $(\Omega_c(M), d)$ are G-split

(2) After (1), the complex C is G-split and since G acts trivially on $\boldsymbol{h}(C)$ and its action on $S(\mathfrak{g}^\vee)$ stabilizes each $S^n(\mathfrak{g}^\vee)$, which is finite dimensional, we can apply Proposition 4.2.5.5. The inclusions $S^n(\mathfrak{g}^\vee)^G \otimes C \supseteq S^n(\mathfrak{g}^\vee)^G \otimes C^G \subseteq (S^n(\mathfrak{g}^\vee) \otimes C)^G$ are then quasi-isomorphisms, and the same for $S^{\geq m}(\mathfrak{g}^\vee) = \bigoplus_{n\geq m} S^n(\mathfrak{g}^\vee)$. The converse is clear since $C = S^0(\mathfrak{g}^\vee) \otimes C$ is the cokernel of the inclusion $S^{\geq 1}(\mathfrak{g}^\vee) \otimes C \subseteq S^{\geq 0}(\mathfrak{g}^\vee) \otimes C$. Details are left to the reader. □

Corollary 4.4.6.2 *Let G be a compact connected Lie group.*[22]

1. *The correspondence* $M \rightsquigarrow (\Omega(M), d, \theta, \iota)$, $f \rightsquigarrow f^*$ *is a contravariant functor from the category of G-manifolds into the category of G-split* $\mathfrak{g}$*-complexes.*
2. *The correspondence* $M \rightsquigarrow (\Omega_c(M), d, \theta, \iota)$, $f \rightsquigarrow f^*$ *is a contravariant functor from the category of G-manifolds and* ***proper*** *maps to the category of G-split* $\mathfrak{g}$*-complexes.*
3. *The correspondence that assigns to a G-manifold M, the spectral sequence* $\mathbb{E}_G(M)$ *associated with the Cartan complex* $(\Omega_G(\_), d_G)$ *(4.3.3.1), is contravariant and functorial on the category* G-Man.

   With a G-equivariant map $f : M \to N$*, the correspondence associates a morphism of spectral sequences*

$$\mathbb{E}_G(f) : \mathbb{E}_G(N) \to \mathbb{E}_G(M),$$

   which converges to the pullback $f^* : H_G(N) \to H_G(M)$ *and which reads at the* $\mathbb{E}_2$ *page level as the homomorphism* $\mathrm{id}_{S(\mathfrak{g}^\vee)} \otimes f^*$*:*

$$\begin{cases} \mathbb{E}_G(N)_2^{p,q} = S^p(\mathfrak{g}^\vee)^G \otimes H^q(N) \Rightarrow H_G^{p+q}(N) \\ \quad \mathbb{E}_G(f)\downarrow \qquad \mathrm{id}\downarrow \qquad f^*\downarrow \qquad f^*\downarrow \\ \mathbb{E}_G(M)_2^{p,q} = S^p(\mathfrak{g}^\vee)^G \otimes H^q(M) \Rightarrow H_G^{p+q}(M) \end{cases}$$

4. *The correspondence that assigns to a G-manifold M, the spectral sequence* $\mathbb{E}_{G,c}(M)$ *associated with the Cartan complex* $(\Omega_{G,c}(\_), d_G)$ *(4.3.3.1), is contravariant and functorial on the category* G-$\mathrm{Man}_{\mathrm{pr}}$.

   With a <u>*proper*</u> *G-equivariant map* $f : M \to N$*, the correspondence associates a morphism of spectral sequences*

$$\mathbb{E}_{G,c}(f) : \mathbb{E}_{G,c}(N) \to \mathbb{E}_{G,c}(M),$$

[22] On the connectedness hypothesis see 4.4.6.3-(1).

which converges to the pullback $f^* : H_{G,c}(N) \to H_{G,c}(M)$ *and which reads at the* $\mathbb{E}_2$ *page level as the homomorphism* $\mathrm{id}_{S(\mathfrak{g}^\vee)} \otimes f^*$*:*

$$\begin{cases} \mathbb{E}_{G,c}(N)_2^{p,q} = S^p(\mathfrak{g}^\vee)^G \otimes H_c^q(N) \Rightarrow H_{G,c}^{p+q}(N) \\ \mathbb{E}_{G,c}(f)\big\downarrow \qquad \mathrm{id}\big\downarrow \qquad f^*\big\downarrow \qquad f^*\big\downarrow \\ \mathbb{E}_{G,c}(M)_2^{p,q} = S^p(\mathfrak{g}^\vee)^G \otimes H_c^q(M) \Rightarrow H_{G,c}^{p+q}(M) \end{cases}$$

Proof The statements are simple consequences of Theorems 4.4.6.1 and 4.3.3.1 interchanging $\mathfrak{g}$ and G, and Propositions 4.3.4.3 and 4.3.4.3.

For the statements (3,4), the only point to check is that the induced morphisms $f^* : (\Omega_{G,?}(N)_{\mathfrak{g}}, d_G) \to (\Omega_{G,?}(M)_{\mathfrak{g}}, d_G)$ respect the filtration by $S(\mathfrak{g}^\vee)$-degrees, which is obvious. □

Comments 4.4.6.3

1. In Theorem 4.4.6.1 and its Corollary 4.4.6.2 connectedness hypothesis on G can be weaken to require simply the action on C of every $g \in G$ to be homotopic to the identity map. This occurs in particular when, G being connected, we are interested in the K-equivariant cohomology $H_K(M)$ of a G-manifold, where K is a closed subgroup of G, connected or not.
2. In view of Exercise 4.3.4.2, we can even completely delete the hypothesis of connectedness of G in Corollary 4.4.6.2, except that in the statements concerning the spectral sequences, will have to write:

$$\mathbb{E}_{G,?}(M)_2^{p,q} = \left(S^p(\mathfrak{g}^\vee)^{\mathfrak{g}} \otimes H_?^q(M)\right)^W \Rightarrow H_{G,?}^{p+q}(M)$$

   where $W := G/G_0$.

4.5 Cohomological Properties of Homotopy Quotients

4.5.1 Local Triviality of G-Spaces[23]

We recall the well-known concept of G-space in order to fix notations and terminology. Given a topological group G, a *left G-space* is a topological space X together with a homomorphism of groups $\rho : G \to \mathrm{Homeom}(X)$ such that the map $G \times X \to X$, $(g, x) \mapsto g \cdot x := \rho(g)(x)$ is continuous. The *G-orbit* of $x \in X$ is the set $G \cdot x := \{g \cdot x \mid x \in X\}$. Two orbits are then either equal or disjoint, so that the

[23]A reference for this section is Hsiang [55] §1.2 Generalities of Fibre Bundles and Free G-Spaces and §1.3 The Existence of Slice and its Consequences on General **G**-Spaces, p. 6–12.

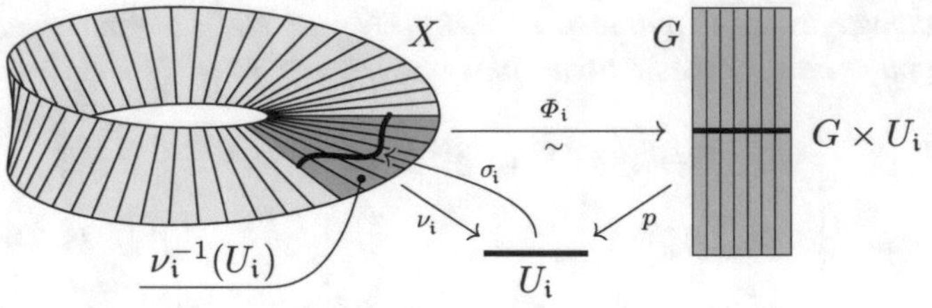

Fig. 4.3 A local slice

orbit decomposition $X = \bigcup_{x \in X} G \cdot x$ can be rewritten as a *partition* parametrized by the set X/G of G-orbits. Endow X/G with the quotient topology and denote by

$$\nu : X \to X/G\,, \tag{4.42}$$

the *projection map* which associates with $x \in X$, its G-orbit $\nu(x) := G \cdot x$.

4.5.2 Slices

When the group G acts freely on X we say that X *is a free G-space*. A classical question is then to find conditions under which (4.42) is a *left principal G-bundle*, which means, by definition, that there exists an open cover $\mathscr{U} := \{U_\mathfrak{i}\}_{\mathfrak{i} \in \mathfrak{I}}$ of X/G, and, for each $\mathfrak{i} \in \mathfrak{I}$, a trivializing homeomorphism

$$\begin{array}{ccccc} X & \xleftarrow{\ \supseteq\ } & \nu^{-1}(U_\mathfrak{i}) & \xrightarrow[\sim]{\ \Phi_\mathfrak{i}\ } & G \times U_\mathfrak{i} \\ \nu \big\downarrow & & \nu_\mathfrak{i} \big\downarrow & \oplus & \big\downarrow p \\ X/G & \xleftarrow{\ \supseteq\ } & U_\mathfrak{i} & = & U_\mathfrak{i} \end{array} \qquad \text{where} \quad p(g, z) = z.$$

which is G-equivariant if we endow $G \times U_\mathfrak{i}$ with the obvious left G-action defined by $g \cdot (h, z) := (gh, z)$, for all $g, h \in G$ and $z \in U_\mathfrak{i}$. A section of ν above $U_\mathfrak{i}$, for example $\sigma_\mathfrak{i}(z) := \Phi_\mathfrak{i}^{-1}(1_G, z)$, is called *a slice* (see Fig. 4.3).

To give a structure of *topological left principal G-bundle* is therefore equivalent to giving a family of slices $\{\sigma_\mathfrak{i} : U_\mathfrak{i} \to X\}_{\mathfrak{i} \in \mathfrak{I}}$ such that the G-equivariant maps

$$\Psi_\mathfrak{i} : G \times U_\mathfrak{i} \to X\,, \quad m_\mathfrak{i}(g, z) := g \cdot \sigma_\mathfrak{i}(z)\,,$$

are open embeddings, for all $\mathfrak{i} \in \mathfrak{I}$.

All these concepts can be given, *mutatis mutandis*, for right actions in which case we obtain the notion of *right principal G-bundle*. We can also work with G-manifolds simply by adding the usual differentiability conditions in the constructions.

4.5.3 Existence of Slices

This question has a positive answer when G is a compact Lie group acting on a manifold M.[24] In that case differentiable local slices always exist, and, beyond being mere locally closed subspaces, they are locally closed submanifolds, which allows us to endow the quotient topological space M/G of a canonical structure of manifold (see 4.5.3.1-(1)).

The idea of the construction as it appears in the paper by Koszul [69],[25] consists of endowing M with a G-invariant Riemannian metric $(.,.)_M$ (possible since G is a compact Lie group and the action is differentiable),[26] and then defining a slice at a point $x \in M$ as the image by the exponential $\exp_x : T_x(M) \to M$ of a small open ball U centered at the origin of the normal space $N_x(G \cdot x) \subseteq T_x(M)$. Compactness of $G \cdot x$ and the fact that G acts by isometries then allow to show that U can be chosen (small enough) so that the map

$$G \times U \to M\,, \quad (u, z) \mapsto g \cdot \exp(z)\,,$$

is a diffeomorphism, G-equivariant by construction.

The map $\exp_x : U \to M$ is then a *slice centered at x defined by the metric* $(.,.)_M$.

In this construction slices are completely determined by the Riemannian metric. The observation is particularly useful when considering an equivariant closed embedding of G-manifolds $N \hookrightarrow M$. In that case, a G-invariant Riemannian metric $(.,.)_N$ in N can always be extended (with the help of partitions of unity) to a Riemannian metric $(.,.)_M$ in M, which can then be chosen to be G-invariant by G-averaging, hence without modifying $(.,.)_N$. As a consequence, if N and M are free G-manifolds, the differentiable slices of $(N, (.,.)_N)$ can be extended as Riemannian submanifolds of differentiable slices of $(M, (.,.)_M)$. This shows, in particular, that the induced map $N/G \hookrightarrow M/G$ is a closed embedding of manifolds. We gather all these observations in a single proposition.

Proposition 4.5.3.1 (Existence of Differentiable Slices) *Let G be a compact Lie group. (G acts either on the left or on the right.)*

1. *For every free G-manifold M, the topological quotient M/G admits a canonical structure of manifold such that the projection map*

$$\nu : M \to M/G$$

is a principal G-bundle. Local sections of ν *are then slices.*

---

[24] Koszul [69] and Gleason [45] for M a *completely regular* topological space.

[25] Besides the work of Gleason and Koszul, a number of papers on the same subject were published at about the same time. Among the most cited are those of Montgomery and Yang [77], Mostow [78] and Palais [79], all in 1957.

[26] See Tu [91] §25.4 Equivariant Tubular Neighborhoods, Theorem 25.11, p. 210.

2. *Given a G-equivariant closed embedding $N \hookrightarrow M$ of free G-manifolds, the induced diagram*

$$\begin{array}{ccc} N & \hookrightarrow & M \\ {\scriptstyle \nu_N}\downarrow & \square & \downarrow{\scriptstyle \nu_M} \\ N/G & \hookrightarrow & M/G \end{array} \tag{4.43}$$

is a morphism of principal G-bundles and a Cartesian diagram where horizontal arrows represent closed embeddings of manifolds.

Furthermore, every principal G-bundle connection for ν_N extends to a principal G-bundle connection for ν_M.

Proof

(1) We must first check that the manifold structure is independent of the choice of the invariant Riemannian metric. Given two slices S_1 and S_2 meeting the same orbit $G \cdot x \subseteq M$ and small enough such that $G \cdot S_1 = C \cdot S_2$, we must show that the map $\Phi : S_1 \to S_2$ which associates with $z \in S_1$ the only $\Phi(z) \in S_2$ such that $\Phi(z) \in G \cdot z$, is a differentiable map, in which case the map $\mathrm{id} : \nu(S_1) \to \nu(S_2)$ is a diffeomorphism and the manifold structures thus defined on the set $\nu(S_i) \subseteq M/G$ coincide.

To see this, we return to the definition of slices which we can assume contained in the same tangent space $T_x(M)$. We have two G-invariant Riemannian metrics $(.,.)_i$ on M, the corresponding exponential maps $\exp_i : T_x(M) \to M$, and, by definition, the slices $S_i := \exp_i(U_i)$, where U_i is a small ball centered at the origin of the orthogonal complement N_i of the tangent space the orbit $T_x(G \cdot x)$ at x, for each of the two induced Euclidean metric on $T_x(M)$.

We thus have decompositions $T_x(M) = T_x(G \cdot x) \oplus N_1 = T_x(G \cdot x) \oplus N_2$, and the map $\mathrm{id} : \nu(S_1) \to \nu(S_2)$ reads, through the exponential maps, as the map $\theta : N_1 \to N_2$ such that, for $z \in N_1$, we have $\theta(z) - \lambda(z)\, z \in T_x(G \cdot x)$ for some $\lambda(z) \in \mathbb{R}_{>0}$, and such that $\exp_1(z)$ and $\exp_2(\theta(z))$ are in the same G-orbit (see Fig. 4.4). Elementary reasons then show that the function $\lambda : N_1 \to \mathbb{R}$ is differentiable, which implies that θ is also differentiable .

We have proved that given a slice $S := \exp_x(U)$ in a free G-manifold M, the map $\varphi_S : \nu(S) \to N \simeq \mathbb{R}^{d_M - d_G}$, defined by the inverse of $\nu \circ \exp_x : U \to \nu(S)$, is a chart $(\nu(S), \varphi_S)$ of M/G, such that, for every $x' \in G \cdot S$, the charts $(\nu(S'), \varphi_{S'})$ defined by slices S' centered at x', all define the same structure of manifold on the neighborhood of $\nu(y)$. In other terms, the collection of charts $\mathfrak{A} := \{(\nu(S), \varphi_S)\}$, indexed by the slices $S \subseteq M$, is a canonical atlas of differentiable manifold for the topological quotient space M/G.

(2) The preliminary discussion already explained how to extend slices from N to M. The canonicity of the associated manifold structures on N/G and M/N, established in (1), then immediately justifies the existence of the commutative diagram of closed embeddings (4.43).

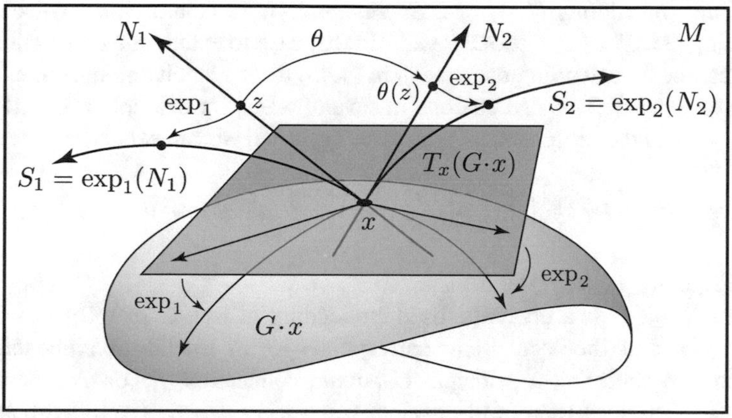

Fig. 4.4 Exponential map

To show that (4.43) is Cartesian, it suffices to note that, after (1) we can endow N with a Riemannian metric induced by M, in which case if S_x is a slice of M centered at $x \in N$, then $S'_x := S_x \cap N$ is automatically a slice of N. The restriction of (4.43) to the open subspace $\nu_M(S_x) \subseteq M/G$ is then easily seen to be isomorphic to the diagram of trivial principal G-bundles

$$\begin{array}{ccc} G \times S'_x & \hookrightarrow & G \times S_x \\ {\scriptstyle p_2}\downarrow & \square & \downarrow{\scriptstyle p_2} \\ S'_x & \hookrightarrow & S_x \end{array} \tag{4.44}$$

which is clearly Cartesian.

We can now address the question of the existence of extensions of principal G-bundles connections. A principal G-bundle connection for ν_N is a G-invariant 1-form $f_N \in (\Omega^1(N) \otimes \mathfrak{g})^G$ such that

$$\iota(Y)(f_N) = Y\,, \quad \forall Y \in \mathfrak{g}\,.$$

When N is the trivial principal G-bundle $G \times \mathbb{R}^d$, denote by $p_1 : N \to G$ the projection $p_1(g, x) := g$. We have the decomposition

$$\Omega^1(N) \otimes \mathfrak{g} := \big(\Omega^0(N) \cdot p_1^*(\Omega^1(G)) \otimes \mathfrak{g}\big) \oplus \big(\Omega^0(N) \otimes \Lambda^1(\mathbb{R}^d) \otimes \mathfrak{g}\big)$$

and since $\Omega^1(G) \cong \Omega^0(G) \otimes \mathfrak{g}^\vee$ (*cf.* 4.4.2–(4.24)), we can conclude that

$$f_N = \big(\textstyle\sum_i p_1^*(e^i) \otimes e_i\big) + \omega\,, \tag{4.45}$$

for some $\omega \in \Omega^0(N) \otimes \Lambda^1(\mathbb{R}^d) \otimes \mathfrak{g}$, where $\{e_i\}$ is a basis of $\mathfrak{g}$ of dual basis $\{e^i\}$.

Call an embedding $N \subseteq M$ *Euclidean* if it is isomorphic to a canonical embedding $G\times\mathbb{R}^d\times\{0\} \subseteq G\times\mathbb{R}^d\times\mathbb{R}^{d'}$. In that case to extend the connection f_N to M, we can begin extending the term ω in (4.45) to M, which amounts to extending $\Omega^0(N)$ to $\Omega^0(M)$. This can be done in several ways, for example if $q : M \twoheadrightarrow N$ $q : M \twoheadrightarrow N$ is the projection $q(g, x, y) = (g, x, 0)$, we can set

$$f_M := \Big(\sum_i p_1^*(e^i) \otimes e_i\Big) + \int_G g^*(q^*(\omega))\, dg\,,$$

and the equality $f_N = f_M|_N$ is clear.

When $N \subseteq M$ is a general closed embedding of free G-manifolds, we know, after (4.44), that the local view corresponds to an Euclidean embedding. We can therefore state that a principal G-bundle connection f_N on N admits local extensions. The idea then is to consider a covering $\mathscr{U} := \{U_\mathrm{i}\}_{\mathrm{i}\in\mathfrak{I}}$ of M/G by trivializing open subspaces for ν_M of two kinds, either $U_\mathrm{i} \cap N = \emptyset$, in which case we set $f_{M,\mathrm{i}}$ to be any connection on $\nu_M^{-1}(U_\mathrm{i})$, or $V_\mathrm{i} := U_\mathrm{i} \cap N \neq \emptyset$ and the embedding $V_\mathrm{i} \subseteq U_\mathrm{i}$ is Euclidean, in which case we set $f_{M,\mathrm{i}}$ to be any extension of f_N. Then, if $\{\phi_\mathrm{i}\}_{\mathrm{i}\in\mathfrak{I}}$ is a partition of unity subordinate to $\mathscr{U}$, the G-invariant differential form

$$f_M := \int_G g^*\Big(\sum_\mathrm{i} \phi_\mathrm{i}\, f_{M,\mathrm{i}}\Big)\, dg \;\in\; \big(\Omega^1(M) \otimes \mathfrak{g}\big)^G\,,$$

is a well-defined extension of f_N. □

Corollary 4.5.3.2 *Let G be a compact Lie group.*

1. *Given a sequence of G-equivariant closed embeddings of free G-manifolds* $M_0 \hookrightarrow M_1 \hookrightarrow \cdots \hookrightarrow M_n \hookrightarrow \cdots$, *the induced diagram*

$$\begin{array}{ccccccccc}
M_0 & \hookrightarrow & M_1 & \hookrightarrow \cdots \hookrightarrow & M_n & \hookrightarrow \cdots \hookrightarrow & \varinjlim_n M_n \\
\downarrow{\scriptstyle \nu_0} & \square & \downarrow{\scriptstyle \nu_1} & & \downarrow{\scriptstyle \nu_n} & & \downarrow{\scriptstyle \text{lim-ind}\, \nu_n} \\
M_0/G & \hookrightarrow & M_1/G & \hookrightarrow \cdots \hookrightarrow & M_n/G & \hookrightarrow \cdots \hookrightarrow & (\varinjlim_n M_n)/G
\end{array}$$

   is a sequence of closed embeddings of principal G-bundles.

   Furthermore, there exists a family $\{f_n\}_{n\in\mathbb{N}}$, *where* $f_n \in (\Omega^1(M_n) \otimes \mathfrak{g})^G$ *is a principal G-bundle connection, such that* $f_{n+1}|_{M_n} = f_n$, *for all* $n \in \mathbb{N}$.
2. *Let* $\mathbb{E}$ *be a right free G-manifold with quotient manifold* $\mathbb{B} := \mathbb{E}/G$ *and projection map* $\nu_{\mathbb{E}} : \mathbb{E} \to \mathbb{B}$. *For every left G-space X, the* diagonal action *of G on* $\mathbb{E} \times X$, $g\cdot(z, x) := (z \cdot g^{-1}, g \cdot x)$, *is a free action. Denote by* $\mathbb{E} \times_G X$ *the quotient space and by* $\nu_{(\mathbb{E}\times X)} : \mathbb{E} \times X \to \mathbb{E} \times_G X$ *be the corresponding projection map.*

Then, the diagram

$$\begin{array}{ccc} \mathbb{E} \times X & \xrightarrow{\nu_{(\mathbb{E}\times X)}} & \mathbb{E} \times_G X \\ {\scriptstyle p_1}\downarrow & \square & \downarrow{\scriptstyle \overline{p_1}} \\ \mathbb{E} & \xrightarrow{\nu_{\mathbb{E}}} & \mathbb{B} \end{array} \quad \text{with} \quad \begin{cases} p_1(z,x) & := z \\ \overline{p_1}([z,x]) & := [z]\,, \end{cases} \tag{4.46}$$

where we denote $[z, x] := \nu_{(\mathbb{E}\times X)}(z, x)$ *and* $[z] := \nu_{\mathbb{E}}(z)$*, is a Cartesian diagram of fiber bundles of vertical fiber* X *and horizontal fiber* G*.*

3. *The correspondence*

$$\mathbb{E} \times_G (\_) : G\text{-Top} \rightsquigarrow \text{Top}_{\mathbb{B}}$$

which associates with a left G*-space* X *the fiber bundle* $(\mathbb{E} \times_G X, \mathbb{B}, \overline{p_1}, X)$*, and with a* G*-equivariant map* $\alpha : N \to M$*, the map*

$$\mathbb{E} \times_G (\alpha) : \mathbb{E} \times_G N \to \mathbb{E} \times_G M\,, \quad [z, x] \mapsto [z, f(x)]\,,$$

is a covariant functor whose restriction to G-Man *takes its values in the subcategory* $\text{Fib}_{\mathbb{B}}$ (cf. *3.1.8).*

Proof

(1) Corollary of 4.5.3.1-(2). The only new point to check is that the natural map $\text{lim-ind}_n(M_n/G) \to (\text{lim-ind}_n M_n)/G$ is a homeomorphism (see ([28]) p. 146). Since the map is clearly continuous and bijective, we need only justify that it is closed. This means that if $Z \subseteq (\text{lim-ind}_n M_n)/G$ is such that its restrictions $Z_n := Z \cap (M_n/G)$ are all closed, then Z is of the form $Z = \text{lim-ind}_n \nu_n(\tilde{Z})$, where $\tilde{Z} \subseteq \text{lim-ind}_n M_n$ is closed and G-stable. The only possibility is then to take $\tilde{Z} = \text{lim-ind}\,\tilde{Z}_n$ with $\tilde{Z}_n := \nu_n^{-1}(Z_n)$, which clearly fulfills the requirements.

The existence of principal bundle connections is a well-known fact, so that a connection f_0 exists and we can apply 4.5.3.1-(2) to justify, by induction on $n \in \mathbb{N}$, the existence of the family of extensions $\{f_n\}_{n\in\mathbb{N}}$.

(2) The diagonal action of G on $\mathbb{E} \times X$ is free since it is so on $\mathbb{E}$, we can then apply 4.5.3.1-(1) to both projection maps $\nu_{\mathbb{E}}$ and $\overline{p_1}$. If $U \subseteq \mathbb{B}$ is a common trivializing open subspace, the diagram (4.46) becomes

$$\begin{array}{ccccc} (U \times G) \times X & \xrightarrow{\nu_{(\mathbb{E}\times X)}} & (U \times G) \times_G X & = & U \times X \\ {\scriptstyle p_1}\downarrow & & \downarrow{\scriptstyle \overline{p_1}} & & \downarrow{\scriptstyle p_1} \\ U \times G & \xrightarrow{\nu_{\mathbb{E}}} & U & = & U \end{array}$$

where the outer diagram is Cartesian with the announced fibers. The diagram (4.46) is thus locally Cartesian, hence Cartesian by 3.1.6.1–(2).

Statement (3) is clear. □

4.6 Constructing Classifying Spaces[27]

Let G be a compact Lie group. Among the different constructions of a *universal fiber bundle* $\mathbb{E}G$ *for* G the two most frequently cited are the following.

4.6.1 The Milnor Construction

Here $\mathbb{E}G$ is the inductive limit[28] of a system of closed embeddings $\{\mathbb{E}G(n) \hookrightarrow \mathbb{E}G(n+1)\}_{n\in\mathbb{N}}$, where $\mathbb{E}G(n)$ is the *join* $G * \cdots * G$ on $(n+1)$ copies of G, equipped with the (free) diagonal right action of G. As a topological space, $\mathbb{E}G(n)$ is compact, oriented and $(n-1)$-connected.[29,30] Moreover, as soon as we fix a triangulation of G, the spaces $\mathbb{E}G(n)$ inherit a canonical structure of CW-complex. The embeddings $\mathbb{E}G(n) \hookrightarrow \mathbb{E}G(n+1)$ are then cellular maps, and the inductive limit $\mathbb{E}G := \text{lim-ind}_n\, \mathbb{E}G(n)$ is a CW-complex, which is then contractible since it is n-connected for all $n \in \mathbb{N}$.[31] The *classifying space of* G is then the topological quotient $\mathbb{B}G := \mathbb{E}G/G$, limit of the inductive system defined by the closed embeddings $\mathbb{B}G(n) \hookrightarrow \mathbb{B}G(n+1)$, where $\mathbb{B}G(n) := \mathbb{E}G(n)/G$.[32]

Although the Milnor construction has undeniable advantages,[33] the fact that it uses CW-complexes which are not manifolds makes it unsuitable for the purposes

[27] See also Tu [91] Appendix §A.10 *Approximation of* $\mathbb{E}G$, *p. 271.*

[28] Given an inductive system of topological spaces $\{f_{j,i} : X_i \to X_j\}_{i\preceq j}$, the topology of $X_{\infty} :=$ lim-ind$_i$ X_i is the smallest topology such that the maps $f_i : X_i \to X_\infty$ are continuous, i.e. $Z \subseteq X_\infty$ is closed (resp. open) if and only if $f_i^{-1}(Z) \subseteq X_i$ is closed (resp. open), for all i.

[29] This is Lemma 2.3 in Milnor [75, p. 432].

[30] A space X is said to be *n-connected* if its first n homotopy groups are trivial, i.e. if we have $\pi_i(X) = 0$ for $i \leq n$. By Hurewicz Theorem, an n-connected space X is n-acyclic over any field $\Bbbk$, which means that the n first $\Bbbk$-Betti numbers of X are those of a singleton $\{\bullet\}$. See Hatcher [53] ch. 4.1 Homotopy Theory, Theorem 4.32, p. 366.

[31] A topological space which is n-connected for all $n \in \mathbb{N}$ is called *weakly contractible*. The celebrated Whitehead's Theorem establishes that a CW-complex which is weakly contractible is contractible. See Hatcher [53] ch. 4.1 Homotopy Theory, Theorem 4.5, p. 346.

[32] See Husemöller [56] ch. 4, §11–13, pp. 54–60, for a thorough discussion of Milnor's construction of the universal and of the classifying bundles. Other references are Husemöller et al. [57] ch. 7, pp. 75–81, as well as the original article of Milnor [75].

[33] It is functorial on the category of groups and greatly simplifies the proof of the classification theorem of principal G-bundles, see Hussemöller [56] Theorem 12.2, p. 57.

of the remainder of this chapter, which is to compare the equivariant de Rham cohomology of G-manifolds with the ordinary cohomology of their homotopy quotients. For this reason we prefer the following alternative classical construction of universal bundles for compact Lie groups.

4.6.2 Stiefel Manifolds

Based on the fact that a compact group G can be embedded in the group of *orthogonal matrices* $O(r)$ for r big enough,[34] we consider, for all $n \in \mathbb{N}$, the *Stiefel manifold* $\mathbb{E}G(n) := V_r(\mathbb{R}^{r+n})$, which consist of all the orthonormal r-tuples $(v_1, \ldots, v_r)$ of vectors in the Euclidean space $\mathbb{R}^{r+n}$. By identifying $\mathbb{R}^{r+n} := \mathbb{R}^{r+n} \times \{0\} \subseteq \mathbb{R}^{r+n+1}$, Stiefel manifolds become naturally nested, and we can write $\mathbb{E}G(n) \subseteq \mathbb{E}G(n+1)$. The sequence

$$\mathbb{E}G(0) \subseteq \mathbb{E}G(1) \subseteq \mathbb{E}G(2) \subseteq \cdots \subseteq \mathbb{E}G(\infty) := \mathbb{E}G\,, \tag{4.47}$$

is then an equivariant sequence of compact, connected, oriented G-manifolds.

The classical way to understand the Stiefel manifold $V_r(\mathbb{R}^{r+n})$ is as homogeneous space. Indeed, the group $O(r+n)$ acts on the left of $V_r(\mathbb{R}^{r+n})$ by its standard representation on $\mathbb{R}^{r+n}$. The action is easily seen to be transitive, and the stabilizer of the r-tuple $(e_1, \ldots, e_r)$, canonical basis of $\mathbb{R}^r \oplus \{0\} \subseteq \mathbb{R}^r \times \mathbb{R}^n$, is the subgroup $\{\mathbf{1}_r\} \times O(n) \subseteq O(r+n)$. We can thus write:

$$V_r(\mathbb{R}^{r+n}) \simeq O(r+n)\Big/_{\{\mathrm{id}_r\}\times O(n)}\,, \tag{4.48}$$

which shows that $\mathbb{E}G(n) := V_r(\mathbb{R}^{r+n})$ has a natural structure of left $O(r+n)$-space, and *right free* $O(r)$-space. Moreover, the inclusions $\mathbb{E}G(n) \subseteq \mathbb{E}G(n+1)$ are compatible with these structures in the obvious way.

The description (4.48) is usually used to prove by induction on $n \in \mathbb{N}$ that the space $\mathbb{E}G(n)$ is $(n-1)$-connected.[35] Furthermore, the manifolds $\mathbb{E}G(n)$ can be equipped with canonical structures of CW-complexes making the embeddings $\mathbb{E}G(n) \subseteq \mathbb{E}G(n+1)$ cellular maps.[36] All the remarks on Milnor construction

---

[34] A corollary of the Completeness Theorem of Peter-Weyl states that every compact topological group has a faithful finite dimensional representation. See Hsiang [55] ch. 1, §1-(D), p. 5.

[35] See Husemöller [56] ch. 8 §1–7, Theorem 6.1, p. 95.

[36] The quotient $G_r(\mathbb{R}^{r+n}) := V_r(\mathbb{R}^{r+n})/O(r)$ is the well-known *Grassmann manifold*, whose points parametrize the vector subspaces of dimension r in $\mathbb{R}^{r+n}$. In Hatcher [54], §1.2 Cell Structures on Grassmannians, p. 31, there is a thorough description of their *Schubert decomposition* which endows these spaces of a canonical structure of CW-complex. The inclusions $G_r(\mathbb{R}^{r+n}) \subseteq G_r(\mathbb{R}^{r+n+1})$ are then cellular maps, and $G_r(\mathbb{R}^\infty) := \text{lim-ind}_n\, G_r(\mathbb{R}^{r+n})$, which is a realization of the classifying space $\mathbb{B}O(r)$, has a natural structure of CW-complex. By fixing a triangulation of the Lie group $O(r)$, this structure lifts to $\mathbb{E}O(r)$.

then apply and the inductive limit $\mathbb{E}G := \text{lim-ind}_n\, \mathbb{E}G(n)$ is a contractible CW-complex endowed with a right free action of $O(r)$, hence of G.

The *classifying space of* G is then the topological quotient $\mathbb{B}G := \mathbb{E}G/G$, and is also the inductive limit of the family of closed embeddings (4.5.3.2-(1))

$$\mathbb{B}G(0) \longrightarrow \cdots \longrightarrow \mathbb{B}G(n) \xrightarrow{\beta_n} \mathbb{B}G(n+1) \longrightarrow \cdots \longrightarrow \mathbb{B}G(\infty) := \mathbb{B}G \tag{4.49}$$

where the $\mathbb{B}G(n)$'s are now nice compact manifolds, oriented and furthermore simply connected if G is connected.

4.6.3 Convention

From now on, the notations $\mathbb{E}G(.)$ and $\mathbb{B}G(.)$ will refer to the realization of these spaces based on Stiefel manifolds unless otherwise stated.

4.7 The Borel Construction

4.7.1 The Homotopy Quotient Functor

Let G be a compact Lie group. As already explained in Sect. 4.1.2, given a G-manifold M, instead of considering the topological quotient M/G, which generally lacks of good properties when the action of G is not free, one replaces M by the G-space $\mathbb{E}G\times M$ on which G acts by the *diagonal action*, i.e. $g\cdot(z,x) := (z\cdot g^{-1}, g\cdot x)$. One thus replaces the G-space M by the homotopy equivalent G-space $\mathbb{E}G \times M$, on which the action of G is free. The topological quotient space, called the *homotopy quotient of* M, and sometimes *the Borel Construction of* M, is denoted by (*cf*. Sect. 4.1.2.1):

$$M_G := \mathbb{E}G \times_G M := (\mathbb{E}G \times M)/G\,.$$

The natural map:

$$\begin{array}{ccc} M_G := \mathbb{E}G \times_G M & \overset{\pi_M}{\twoheadrightarrow} & \mathbb{E}G/G =: \mathbb{B}G \\ [z,x] & \longmapsto & [z] \end{array}$$

defines a fiber bundle $(M_G, \mathbb{B}G, \pi_M, M)$, after 4.5.3.2-(2).

Furthermore, if $f : M \to N$ is a G-equivariant map, then the induced map

$$f_G : M_G \to N_G\,, \quad [z,m] \mapsto [z, f(m)]\,,$$

is a well-defined morphism in the category $\mathrm{Fib}_{\mathbb{B}G}$, after 4.5.3.2-(3).

The correspondence thus defined, $M \rightsquigarrow M_G$, $f \rightsquigarrow f_G$, from the category of G-manifolds to the category $\mathrm{Fib}_{\mathbb{B}G}$, is clearly covariant and functorial.

Definition 4.7.1.1 The functor

$$(\_)_G : G\text{-Man} \rightsquigarrow \mathrm{Fib}_{\mathbb{B}G}\,, \quad M \rightsquigarrow M_G,\ f \rightsquigarrow f_G\,,$$

is the *homotopy quotient functor*, or *Borel construction functor*.

Exercise 4.7.1.2 Recall that the connected component G_0 of the identity element $e \in G$ is a normal subgroup of G and that the quotient space $W := G/G_0$ is a discrete group.

1. Show that for every G-manifold M, the group W acts naturally on M_{G_0} and the canonical surjection $\epsilon(M) : M_{G_0} \twoheadrightarrow M_G$ then induces a homeomorphism $\bar{\epsilon}(M) : M_{G_0}/W \simeq M_G$ which is functorial for $M \in G$-Man.
2. Show that if W is finite (e.g. G is compact) and if $\Bbbk$ is a field of characteristic prime to $|W|$, then there exists a canonical isomorphism of functors

$$H((\_)_G; \Bbbk) \simeq H((\_)_{G_0}; \Bbbk)^W\,.$$

4.7.2 On the Cohomology of the Homotopy Quotient

By Corollary 4.5.3.2, the space M_G is the inductive of the manifolds $M_G(n) := \mathbb{E}G(n) \times_G M$, for which we denote by $\nu_{(n,M)} : \mathbb{E}G(n) \times M \twoheadrightarrow M_G(n)$ the corresponding projection map. We have therefore, for each $n \in \mathbb{N}$, a Cartesian diagram (after Exercise 3.1.4.3) of fiber bundles of manifolds

$$\begin{array}{ccc} \mathbb{E}G(n) \times M & \xrightarrow[\;[G]\;]{\nu_{(n,M)}} & M_G(n) \\ {\scriptstyle [M]}\big\downarrow{\scriptstyle \mathrm{id}\times c_M} & \square & {\scriptstyle [M]}\big\downarrow{\scriptstyle \pi_n} \\ \mathbb{E}G(n) \times \{\bullet\} & \xrightarrow[\;[G]\;]{\nu_{(n,\{\bullet\})}} & \mathbb{B}G(n) \end{array} \tag{4.50}$$

where fibers are shown in brackets.

These constructions, especially diagram (4.50), are functorial in both entries, $\mathbb{E}G(n)$ in the category of right free G-spaces, and in M in the category of left G-manifolds. In particular, when applied to the G-equivariant sequence

$$\mathbb{E}G(0) \subseteq \mathbb{E}G(1) \subseteq \mathbb{E}G(2) \subseteq \cdots \subseteq \mathbb{E}G(n) \subseteq \cdots \subseteq \mathbb{E}G(\infty) := \mathbb{E}G\,,$$

we obtain the inductive system of fiber bundles with fiber M

$$\begin{array}{ccccccccc} \cdots \longrightarrow & M_G(n) & \xrightarrow{\mu_n} & M_G(n{+}1) & \longrightarrow \cdots\cdots \longrightarrow & M_G(\infty) & = & M_G \\ & {\scriptstyle [M]}\big\downarrow{\scriptstyle \pi_n} & \square & {\scriptstyle [M]}\big\downarrow{\scriptstyle \pi_{n+1}} & & {\scriptstyle [M]}\big\downarrow{\scriptstyle \pi_\infty} & & {\scriptstyle [M]}\big\downarrow{\scriptstyle \pi} \\ \cdots \longrightarrow & I\!\!B G(n) & \xrightarrow{\beta_n} & I\!\!B G(n{+}1) & \longrightarrow \cdots\cdots \longrightarrow & I\!\!B G(\infty) & = & I\!\!B G \end{array} \tag{4.51}$$

where the horizontal arrows are closed embeddings.

We are thus lead to consider the projective systems of complexes of Alexander-Spanier cochains with coefficients in a field $\Bbbk$

$$\begin{cases} \text{(i) } \Omega(I\!\!B G;\Bbbk) \to \varprojlim_{n\in\mathbb{N}} \left\{\Omega^*(I\!\!B G(n{+}1);\Bbbk) \xrightarrow{\beta_n^*} \Omega^*(I\!\!B G(n);\Bbbk)\right\} \\ \text{(ii) } \Omega(M_G;\Bbbk) \to \varprojlim_{n\in\mathbb{N}} \left\{\Omega^*(M_G(n{+}1);\Bbbk) \xrightarrow{\mu_n^*} \Omega^*(M_G(n);\Bbbk)\right\} \\ \text{(iii) } \Omega_{\mathrm{cv}}(M_G;\Bbbk) \to \varprojlim_{n\in\mathbb{N}} \left\{\Omega^*_{\mathrm{cv}}(M_G(n{+}1);\Bbbk) \xrightarrow{\mu_n^*} \Omega^*_{\mathrm{cv}}(M_G(n);\Bbbk)\right\} \end{cases} \tag{4.52}$$

where the horizontal arrows are the restriction morphisms (see Exercise 4.7.2.1).

(Notice that the first line is a particular case of the second since $I\!\!B G = \{\bullet\}_G$. Also notice that in the third line we can replace $\Omega^*_{\mathrm{cv}}(M_G(.);\Bbbk)$ by $\Omega^*_{\mathrm{c}}(M_G(.);\Bbbk)$, since $I\!\!B G(.)$ is compact.)

Exercise 4.7.2.1 Show that in (4.52), the morphisms (i,ii) are always isomorphisms, while (iii) is injective and generally not surjective. (🕮, p. 349)

Theorem 4.7.2.2 *Let G be a compact, connected, Lie group, and let M be a G-manifold. (Cohomology is Alexander-Spanier's with coefficients in a field $\Bbbk$.)*

In the following, $n \in \mathbb{N}$ and $m \in \mathbb{N} \cup \{+\infty\}$, are such that $n \leq m$.

We denote by $\mu_{m,n} : M_G(n) \to M_G(m)$ and $\beta_{m,n} : I\!\!B G(n) \to I\!\!B G(m)$ the maps defined by the projective system (4.51).

1. *Let $I\!\!E(\pi_{m*}) \Rightarrow H(M_G(m))$ denote the Leray spectral sequence associated with $\pi_m : M_G(m) \to I\!\!B G(m)$. The Cartesian diagram*

$$\begin{array}{ccc} M_G(n) & \xrightarrow{\mu_{m,n}} & M_G(m) \\ {\scriptstyle [M]}\big\downarrow{\scriptstyle \pi_n} & \square & {\scriptstyle [M]}\big\downarrow{\scriptstyle \pi_m} \\ I\!\!B G(n) & \xrightarrow{\beta_{m,n}} & I\!\!B G(m) \end{array} \tag{4.53}$$

induces a morphism of associated Leray spectral sequences which reads at the $I\!\!E_2$ pages level as the homomorphism $\beta^*_{m,n} \otimes \mathrm{id}^*_M$.

$$\begin{array}{ccccc} \mathbb{E}(\pi_{m*})_2^{p,q} & = H^p(\mathbb{B}G(m)) \otimes H^q(M) & \Longrightarrow & H^{p+q}(M_G(m)) \\ & \beta^*_{m,n}\downarrow \qquad \mathrm{id}^*_M \downarrow & & \downarrow \mu^*_{m,n} \\ \mathbb{E}(\pi_{n*})_2^{p,q} & = H^p(\mathbb{B}G(n)) \otimes H^q(M) & \Rightarrow & H^{p+q}(M_G(n)) \end{array}$$

2. *Let $\mathbb{E}(\pi_{m!}) \Rightarrow H_{\mathrm{cv}}(M_G(n))$ denote the Leray spectral sequence associated with $\pi_m : M_G(m) \to \mathbb{B}G(m)$. The Cartesian diagram (4.53) induces a morphism of associated Leray spectral sequences which reads at the $\mathbb{E}_2$ pages level as the homomorphism $\beta^*_{m,n} \otimes \mathrm{id}^*_M$.*

$$\begin{array}{ccccc} \mathbb{E}(\pi_{m!})_2^{p,q} & = H^p(\mathbb{B}G(m) \otimes H^q_{\mathrm{c}}(M) & \Longrightarrow & H^{p+q}_{\mathrm{cv}}(M_G(m)) \\ & \beta^*_{m,n}\downarrow \qquad \mathrm{id}^*_M \downarrow & & \downarrow \mu^*_{m,n} \\ \mathbb{E}(\pi_{n!})_2^{p,q} & = H^p(\mathbb{B}G(n)) \otimes H^q_{\mathrm{c}}(M) & \Rightarrow & H^{p+q}_{\mathrm{cv}}(M_G(n)) \end{array}$$

3. *Given $i \in \mathbb{N}$, the projective systems (4.52) induce isomorphisms*

$$\left\{ \begin{array}{l} H^i(M_G(m)) \xrightarrow[\simeq]{\mu^*_{m,n}} H^i(M_G(n)) \,, \\ H^i_{\mathrm{cv}}(M_G(m)) \xrightarrow[\simeq]{\mu^*_{m,n}} H^i_{\mathrm{cv}}(M_G(n)) \,, \end{array} \right. \qquad \forall m \geq n > i \,. \tag{4.54}$$

4. *The morphisms (4.52) are quasi-isomorphism and induce isomorphisms*

$$\left\{ \begin{array}{ll} \text{(i)} & H(\mathbb{B}G) \underset{\simeq}{\to} \mathbf{h}\big(\varprojlim_n \Omega(\mathbb{B}G(n))\big) \underset{\simeq}{\to} \varprojlim_n H(\mathbb{B}G(n)) \,, \\ \text{(ii)} & H(M_G) \underset{\simeq}{\to} \mathbf{h}\big(\varprojlim_n \Omega(M_G(n))\big) \underset{\simeq}{\to} \varprojlim_n H(M_G(n)) \,, \\ \text{(iii)} & H_{\mathrm{cv}}(M_G) \underset{\simeq}{\to} \mathbf{h}\big(\varprojlim_n \Omega_{\mathrm{cv}}(M_G(n))\big) \underset{\simeq}{\to} \varprojlim_n H_{\mathrm{cv}}(M_G(n)) \,. \end{array} \right. \tag{4.55}$$

5. *Statements (3,4) are true when G is not connected and the characteristic of the underlying field $\Bbbk$ is prime to the cardinality of G/G_0* (cf. *Exercise 4.7.1.2-(2)).*

Proof (1, 2) Both statements are related to the functoriality of Leray spectral sequences, which we first justify.

Lemma A *Let (E', B', π, M) and (E, B, π, M) be fiber bundles and let*

$$\begin{array}{ccc} E' & \xrightarrow{\mu} & E \\ \pi'\downarrow & \square & \downarrow\pi \\ B' & \xrightarrow{\beta} & B \end{array} \tag{4.56}$$

be a Cartesian diagram, with $\mu : E' \to E$ and $\beta : B' \to B$ closed embeddings.

A.1. *For any sheaf* $\mathcal{F} \in \mathrm{Sh}(B'; \Bbbk)$*, Godement's resolution* $\mathcal{F} \to C_\star(\mathcal{F})$ *commutes with the direct image functor* $\beta_* : \mathrm{Sh}(B'; \Bbbk) \rightsquigarrow \mathrm{Sh}(B; \Bbbk)$*, i.e. we have a natural identification*

$$\big(\beta_*\mathcal{F} \to C_\star(\beta_*\mathcal{F})\big) = \beta_*\big(\mathcal{F} \to C_\star(\mathcal{F})\big).$$

A.2. *The natural restriction morphism of complexes of sheaves of Alexander-Spanier cochains* $\underline{\underline{\Omega}}^*_E \to \beta_*(\underline{\underline{\Omega}}^*_{E'})$ *induces morphisms of bicomplexes*

$$\begin{array}{ccc} \pi_*\underline{\underline{\Omega}}^*_E & \longrightarrow & \beta_*\big(\pi'_*\underline{\underline{\Omega}}^*_{E'}\big) \\ \downarrow & & \downarrow \\ C_\star(\pi'_*\underline{\underline{\Omega}}^*_E) & \longrightarrow & \beta_*\big(C_\star(\pi_*\underline{\underline{\Omega}}^*_{E'})\big) \end{array} \qquad \begin{array}{ccc} \pi_!\underline{\underline{\Omega}}^*_E & \longrightarrow & \beta_*\big(\pi'_!\underline{\underline{\Omega}}^*_{E'}\big) \\ \downarrow & & \downarrow \\ C_\star(\pi_!\underline{\underline{\Omega}}^*_E) & \longrightarrow & \beta_*\big(C_\star(\pi'_!\underline{\underline{\Omega}}^*_{E'})\big) \end{array} \tag{4.57}$$

where the vertical arrows are flasque resolutions is $\mathrm{Sh}(B; \Bbbk)$.

Proof of Lemma

(*A*.1) The main point is that, since β is a closed embedding, the direct image β_* preserves stalks (hence is exact). The identification $C_0\beta_*\mathcal{F} = \beta_*C_0\mathcal{F}$ is then almost tautological after the definition of Godement's flasque resolution.[37]

The exactness of the functors $\beta_*(\_)$ and $C_0(\_)$, then identifies the cokernels

$$\begin{array}{ccccccccc} 0 & \longrightarrow & \beta_*\mathcal{F} & \xrightarrow{\epsilon} & \beta_*C_0\mathcal{F} & \longrightarrow & \beta_*\,\mathrm{coker}(\epsilon) & \longrightarrow & 0 \\ & & \| & & \| & & \vdots & & \\ 0 & \longrightarrow & \beta_*\mathcal{F} & \xrightarrow{\beta_*\epsilon} & C_0\beta_*\mathcal{F} & \longrightarrow & \mathrm{coker}(\beta_*\epsilon) & \longrightarrow & 0 \end{array}$$

on which the Godement Resolution procedure can be iterated.

(*A*.2) By definition of Alexander-Spanier cochains,[38] the set $\Gamma(U; \pi_*\underline{\underline{\Omega}}^d_E)$ consists of the set-theoretic maps $f : \pi^{-1}(U)^{d+1} \to \Bbbk$, and $\Gamma(U; \beta_*\pi_*\underline{\underline{\Omega}}^d_{E'})$ of $f' : \pi^{-1}(\beta^{-1}(U))^{d+1} \to \Bbbk$. A map $f \in \Gamma(U; \pi_*\underline{\underline{\Omega}}^d_E)$ composed with the embedding $\beta^d : (U \cap B')^d \hookrightarrow U^d$ then defines a map $f' \in \Gamma(U; \beta_*\pi'_*\underline{\underline{\Omega}}^d_{E'})$. This correspondence $f \mapsto f'$ is easily seen to be a morphism of complexes, it is the announced morphism $\pi_*\underline{\underline{\Omega}}^d_E \to \beta_*\pi_*\underline{\underline{\Omega}}^d_{E'}$.

The morphism for proper supports $\pi_!\underline{\underline{\Omega}}^d_E \to \beta_*\pi'_!\underline{\underline{\Omega}}^d_{E'}$ is defined as the restriction of $\pi_*\underline{\underline{\Omega}}^d_E \to \beta_*\pi_*\underline{\underline{\Omega}}^d_{E'}$. The fact that is is well-defined results from the properness of μ (we do not even need the diagram (4.56) to be Cartesian), the details are the same as in Proposition 3.1.10.2-(2).

---

[37] See Godement [46] §4.3 Resolution canonique d'un faisceau, p. 167, or [20] §II.2 The canonical resolution and sheaf cohomology, p. 36.

[38] See Godement [46] Ex. 2.5.2, p. 134, or Bredon [20] Alexander-Spanier cohomology p. 24.

The diagrams (4.57) then follow applying (A.1) and recalling that the direct image functor β_* is exact (since β is a closed embedding) which transforms flasque sheaves in flasque sheaves (as do any direct image functor). ⊟

We can now prove the statement (1) of the Proposition.

Identify the notations of corresponding terms in the two Cartesian diagrams:

$$\begin{array}{ccc} M_G(n) & \xrightarrow{\mu_{m,n}} & M_G(m) \\ {\scriptstyle [M]}\downarrow{\scriptstyle \pi_n} & \square & {\scriptstyle [M]}\downarrow{\scriptstyle \pi_m} \\ \mathbb{B}G(n) & \xrightarrow{\beta_{m,n}} & \mathbb{B}G(m) \end{array} \quad = \quad \begin{array}{ccc} E' & \xrightarrow{\mu} & E \\ {\scriptstyle \pi'}\downarrow & \square & \downarrow{\scriptstyle \pi} \\ B' & \xrightarrow{\beta} & B \end{array}$$

By filtering the morphism of bicomplexes

$$\Gamma\big(B; C_\star(\pi_*\underline{\underline{\Omega}}{}^*_E)\big) \to \Gamma\big(B; \beta_*(C_\star(\pi'_*\underline{\underline{\Omega}}{}^*_{E'}))\big) = \Gamma\big(B'; C_\star(\pi'_*\underline{\underline{\Omega}}{}^*_{E'})\big)$$

by the (regular) decreasing $\star$-filtration, we get a morphism of spectral sequences which reads in the $\mathbb{E}_1$ page as

$$\Gamma\big(B; C^p(\mathcal{H}^q(\pi_*))\big) \to \Gamma\big(B'; C^p(\mathcal{H}^q(\pi'_*))\big)\,, \tag{4.58}$$

where $\mathcal{H}^q(\pi_*)$ denotes the q'th cohomology sheaf of the complex $\pi_*\underline{\underline{\Omega}}_E$, and *mutatis mutandis* for $\mathcal{H}^q(\pi'_*)$. These sheaves are a priori locally constant with fiber $H(M; \mathbb{k})$ (3.1.10.4-(2)), but they are also *globally* constant since the base spaces $\mathbb{B}G(.)$, being the quotients of the simply connected spaces $\mathbb{E}G(.)$ by the free action of a *connected* group G, are themselves simply connected. As a consequence, the morphism of the $\mathbb{E}_2$ pages of the associated spectral sequences induced by (4.58) reads like the usual restriction

$$H^p(B; H^q(M; \mathbb{k})) \to H^p(B'; H^q(M; \mathbb{k}))\,. \tag{4.59}$$

To finish, we recall that if V is any $\mathbb{k}$-vector space, the exactness of the functor $\mathrm{Hom}_{\mathbb{k}}(\_; V)$ induces an isomorphism

$${}_SH^p(B; V) = \mathrm{Hom}_{\mathbb{k}}({}_SH_p(B; \mathbb{k}), V)\,,$$

where ${}_SH$ refers to *singular* homology or cohomology.

In the forthcoming proof of statement (3), we will show that the $\mathbb{k}$ Betti numbers of $B := \mathbb{B}G$ are finite, so that we will then be able to write

$$\begin{aligned} {}_SH^p(B; H^q(M; \mathbb{k})) &= \mathrm{Hom}_{\mathbb{k}}({}_SH_p(B; \mathbb{k}), H^q(M; \mathbb{k})) \\ &= \mathrm{Hom}_{\mathbb{k}}({}_SH_p(B; \mathbb{k})) \otimes H^q(M; \mathbb{k}) \\ &= {}_SH^p(B; \mathbb{k}) \otimes H^q(M; \mathbb{k}) \end{aligned}$$

and the statement (1) will thus follow by the well-known equivalence between Singular and Alexander-Spanier cohomologies on CW-complexes.[39]

The proof of (2) is the same, replacing π_* by $\pi_!$ and $H(M;\Bbbk)$ by $H_c(M;\Bbbk)$.

(3) We start with $M := \{\bullet\}$, in which case $M_G(m) = \mathbb{B}G(m)$.

We apply Corollary 4.5.3.2-(2) to $\mathbb{E}G \times \mathbb{E}G(m)$, where we must be careful since $\mathbb{E}G(m)$ is, both, a *left* G-space and a *right free* G-space (*cf*. 4.6.2). The action for the identification $\mathbb{B}G(m) = \mathbb{E}G(m)/G$ is the *right free action*.

We thus endow (temporally) $\mathbb{E}G(m)$ with a left action of G by setting $g * z := z \cdot g^{-1}$, and we define $\mathbb{E}G \times_G \mathbb{E}G(m)$ as in 4.5.3.2-(2), as the space of equivalence classes for the relation $(z, x) \sim (z \cdot g^{-1}, g * x)$. Since both $\mathbb{E}G$ and $\mathbb{E}G(m)$ are free G-spaces, we can apply the Corollary in two symmetric ways obtaining two locally trivial fibrations

$$\begin{array}{ccccc} & & \mathbb{E}G \times_G \mathbb{E}G(m) & & \\ & {\scriptstyle [\mathbb{E}G(m)]}\swarrow {\scriptstyle \overline{p}_1} & & {\scriptstyle \overline{p}_2}\searrow {\scriptstyle [\mathbb{E}G]} & \\ \mathbb{B}G & & & & \mathbb{B}G(m) \end{array}$$

where the fibers are shown in brackets.

Notice that the contractibility of fibers of $\overline{p}_2$ immediately implies that the pullback $\overline{p}_2^* : H(\mathbb{B}G(m)) \to H(\mathbb{E}G \times_G \mathbb{E}G(m))$ is an isomorphism. The study of the restriction $H(\mathbb{B}G) \to H(\mathbb{B}G(m))$ is then equivalent to the study of the pullback morphism

$$\overline{p}_1^* : H^i(\mathbb{B}G) \to H^i(\mathbb{E}G \times_G \mathbb{E}G(m)), \tag{4.60}$$

associated with the morphism of fiber bundles above $\mathbb{B}G$

$$\begin{array}{ccc} \mathbb{E}G \times_G \mathbb{E}G(m) & \xrightarrow{\ \overline{p}_1\ } & \mathbb{B}G \\ {\scriptstyle [\mathbb{E}G(m)]}\searrow {\scriptstyle \overline{p}_1} & & {\scriptstyle \mathrm{id}}\swarrow {\scriptstyle [\{\bullet\}]} \\ & \mathbb{B}G & \end{array}$$

for which we know after Lemma A, that $\overline{p}_1^*$ is the abutment of a morphism of Leray spectral sequences which reads at the $\mathbb{E}_2$ pages level as the following homomorphism $\xi(m)$ induced by the pullback $c^* : H(\{\bullet\}) \to H(\mathbb{E}G(m))$ associated with the constant map $c : \mathbb{E}G(m) \to \{\bullet\}$:

$$\begin{array}{ccc} H^p(\mathbb{B}G; H^q(\{\bullet\})) & \Longrightarrow & H^{p+q}(\mathbb{B}G) \\ {\scriptstyle \xi(m)_{p,q}}\big\downarrow & & \big\downarrow{\scriptstyle \overline{p}_1^*} \\ H^p(\mathbb{B}G; H^q(\mathbb{E}G(m))) & \Longrightarrow & H^{p+q}(\mathbb{E}G \times_G \mathbb{E}G(m)) \end{array}$$

[39]On CW-complexes, Čech, Sheaf, Singular and Alexander-Spanier cohomologies coincide. See Bredon [20], ch. III, Comparison with other cohomology theories, Theorem 2.1, p. 187.

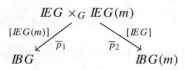

When $i = p + q$, we conclude that $\xi(m)_{p,q}$ is an isomorphism for all $m > i$, since $\mathbb{E}G(m)$ is $(m-1)$-connected. The equivalent to statement (3) for the family (4.60), hence for $M = \{\bullet\}$, is thus proved, and, at the same time, this ends the proof of (1) and (2).

The statement (3) for general M is now immediate by the spectral sequences in (1) and (2) since we now know that the $\beta^*_{m,n} : H^i(\mathbb{B}G(m)) \to H^i(\mathbb{B}G(n))$ become isomorphisms as $n \mapsto \infty$.

(4) (i) is a particular case of (ii). For (ii) consider the natural diagram of morphisms of graded vector spaces

$$\begin{array}{ccc} H(M_G) & \xrightarrow{(1)} & \boldsymbol{h}\big(\varprojlim_n \Omega(M_G(n))\big) \\ & {\scriptstyle (3)}\searrow \quad \swarrow{\scriptstyle (2)} & \\ & \varprojlim_n H(M_G(n)) & \end{array}$$

Thanks to (3), we can immediately claim that (iii) is an isomorphism, and also that, for each $i \in \mathbb{N}$, the projective system $\{H^i(M_G(n+1)) \to H^i(M_G(n))\}_{n\in\mathbb{N}}$ verifies Mittag-Leffler condition.[40] Besides, this same condition is also verified at the cochain level, since, the the maps $\mu_{m,n} : M(n) \to M(m)$ being closed embeddings, the restrictions $\Omega(M_G(m)) \to \Omega(M_G(n))$ are all surjective maps. When these Mittag-Leffler conditions are satisfied, it is well-known that *the cohomology of a projective limit is the projective limit of the cohomologies*,[41] which immediately implies that (2) is an isomorphism. The fact that (1) is an isomorphism too then results from the equality $(3) = (2) \circ (1)$.

The isomorphisms in (iii) are proved by exactly that same arguments.

(5) Statements (3,4) are true for the connected component G_0 of the identity element $e \in G$. But then, following Exercise 4.7.1.2 (see also proof of 4.7.3.1), the group $W := G/G_0$ acts naturally on $\mathbb{E}G(m)\times_{G_0}(N)$, for every $m \in \mathbb{N} \cup \{+\infty\}$ and every G-manifold N. The complexes $\Omega(M_{G_0}(m))$, $\Omega_{\mathrm{cv}}(M_{G_0}(m))$ are W-dg-modules, and all the morphisms are compatible with the action of W. On the other hand, as $(|W|, \mathrm{char}(\Bbbk)) = 1$, the functor of *$W$-invariants* $(\_)^W$ is exact. This implies that it commutes with cohomology, giving a canonical isomorphism of functors: $H((\_)_G; \Bbbk) \simeq H((\_)_{G_0}; \Bbbk)^W$ (*cf*. 4.7.1.2-(2)), but also that it preserves Mittag-Leffler conditions at complexes and cohomology levels, and that it commutes with projective limits. The statements (3,4) for G then follow. □

[40] A projective system of vector spaces $\{\rho_{n,m} : V_m \to V_n\}_{m\geq n\in\mathbb{N}}$ is said to satisfy Mittag-Leffler condition if for any $n \in \mathbb{N}$, the decreasing sequence $\{\rho_{n,m}(V_m)\}_{m\geq n}$ of subspaces of V_n is stationary. The condition is trivially verified when the vector spaces V_n are all finite dimensional, and also when $V_m \to V_n$ is an isomorphism for $n \gg 0$. See Kashiwara-Schapira [61] ch. I §1.12 The Mittag-Leffler condition, p. 64, or Weibel [95] §3.5 Derived Functors of the Inverse Limit, Definition 3.5.6, p. 82.

[41] See Kashiwara-Schapira [61] Proposition 1.12.4, p. 67, or Weibel [95] Theorem 3.5.8, p. 83.

4.7.3 Orientability of the Homotopy Quotient

Proposition 4.7.3.1 *Let G be a compact Lie group and let M be a G-manifold. If M is orientable and the action of G preserves the orientation, then the fiber bundle $M_G(n) := (M_G(n), \mathbb{B}G(n), \pi_M, M)$ is orientable. This arrives, for example, when G is connected.*[42]

Proof It suffices to consider the case where M is equidimensional.

Denote by G_0 the connected component of G containing 1_G. The connectedness of G_0 implies that $\mathbb{B}G_0(n)$ is simply connected, in which case the fiber bundle $M_{G_0}(n) := (M_{G_0}(n), \mathbb{B}G_0(n), \pi_M, M)$ is automatically orientable after Corollary 3.1.7.2-(3c). The quotient $W := G/G_0$ is a finite group which acts naturally on $M_{G_0}(n) := \mathbb{E}G(n) \times_{G_0} M$ by $w \cdot [z, x] = [z \cdot w^{-1}, w \cdot x]$. We then have the following diagram of natural morphisms of fiber bundles

$$\begin{array}{ccccc} \nu_M : M_{G_0}(n) & \xrightarrow[{[W]}]{} & M_{G_0}(n)/W & \xrightarrow{\simeq} & M_G(n) \\ {\scriptstyle [M]}\Big\downarrow{\scriptstyle \pi_M} & \text{(I)} & {\scriptstyle [M]}\Big\downarrow{\scriptstyle \pi_M} & & {\scriptstyle [M]}\Big\downarrow{\scriptstyle \pi_M} \\ \nu_{\mathbb{B}} : \mathbb{B}G_0(n) & \xrightarrow[{[W]}]{} & \mathbb{B}G_0(n)/W & \xrightarrow[\simeq]{} & \mathbb{B}G(n) \end{array}$$

where the subdiagram (I) is Cartesian after Exercise 3.1.4.3.

An orientation of the fiber bundle $(M_G(n), \mathbb{B}G(n), \pi_M, M)$ over a field $\Bbbk$ is given by a nowhere vanishing global section of the cohomology sheaf

$$\mathcal{H}^{d_M}(\pi_!) := \mathcal{H}^{d_M}\left(\pi_{M,!}\, \underline{\underline{\Omega}}(M_{G_0}(n); \Bbbk)\right).$$

When M is orientable over $\Bbbk$, the fibers of $\mathcal{H}^{d_M}(\pi_!)$ are isomorphic to $\Bbbk$. The space $\mathbb{B}G_0(n)$, being simply connected, the sheaf $\mathcal{H}^{d_M}(\pi_!)$ is the constant sheaf $\underline{\underline{\Bbbk}} \cdot \zeta_M$, where ζ_M is the fundamental class of M. We can therefore define a global section $o \in \Gamma(\mathbb{B}G_0; \mathcal{H}^{d_M}(\pi_!))$ by $o : x \mapsto \zeta_M$, which is W invariant since we assume ζ_M invariant under the action of G. On the other hand, the projection $\nu_{\mathbb{B}} : \mathbb{B}G_0(n) \to \mathbb{B}G_0/W$, being a group covering, induces an isomorphism $\nu_{\mathbb{B}}^* : \Gamma(\mathbb{B}G; \mathcal{H}(\pi_!)) \to \Gamma(\mathbb{B}G_0; \mathcal{H}(\pi_!))^W$, so we can transfer the orientation of the fibers of M_{G_0} to $M_G(n)$.

When $\Bbbk := \mathbb{R}$ we can use differential forms to show orientability of $M_G(n)$. Indeed, by 3.1.7.1 and 3.1.7.2-(3c), there exists $\omega_\pi \in \Omega^{d_M}(M_{G_0}(n))$ whose restrictions to fibers of $\pi_M : M_{G_0}(n) \to \mathbb{B}G_0(n)$ are nowhere vanishing. For all $g \in G$, the differential form $g^*(\omega_\pi)$ defines the same orientation on the fibers of

[42] The same is true for $\Bbbk$-orientations over any field $\Bbbk$.

π_M since we assume G to preserve the orientation of M. As a consequence, the *G-average*[43]

$$\tilde{\omega}_\pi := \frac{1}{|W|}\sum\nolimits_{w\in W} w^*\Big(\int_{G_0} g^*(\omega_\pi)\,dg\Big)\,,$$

still defines the same orientation on the fibers of π_M, and is also W invariant. We can then transfer $\tilde{\omega}_\pi$ to $H^{d_M}(M_G)$ through the pullback homomorphism induced by the group covering $\nu_M : M_{G_0}(n) \to M_G(n)$ which induces an isomorphism $\nu_M^* : H^{d_M}(M_G) \to H^{d_M}(M_{G_0})^W$ whose restrictions to fibers are nowhere vanishing. Hence the orientation of $\big(M_G(n), \mathbb{B}G(n), \pi, M\big)$. □

4.8 Equivariant de Rham Comparison Theorems

Using Cartan's method (*cf*. Sect. 4.1.1), we prove in this section and in Sect. 4.10, three extensions to the classical de Rham Theorem, which compare equivariant cohomologies of a manifold M with corresponding ordinary cohomologies of the homotopy quotient M_G. For G compact, not necessarily connected, we give canonical isomorphisms[44]

$$\text{(i) } H_G(M) \simeq H(M_G;\mathbb{R}) \quad \text{(ii) } H_{G,\mathrm{c}}(M) \simeq H_{\mathrm{cv}}(M_G;\mathbb{R}) \tag{4.61}$$

$$\text{(iii) } H_{G,N}(M) \simeq H_{N_G}(M_G;\mathbb{R})\,,$$

where (iii) concerns *local cohomology*, to be discussed in Sect. 4.10.

The strategy to prove equivalences (i,ii)-(4.61) follows the approach we used in Sect. 4.7.2, which is based on the fact that $\mathbb{E}G = \text{lim-ind}_{n\in\mathbb{N}}\, \mathbb{E}G(n)$.

In these preliminaries, we assume the compact Lie group G to be **connected**.

Given a G-manifold M, we have, in the category of topological G-spaces,

$$\begin{array}{ccccc} \big(\varinjlim_{n\in\mathbb{N}} \mathbb{E}G(n)\times M\big) & = & \mathbb{E}G\times M & \xrightarrow{p_2} & M \\ {\scriptstyle [G]}\big\downarrow{\scriptstyle \nu_{(n,M)}} & & {\scriptstyle [G]}\big\downarrow{\scriptstyle \nu_M} & & \\ \big(\varinjlim_{n\in\mathbb{N}} \mathbb{E}G(n)\times_G M\big) & = & \mathbb{E}G\times_G M =: M_G & & \end{array} \tag{4.62}$$

where the projection map $p_2(z,x) := x$ is a homotopy equivalence. An important point here is that, in the top row, the spaces $\mathbb{E}G(n)\times M$ are *free* G-manifolds, whether the action on M is free or not.

[43] See Tu [91] §13.2 Integrating over a Compact connected Lie Group, p. 105.

[44] For (i), see Tu [91], §A.9 *Proof of the Equivariant de Rham Theorem in General*, p. 269, and Guillemin-Sternberg [50], §2.5 The Equivariant de Rham Theorem, p. 28.

From (4.62), we deduce the following diagrams of homomorphisms

$$\begin{array}{ccc} H_G(M) & \xrightarrow[(1)]{} & \varprojlim_n H_G(\mathbb{E}G(n)\times M) \\ {\scriptstyle (4)}\Updownarrow & & {\scriptstyle (2)}\Updownarrow \simeq \\ H(M_G;\mathbb{R}) & \xrightarrow[(3)]{\simeq} & \varprojlim_n H(\mathbb{E}G(n)\times_G M;\mathbb{R}) \end{array} \tag{4.63}$$

and

$$\begin{array}{ccc} H_{G,c}(M) & \xrightarrow[(1)]{} & \varprojlim_n H_{G,c}(\mathbb{E}G(n)\times M) \\ {\scriptstyle (4)}\Updownarrow & & {\scriptstyle (2)}\Updownarrow \simeq \\ H_{cv}(M_G;\mathbb{R}) & \xrightarrow[(3)]{\simeq} & \varprojlim_n H_c(\mathbb{E}G(n)\times_G M;\mathbb{R}) \end{array} \tag{4.64}$$

where the arrows (3) are the isomorphisms in Alexander-Spanier cohomology of Theorem 4.7.2.2-(4), and the arrows (2) recall the existence, for each $n \in \mathbb{N}$, of a canonical isomorphism which we will later make specific (*cf*. 4.8.1.2).

To show that we have canonical isomorphisms (4), the idea is then to show that:

Iso 1: the isomorphisms (2) define an isomorphism of projective systems;

Iso 2: for each $i \in \mathbb{N}$, the pullback morphism

$$p_2^* : H^i_G(M) \to H^i_G(\mathbb{E}G(n)\times M)$$

induced by the projection $p_2 : \mathbb{E}G(n)\times M \to M$, $p_2(z,x) := x$, is an isomorphism for all $n > i$.

We will address these questions in the following two sections.

4.8.1 Question Iso 1

The notations are those introduced in 4.7.2, where, for every $n \in \mathbb{N}$, we considered the Cartesian diagram of locally trivial fibrations of manifolds 4.7.2-(4.50)

$$\begin{array}{ccc} \mathbb{E}G(n)\times M & \overset{\nu_{(n,M)}}{\underset{[G]}{\twoheadrightarrow}} & \mathbb{E}G(n)\times_G M \\ {\scriptstyle [M]}\Big\downarrow {\scriptstyle \mathrm{id}\times c_M} & \square & {\scriptstyle [M]}\Big\downarrow {\scriptstyle \pi_M} \\ \mathbb{E}G(n)\times\{\bullet\} & \overset{\nu_{(n,\{\bullet\})}}{\underset{[G]}{\twoheadrightarrow}} & \mathbb{E}G(n)\times_G\{\bullet\} := \mathbb{B}G(n) \end{array} \tag{4.65}$$

to which we apply the approach of Cartan's lectures, as recalled in Sect. 4.1.1.

For the sequel, it will be convenient to shrink notations. We set:

$$\boxed{\mathbb{E}M_n := \mathbb{E}G(n)\times M\,,\quad \overline{\mathbb{E}M_n} := \mathbb{E}G(n)\times_G M \quad \text{and} \quad \nu_n := \nu_{n,M}}$$

Since $\nu_n : \mathbb{E}M_n \to \overline{\mathbb{E}M_n}$ is a projection of principal G-bundles, the corresponding pullback morphism ν_n^* identifies (*since G is connected*)

$$\Omega(\overline{\mathbb{E}M_n}) \xrightarrow[\simeq]{\nu_n^*} \Omega(\mathbb{E}M_n)^{\mathrm{bas}}$$

and, applying Proposition 3.1.10.2-(2), which we can do since the diagram (4.65) is Cartesian, we also have $\nu_n^*(\Omega_{\mathrm{c}}(\overline{\mathbb{E}M_n})) = \Omega_{\mathrm{c}}(\mathbb{E}M_n)^{\mathrm{bas}}$, as G is compact.

Hence, we deduce the following commutative diagram

$$\begin{array}{ccc} \Omega_{\mathrm{c}}(\overline{\mathbb{E}M_n}) & \xrightarrow[\simeq]{\nu_n^*} & \Omega_{\mathrm{c}}\big(\mathbb{E}M_n\big)^{\mathrm{bas}} \\ \downarrow & \oplus & \downarrow \\ \Omega(\overline{\mathbb{E}M_n}) & \xrightarrow[\simeq]{\nu_n^*} & \Omega\big(\mathbb{E}M_n\big)^{\mathrm{bas}} \end{array} \tag{4.66}$$

where, we recall, $\Omega_{\mathrm{c}}(\overline{\mathbb{E}M_n})$ and $\Omega_{\mathrm{c}}(\mathbb{E}M_n)$ denote the complexes of differential forms respectively with π_m and $(\mathrm{id}\times c_M)$-proper supports in (4.65). (Notice that the horizontal arrows are actual isomorphisms and not just quasi-isomorphisms.)

Given a connection f_n for the principal G-bundle $\nu_n : \mathbb{E}M_n \to \overline{\mathbb{E}M_n}$, let

$$f_n : W(\mathfrak{g}) \to \Omega\big(\mathbb{E}M_n\big)\,, \tag{4.67}$$

be the corresponding Weil morphism of $\mathfrak{g}$-dga's and consider the analogue to the diagram 4.1.1.1-(4.3) in the present context, i.e. the commutative diagram of Cartan-Weil morphisms

$$\begin{array}{ccccc} (\Omega_{\mathrm{c}}(\mathbb{E}M_n))^{\mathrm{bas}} & \xrightarrow[\text{q.i.}]{\bar{\mathfrak{i}}} & (W(\mathfrak{g})\otimes\Omega_{\mathrm{c}}(\mathbb{E}M_n))^{\mathrm{bas}} & \xrightarrow[\text{q.i.}]{\bar{\mathfrak{f}}} & (\Omega_{\mathrm{c}}(\mathbb{E}M_n))^{\mathrm{bas}} \\ \downarrow & \oplus & \downarrow & \oplus & \downarrow \\ (\Omega(\mathbb{E}M_n))^{\mathrm{bas}} & \xrightarrow[\text{q.i.}]{\bar{\mathfrak{i}}} & (W(\mathfrak{g})\otimes\Omega(\mathbb{E}M_n))^{\mathrm{bas}} & \xrightarrow[\text{q.i.}]{\bar{\mathfrak{f}}} & (\Omega(\mathbb{E}M_n))^{\mathrm{bas}} \end{array} \tag{4.68}$$

with $\bar{\mathfrak{i}}(\omega) := 1\otimes\omega$ and $\bar{\mathfrak{f}}(\alpha\otimes\omega) := f(\alpha)\wedge\omega$, hence $\bar{\mathfrak{f}}\circ\bar{\mathfrak{i}} = \mathrm{id}$ in each row.

There are two important remarks about this diagram:

- Since $\Omega(\mathbb{E}M_n)$ is a $\mathfrak{g}$-dg-algebra, we can apply Cartan's Theorem 4.1.1.1, and state that, in the bottom row, both morphisms $\bar{\mathfrak{i}}$ and $\bar{\mathfrak{f}}$ are quasi-isomorphisms.
- The same is true in the top row, although some care must be taken since $\Omega_{\mathrm{c}}(\mathbb{E}M_n)$ is not an algebra. In Appendix C, we show that Cartan's theorem is

more generally true for any differential graded ideal of $(\Omega(\mathbb{E}M_n), d)$ stable by $\mathfrak{g}$-derivatives and $\mathfrak{g}$-interior products, all properties clearly satisfied by $\Omega_c(\mathbb{E}M_n)$, since these operators reduce supports.

To compare the diagram (4.68) for different values n, recall that we are working on the inductive system of closed embeddings of fiber bundles (4.5.3.2-(1)):

$$
\begin{array}{ccccccccc}
\cdots & \longrightarrow & \mathbb{E}M_n & \xrightarrow{\epsilon_n} & \mathbb{E}M_{n+1} & \longrightarrow & \cdots & \mathbb{E}G\times M \\
 & & {\scriptstyle [G]}\big\downarrow{\scriptstyle \nu_n} & \square & {\scriptstyle [G]}\big\downarrow{\scriptstyle \nu_{n+1}} & & & {\scriptstyle [G]}\big\downarrow{\scriptstyle \nu_M} \\
\cdots & \longrightarrow & \overline{\mathbb{E}M_n} & \xrightarrow{\mu_n} & \overline{\mathbb{E}M_{n+1}} & \longrightarrow & \cdots & M_G
\end{array}
\tag{4.69}
$$

which leads us to consider the diagram

$$
\begin{array}{ccc}
\Omega(\mathbb{E}M_{n+1})^{\mathrm{bas}} & \xrightarrow{\qquad \epsilon_n^* \qquad} & \Omega(\mathbb{E}M_n)^{\mathrm{bas}} \\
{\scriptstyle \bar{\mathfrak{i}}_{n+1}}\big\downarrow{\scriptstyle \text{q.i.}} & \text{(I)} & {\scriptstyle \bar{\mathfrak{i}}_{n}}\big\downarrow{\scriptstyle \text{q.i.}} \\
\big(W(\mathfrak{g})\otimes\Omega(\mathbb{E}M_{n+1})\big)^{\mathrm{bas}} & \xrightarrow{\mathrm{id}\otimes\epsilon_n^*} & \big(W(\mathfrak{g})\otimes\Omega(\mathbb{E}M_n)\big)^{\mathrm{bas}} \\
{\scriptstyle \bar{\mathfrak{f}}_{n+1}}\big\downarrow{\scriptstyle \text{q.i.}} & \text{(II)} & {\scriptstyle \text{q.i.}}\big\downarrow{\scriptstyle \bar{\mathfrak{f}}_{n}} \\
\Omega(\mathbb{E}M_{n+1})^{\mathrm{bas}} & \xrightarrow{\qquad \epsilon_n^* \qquad} & \Omega(\mathbb{E}M_n)^{\mathrm{bas}}
\end{array}
\tag{4.70}
$$

where

- the subdiagram (I) is commutative by construction, but
- the subdiagram (II) need not be commutative. Indeed, for that we need the family of connections $\mathcal{F} := \{f_n\}_{n\in\mathbb{N}}$ to be compatible with the family of restrictions $\{\epsilon_n\}_{n\in\mathbb{N}}$, i.e. we need the diagrams

$$
\begin{array}{ccc}
\Omega(\mathbb{E}M_{n+1}) & \xrightarrow{\epsilon_n^*} & \Omega(\mathbb{E}M_n) \\
{\scriptstyle f_{n+1}}\nwarrow & \oplus & \nearrow{\scriptstyle f_n} \\
 & W(\mathfrak{g}) &
\end{array}
$$

to be commutative for all $n \in \mathbb{N}$. And this is indeed possible after Corollary 4.5.3.2-(1) which states the existence of such families $\mathcal{F}$, which we call *projective system of connections* for $\{\mathbb{E}M_n\}_{n\in\mathbb{N}}$ (*cf*. fn. (3), p. 110).

To finish, notice that since the subdiagrams in (4.69) are Cartesian, we can exchange $\Omega \leftrightarrow \Omega_c$ in (4.70) and still have a commutative diagram with vertical quasi-isomorphisms.

We can now state the main result in this section.

Proposition 4.8.1.1 *Let G be a compact connected Lie group. For every G-manifold M, the families $\{\bar{\mathfrak{i}}_n\}_{n\in\mathbb{N}}$ of Cartan-Weil quasi-isomorphisms:*

$$\begin{cases} \bar{\mathfrak{i}}_n : \Omega(\mathbb{E}G(n)\times_G M) \to \big(W(\mathfrak{g}) \otimes \Omega(\mathbb{E}G(n)\times M)\big)^{\mathrm{bas}} \\ \bar{\mathfrak{i}}_n : \Omega_{\mathrm{c}}(\mathbb{E}G(n)\times_G M) \to \big(W(\mathfrak{g}) \otimes \Omega_{\mathrm{c}}(\mathbb{E}G(n)\times M)\big)^{\mathrm{bas}} \end{cases} \tag{4.71}$$

are morphisms of projective systems from the projective system

$$\big\{\Omega(\mathbb{E}G(n)\times_G M) \xrightarrow{\mu_n^*} \Omega(\mathbb{E}G(n+1)\times_G M)\big\}_{n\in\mathbb{N}}$$

to the projective system

$$\big\{\big(W(\mathfrak{g}) \otimes \Omega(\mathbb{E}G(n)\times M)\big)^{\mathrm{bas}} \xrightarrow{\mathrm{id}\otimes\epsilon_n^*} \big(W(\mathfrak{g}) \otimes \Omega(\mathbb{E}G(n+1)\times M)\big)^{\mathrm{bas}}\big\}_{n\in\mathbb{N}}$$

Furthermore, if $\mathcal{F} := \{f_n\}_{n\in\mathbb{N}}$ is a projective system of connections for the inductive system $\big\{\mathbb{E}G(n)\times M\big\}_{n\in\mathbb{N}}$ (4.69), then the families $\{\bar{\mathfrak{f}}_n\}_{n\in\mathbb{N}}$ of Cartan-Weil quasi-isomorphisms:

$$\begin{cases} \bar{\mathfrak{f}}_n : \big(W(\mathfrak{g}) \otimes \Omega(\mathbb{E}G(n)\times_G M)\big)^{\mathrm{bas}} \to \Omega(\mathbb{E}G(n)\times_G M) \\ \bar{\mathfrak{f}}_n : \big(W(\mathfrak{g}) \otimes \Omega_{\mathrm{c}}(\mathbb{E}G(n)\times_G M)\big)^{\mathrm{bas}} \to \Omega_{\mathrm{c}}(\mathbb{E}G(n)\times_G M) \end{cases} \tag{4.72}$$

are also morphisms of projective systems.

Proof Since $\Omega(\mathbb{E}G(n)\times_G M) = \Omega(\mathbb{E}G(n)\times M)^{\mathrm{bas}}$ and $\Omega_{\mathrm{c}}(\mathbb{E}G(n)\times_G M) = \Omega_{\mathrm{c}}(\mathbb{E}G(n)\times M)^{\mathrm{bas}}$, the statements are just a rewriting of the commutative diagram (4.70) and of the same diagram replacing Ω with Ω_{c}. □

The following Corollary is now immediate.

Corollary 4.8.1.2 *Let G be a compact connected Lie group. For every G-manifold M, the families $\{H(\bar{\mathfrak{i}}_n)\}_{n\in\mathbb{N}}$ of Cartan-Weil isomorphisms*

$$\begin{cases} H(\bar{\mathfrak{i}}_n) : H(\mathbb{E}G(n)\times_G M) \to H_G(\mathbb{E}G(n)\times M) \\ H(\bar{\mathfrak{i}}_n) : H_{\mathrm{c}}(\mathbb{E}G(n)\times_G M) \to H_{G,\mathrm{c}}(\mathbb{E}G(n)\times M) \end{cases}$$

induce isomorphisms of projective limits

$$\begin{cases} \varprojlim_n H(\bar{\mathfrak{i}}_n) : \varprojlim_n H(\mathbb{E}G(n)\times_G M) \xrightarrow[\simeq]{} \varprojlim_n H_G(\mathbb{E}G(n)\times M) \\ \varprojlim_n H(\bar{\mathfrak{i}}_n) : \varprojlim_n H_{\mathrm{c}}(\mathbb{E}G(n)\times_G M) \xrightarrow[\simeq]{} \varprojlim_n H_{G,\mathrm{c}}(\mathbb{E}G(n)\times M) \end{cases}$$

Furthermore, if $\mathcal{F} := \{f_n\}_{n\in\mathbb{N}}$ is a projective system of connections for the inductive system $\big\{\mathbb{E}G(n)\times M\big\}_{n\in\mathbb{N}}$, then the corresponding family of Cartan-Weil isomorphisms $\{H(\bar{\mathfrak{f}}_n)\}_{n\in\mathbb{N}}$ verifies

$$\varprojlim_n H(\bar{\mathfrak{f}}_n) := \varprojlim_n H(\bar{\mathfrak{i}}_n)^{-1}\,.$$

Remark 4.8.1.3 The classical de Rham Theorem allows replacing in 4.8.1.2 the de Rham cohomologies of the manifolds $\mathbb{E}G(n)\times_G M$, by the corresponding ordinary cohomologies, hence completing the answer of question **Iso 1**.

4.8.2 Question Iso 2

We apply Corollary 4.4.6.2-(3,4) to the G-equivariant proper map

$$p_2(n) : \mathbb{E}G(n)\times M \to M\,,$$

which gives morphisms of spectral sequences converging to the morphisms

$$\begin{cases} p_2(n)^* : \ H_G(M) \ \to \ H_G(\mathbb{E}G(n)\times M) \\ p_2(n)^* : H_{G,\mathrm{c}}(M) \to H_{G,\mathrm{c}}(\mathbb{E}G(n)\times M) \end{cases}$$

whose terms $\mathbb{E}_2^{p,q}$ are respectively

$$\mathbb{E}_G(p_2(n))_2^{p,q} := \begin{cases} S^p(\mathfrak{g}^\vee)^{\mathfrak{g}} \otimes H^q(M) \Rightarrow H_G^{p+q}(M) \\ \quad \mathrm{id}\downarrow \qquad p_2(n)^*\downarrow \qquad p_2(n)^*\downarrow \\ S^p(\mathfrak{g}^\vee)^{\mathfrak{g}} \otimes H^q(\mathbb{E}G(n)\times M) \Rightarrow H_G^{p+q}(\mathbb{E}G(n)\times M) \end{cases}$$

and

$$\mathbb{E}_{G,\mathrm{c}}(p_2(n))_2^{p,q} := \begin{cases} S^p(\mathfrak{g}^\vee)^{\mathfrak{g}} \otimes H_{\mathrm{c}}^q(M) \Rightarrow H_{G,\mathrm{c}}^{p+q}(M) \\ \quad \mathrm{id}\downarrow \qquad p_2(n)^*\downarrow \qquad p_2(n)^*\downarrow \\ S^p(\mathfrak{g}^\vee)^{\mathfrak{g}} \otimes H_{\mathrm{c}}^q(\mathbb{E}G(n)\times M) \Rightarrow H_{G,\mathrm{c}}^{p+q}(\mathbb{E}G(n)\times M) \end{cases}$$

where

$$\begin{cases} H^q(\mathbb{E}G(n)\times M) = \bigoplus_{a+b=q} H^a(\mathbb{E}G(n)) \otimes H^b(M) \\ H_{\mathrm{c}}^q(\mathbb{E}G(n)\times M) = \bigoplus_{a+b=q} H^a(\mathbb{E}G(n)) \otimes H_{\mathrm{c}}^b(M) \end{cases}$$

But then, if $n > (p+q)$, we have $n > a$, in which case $H^a(\mathbb{E}G(n)) = 0$ since $\mathbb{E}G(n)$ is n-connected (*cf*. fn. ([30]), p. 146), and the morphisms $\mathbb{E}_G(p_2(n))_r$ and $\mathbb{E}_{G,\mathrm{c}}(p_2(n))_r$ are isomorphisms, for all $r \geq 2$.

We have thus proved the following analogue to Theorem 4.7.2.2-(3), which answers question **Iso 2**.

Theorem 4.8.2.1 *Let G be a compact connected Lie group. For every G-manifold M, given $i \in \mathbb{N}$, the homomorphisms*

$$\begin{cases} p_2(n)^* : \ H_G^i(M) \ \to H_G^i(\mathbb{E}G(n)\times M)\,, \\ p_2(n)^* : H_{G,\mathrm{c}}^i(M) \to H_{G,\mathrm{c}}^i(\mathbb{E}G(n)\times M)\,, \end{cases} \tag{4.73}$$

are isomorphisms for all $m \geq n > i$. Furthermore, the induced homomorphisms

$$\begin{cases} H_G(M) \xrightarrow{\lim_n p_2(n)^*} \varprojlim_n H_G(\mathbb{E}G(n)\times M) \leftarrow h\big(\varprojlim_n \Omega_G(\mathbb{E}G(n)\times M)\big)\,, \\ H_{G,\mathrm{c}}(M) \xrightarrow{\lim_n p_2(n)^*} \varprojlim_n H_{G,\mathrm{c}}(\mathbb{E}G(n)\times M) \leftarrow h\big(\varprojlim_n \Omega_{G,\mathrm{c}}(\mathbb{E}G(n)\times M)\big)\,, \end{cases}$$

are isomorphisms.

In particular, the natural morphisms

$$\begin{cases} \varprojlim_n p_2(n)^* : \Omega_G(M) \to \varprojlim_n \Omega_G(\mathbb{E}G(n)\times M)\,, \\ \varprojlim_n p_2(n)^* : \Omega_{G,\mathrm{c}}(M) \to \varprojlim_n \Omega_{G,\mathrm{c}}(\mathbb{E}G(n)\times M)\,, \end{cases} \tag{4.74}$$

are quasi-isomorphisms.

Proof The isomorphisms (4.73) are already justified. For the rest, the proof is the same as 4.7.2.2-(4). Consider the natural diagram of morphisms of graded vector spaces

$$\begin{array}{ccc} H_G(M) & \xrightarrow{\ (1)\ } & \mathbf{h}\big(\varprojlim_n \Omega_G(\mathbb{E}G(n)\times M)\big) \\ & {\scriptstyle (3)}\searrow \quad \swarrow{\scriptstyle (2)} & \\ & \varprojlim_n H_G(\mathbb{E}G(n)\times M) & \end{array}$$

The isomorphisms (4.73), immediately imply that (3) is an isomorphism, and also that $\{H_G(\mathbb{E}G(n+1)\times M) \to H_G(\mathbb{E}G(n)\times M)\}_{n\subset\mathbb{N}}$ verifies Mittag-Leffler condition. We can therefore conclude as in 4.7.2.2-(4), showing that the projective system $\{\Omega_G(\mathbb{E}G(n+1)\times M) \to \Omega_G(\mathbb{E}G(n)\times M)\}_{n\in\mathbb{N}}$ verifies M-L condition too. But this is clear since, $\epsilon_n : (\mathbb{E}G(n)\times M) \to (\mathbb{E}G(n+1)\times M)$ being closed embeddings of manifolds, the morphisms

$$\mathrm{id}\otimes\epsilon_n^* : S(\mathfrak{g}^\vee)\otimes\Omega(\mathbb{E}G(n+1)\times M) \to S(\mathfrak{g}^\vee)\otimes\Omega(\mathbb{E}G(n)\times M)$$

are surjective, as are also their restrictions to G-invariants

$$\mathrm{id}\otimes\epsilon_n^* : \big(S(\mathfrak{g}^\vee)\otimes\Omega(\mathbb{E}G(n+1)\times M)\big)^G \to \big(S(\mathfrak{g}^\vee)\otimes\Omega(\mathbb{E}G(n)\times M)\big)^G\,,$$

as we dispose of the averaging operator $\int_G$, both on $S(\mathfrak{g}^\vee)$ and on $\Omega(\_)$. □

4.8.3 Equivariant Cohomology Comparison Theorem

With the last proposition, the strategy proposed at the preliminary discussion in Sect. 4.8 is fully confirmed. We can therefore state our first comparison theorem.

Equivariant de Rham Theorem 4.8.3.1 *Let G be a compact Lie group. For every G-manifold M, we have canonical isomorphisms*

$$H_G(M) \simeq H(M_G;\mathbb{R}) \quad \textit{and} \quad H_{G,\mathrm{c}}(M) \simeq H_{\mathrm{cv}}(M_G;\mathbb{R}), \tag{4.75}$$

coming from the following diagrams of canonical quasi-isomorphisms

$$\begin{array}{ccc} \Omega_G(M) & \xrightarrow[(1)]{} & \varprojlim_n \Omega_G(\mathbb{E}G(n)\times M) \\ \wr| & & (2)\ \uparrow \bar{\mathfrak{i}}_n \downarrow \bar{\mathfrak{f}}_n \\ \Omega(M_G;\mathbb{R}) & \xrightarrow[(3)]{} & \varprojlim_n \Omega(\mathbb{E}G(n)\times_G M;\mathbb{R}) \end{array} \tag{4.76}$$

and

$$\begin{array}{ccc} \Omega_{G,\mathrm{c}}(M) & \xrightarrow[(1)]{} & \varprojlim_n \Omega_{G,\mathrm{c}}(\mathbb{E}G(n)\times M) \\ \wr| & & (2)\ \uparrow \bar{\mathfrak{i}}_n \downarrow \bar{\mathfrak{f}}_n \\ \Omega_{\mathrm{cv}}(M_G;\mathbb{R}) & \xrightarrow[(3)]{} & \varprojlim_n \Omega_{\mathrm{c}}(\mathbb{E}G(n)\times_G M;\mathbb{R}) \end{array} \tag{4.77}$$

where

- (1) *is* lim-proj$_n(p_2(n)^*)$, *from Theorem 4.8.2.1.*
- (2) *are* lim-proj$_n(\bar{\mathfrak{i}}_n)$ *and* lim-proj$_n(\bar{\mathfrak{f}}_n)$, *from Corollary 4.8.1.2.*
- (3) *comes from Theorem 4.7.2.2-(4).*

Furthermore, the diagram (4.76) is functorial in the category G-Man *and the diagram (4.77) is functorial in the category* G-Man$_{\mathrm{pr}}$.

Proof For G connected, the theorem is an easy consequence of Theorem 4.7.2.2 together with the conclusions of questions **Iso 1, 2** in Corollary 4.8.1.2 and Theorem 4.8.2.1. Functoriality is clear since all the constructions were so.

For G non-connected, let G_0 be the connected component of the identity element $e \in G$, set $W := G/G_0$, and use the functorial quasi-isomorphisms

$$\begin{cases} \Omega_G(\_) \simeq \Omega_{G_0}(\_)^W \\ \Omega_{G,\mathrm{c}}(\_) \simeq \Omega_{G_0,\mathrm{c}}(\_)^W \end{cases} \quad \text{and} \quad \begin{cases} \Omega(\_G;\mathbb{R}) \simeq \Omega(\_{G_0};\mathbb{R})^W \\ \Omega_{\mathrm{cv}}(\_G;\mathbb{R}) \simeq \Omega_{\mathrm{cv}}(\_{G_0};\mathbb{R})^W \end{cases}$$

established in Proposition 4.4.5.2 and Exercise 4.7.1.2. □

Remark 4.8.3.2 Theorem 4.8.2.1 gives us an isomorphism:

$$H_G(M) \to \varprojlim_n H_G(\mathbb{E}G(n)\times M) \tag{4.78}$$

where, on the right-hand side, the spaces $\mathbb{E}G(n)\times M$ are all free G-manifolds. We can therefore apply Corollary 4.8.1.2 which gives the projective family of isomorphisms

$$\big\{H(\mathfrak{f}_n) : H_G(\mathbb{E}G(n)\times M) \to H(\mathbb{E}G(n)\times_G M)\big\}_{n\in\mathbb{N}},$$

and, composing (4.78) with lim-proj$_n$ $H(\mathfrak{f}_n)$, we obtain a canonical isomorphism

$$\boxed{H_G(M) \to \varprojlim_n H(M_G(n))} \tag{4.79}$$

which is functorial on $M \in G$-Man.

Expression (4.79) can therefore be retained as an alternative definition of equivariant cohomology theory of G-manifolds as limit theory of de Rham cohomology theory on the approximations of the homotopy quotient.

4.9 Cohomology of Classifying Spaces

4.9.1 Canonicity of the Cohomology of Classifying Spaces

In Sect. 4.6 we recalled two well-known constructions of universal fiber bundles $\mathbb{E}$ for a given compact Lie group G. The resulting topological spaces share the properties of being contractible CW-complexes (*cf*. fn. ([31]), p. 146), free G-spaces and such that the projection $\mathbb{E} \twoheadrightarrow \mathbb{E}/G$ is a locally trivial fibration. In 4.6.3 we decided to work with some fixed universal fiber bundle $\mathbb{E}G$ which is the inductive limit of Stiefel manifolds in order to prove the Equivariant de Rham comparison theorems. However, in practice, we sometimes need to change the choice of the space $\mathbb{E}G$, and although the classification theorem of principal fiber bundles[45] tells us already that different constructions of $\mathbb{B}G$ are homotopy-equivalent, it is interesting to return to Cartan's remark in the very last lines of [27], on an alternative a priori justification of the canonicity of the cohomology of classifying spaces based on Cartan's theorem 4.1.1.1 and the equivariant de Rham Theorem (4.8.3.1).

Corollary 4.9.1.1 *Let G be a compact Lie group. For every weakly contractible G-principal fiber bundle* $\mathbb{E}$ (cf. *fn.* ([31]), *p. 146), there exists a canonical isomorphism*

$$H(\mathbb{E}/G;\mathbb{R}) \simeq H_G(\{\bullet\}) = S(\mathfrak{g}^\vee)^G\,.$$

Proof Let $\mathbb{E}G$ denote the universal fiber bundle inductive limit of Stiefel manifolds (4.6.3). Let G act on the right of $\mathbb{E}G\times\mathbb{E}$ by $(x,y)\cdot g := (x\cdot g, y\cdot g)$. We then have the two locally trivial fibrations

$$\mathbb{B}G \xleftarrow[\;[\mathbb{E}G]\;]{\overline{p}_1} (\mathbb{E}G\times\mathbb{E})/G \xrightarrow[\;[\mathbb{E}]\;]{\overline{p}_2} \mathbb{E}/G\,,$$

[45]See Steenrod [84], §19.3 Classification Thm., p. 101, or Hussemöller [56] Thm. 12.2, p. 57.

where the fibers are shown in brackets. Since these fibers are weakly contractible, the Leray spectral sequences associated with the projections $\overline{p}_i$ degenerate at the second page, which implies that the pullbacks $\overline{p}_i^*$ are isomorphisms

$$H(\mathbb{B}G;\Bbbk) \xrightarrow[\simeq]{\overline{p}_1^*} H((\mathbb{E}G\times\mathbb{E})/G;\Bbbk) \xleftarrow[\simeq]{\overline{p}_2^*} H(\mathbb{E}/G;\Bbbk)\,.$$

We can then conclude, applying the isomorphism $H(\mathbb{B}G;\mathbb{R})\simeq H_G(\{\bullet\}) = S(\mathfrak{g}^\vee)^G$ given by the equivariant de Rham theorem (4.8.3.1). □

4.9.2 Formality of Classifying Spaces

In Sect. 3.4.1, we recalled the definition of $\mathbb{R}$-*formal topological spaces*, which, for classifying space $\mathbb{B}G$, says that there exists a diagram

$$\Omega(\mathbb{B}G;\mathbb{R}) \to (A_1,d) \leftarrow (A_2,d) \to \cdots \leftarrow (A_{n-1},d) \to (A_n,d) \leftarrow H(\mathbb{B}G,\mathbb{R})\,,$$

where the arrows represent quasi-isomorphic morphisms of dg-algebras.

Theorems 4.8.2.1 and 4.7.2.2-(4) established that the morphisms of dg-algebras

$$\begin{cases} \Omega_G(\{\bullet\}) \xrightarrow[\text{q.i.}]{} \lim\limits_{\longleftarrow n} \Omega_G(\mathbb{E}G(n)) \\ \Omega(\mathbb{B}G;\mathbb{R}) \xrightarrow[\text{q.i.}]{} \lim\limits_{\longleftarrow n} \Omega(\mathbb{B}G(n);\mathbb{R}) \end{cases} \tag{4.80}$$

are quasi-isomorphism, and we also know that

$$\Omega_G(\{\bullet\}) = \boldsymbol{h}(\Omega_G(\{\bullet\})) \simeq H(\mathbb{B}G;\mathbb{R})\,, \tag{4.81}$$

where $(=)$ is simply because $\Omega_G(\{\bullet\}) = (S(\mathfrak{g}^\vee),0)$, and $(\simeq)$ is the canonical isomorphism given by the equivariant de Rham Theorem 4.8.3.1. Therefore, to prove that $\mathbb{B}G$ is $\mathbb{R}$-formal, we need only show that the projective limits in (4.80) can be joined by a zig-zag ($\leftrightsquigarrow$) of quasi-isomorphic morphisms of dga's, and, moreover, since the corresponding projective systems verify Mittag-Leffler conditions at the levels both of complexes and cohomology, we need only show (see ([41]), p. 155) that there exists a family of ziz-zags $\{\Omega_G(\mathbb{E}G(n)) \leftrightsquigarrow \Omega(\mathbb{B}G(n);\mathbb{R})\}_{n\in\mathbb{N}}$, such that the diagrams

$$\begin{array}{ccc} \Omega_G(\mathbb{E}G(n{+}1)) & \longrightarrow & \Omega_G(\mathbb{E}G(n)) \\ \updownarrow\wr & \oplus & \updownarrow\wr \\ \Omega(\mathbb{B}G(n{+}1);\mathbb{R}) & \longrightarrow & \Omega(\mathbb{B}G(n);\mathbb{R})\,, \end{array}$$

where the horizontal arrows are the usual restrictions, are commutative.

For this, the easiest way is to work with sheaves. Indeed, given a manifold M, let $\underline{\underline{\Omega}}_M$ and $\underline{\underline{\mathit{\Omega}}}_M$ denote the complexes of sheaves on M respectively of de Rham differential forms and of Alexander-Spanier cochains with coefficients in $\mathbb{R}$.

Since M is locally contractible, the cohomology sheaves are concentrated in degree 0 where we have $\mathcal{H}^0(\underline{\underline{\Omega}}_M) = \mathcal{H}^0(\underline{\underline{\mathit{\Omega}}}_M) = \underline{\underline{\mathbb{R}}}_M$.

The two natural morphisms of sheaves of dg-algebras

$$\begin{cases} \lambda_M : \underline{\underline{\Omega}}_M \underset{\text{q.i.}}{\longrightarrow} (\underline{\underline{\Omega}}_M \otimes \underline{\underline{\mathit{\Omega}}}_M)\,, & s \mapsto s \otimes 1\,, \\ \rho_M : \underline{\underline{\mathit{\Omega}}}_M \underset{\text{q.i.}}{\longrightarrow} (\underline{\underline{\Omega}}_M \otimes \underline{\underline{\mathit{\Omega}}}_M)\,, & s \mapsto 1 \otimes s\,, \end{cases}$$

are then easily seen to be quasi-isomorphic by Künneth's theorem at stalks level.

Applying these considerations to a closed embedding of manifolds $\iota : N \to M$, leads us to consider the following commutative diagram:

$$\begin{array}{ccc} \underline{\underline{\Omega}}_M & \longrightarrow & \iota_* \underline{\underline{\Omega}}_N \\ {\scriptstyle \lambda_M}\downarrow{\scriptstyle \text{q.i.}} & \oplus & {\scriptstyle \text{q.i.}}\downarrow{\scriptstyle \iota_*(\lambda_N)} \\ \underline{\underline{\Omega}}_M \otimes \underline{\underline{\mathit{\Omega}}}_M & \longrightarrow & \iota_*(\underline{\underline{\Omega}}_N \otimes \underline{\underline{\mathit{\Omega}}}_N) \\ {\scriptstyle \rho_M}\uparrow{\scriptstyle \text{q.i.}} & \oplus & {\scriptstyle \text{q.i.}}\uparrow{\scriptstyle \iota_*(\rho_N)} \\ \underline{\underline{\mathit{\Omega}}}_M & \longrightarrow & \iota_* \underline{\underline{\mathit{\Omega}}}_N\,, \end{array} \tag{4.82}$$

where the horizontal arrows in the top and bottom rows are induced by the usual restrictions of differential forms and of Alexander-Spanier cochains respectively. The arrow in the middle row is then determined by the commutativity of the diagram, which, at the level of presheaves, associates $\omega \otimes \varpi$ with $\omega|_N \otimes \varpi|_N$.

A key point in diagram (4.82) is that, by the acyclicity theorem B.6.3.4, the sheaves $\underline{\underline{\Omega}}^i_M$, $\underline{\underline{\mathit{\Omega}}}^j_M$ and $\underline{\underline{\Omega}}^i_M \otimes \underline{\underline{\mathit{\Omega}}}^j_M$ are all $\Gamma(M, \_)$-acyclic, the first since it is an $\underline{\underline{\Omega}}^0_M$-module and the two others, since they are $\underline{\underline{\mathit{\Omega}}}^0_M$-modules. The same is obviously true for N. Consequently, applying the functor $\Gamma(M, \_)$ to (4.82) gives the following diagram of morphisms of dg-algebras, where the vertical arrows are still quasi-isomorphisms:[46]

$$\begin{array}{ccc} \Omega(M) & \longrightarrow\!\!\!\!\!\rightarrow & \Omega(N) \\ {\scriptstyle \lambda_M}\downarrow{\scriptstyle \text{q.i.}} & & {\scriptstyle \text{q.i.}}\downarrow{\scriptstyle \lambda_N} \\ \Gamma(M; \underline{\underline{\Omega}}_M \otimes \underline{\underline{\mathit{\Omega}}}_M) & \longrightarrow\!\!\!\!\!\rightarrow & \Gamma(N; \underline{\underline{\Omega}}_N \otimes \underline{\underline{\mathit{\Omega}}}_N) \\ {\scriptstyle \rho_M}\uparrow{\scriptstyle \text{q.i.}} & \oplus & {\scriptstyle \text{q.i.}}\uparrow{\scriptstyle \rho_N} \\ \mathit{\Omega}(M; \mathbb{R}) & \longrightarrow\!\!\!\!\!\rightarrow & \mathit{\Omega}(N; \mathbb{R})\,. \end{array} \tag{4.83}$$

[46]The reader will have recognized in these lines the proof of the classical de Rham's theorem using tools of Sheaf Theory.

Furthermore, the horizontal rows are surjective. Indeed, while the surjectivity of the top and bottom arrows are standard features for a closed embedding $N \subseteq M$, the surjectivity of the middle arrow is a bit more subtle because it comes from the surjectivity of the morphism of sheaves $(\underline{\underline{\Omega}}_N \otimes \underline{\underline{\Omega}}_N) \twoheadrightarrow \iota_*(\underline{\underline{\Omega}}_N \otimes \underline{\underline{\Omega}}_N)$ in (4.82) (obvious at stalks), and the fact that, being a morphism of Ω^0_M-modules, the Ω^0_M-modules, the global section functor $\Gamma(M;\_) = \mathrm{GM}(\underline{\underline{\Omega}}^0_M) \to \mathrm{GM}(\Omega^0(M;\mathbb{R}))$, which is an equivalence of categories after B.4.1-(3), preserves its surjectivity.

If we now apply (4.83) to $M := \mathbb{B}G(n{+}1)$ and $N := \mathbb{B}G(n)$, for all $n \in \mathbb{N}$, we obtain the morphisms of projective systems:

$$\begin{array}{ccc} \{\Omega(\mathbb{B}G(n))\}_{n\in\mathbb{N}} & & \{\Omega(\mathbb{B}G(n))\}_{n\in\mathbb{N}} \\ {\scriptstyle\{\lambda\}_n}\searrow & & \swarrow{\scriptstyle\{\rho\}_n} \\ & \{\Gamma(\mathbb{B}G(n); \underline{\underline{\Omega}}_{\mathbb{B}G(n)} \otimes \underline{\underline{\Omega}}_{\mathbb{B}G(n)})\}_{n\in\mathbb{N}} & \end{array} \tag{4.84}$$

which satisfies Mittag-Leffler conditions, as in (4.80), hence inducing quasi-isomorphic morphisms of dg-algebras at projective limits.

We have thus established the existence of a zig-zag of quasi-isomorphic morphisms of dg-algebras:

$$\varprojlim_n \Omega(\mathbb{B}G(n)) \leftrightsquigarrow \varprojlim_n \Omega(\mathbb{B}G(n);\mathbb{R})\,. \tag{4.85}$$

One last step can be achieved by considering the following sequence of quasi-isomorphic morphisms of dg-algebras:

$$\begin{array}{ccc} \Omega(\mathbb{B}G(n)) \xrightarrow[\text{q.i.}]{\nu_n^*} \Omega(\mathbb{E}G(n))^{\mathrm{bas}} & & \\ \text{q.i.}\big\downarrow \mathrm{i}_n & & \\ \Omega(W(\mathfrak{g}) \otimes \mathbb{E}G(n))^{\mathrm{bas}} \xrightarrow[\text{q.i.}]{\Xi_n} \Omega_G(\mathbb{E}G(n)) & & \end{array} \tag{4.86}$$

where

- $\nu_n : \mathbb{E}G(n) \to \mathbb{B}G(n)$ is the canonical projection.
- i_n is the Cartan-Weil morphism 4.1.1.1.
- Ξ_n is the Cartan isomorphism 4.1.1-4.4 (see also 4.3.2-(4.16)).

It is easy to see that the sequences (4.86), being functorial relative to the closed embeddings $\mathbb{E}G(n) \hookrightarrow \mathbb{E}G(n{+}1)$, induce a morphism of projective systems from $\{\Omega(\mathbb{B}G(n))\}_{n\in\mathbb{N}}$ to $\{\Omega_G(\mathbb{E}G(n))\}_{n\in\mathbb{N}}$, and since these systems satisfy Mittag-Leffler conditions at both complexes and cohomology levels, we obtain a quasi-isomorphic morphism of dg-algebras

$$\varprojlim_n \Omega(\mathbb{B}G(n)) \xrightarrow[\text{q.i.}]{} \varprojlim_n \Omega_G(\mathbb{E}G(n))\,. \tag{4.87}$$

We can now prove the main result of this section.

Theorem 4.9.2.1 *Let G be a compact Lie group.*

1. *For every M be a G-manifold, there exists a zig-zag of quasi-isomorphic morphisms of dg-algebras:*

$$\Omega_G(M) \leftrightsquigarrow \Omega(M_G; \mathbb{R}) \tag{4.88}$$

2. *The classifying space $I\!\!BG$ is an $\mathbb{R}$-formal space.*

Proof

(1) If we put aside equality (4.81), the preliminary discussion in Sect. 4.9.2 is valid for any G-manifold M in lieu of the singleton $\{\bullet\}$. Consequently, for G connected, the concatenation of (4.80), (4.85) and (4.87) gives a canonical zig-zag $\Omega_G(M) \leftrightsquigarrow \Omega(M_G; \mathbb{R})$ of quasi-isomorphic morphisms of dg-algebras which is functorial on $M \in G$-Man.

When G is not connected, we use the fact the finite group $W := G/G_0$, acts on M_{G_0} (*cf*. Exercise 4.7.1.2) in a way that if we denote by $\pi : M_{G_0} \to M_{G_0}/W = M_G$ the canonical projection, then the pullback π^* induces algebra isomorphisms

$$\pi^* : \Omega(M_G; \mathbb{R}) \xrightarrow[\simeq]{} \Omega(M_{G_0}; \mathbb{R})^W \quad \text{and} \quad \pi^* : H(M_G; \mathbb{R}) \xrightarrow[\simeq]{} H(M_{G_0}; \mathbb{R})^W ,$$

which are functorial on $M \in G$-Man, by 4.4.5.2 and 4.7.1.2.

Analogously, as explained in Sect. *cf*. 4.4.5, the group W acts on the complex of G_0-equivariant differential forms $\Omega_{G_0}(M)$, and the restriction maps

$$\Omega_G(M) \xrightarrow[\simeq]{} \Omega_{G_0}(M)^W \quad \text{and} \quad H_G(M) \xrightarrow[\simeq]{} H_{G_0}(M)^W ,$$

are also isomorphisms (*cf*. Proposition 4.4.5.2).

Therefore, to establish ((1)), we need only show that all the morphisms of dg-algebras in the zig-zag $\Omega_G(M) \leftrightsquigarrow \Omega(M_G; \mathbb{R})$ are W-equivariant. This is elementary in (4.80) and in (4.84), hence in (4.85). The W-equivariance of (4.87) results from that of (4.86) which was established in 4.4.5.2.

(2) Apply (1) to $M := \{\bullet\}$, in which case $\Omega_G(\{\bullet\}) \simeq H(I\!\!BG; \mathbb{R})$ (see (4.81)). □

4.10 Local Equivariant Cohomology

Let X be a mild topological space (*cf*. B.1). Given a closed subspace $Z \subseteq X$, let $U := X \smallsetminus Z$, and denote by $i : Z \subseteq X$ and $j : U \subseteq X$ the corresponding inclusion maps. One has an exact triangle of functors in the category $\mathcal{D}(M; \Bbbk)$

$$I\!\!R\,\Gamma_Z(X; \_) \to I\!\!R\,\Gamma(X; \_) \to I\!\!R\,\Gamma(U; \_) \to \tag{4.89}$$

where $\Gamma_Z(X; \_)$ is the functor of sections with support in Z.

The cohomology of the right derived functor $I\!R\,\Gamma_Z(X;\,\_)$ is called the *local cohomology of X, relative to X*.[47]

In the case of the constant sheaf $\underline{\underline{\Bbbk}}_X$, the triangle (4.89) gives rise to the *long exact sequence of local cohomology*:[48]

$$\to H^i_Z(X;\Bbbk) \to H^i(X;\Bbbk) \to H^i(U;\Bbbk) \to \tag{4.90}$$

where we see that local cohomology is a relative cohomology. Indeed, we have a canonical isomorphism

$$H^i_Z(X;\Bbbk) \simeq H^i(X, X \smallsetminus Z;\Bbbk)\,,$$

as well as a canonical isomorphism in derived category

$$I\!R\,\Gamma_Z(X;\,\_) \simeq \hat{c}\big(\, I\!R\,\Gamma(X;\,\_) \to I\!R\,\Gamma(X;\,\_)\big)[-1]$$

where $\hat{c}(\_)$ denotes the mapping cone of a morphism of complexes (*cf.* A.1.3).

We will apply these consideration to the case of a closed inclusion $i : N \subseteq M$ of G-manifolds, and to each column of the following induced inclusion of inductive systems (*cf.* 4.5.3.1-(2))

$$\begin{array}{ccccccc}
\cdots\hookrightarrow & N_G(n) & \overset{\nu_n}{\hookrightarrow} & N_G(n{+}1) & \hookrightarrow\cdots\cdots\hookrightarrow & N_G \\
& \downarrow{\scriptstyle i_{G,n}} & \square & \downarrow{\scriptstyle i_{G,n+1}} & & \downarrow{\scriptstyle i_G} \\
\cdots\longrightarrow & M_G(n) & \overset{\mu_n}{\hookrightarrow} & M_G(n{+}1) & \hookrightarrow\cdots\cdots\hookrightarrow & M_G
\end{array} \tag{4.91}$$

Let $U := M \smallsetminus N$ and denote by $j_G : U_G \to M_G$ and $j_{G,n} : U_G(n) \subseteq M_G(n)$ the corresponding open inclusions.

Since all the spaces in consideration are mild spaces, diagram (4.91) gives rise to the following projective system of (vertical) long exact sequences of local (ordinary) cohomology with coefficients in a field $\Bbbk$:

$$\begin{array}{ccccccc}
\downarrow & & \downarrow & & \downarrow & \\
H^i_{N_G}(M_G) & \to\cdots\cdots\to & H^i_{N_G(n+1)}(M_G(n{+}1)) & \to & H^i_{N_G(n)}(M_G(n)) & \to\cdots\cdots \\
\downarrow & & \downarrow & & \downarrow & \\
H^i(M_G) & \to\cdots\cdots\longrightarrow & H^i(M_G(n{+}1)) & \longrightarrow & H^i(M_G(n)) & \longrightarrow\cdots\cdots \\
\downarrow & & \downarrow & & \downarrow & \\
H^i(U_G) & \to\cdots\cdots\longrightarrow & H^i(U_G(n{+}1)) & \longrightarrow & H^i(U_G(n)) & \longrightarrow\cdots\cdots \\
\downarrow & & \downarrow & & \downarrow &
\end{array} \tag{4.92}$$

---

[47] The terminology is due to Grothendieck, see Hartshorne [52] §1, pp. 1–15.

[48] See Hartshorne [52] §1, Proposition 1.9, p. 9.

where we apply Theorem 4.7.2.2-(3) and claim that the restrictions

$$H^i(M_G; \Bbbk) \to H^i(M_G(n); \Bbbk) \quad \text{and} \quad H^i(U_G; \Bbbk) \to H^i(U_G(n); \Bbbk)$$

are isomorphisms for all $n > i$. But then, by the Five Lemma, the same is true for local (ordinary) cohomology, i.e. we have isomorphisms

$$\boxed{H^i_{N_G}(M_G; \Bbbk) \underset{\simeq}{\to} H^i_{N_G(n)}(M_G(n); \Bbbk)\,, \quad \forall n > i\,,} \tag{4.93}$$

whence, an isomorphism

$$\boxed{H_{N_G}(M_G; \Bbbk) \to \varprojlim_n H_{N_G(n)}(M_G(n); \Bbbk)} \tag{4.94}$$

We can proceed in the same way with equivariant cohomology using the expression 4.8.3.2-(4.79) which gives the projective system of (vertical) long exact sequences of local (de Rham) cohomology:

$$\begin{array}{ccccccc}
\downarrow & & & & \downarrow & & \downarrow \\
(?) & \longrightarrow & \cdots\cdots & \to & H^i_{N_G(n+1)}(M_G(n+1)) & \to & H^i_{N_G(n)}(M_G(n)) & \to \cdots\cdots \\
\downarrow & & & & \downarrow & & \downarrow \\
H^i_G(M) & \to & \cdots\cdots & \longrightarrow & H^i(M_G(n+1)) & \longrightarrow & H^i(M_G(n)) & \longrightarrow \cdots\cdots \\
\downarrow & & & & \downarrow & & \downarrow \\
H^i_G(U) & \to & \cdots\cdots & \longrightarrow & H^i(U_G(n+1)) & \longrightarrow & H^i(U_G(n)) & \longrightarrow \cdots\cdots \\
\downarrow & & & & \downarrow & & \downarrow
\end{array} \tag{4.95}$$

where we now apply Theorem 4.8.2.1 and claim that the restrictions

$$H^i_G(M) \to H^i(M_G(n)) \quad \text{and} \quad H^i_G(U) \to H^i(U_G(n))$$

are isomorphisms for all $n > i$. But then, by the Five Lemma, we have isomorphisms for local (de Rham) cohomology

$$H^i_{N_G(m)}(M_G(m)) \underset{\simeq}{\to} H^i_{N_G(n)}(M_G(n))\,, \quad \forall m \geq n > i\,.$$

We are thus lead to <u>define</u> *the G-equivariant cohomology of M with supports in N*, or *the local equivariant cohomology of M relative to N*. (*cf*. 4.8.3.2)

$$\boxed{H_{G,N}(M) := \varprojlim_n H_{N_G(n)}(M_G(n))} \tag{4.96}$$

which clearly verifies, for fixed $i \in \mathbb{N}$, that the "restriction" homomorphisms

$$\boxed{H^i_{G,N}(M) \underset{\simeq}{\to} H^i_{NG(n)}(M_G(n))} \tag{4.97}$$

are isomorphisms for all $n \geq i$ (likewise (4.93)).

4.10.1 *The Long Exact Sequence of Local Equivariant Cohomology*

The definition (4.96) fills the missing term (?) in (4.95), giving the *long exact sequence of local equivariant de Rham cohomology for G-manifolds*:

$$\boxed{\to H^i_{G,N}(M) \to H^i_G(M) \to H^i_G(U) \to} \tag{4.98}$$

All these facts lead us to propose the following definition motivated by the fact that local cohomology is the relative cohomology of the pair (U, M), as recalled in the preamble of 4.10. The definition completes Definition 4.4.4.

Definition 4.10.1.1 Let G be a compact Lie group. Let $N \subseteq M$ be a closed inclusion of G-manifolds. The *complex of G-equivariant differential forms of M with supports in N*, is, by definition, the complex

$$\big(\Omega_{G,N}(M), d_G\big) := \hat{c}\big(\Omega_G(M) \to \Omega_G(M \smallsetminus N)\big)[-1]\,. \tag{4.99}$$

where $\hat{c}(\_)$ denotes the mapping cone of a morphism of complexes (*cf*. A.1.3).

The following theorem ends the chapter showing that this definition is well-founded and in agreement with the previous observations.

Local Equivariant de Rham Theorem 4.10.1.2 *Let G be a compact Lie group, and let $N \subseteq M$ be a closed inclusion of G-manifolds. Then, there exist canonical isomorphisms*

$$\mathbf{h}\big(\Omega_{G,N}(M), d_G\big) \simeq H_{G,N}(M) \simeq H_{N_G}(M_G; \mathbb{R})\,.$$

Proof Set $U := M \smallsetminus N$, and let $j : U \to M$ denote the inclusion map. for $n \in \mathbb{N}$, denote by $p_2 : \mathbb{E}G(n) \times (\_) \to (\_)$ the canonical projection (as in Sect. 4.8.2).

- $H\big(\Omega_{G,N}(M), d_G\big) \simeq H_{G,N}(M)$. The commutative diagram (I) of Cartan complexes, canonically extends to a morphism of exact triangles defined by the mapping cones of $j_G^* : \Omega_G(M) \to \Omega_G(N)$ and $j(n)^* : \Omega(M_G(n)) \to \Omega(N_G(n))$:

$$\begin{array}{cccccc}
\Omega_{G,N}(M) & \dashrightarrow & \Omega_G(M) & \xrightarrow{\ j_G^*\ } & \Omega_G(U) & \dashrightarrow \\
{\scriptstyle q^*}\downarrow & & {\scriptstyle p_2^*}\downarrow & (\mathrm{I}) & {\scriptstyle p_2^*}\downarrow & \\
\Omega_{N_G(n)}(M_G(n)) & \dashrightarrow & \Omega(M_G(n)) & \xrightarrow{\ j(n)^*\ } & \Omega(U_G(n)) & \dashrightarrow
\end{array}$$

which gives rise to a morphism of long exact sequences

$$\begin{array}{ccccccc}
\longrightarrow & h^i\big(\Omega_{G,N}(M)\big) & \longrightarrow & H^i_G(M) & \longrightarrow & H^i_G(U) & \longrightarrow \\
 & \simeq\big\downarrow{\scriptstyle (1)} & & \simeq\big\downarrow{\scriptstyle (2)} & & \simeq\big\downarrow{\scriptstyle (3)} & \\
\longrightarrow & H^i_{N_G(n)}(M(n)) & \longrightarrow & H^i(M_G(n)) & \longrightarrow & H^i(U_G(n)) & \longrightarrow
\end{array}$$

where the arrow (1) is an isomorphism for all $n > i$ since this is so of arrows (2, 3) after Theorem 4.8.2.1 and Remark 4.8.3.2. Hence, the isomorphism

$$h\big(\Omega_{G,N}(M)\big) \xrightarrow[\simeq]{} \varprojlim_n H_{N_G(n)}(M(n)) =: H_{G,N}(M)\,,$$

after the definition 4.96.

- $H_{G,N}(M) \simeq H_{N_G}(M_G; \mathbb{R})$. This isomorphism follows in the same way as Theorem 4.8.3.1. The ingredients of the proof, i.e. formulas (4.93), (4.94), (4.96) and (4.97), have already been justified. □

Exercise 4.10.1.3 *A spectral sequence for local equivariant cohomology.*

1. Let $\alpha : (C_1, d, \theta, \iota) \to (C_2, d, \theta, \iota)$ be a morphism of positively graded $\mathfrak{g}$-dgm's. Extend Theorem 4.3.3.1-(3) to construct a convergent spectral sequence

$$\Big(\mathbb{E}_2^{p,q} = S^p(\mathfrak{g}^\vee)^{\mathfrak{g}} \otimes H^q(\hat{c}(\alpha))\,,\ d_2 = \textstyle\sum_i e^i \otimes \iota(e_i)\Big) \Rightarrow H_{\mathfrak{g}}^{p+q}(\hat{c}(\alpha))\,.$$

2. Let G be a compact Lie group, and let $N \subseteq M$ be a closed inclusion of G-manifolds. Use (1) to construct a spectral sequence $\mathbb{E}_{G,N,r}(M))$, such that:

$$\mathbb{E}_{G,N,2}(M)^{p,q} := \big(S^p(\mathfrak{g}^\vee)^{\mathfrak{g}} \otimes H^q_N(M)\big)^W \Rightarrow H^{p+q}_{G,N}(M)\,,$$

where $W := G/G_0$.
3. Show that there exists a long exact sequence of spectral sequences

$$\longrightarrow \mathbb{E}_{G,N,r}(M) \longrightarrow \mathbb{E}_{G,r}(M) \longrightarrow \mathbb{E}_{G,r}(M \smallsetminus N) \longrightarrow$$

where $\mathbb{E}_{G,r}(\_)$ denotes the spectral sequence for equivariant cohomology in Corollary 4.4.6.2-(3). *Hint. Exercise 5.4.4 in Weibel [95], page 135.* (☝, p. 349)

Comment 4.10.1.4 We will see in Sect. 6.4.6, which introduces the Gysin long exact sequence associated with a closed inclusion of oriented G-manifolds $N \subseteq M$, the existence of the canonical isomorphism (6.34)

$$H_{G,N}(M) \simeq H_G(N)[d_N - d_M]\,.$$

Chapter 5
Equivariant Poincaré Duality

5.1 Differential Graded Modules over a Graded Algebra

We begin extending the definitions and terminology of Sect. 2.1, by replacing the field $\Bbbk$ by a graded $\Bbbk$-algebra $\boldsymbol{A}$ (2.1.5).

5.1.1 Graded Modules and Algebras over Graded Algebras

Let $\boldsymbol{A}$ be a graded $\Bbbk$-algebra. An *$\boldsymbol{A}$-graded module*, *$\boldsymbol{A}$-gm* in short, is a graded space $\boldsymbol{V} \in \mathrm{GV}(\Bbbk)$ together with a morphism $\boldsymbol{A} \to \mathrm{Endgr}_{\Bbbk}^{*}(\boldsymbol{V})$ of graded $\Bbbk$-algebras. Given two $\boldsymbol{A}$-gm's $\boldsymbol{V}$ and $\boldsymbol{W}$, a *graded homomorphism of $\boldsymbol{A}$-gm's of degree d from $\boldsymbol{V}$ to $\boldsymbol{W}$* is a graded homomorphism of graded spaces $\boldsymbol{\alpha} : \boldsymbol{V} \to \boldsymbol{W}$ of degree d (2.1.3), compatible with the action of $\boldsymbol{A}$, i.e. $\boldsymbol{\alpha}(\boldsymbol{a} \cdot \boldsymbol{v}) = \boldsymbol{a} \cdot \boldsymbol{\alpha}(\boldsymbol{v})$. The space of such homomorphisms is denoted by $\mathrm{Homgr}_{\boldsymbol{A}}^{d}(\boldsymbol{V}, \boldsymbol{W})$. The graded space of *graded homomorphisms of $\boldsymbol{A}$-gm's* from $\boldsymbol{V}$ to $\boldsymbol{W}$ is then

$$\mathrm{Homgr}_{\boldsymbol{A}}^{*}(\boldsymbol{V}, \boldsymbol{W}) = \left\{ \mathrm{Homgr}_{\boldsymbol{A}}^{d}(\boldsymbol{V}, \boldsymbol{W}) \right\}_{d \in \mathbb{Z}}, \tag{5.1}$$

which is again an $\boldsymbol{A}$-graded module.

5.1.1.1 The the $\boldsymbol{A}$-gm's are the objects of the *category* $\mathrm{GM}(\boldsymbol{A})$ *of $\boldsymbol{A}$-graded modules*. The *morphisms* in $\mathrm{GM}(\boldsymbol{A})$ are the graded homomorphisms of degree 0, hence we set $\mathrm{Mor}_{\mathrm{GM}(\boldsymbol{A})}(\boldsymbol{V}, \boldsymbol{W}) := \mathrm{Homgr}_{\boldsymbol{A}}^{0}(\boldsymbol{V}, \boldsymbol{W})$. The category $\mathrm{GM}(\boldsymbol{A})$ is abelian. The full subcategories of bounded graded modules $\mathrm{GM}^{b}(\boldsymbol{A})$, $\mathrm{GM}^{+}(\boldsymbol{A})$, $\mathrm{GM}^{\geq \ell}(\boldsymbol{A})$, $\mathrm{GM}^{-}(\boldsymbol{A})$ and $\mathrm{GM}^{\leq \ell}(\boldsymbol{A})$ are as in Sect. 2.1.4.

5.1.1.2 An *$\boldsymbol{A}$-graded algebra* is an $\boldsymbol{A}$-graded module $\boldsymbol{B} \in \mathrm{GM}(\boldsymbol{A})$ together with a *multiplication*, i.e. a family of $\boldsymbol{A}$-bilinear maps $\{\cdot : B^{a} \times B^{b} \to B^{a+b}\}_{a,b \in \mathbb{Z}}$, such that the triple $(\boldsymbol{B}, 0, +, \cdot)$ verifies the axioms of an $\boldsymbol{A}$-algebra.

A. Arabia, *Equivariant Poincaré Duality on G-Manifolds*, Lecture Notes in Mathematics 2288, https://doi.org/10.1007/978-3-030-70440-7_5

The algebra is said to be

- *positively graded*: if $B^m = 0$ for all $m < 0$,
- *evenly graded*: if $B^m = 0$, for all odd m,
- *anticommutative*:[1] if $b_1 \cdot b_2 := (-1)^{d_1 d_2} b_2 \cdot b_1$, for all $b_1 \in B^{d_1}$, $b_2 \in B^{d_2}$.

A *morphism of* $\boldsymbol{A}$*-graded (resp. unital) algebras* $\boldsymbol{\alpha} : (\boldsymbol{B}, 0, +, \cdot) \to (\boldsymbol{B}', 0, +, \cdot)$ is a morphism of $\boldsymbol{A}$-graded modules $\boldsymbol{\alpha} \in \mathrm{Homgr}^0_{\boldsymbol{A}}(\boldsymbol{B}, \boldsymbol{B}')$ that is compatible with the multiplication operation, i.e. $\boldsymbol{\alpha}(x \cdot y) = \boldsymbol{\alpha}(x) \cdot \boldsymbol{\alpha}(y)$ (resp. $\boldsymbol{\alpha}(\mathbf{1}_{\boldsymbol{A}}) = \mathbf{1}_{\boldsymbol{B}}$).

Examples 5.1.1.3

(a) Given a G-manifold M, the complexes $\Omega_G(M)$ and $\Omega_{G,c}(M)$ are positively $S(\mathfrak{g}^\vee)^G$-graded anticommutative algebras.
(b) For any $\boldsymbol{V} \in \mathrm{GM}(\boldsymbol{A})$, the $\boldsymbol{A}$-graded module $(\mathrm{Endgr}^*_{\boldsymbol{A}}(\boldsymbol{V}), 0, +, \mathrm{id}, \circ)$ of graded endomorphisms is a noncommutative unital graded $\boldsymbol{A}$-algebra.

5.1.2 The Category of Ω_G-Graded Modules

We now turn to the graded algebra (*cf.* 4.4.4.1),

$$\boxed{\Omega_G := \Omega_G(\{\bullet\}) = S(\mathfrak{g}^\vee)^G}$$

where G is a compact Lie group. The *category of* Ω_G*-graded modules* is denoted by $\mathrm{GM}(\Omega_G)$, where $\mathrm{Mor}_{\mathrm{GM}(\Omega_G)}(\_, \_) := \mathrm{Hom}_{\Omega_G}(\_, \_)$.

Definition 5.1.2.1 A direct sum $\bigoplus_{\mathrm{i} \in \mathfrak{I}} \Omega_G[m_{\mathrm{i}}]$, with $m_{\mathrm{i}} \in \mathbb{Z}$, is called *a free* Ω_G*-graded module.*

Proposition 5.1.2.2

1. *An object* $\boldsymbol{V} \in \mathrm{GM}(\Omega_G)$ *is* projective (resp. injective) *if and only if the functor* $\mathrm{Homgr}^*(\boldsymbol{V}, \_) : \mathrm{GM}(\Omega_G) \rightsquigarrow \mathrm{GM}(\Omega_G)$ (resp. $\mathrm{Homgr}^*(\_, \boldsymbol{V})$) *is exact.*
2. [2] *The category* $\mathrm{GM}(\Omega_G)$ *is an abelian category with enough injective and projective objects. A graded module* $\boldsymbol{V} \in \mathrm{GM}^{\geq \ell}(\Omega_G)$ *admits a free resolution*

$$\cdots \xrightarrow{d_2} \boldsymbol{F}_2 \xrightarrow{d_1} \boldsymbol{F}_1 \xrightarrow{d_0} \boldsymbol{F}_0 \overset{\epsilon}{\twoheadrightarrow} \boldsymbol{V} \to 0$$

with $\boldsymbol{F}_i \in \mathrm{GM}^{\geq \ell}(\Omega_G)$*. In particular,* $\mathrm{GM}^{\geq \ell}(\Omega_G)$ *and* $\mathrm{GM}^+(\Omega_G)$ *are abelian categories with enough projective objects.*
3. *The homological dimension*[3] *of* $\mathrm{GM}(\Omega_G)$ *is finite.*

---

[1] A synonym of *graded commutative*, frequently used in the works of Cartan [25, 26].

[2] See Grothendieck [49], chapter **I**, Thm. 1.10, p. 135.

[3] See Weibel [95], Global Dimension Theorem 4.1.2, p. 91, or Jacobson [59], 6.12 Homological Dimension, p. 375.

Proof

(1) immediate since $\mathrm{Homgr}^*_{\Omega_G}(\_,\_)$ can be seen as the product

$$\mathrm{Homgr}^*_{\Omega_G}(\_,\_) = \prod\nolimits_{m\in\mathbb{Z}} \mathrm{Homgr}^0_{\Omega_G}(\_,\_[m]) = \prod\nolimits_{m\in\mathbb{Z}} \mathrm{Homgr}^0_{\Omega_G}(\_[m],\_)\,,$$

where the shift functor $[m]$ is exact.

(2) *Enough Projectives.* Let $\{v_\mathfrak{i}\}_{\mathfrak{i}\in\mathfrak{I}}$ be a family of *homogeneous* (2.1.3.1) generators for $\boldsymbol{V} \in \mathrm{GM}(\Omega_G)$ and consider, for each $\mathfrak{i} \in \mathfrak{I}$, the map $\gamma_\mathfrak{i} : \Omega_G[-d_\mathfrak{i}] \to \boldsymbol{V}, x \mapsto x v_\mathfrak{i}$ which is clearly a morphism in $\mathrm{GM}(\Omega_G)$. The sum

$$\sum\nolimits_{\mathfrak{i}\in\mathfrak{I}} \gamma_\mathfrak{i} : \bigoplus\nolimits_{\mathfrak{i}\in\mathfrak{I}} \Omega_G[-d_\mathfrak{i}] \twoheadrightarrow \boldsymbol{V} \tag{5.2}$$

represents $\boldsymbol{V}$ as the quotient in $\mathrm{GM}(\Omega_G)$ of a free, and thus projective, Ω_G-gm.

Enough Injectives.[4] The correspondence

$$\boldsymbol{V} \rightsquigarrow \widehat{\boldsymbol{V}} := \mathrm{Homgr}^*_{\mathbb{Z}}(\boldsymbol{V}, (\mathbb{Q}/\mathbb{Z})[0]) \tag{5.3}$$

is an additive contravariant functor from the category of *left* (resp. *right*) Ω_G-gm to the category of *right* (resp. *left*) Ω_G-gm,[5] and is exact, by (1), since

$$\mathrm{Homgr}^0_{\mathbb{Z}}(\_, (\mathbb{Q}/\mathbb{Z})[0]) = \mathrm{Hom}_{\mathbb{Z}}((\_)^0, \mathbb{Q}/\mathbb{Z})$$

and since $\mathbb{Q}/\mathbb{Z}$ is an injective $\mathbb{Z}$-module.

Lemma 1 *The map* $\nu(\boldsymbol{V}) : \boldsymbol{V} \to \widehat{\widehat{\boldsymbol{V}}}$, $v \mapsto (\gamma \mapsto \gamma(v))$ *is an* injective *morphism.*

Proof of Lemma 1 Since $\nu(\boldsymbol{V})$ is clearly a morphism of graded modules, it is injective if and only if it does not kill any homogeneous nonzero element. If $0 \neq v \in \boldsymbol{V}^d$, the subgroup $\mathbb{Z}\cdot v \subseteq \boldsymbol{V}^d$ is isomorphic to some $\mathbb{Z}/n\mathbb{Z}$ for $n \neq \pm 1$, and there exists a nonzero homomorphism $\gamma'' : \mathbb{Z}\cdot v \to \mathbb{Q}/\mathbb{Z}$ (exercise), restriction of some $\gamma' : \boldsymbol{V}^d \to \mathbb{Q}/\mathbb{Z}$ (thanks to the injectivity of $\mathbb{Q}/\mathbb{Z}$). Extend this γ' to the whole of $\boldsymbol{V}$, assigning the value 0 on the homogeneous factors $\boldsymbol{V}^e$ when $e \neq d$.

This last extension, denoted by $\gamma : \boldsymbol{V} \to \mathbb{Q}/\mathbb{Z}$, is a graded morphism of degree $-d$ and verifies $\nu(\boldsymbol{V})(v)(\gamma) = \gamma(v) \neq 0$ by construction, so that $\nu(\boldsymbol{V})(v) \neq 0$, which completes the proof of Lemma 1. ⊟

Lemma 2 *For any free* right Ω_G*-gm* $\boldsymbol{F}$*, the* left Ω_G*-gm* $\widehat{\boldsymbol{F}}$ *is injective.*

[4]We reproduce the proof in Godement [46] Thm. 1.4.1, p. 6, in the graded framework.

[5]If $\boldsymbol{N}$ is a *right* Ω_G-gm, the structure of *left* Ω_G-module of $\mathrm{Homgr}^*_{\mathbb{Z}}(\boldsymbol{N}, (\mathbb{Q}/\mathbb{Z})[0])$ is given by $(x\cdot\gamma)(y) := \gamma(yx)$ for all $x \in \Omega_G$ and $y \in \boldsymbol{N}$. If $\boldsymbol{N}$ is a *left* Ω_G-gm, the structure of *right* Ω_G-module of $\mathrm{Homgr}^*_{\mathbb{Z}}(\boldsymbol{N}, (\mathbb{Q}/\mathbb{Z})[0])$ is given by $(\gamma\cdot x)(y) := \gamma(xy)$ for all $x \in \Omega_G$ and $y \in \boldsymbol{N}$.

Proof of Lemma 2 We recall (*cf*. [19] Chap. II, §4, Prop. 1) that for any left Ω_G-dgm $\boldsymbol{N}$, the maps

$$\begin{array}{ccc} \mathrm{Homgr}^*_{\Omega_G}(\boldsymbol{N}, \mathrm{Homgr}^*_{\mathbb{Z}}(\Omega_G, (\mathbb{Q}/\mathbb{Z})[0])) & \rightleftarrows & \mathrm{Homgr}^*_{\mathbb{Z}}(\boldsymbol{N}, (\mathbb{Q}/\mathbb{Z})[0]) \\ \gamma & \longmapsto & \big(v \mapsto \gamma(v)(1)\big) \\ \big(v \mapsto (x \mapsto \bar{\xi}(xv))\big) & \longleftarrow & \xi \end{array}$$

are isomorphisms inverse of each other. It follows that $\mathrm{Homgr}^*_{\mathbb{Z}}(\Omega_G, (\mathbb{Q}/\mathbb{Z})[0]))$ is an injective left Ω_G-gm if and only if $\mathrm{Homgr}^*_{\mathbb{Z}}(\_, (\mathbb{Q}/\mathbb{Z})[0])$ is exact, but this is equivalent, by (1), to the exactness of the functor $\mathrm{Homgr}^0_{\mathbb{Z}}(\_, (\mathbb{Q}/\mathbb{Z})[0]) = \mathrm{Hom}_{\mathbb{Z}}((\_)^0, \mathbb{Q}/\mathbb{Z})$, which is clear since $\mathbb{Q}/\mathbb{Z}$ is an injective $\mathbb{Z}$-module. ⊟

Let $\boldsymbol{V}$ be a left Ω_G-gm. Fix some epimorphism of right Ω_G-gm $\pi : \boldsymbol{F} \twoheadrightarrow \widehat{\boldsymbol{V}}$ where $\boldsymbol{F}$ is free as in (5.2). The morphism $\widehat{\pi} : \widehat{\widehat{\boldsymbol{V}}} \to \widehat{\boldsymbol{F}}$ is injective and, composed with $\nu(\boldsymbol{V}) : \boldsymbol{V} \to \widehat{\widehat{\boldsymbol{V}}}$, which is also injective by Lemma 1, gives an injective morphism $\boldsymbol{V} \hookrightarrow \widehat{\boldsymbol{F}}$ of left Ω_G-gm, where $\widehat{\boldsymbol{F}}$ is an injective left Ω_G-gm by Lemma 2. This completes the proof of the existence of enough injective objects in $\mathrm{GM}(\Omega_G)$.

The fact that $\mathrm{GM}^{\geq \ell}(\Omega_G)$ and $\mathrm{GM}^+(\Omega_G)$ are abelian categories is obvious.

The proof of the existence of enough projective modules also shows that a graded module $\boldsymbol{V} \in \mathrm{GM}^{\geq \ell}(\Omega_G)$ is a quotient $\epsilon : \boldsymbol{F}_0 \twoheadrightarrow \boldsymbol{V}$ of a free Ω_G-graded module $\boldsymbol{F}_0 \in \mathrm{GM}^{\geq \ell}(\Omega_G)$, and since $\ker(\epsilon) \in \mathrm{GM}^{\geq \ell}(\Omega_G)$, we can iterate the procedure.

(3) When G is connected, Ω_G is a polynomial algebra in $\mathrm{rk}(G)$ variables[6] and we can apply Hilbert's Syzygy Theorem.[7] Otherwise $\Omega_G = (\Omega_{G_0})^{G/G_0}$, and $\dim_{\mathrm{h}}(\Omega_G) \leq \dim_{\mathrm{h}}(\Omega_{G_0})$ since $\Omega_{G_0} = S(\mathfrak{g}^\vee)^{\mathfrak{g}}$ is a polynomial algebra over a field of characteristic 0 and that G/G_0 is a finite group. (☡, p. 349) □

Exercise 5.1.2.3 Let $\boldsymbol{A}$ be a graded $\mathbb{R}$-algebra which is an integral domain.

1. Let S denote the multiplicative system of homogeneous nonzero elements of $\boldsymbol{A}$. Show that $S^{-1}\boldsymbol{A}$ is an injective object of $\mathrm{GM}(\boldsymbol{A})$ (see ch. D). Prove that the canonical inclusion $\boldsymbol{A} \hookrightarrow S^{-1}\boldsymbol{A}$ is an *injective envelope* for $\boldsymbol{A}$.[8]
2. Show that when $\mathrm{rk}(G) > 0$, the degrees of a non trivial injective object of $\mathrm{GM}(\Omega_G)$ cannot be bounded below (2.1.4).

The next two sections are straightforward generalizations of Sects. 2.1.6 and 2.1.8 from graded vector spaces to Ω_G-graded modules.

---

[6]This is Theorem 19.1, p. 171, in Armand Borel's Ph.D. thesis [13]. The result is based on the identification $S(\mathfrak{g}^\vee)^{\mathfrak{g}} = S(\mathfrak{t}^\vee)^W$, where $W := N_G(T)/T$ is the Weyl group of the pair (G, T), and the celebrated Chevalley's theorem stating that $S(\mathfrak{t}^\vee)^W$ is a polynomial algebra in $\mathrm{rk}(G)$ variables, see Chevalley [30].

[7]See Jacobson [59] Hilbert's Theorem in p. 385, Corollary, p. 386, and Ex. 2, p. 387.

[8]The injective envelope of an object $\boldsymbol{O}$, also called its *injective hull*, is the '*smallest*' injective object containing $\boldsymbol{O}$. More precisely, it is an injective object $\boldsymbol{I}$ together with a monomorphism $\epsilon : \boldsymbol{O} \hookrightarrow \boldsymbol{I}$ such that any other monomorphism $\eta : \boldsymbol{O} \hookrightarrow \boldsymbol{J}$ where $\boldsymbol{J}$ is injective, factors through ϵ in a monomorphism $\iota : \boldsymbol{I} \hookrightarrow \boldsymbol{J}$, i.e. $\eta = \iota \circ \epsilon$.

5.2 The Category of Ω_G-Differential Graded Modules

5.2.1 *Definition*

An Ω_G*-differential graded module*, Ω_G*-dgm* in short, is a pair $(\boldsymbol{V}, \boldsymbol{d})$ with $\boldsymbol{V} \in \mathrm{GM}(\Omega_G)$ and $\boldsymbol{d} \in \mathrm{Endgr}^1_{\Omega_G}(\boldsymbol{V})$, *the differential*, such that $(\boldsymbol{V}, \boldsymbol{d})$ is a complex of vector spaces, i.e. $\boldsymbol{d}^2 = 0$.

A *morphism of* Ω_G*-dgm* $\boldsymbol{\alpha} : (\boldsymbol{V}, \boldsymbol{d}) \to (\boldsymbol{V}', \boldsymbol{d}')$ is a morphism of Ω_G-gm's and a morphism of complexes, i.e. $\boldsymbol{d}' \circ \boldsymbol{\alpha} = \boldsymbol{\alpha} \circ \boldsymbol{d}$.

The Ω_G-dgm's and their morphisms constitute the *category* $\mathrm{DGM}(\Omega_G)$ *of* Ω_G*-differential graded modules*. The category $\mathrm{DGM}(\Omega_G)$ is an abelian category.

Comment 5.2.1.1 This definition of differential graded modules is specific to evenly graded algebras with zero differential, as is Ω_G. In the general definition of differential graded modules over a differential graded algebra $(\boldsymbol{A}, \boldsymbol{d})$, the anticommutativity of the differentiation operation is required, i.e. we also need to satisfy the Koszul sign rule by which $d(a \cdot x) = d(a) \cdot x + (-1)^{[a]} a \cdot d(x)$, for homogeneous $a \in \boldsymbol{A}$, in which case the action of $\boldsymbol{d}$ does not necessarily commute with the action of $\boldsymbol{A}$ (*cf*. 2.1.9).

5.2.2 *The* $\mathbf{Hom}^\bullet_{\Omega_G}(\_, \_)$ *and* $(\_ \otimes_{\Omega_G} \_)^\bullet$ *Bifunctors on* $\mathrm{DGM}(\Omega_G)$

Given two Ω_G-dgm's $(\boldsymbol{V}, \boldsymbol{d})$ and $(\boldsymbol{V}', \boldsymbol{d}')$, we define the two Ω_G-dgm's:

$$\left(\mathbf{Hom}^\bullet_{\Omega_G}(\boldsymbol{V}, \boldsymbol{V}'), \boldsymbol{D}_\bullet\right) \quad \text{and} \quad \left((\boldsymbol{V} \otimes_{\Omega_G} \boldsymbol{V}')^\bullet, \boldsymbol{\Delta}_\bullet\right).$$

As Ω_G-graded modules, defined by

$$m \mapsto \begin{cases} \mathbf{Hom}^m_{\Omega_G}(\boldsymbol{V}, \boldsymbol{V}') & := \mathrm{Homgr}^0_{\Omega_G}(\boldsymbol{V}, \boldsymbol{V}'[m]) = \mathrm{Homgr}^0_{\Omega_G}(\boldsymbol{V}[-m], \boldsymbol{V}') \\ (\boldsymbol{V} \otimes_{\Omega_G} \boldsymbol{V}')^m & := \pi\left((\boldsymbol{V} \otimes_{\mathbb{R}} \boldsymbol{V}')^m\right) \end{cases} \tag{5.4}$$

where $\pi : \boldsymbol{V} \otimes_{\mathbb{R}} \boldsymbol{V}' \twoheadrightarrow \boldsymbol{V} \otimes_{\Omega_G} \boldsymbol{V}'$, $v \otimes v' \mapsto [v \otimes v']$, is the canonical (graded) surjection (see remark 5.2.2.1-(2)). The differentials $\boldsymbol{D}_\bullet$ and $\boldsymbol{\Delta}_\bullet$ are:

$$\begin{cases} D_m(f) = \boldsymbol{d}' \circ f - (-1)^m f \circ \boldsymbol{d} \\ \boldsymbol{\Delta}_m([v \otimes v']) = [\boldsymbol{d}(v) \otimes v'] + (-1)^a [v \otimes \boldsymbol{d}'(v')] \end{cases} \tag{5.5}$$

where $v \otimes v' \in V^a \otimes V'^b$ and $a + b = m$. Notice that the two differentials are Ω_G-linear.

5.2.2.1 As in Definition 2.1.8.1, these constructions are natural w.r.t. each entry. They define the two bifunctors:

$$\begin{cases} \mathbf{Hom}^{\bullet}_{\Omega_G}(\_,\_) : \mathrm{DGM}(\Omega_G) \times \mathrm{DGM}(\Omega_G) \rightsquigarrow \mathrm{DGM}(\Omega_G)\,, \\ (\_ \otimes_{\Omega_G} \_)^{\bullet} : \mathrm{DGM}(\Omega_G) \times \mathrm{DGM}(\Omega_G) \rightsquigarrow \mathrm{DGM}(\Omega_G)\,. \end{cases} \tag{5.6}$$

The '$\mathbf{Hom}^{\bullet}_{\Omega_G}$' functor is contravariant left exact on the first entry, and covariant left exact on the second entry. The '$\otimes$' functor is covariant right exact on each entry (compare with Definition 2.1.8.1).

Comments 5.2.2.1

1. The formulas (5.5), by which $\boldsymbol{D}$ and $\boldsymbol{\Delta}$ are Ω_G-linear, apply only where Ω_G is evenly graded with zero differential, otherwise the formulas must include additional signs and terms related to graded commutativity and graded derivability (*cf.* 5.2.1.1).
2. Some care must be taken with the tensor product, which can conceal some subtleties. A good way to understand this is to note that $\boldsymbol{V} \otimes_{\Omega_G} \boldsymbol{V}'$ is the quotient of the graded vector space $\boldsymbol{V} \otimes_{\mathbb{R}} \boldsymbol{V}'$ by the subspace $\boldsymbol{W}$ spanned by the tensors $Pv \otimes v\prime - v \otimes Pv\prime$ with $P \in \Omega_G$ and $(v, v\prime) \in \boldsymbol{V} \times \boldsymbol{V}'$ *homogeneous*. $\boldsymbol{W}$ is a graded subcomplex of $(\boldsymbol{V} \otimes_{\mathbb{R}} \boldsymbol{V}', \boldsymbol{\Delta})$ and the canonical surjection $\pi : (\boldsymbol{V} \otimes_{\mathbb{R}} \boldsymbol{V}', \boldsymbol{\Delta}) \twoheadrightarrow (\boldsymbol{V} \otimes_{\mathbb{R}} \boldsymbol{V}', \boldsymbol{\Delta})/\boldsymbol{W}$ is an epimorphism of graded complexes, therefore inducing over $\boldsymbol{V} \otimes_{\Omega_G} \boldsymbol{V}'$ a structure of Ω_G-dgm. Again, this setting is specific to the fact that Ω_G is evenly graded.

5.2.3 *The Duality Functor on* DGM(Ω_G)

In 2.1.11, we introduced the duality functor $\mathbf{Hom}^{\bullet}_{\Bbbk}(\_, \Bbbk) : \mathrm{DGV}(\Bbbk) \rightsquigarrow \mathrm{DGV}(\Bbbk)$ and noted that it was exact (2.1.11.2). In the framework of Ω_G-dgm's, the corresponding functor is the Ω_G*-duality functor*

$$\mathbf{Hom}^{\bullet}_{\Omega_G}(\_, \Omega_G) : \mathrm{DGM}(\Omega_G) \rightsquigarrow \mathrm{DGM}(\Omega_G) \tag{5.7}$$

which is contravariant left exact, but not exact.

5.2.4 *The Forgetful Functor*

If we disregard differentials, an Ω_G-dgm is simply as an Ω_G-gm, and likewise for morphisms. This is the action of the *the forgetful functor*, which we denote by

$$o : \mathrm{DGM}(\Omega_G) \rightsquigarrow \mathrm{GM}(\Omega_G)\,.$$

It is an exact functor, which will often be implicit in our considerations.

5.2.5 *On the Exactness of* $\mathbf{Hom}^\bullet(\_, \_)$ *and* $(\_ \otimes \_)^\bullet$

The most relevant difference between Sect. 2.1.8 and 5.2.2 is the change of the coefficients ring A from a field $\Bbbk$ to the graded ring Ω_G. When $A = \Bbbk$, the categories $\mathrm{Vec}(\Bbbk)$ and $\mathrm{GV}(\Bbbk)$ are both *split* ((3), p. 13), which implies that additive functors on these categories always induce exact functors on the categories $C(\Bbbk)$ and $C(\mathrm{GV}(\Bbbk))$, where, in particular, they preserve acyclicity and quasi-isomorphism (*cf*. Exercise 2.1.6.1). By contrast, when $A = \Omega_G$, the category $\mathrm{GM}(\Omega_G)$ is no longer split, and additive functors are no longer necessarily exact on the categories $C(\mathrm{GM}(\Omega_G))$ and $\mathrm{DGM}(\Omega_G)$ (*cf*. 5.2.5.1-(1)).

The functor $\mathbf{Hom}^\bullet_{\Bbbk}(\_, \Bbbk)$ is central to proving nonequivariant Poincaré duality and its exactness on $C(\mathrm{GV}(\Bbbk))$ is crucial for the proof to work. When we follow the same approach in the equivariant setting, we are lead to consider $\mathbf{Hom}^\bullet_{\Omega_G}(\_, \Omega_G)$ which is not exact on $\mathrm{DGM}(\Omega_G)$ (*cf*. 5.2.5.1-(2)), causing the attempt at establishing equivariant Poincaré duality to fail.

This failure of exactness is the main reason for introducing the *derived categories* $\mathcal{D}(\mathrm{GM}(\Omega_G))$ *and* $\mathcal{D}(\mathrm{DGM}(\Omega_G))$ and the corresponding *derived duality* functor $I\!R\,\mathbf{Hom}^\bullet_{\Omega_G}(\_, \Omega_G)$, which we will address in Sects. 5.4.5 and 5.4.6 below. We will see that replacing $\mathbf{Hom}^\bullet_{\Omega_G}(\_, \Omega_G)$ by $I\!R\,\mathbf{Hom}^\bullet_{\Omega_G}(\_, \Omega_G)$ suffices to overcome the obstructions in the proof of equivariant Poincaré duality, justifying *a posteriori* the use of derived categories.

Exercises 5.2.5.1

1. Let A be a noetherian ring.
   a. Show that the simple objects[9] in the category $\mathrm{Mod}(A)$ of A-modules are the fields $A/\mathfrak{M}$, where $\mathfrak{M}$ is a maximal ideal in A.
   b. Show that $\mathrm{Mod}(A)$ is split (fn. (3), p. 13) if and only if A is a *finite* product of fields. Conclude that $\mathrm{Mod}(\Omega_G)$ is not split if $\dim \mathfrak{g} > 0$.
   c. Show that $\mathrm{Mod}(A)$ is split if and only if the functor $\mathrm{Hom}_A(S, \_) : \mathrm{Mod}(A) \rightsquigarrow \mathrm{Mod}(A)$ is exact for every simple A-module S. (*cf*. Exercise 2.1.6.1).
   d. Same as (1c) for the functors $\mathrm{Hom}_A(\_, S)$ and $S \otimes_A (\_)$.
2. Show that if $\mathrm{Mod}(A)$ is not split, then the A-duality functor does not preserve quasi-isomorphisms of complexes. Apply to the Ω_G-duality functor (5.7).

[9] An object S in an abelian category Ab is called *simple* if $S \neq 0$ and if 0 and S are its only subobjects.

5.3 Comparing the Categories $C(\mathrm{GM}(\Omega_G))$ and $\mathrm{DGM}(\Omega_G)$

This section is quite technical, so first-time readers might choose to skip it. It examines the differences between the categories of complexes of Ω_G-graded modules and of differential graded modules over Ω_G.

In an abelian category Ab the *category* $C(\mathrm{Ab})$ *of complexes of* Ab is defined in the same way as for vector spaces. A *complex of objects of* Ab is a sequence

$$(\boldsymbol{X}, \mathbf{d}) := \big(\cdots \xrightarrow{d_{i-2}} X_{i-1} \xrightarrow{d_{i-1}} X_i \xrightarrow{d_i} X_{i+1} \xrightarrow{d_{i+1}} \cdots\big). \tag{5.8}$$

of objects $X_i \in \mathrm{Ob(Ab)}$ where $d_i \in \mathrm{Mor}_{\mathrm{Ab}}(X_i, X_{i+1})$ and $d_{i+1} \circ d_i = 0$. A *morphism of complexes* $\boldsymbol{\alpha} : (\boldsymbol{X}, \mathbf{d}) \to (\boldsymbol{X}', \mathbf{d}')$ is then a family of morphisms $\alpha_i \in \mathrm{Mor(Ab)}(C_i, C_i')$ such that $d'_{i+1} \circ \alpha_i = \alpha_{i+1} \circ d_i$.

In this presentation, the objects in the complex have no grading, but if we use the subscripts $i \in \mathbb{Z}$ as degrees, we get a sequence (*cf*. 2.1.7-(S-1))

$$\big(\cdots \xrightarrow{d_{i-2}} X_{i-1}[-i+1] \xrightarrow{d_{i-1}} X_i[-i] \xrightarrow{d_i} X_{i+1}[-i-1] \xrightarrow{d_{i+1}} \cdots\big), \tag{5.9}$$

where, now, the differentials d_i are of degree 1. The family $\{X_i[-i]\}_{i\in\mathbb{Z}}$ belongs then to the category $\mathrm{GO(Ab)} := \mathrm{Ab}^{\mathbb{Z}}$ of graded objects of Ab, and the category $C(\mathrm{Ab})$ is then (tautologically) equivalent to the category DGO(Ab) of *differential graded objects* of Ab. This was also the case for vector spaces where we had $C(\mathrm{Vec}(\Bbbk)) = \mathrm{DGV}(\Bbbk)$ (2.1.6).

5.3.1 *The* Tot *Functors*

When Ab is the category $\mathrm{GM}(\Omega_G)$, we use the same recipe (5.9) and get objects with two gradings, one coming from the grading as Ω_G-gm and the other from the subscript index in the complex. Hence, the two dimensional ladder (5.10). We can then associate with a complex of Ω_G-gm's $(\boldsymbol{X}, \boldsymbol{d})$ a differential graded module in $\mathrm{DGM}(\Omega_G)$ in ways that depend on how we collect terms of equal *total degree*[10] in the bigraded ladder (5.10). In the present case where there are only two degrees, we can do this either by taking the direct sum '$\oplus$' or the direct product '$\boldsymbol{\Pi}$' of the diagonals of constant total degree.

[10] The *total degree* of a multi-degree object is the sum of its degrees. Here, it is the sum of the graded module degree plus the index number in the complex.

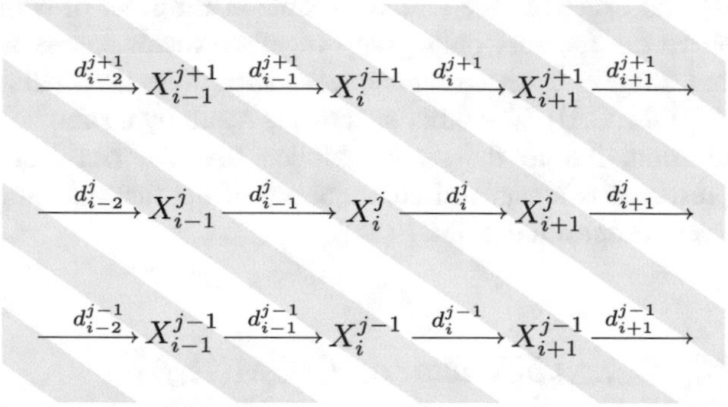

$$\xrightarrow{d^{j+1}_{i-2}} X^{j+1}_{i-1} \xrightarrow{d^{j+1}_{i-1}} X^{j+1}_i \xrightarrow{d^{j+1}_i} X^{j+1}_{i+1} \xrightarrow{d^{j+1}_{i+1}}$$

$$\xrightarrow{d^{j}_{i-2}} X^{j}_{i-1} \xrightarrow{d^{j}_{i-1}} X^{j}_i \xrightarrow{d^{j}_i} X^{j}_{i+1} \xrightarrow{d^{j}_{i+1}} \tag{5.10}$$

$$\xrightarrow{d^{j-1}_{i-2}} X^{j-1}_{i-1} \xrightarrow{d^{j-1}_{i-1}} X^{j-1}_i \xrightarrow{d^{j-1}_i} X^{j-1}_{i+1} \xrightarrow{d^{j-1}_{i+1}}$$

Fig. 5.1 A complex in $C(\mathrm{GM}(\Omega_G))$

For direct sums, we define:[11]

- $\mathbf{Tot}^m_\oplus X := \bigoplus_{i \in \mathbb{Z}} X^{m-i}_i$, and

$$\mathbf{Tot}_\oplus(X) := \{\mathbf{Tot}^m_\oplus X\}_{m \in \mathbb{Z}} = \bigoplus_{i \in \mathbb{Z}} X_i[-i] \in \mathrm{GM}(\Omega_G). \tag{5.11}$$

- The differential $\mathbf{Tot}_\oplus d : \mathbf{Tot}_\oplus X \to \mathbf{Tot}_\oplus X[1]$ is the morphism of Ω_G-gm's:

$$d_m(\oplus_{i \in \mathbb{Z}} x_i) := i(\oplus_{i \in \mathbb{Z}} d_i(x_i)). \tag{5.12}$$

- If $\alpha : (X, d) \to (X', d')$ is a morphism in $C(\mathrm{GM}(\Omega_G))$, we define

$$\mathbf{Tot}_\oplus \alpha : \mathbf{Tot}_\oplus X \to \mathbf{Tot}_\oplus X', \quad \alpha_m(\oplus_{i \in \mathbb{Z}} x_i) := \big(\oplus_{i \in \mathbb{Z}} \alpha_i(x_i)\big).$$

For direct products, we define $\mathbf{Tot}_\Pi$ *mutatis mutandis* replacing $\oplus$ by Π.

Proposition 5.3.1.1 *The previous settings define faithful exact functors*

$$\begin{cases} \mathbf{Tot}_\oplus : C(\mathrm{GM}(\Omega_G)) \rightsquigarrow \mathrm{DGM}(\Omega_G), \\ \mathbf{Tot}_\Pi : C(\mathrm{GM}(\Omega_G)) \rightsquigarrow \mathrm{DGM}(\Omega_G). \end{cases}$$

which coincide on $C^b(\mathrm{GM}(\Omega_G))$. They preserve cones (Sect. A.1.3), homotopies (Sect. A.2.2) and commute with cohomology:

$$h \circ \mathbf{Tot}_\oplus = \mathbf{Tot}_\oplus \circ h \quad and \quad h \circ \mathbf{Tot}_\Pi = \mathbf{Tot}_\Pi \circ h.$$

In particular they preserve acyclicity and quasi-isomorphisms.

---

[11] We will come back to this in Sect. 5.4.2, where spectral sequences will come on stage.

Proof The additivity and exactness are already visible at the level of complexes of vector spaces where the functors only group together vector spaces with no modification. One has $\mathbf{Tot}_{\oplus} = \mathbf{Tot}_{\Pi}$ on bicomplexes with finite length diagonals, hence the category $C^b(\mathrm{GM}(\Omega_G))$. The other statements result by a straightforward albeit tedious verification, left to the reader. Notice that the fact that quasi-isomorphisms are preserved is a standard consequence of the fact that cones and acyclicity are preserved (Proposition A.1.4.1-(3c)). □

5.3.2 *The* $\mathbf{Hom}^{\bullet}_{\Omega_G}(\_,\_)$ *bifunctor on* $C(\mathbf{GM}(\Omega_G))$

We start by recalling that in any abelian category Ab with countable products, standard practice is to consider the bifunctor

$$\big(\mathbf{Hom}^{\bullet}_{\mathrm{Ab}}(\_,\_), \boldsymbol{D}_{\bullet}\big) : C(\mathrm{Ab}) \times C(\mathrm{Ab}) \rightsquigarrow C(\mathrm{Mod}(\mathbb{Z}))\,, \tag{5.13}$$

(as we did for the category $\mathrm{Vec}(\Bbbk)$ (Sect. 2.1.8)) by setting

$$\mathbf{Hom}^{i}_{\mathrm{Ab}}\big((\boldsymbol{X}, \boldsymbol{d}), (\boldsymbol{X}', \boldsymbol{d}')\big) := \Big(\prod\nolimits_n \mathrm{Hom}_{\mathrm{Ab}}(X_n, X'_{n+i})\Big) \tag{5.14}$$

where 'i' denotes the index in the resulting complex, and

$$D_i(\boldsymbol{\alpha}) := \boldsymbol{d}' \circ \boldsymbol{\alpha} - (-1)^i \boldsymbol{\alpha} \circ \boldsymbol{d}\,. \tag{5.15}$$

However, in the case of graded categories, for example $\mathrm{Ab} := \mathrm{GM}(\Omega_G)$, this definition would give

$$\mathbf{Hom}^{i}_{\mathrm{GM}(\Omega_G)}\big((\boldsymbol{X}, \boldsymbol{d}), (\boldsymbol{X}', \boldsymbol{d}')\big) := \prod\nolimits_n \mathrm{Homgr}^{0}_{\Omega_G}(X_n, X'_{n+i}) \in \mathrm{Vec}(\Bbbk)\,,$$

which is not what we are interested in for several reasons, among which is that it would give us complexes of vector spaces rather than Ω_G-graded modules.

This leads us naturally to define instead

$$\mathbf{Hom}^{\bullet}_{\Omega_G}(\_,\_) : C(\mathrm{GM}(\Omega_G)) \times C(\mathrm{GM}(\Omega_G)) \rightsquigarrow C(\mathrm{GM}(\Omega_G))\,, \tag{5.16}$$

$$\mathbf{Hom}^{i}_{\Omega_G}\big((\boldsymbol{X}, \boldsymbol{d}), (\boldsymbol{X}', \boldsymbol{d}')\big) := \prod\nolimits_n \mathrm{Homgr}^{*}_{\Omega_G}(X_n, X'_{n+i}) \in \mathrm{GM}^*(\Omega_G)\,, \tag{5.17}$$

where '$*$' denotes the grading of $\mathrm{GM}(\Omega_G)$, with the same differential (5.15).

As in Sect. 5.2.2.1, the '$\mathbf{Hom}^{\bullet}_{\Omega_G}$' functor (5.17) is contravariant left exact on the first entry, and covariant left exact on the second entry.

Proposition 5.3.2.1 *The compatibility between the* $\mathbf{Hom}^\bullet_{\Omega_G}(\_,\_)$ *functors on* $C(\mathrm{GM}(\Omega_G))$ *and on* $\mathrm{DGM}(\Omega_G)$*, is given by the following commutative diagram.*

$$\begin{array}{ccccc} C(\mathrm{GM}(\Omega_G)) & \times & C(\mathrm{GM}(\Omega_G)) & \xrightarrow{\mathbf{Hom}^\bullet_{\Omega_G}(\_,\_)} & C(\mathrm{GM}(\Omega_G)) \\ \wr\downarrow \mathbf{Tot}_\oplus & & \wr\downarrow \mathbf{Tot}_\Pi & & \wr\downarrow \mathbf{Tot}_\Pi \\ \mathrm{DGM}(\Omega_G) & \times & \mathrm{DGM}(\Omega_G) & \xrightarrow{\mathbf{Hom}^\bullet_{\Omega_G}(\_,\_)} & \mathrm{DGM}(\Omega_G) \end{array} \tag{5.18}$$

Furthermore, if we restrict to the full subcategory $C^b(\mathrm{GM}(\Omega_G)) \times C^b(\mathrm{GM}(\Omega_G))$*, then the* $\mathbf{Tot}_\Pi$ *functors can be replaced by* $\mathbf{Tot}_\oplus$*.*

Proof Since the differentials are not essential to the statement, the proof is elementary and based on the following equalities. Let $\boldsymbol{X} := (X_i)$, $\boldsymbol{X}' := (X'_i)$ be complexes of Ω_G-gm's.

- $\mathbf{Tot}_\Pi \circ \mathbf{Hom}^\bullet_{\Omega_G}(\_,\_)$. The i-th term in the complex $\mathbf{Hom}^\bullet_{\Omega_G}(\boldsymbol{X},\boldsymbol{X}')$, given by the definition (5.17), is

$$\mathbf{Hom}^i_{\Omega_G}(\boldsymbol{X},\boldsymbol{X}')^* = \Big(\prod\nolimits_n \mathrm{Homgr}^*_{\Omega_G}(X_n, X'_{n+i})\Big) \in \mathrm{GM}^*(\Omega_G)\,, \tag{5.19}$$

  and if we apply $\mathbf{Tot}_\Pi$ to $\mathbf{Hom}^\bullet_{\Omega_G}(\boldsymbol{X},\boldsymbol{X}')$, we get the Ω^*_G-dgm

$$\prod\nolimits_i \prod\nolimits_n \mathrm{Homgr}^0_{\Omega_G}(X_n, X'_{n+i}[*-i])\,. \tag{5.20}$$

- $\mathbf{Hom}^\bullet_{\Omega_G}(\_,\_) \circ (\mathbf{Tot}_\oplus \times \mathbf{Tot}_\Pi)$. In the Ω^*_G-gm $\mathbf{Hom}^\bullet_{\Omega_G}(\mathbf{Tot}_\oplus \boldsymbol{X}, \mathbf{Tot}_\Pi \boldsymbol{X}')$, the term of degree '$*$' is:

$$\begin{aligned} &\mathrm{Homgr}^*_{\Omega_G}\big(\textstyle\bigoplus_n X_n[-n],\ \prod_m X'_m[-m]\big) = \\ &\qquad = \prod\nolimits_{n,m} \mathrm{Homgr}^0_{\Omega_G}\big(X_n, X'_m[n-m+*]\big)\,, \end{aligned} \tag{5.21}$$

  equal to (5.20), following an obvious change of subscripts.

For the last statement, we note that if $\boldsymbol{X} \in C^b(\mathrm{GM}(\Omega_G))$, then, for each i, there is only a finite number of n such that $X_n \neq 0$, in which case the '$\prod_n$' symbol in (5.19) can be changed to '$\bigoplus_n$'. After that, if, in addition, $\boldsymbol{X}' \in C^b(\mathrm{GM}(\Omega_G))$, then the set of couples (n,i), for which $X'_{n+i} \neq 0$, is also finite, and the '$\prod_i$' symbol in (5.20) can be changed to '$\bigoplus_i$'. Hence,

$$\mathbf{Tot}_\Pi\, \mathbf{Hom}^\bullet_{\Omega_G}(\boldsymbol{X},\boldsymbol{X}') = \mathbf{Tot}_\oplus\, \mathbf{Hom}^\bullet_{\Omega_G}(\boldsymbol{X},\boldsymbol{X}')\,.$$

On the other hand, for $\boldsymbol{X}' \in C^b(\mathrm{GM}(\Omega_G))$, we have $\mathbf{Tot}_{\prod}\boldsymbol{X}' = \mathbf{Tot}_{\oplus}\boldsymbol{X}'$ and the '$\prod_m$' in (5.21) can be replaced by '$\bigoplus_m$'. Hence,

$$\mathbf{Hom}^{\bullet}_{\Omega_G}(\mathbf{Tot}_{\oplus}\boldsymbol{X}, \mathbf{Tot}_{\prod}\boldsymbol{X}') = \mathbf{Hom}^{\bullet}_{\Omega_G}(\mathbf{Tot}_{\oplus}\boldsymbol{X}, \mathbf{Tot}_{\oplus}\boldsymbol{X}'),$$

which ends the proof of the proposition. □

5.3.3 The $(\_ \otimes_{\Omega_G} \_)^{\bullet}$ Bifunctor on $C(\mathrm{GM}(\Omega_G))$

Given $(\boldsymbol{X}, \boldsymbol{d})$, $(\boldsymbol{X}', \boldsymbol{d}') \in C(\mathrm{GM}(\Omega_G))$, we define the complex

$$\big((\boldsymbol{V} \otimes_{\Omega_G} \boldsymbol{V}')^{\bullet}, \boldsymbol{\Delta}\big) \in C(\mathrm{GM}(\Omega_G)),$$

with i-th term

$$(\boldsymbol{V} \otimes_{\Omega_G} \boldsymbol{V}')_i := \sum\nolimits_{a+b=i} \boldsymbol{X}_a \otimes_{\Omega_G} \boldsymbol{X}'_b,$$

where $\boldsymbol{X}_a \times_{\Omega_G} \boldsymbol{X}'_b$ is endowed with the grading induced by the *total degree* $a+b$ of $\boldsymbol{X}_a \otimes_{\mathbb{R}} \boldsymbol{X}'_b$, as in (5.2.2–(5.4)). The differential Δ_i on $\boldsymbol{X}_a \otimes \boldsymbol{X}'_b$ is

$$\Delta_i(v \otimes v\prime) := d_a(v) \otimes v\prime + (-1)^a v \otimes d_b(v\prime). \tag{5.22}$$

In this way we get a bifunctor covariant additive and right exact on each entry

$$(\_ \otimes_{\Omega_G} \_)^{\bullet} : C(\mathrm{GM}(\Omega_G)) \times C(\mathrm{GM}(\Omega_G)) \rightsquigarrow C(\mathrm{GM}(\Omega_G)). \tag{5.23}$$

Proposition 5.3.3.1 *The compatibility of* $(\_ \otimes_{\Omega_G} \_)^{\bullet}$ *on* $C(\mathrm{GM}(\Omega_G))$ *and on* $\mathrm{DGM}(\Omega_G)$*, is described by the following commutative diagram*

$$\begin{array}{ccc}
C(\mathrm{GM}(\Omega_G)) \times C(\mathrm{GM}(\Omega_G)) & \xrightarrow{(_\otimes_{\Omega_G}_)^{\bullet}} & C(\mathrm{GM}(\Omega_G)) \\
\wr\, \mathbf{Tot}_{\oplus} \downarrow \qquad \wr\, \mathbf{Tot}_{\oplus} \downarrow & & \wr\, \downarrow \mathbf{Tot}_{\oplus} \\
\mathrm{DGM}(\Omega_G) \times \mathrm{DGM}(\Omega_G) & \xrightarrow{(_\otimes_{\Omega_G}_)^{\bullet}} & \mathrm{DGM}(\Omega_G)
\end{array} \tag{5.24}$$

Sketch of Proof Results by an elementary check, much simpler than for Proposition 5.3.2.1. Details are left to the reader. □

5.4 Derived Functors in the Category GM(Ω_G)

Readers with little experience in these subjects may find it helpful to read Appendix A first, as it gives an overview of Derived Categories and Derived Functors. This has resulted in some duplication, for which we apologize.

We showed in Proposition 5.1.2.2 that the abelian category GM(Ω_G) has enough projective and injective objects. We now recall the action on GM(Ω_G) of the *derived functors* of an *additive* functor

$$\boldsymbol{F} : \mathrm{GM}(\Omega_G) \to \mathrm{Ab} \tag{5.25}$$

where Ab is an abelian category for which we denote by $C(\mathrm{Ab})$ its category of complexes and by $\mathcal{K}(\mathrm{Ab})$ its homotopy category (*cf*. A.1.1, A.1.5, A.2.2).

The *right and left derived functors of* $\boldsymbol{F}$ are respectively denoted by

$$\begin{cases} \mathbb{R}\,\boldsymbol{F} : \mathrm{GM}(\Omega_G) \rightsquigarrow \mathcal{K}(\mathrm{Ab}) \\ \mathbb{L}\boldsymbol{F} \; : \mathrm{GM}(\Omega_G) \rightsquigarrow \mathcal{K}(\mathrm{Ab})\,. \end{cases} \tag{5.26}$$

Notice that while the target category of $\boldsymbol{F}$ is Ab, the target category of its derived functors is the homotopy category $\mathcal{K}(\mathrm{Ab})$.

For $\boldsymbol{V},\ \boldsymbol{W} \in \mathrm{GM}(\Omega_G)$ the complexes $\mathbb{R}\,\boldsymbol{F}(\boldsymbol{V}),\ \mathbb{L}\boldsymbol{F}(\boldsymbol{W}) \in \mathcal{K}(\mathrm{Ab})$ are defined by the following recipe.

– Choose a projective (resp. injective) resolution of $\boldsymbol{V}$ (resp. $\boldsymbol{W}$) in $C(\mathrm{GM}(\Omega_G))$.

$$\begin{array}{c} \cdots \xrightarrow{d_{-2}} \mathcal{P}_{-2} \xrightarrow{d_{-1}} \mathcal{P}_{-1} \xrightarrow{d_0} \mathcal{P}_0 \xrightarrow{\epsilon} \boldsymbol{V} \longrightarrow 0 \\ 0 \longrightarrow \boldsymbol{W} \xrightarrow{\epsilon} \mathcal{I}_0 \xrightarrow{d_0} \mathcal{I}_1 \xrightarrow{d_1} \mathcal{I}_2 \xrightarrow{d_2} \cdots \end{array} \tag{5.27}$$

– Set

$$\left\{ \begin{array}{l} \mathcal{P}_\star \boldsymbol{V} := (\cdots \xrightarrow{d_{-1}} \mathcal{P}_{-1} \xrightarrow{d_0} \mathcal{P}_0 \to 0) \\ \mathcal{I}_\star \boldsymbol{W} := (0 \to \mathcal{I}_0 \xrightarrow{d_0} \mathcal{I}_1 \xrightarrow{d_1} \cdots) \end{array} \right\} \in C^\star(\mathrm{GM}(\Omega_G))$$

and define

$$\begin{cases} \mathbb{L}\boldsymbol{F}(\boldsymbol{V}) := \boldsymbol{F}(\mathcal{P}_\star \boldsymbol{V}) \in \mathcal{K}(\mathrm{GM}(\Omega_G))\,, \\ \mathbb{R}\,\boldsymbol{F}(\boldsymbol{V}) := \boldsymbol{F}(\mathcal{I}_\star \boldsymbol{V}) \in \mathcal{K}(\mathrm{GM}(\Omega_G))\,. \end{cases} \tag{5.28}$$

Notice that, since projective and injective resolutions of objects are always homotopic, these complexes are homotopically independent of the chosen resolutions, so that they are *well-defined objects* in the homotopy category $\mathcal{K}(\mathrm{Ab})$. Their cohomologies are respectively denoted by

$$(I\!R^i\, \boldsymbol{F})(\_) := H^i(I\!R\, \boldsymbol{F}(\_)) \quad \text{and} \quad (I\!L^i \boldsymbol{F})(\_) := H^{-i}(I\!L \boldsymbol{F}(\_))\,.$$

5.4.1 Augmentations

We easily see from the above definitions that the *augmentation morphisms* of complexes $\epsilon : \boldsymbol{V}[0] \to \mathcal{I}_\star \boldsymbol{V}$ and $\epsilon : \mathcal{P}_\star \boldsymbol{V} \to \boldsymbol{V}[0]$, give rise to natural morphisms of complexes $\boldsymbol{F}(\epsilon) : \boldsymbol{F}(\boldsymbol{V}[0]) \to I\!R\, \boldsymbol{F}(\boldsymbol{V})$ and $\boldsymbol{F}(\epsilon) : I\!L \boldsymbol{F}(\boldsymbol{V}) \to \boldsymbol{F}(\boldsymbol{V}[0])$, inducing canonical morphisms

$$\boldsymbol{F}(\boldsymbol{V}) \to (I\!R^0\, \boldsymbol{F})(\boldsymbol{V}) \quad \text{and} \quad (I\!L^0 \boldsymbol{F})(\boldsymbol{V}) \to \boldsymbol{V}\,.$$

These are isomorphisms if $\boldsymbol{F}$ is respectively left and right exact (see A.2.3.6).

Before going further into the subject of derived functors we need to recall some basics of bicomplexes of an abelian category Ab.

5.4.2 Simple Complex Associated with a Bicomplex

Since the category $C^\natural(\mathrm{Ab})$ of complexes of an abelian category Ab is additive (in fact abelian), we can consider the category $C^{\star,\natural}(\mathrm{Ab}) := C^\star(C^\natural(\mathrm{Ab}))$ of complexes of $C^\natural(\mathrm{Ab})$ also called *double complexes, or bicomplexes, of* Ab.

A bicomplex $\boldsymbol{N}^{\star,\natural} := (\boldsymbol{N}^{\star,\natural}, \delta_{\star,\natural}, d_{\star,\natural}) \in C^{\star,\natural}(\mathrm{Ab})$ is generally represented as a two dimensional ladder (5.29) all of whose subdiagrams are commutative.

We extend the definition of the $\mathbf{Tot}_\oplus$ functor of 5.3.1 to $C^{\star,\natural}(\mathrm{Ab})$.

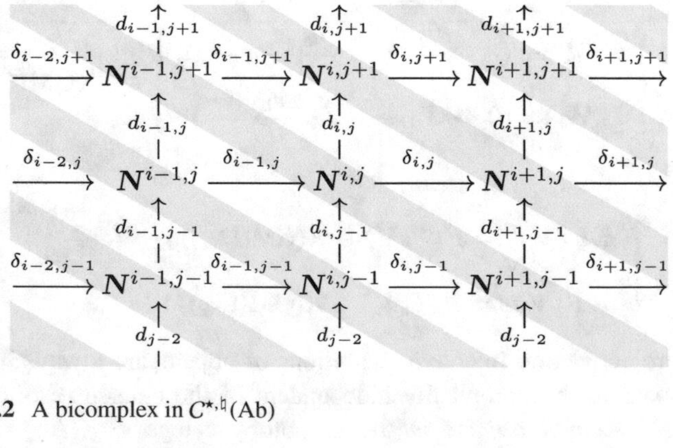

(5.29)

Fig. 5.2 A bicomplex in $C^{\star,\natural}(\mathrm{Ab})$

The *simple (or total) complex* associated with $N^{\star,\natural}$ is the complex denoted by $(\mathbf{Tot}^{\circ}_{\oplus}(N^{\star,\natural}), D_{\circ})$, where, for $m \in \mathbb{Z}$,

$$\begin{cases} \mathbf{Tot}^{m}_{\oplus}(N^{\star,\natural}) := \bigoplus_{m=i+j} N^{i,j} \\ D_m(n_{i,j}) := d_{i,j}(n_{i,j}) + (-1)^j \delta_{i,j}(n_{i,j}) \end{cases} \qquad \begin{array}{ccc} N^{i,j+1} & & \\ {\scriptstyle d_{i,j}}\uparrow & & \\ N^{i,j} & \xrightarrow{(-1)^j \delta_{i,j}} & N^{i+1,j} \end{array}$$

In this way, we get an exact functor

$$\mathbf{Tot}^{\circ}_{\oplus} := C^{\star,\natural}(\mathrm{Ab}) \rightsquigarrow C^{\circ}(\mathrm{Ab}),$$

where '$\circ := \star + \natural$'.

5.4.2.1 Spectral Sequences Associated with Bicomplexes

The bicomplex $N^{\star,\natural}$ is said to be *of the first quadrant* if $N^{i,j} = 0$ for all $(i, j) \notin \mathbb{N} \times \mathbb{N}$. As explained in Godement [46] (§4.8, p. 86), we assign to such bicomplex, two *regular* decreasing filtrations of $(\mathbf{Tot}^{\circ}_{\oplus}(N^{\star,\natural}), D_{\circ})$. The first is relative to the *row* $\natural$*-filtration* $\mathbf{Tot}^{\circ}_{\oplus}(N^{\star,\natural})_{\ell} := \mathbf{Tot}^{\circ}_{\oplus}(N^{\star,\natural\geq\ell})$, and the second to the *column* $\star$*-filtration* $\mathbf{Tot}^{\circ}_{\oplus}(N^{\star,\natural})_{c} := \mathbf{Tot}^{\circ}_{\oplus}(N^{\star\geq c,\natural})$. The same is true for bicomplexes $N^{\star,\natural}$ of the *third quadrant* or with finite number of nonzero rows or columns.

In those cases, each filtration gives rise to a spectral sequence which converges to the cohomology of $(\mathbf{Tot}^{\circ}_{\oplus} N^{\star,\natural}, D_{\circ})$, respectively

$$\begin{cases} {}'\mathbb{E}_2^{p,q} := h^p_{\natural}\, h^q_{\star}(N^{\star,\natural}) \Rightarrow h^{p+q}_{\circ}(\mathbf{Tot}^{\circ}_{\oplus} N^{\star,\natural}, D_{\circ}), \\ {}''\mathbb{E}_2^{p,q} := h^p_{\star}\, h^q_{\natural}(N^{\star,\natural}) \Rightarrow h^{p+q}_{\circ}(\mathbf{Tot}^{\circ}_{\oplus} N^{\star,\natural}, D_{\circ}), \end{cases}$$

where $h_{\star}$ (resp. $h_{\natural}$) is the cohomology w.r.t. $\delta_{\star}$ (resp. $d_{\natural}$).

We now return to the action of derived functors on the category GM(Ω_G). Note that the category Ab in the previous sections, now becomes the category GM$^*(\Omega_G)$ whose objects have a grading '$*$', and that the bicomplexes in $C^{\star,\natural}(\mathrm{GM}^*(\Omega_G))$ will thereby be equipped with three degrees: the column degree '$\star$', the row degree '$\natural$' and the Ω_G-graded module degree '$*$'. Consequently, when using **Tot** functors we must be careful which pair of degrees we are dealing with.

5.4.3 The $\mathbb{R}\,\mathbf{Hom}^{\bullet}_{\Omega_G}(\_,\_)$ and $(\_)\otimes^{\mathbb{L}}_{\Omega_G}(\_)$ Bifunctors on GM(Ω_G)

Given $V, V', W \in \mathrm{GM}(\Omega_G)$, the four functors

$$\mathbf{Hom}^{\bullet}_{\Omega_G}(V,\_)\,, \quad \mathbf{Hom}^{\bullet}_{\Omega_G}(\_,W)\,, \quad V \otimes_{\Omega_G}(\_)\,, \quad (\_)\otimes_{\Omega_G} V'\,, \tag{5.30}$$

where the first two are left exact and the other two are right exact.

To shrink notations, we will now write '$\mathbf{Hom}^\bullet$' for '$\mathbf{Hom}^\bullet_{\Omega_G}$', and '$\otimes$' for '$\otimes_{\Omega_G}$'.

Given projective resolutions $\mathcal{P}_\natural(\boldsymbol{V}) \to \boldsymbol{V}$, $\mathcal{P}_\natural(\boldsymbol{V}') \to \boldsymbol{V}'$ and an injective resolution $\boldsymbol{W} \to \mathcal{I}_\star(\boldsymbol{W})$ in $\mathrm{GM}(\Omega_G)$, we have natural morphisms of bicomplexes

$$\begin{array}{ccccc} \mathbf{Hom}^\bullet(\mathcal{P}_\natural(\boldsymbol{V}), \boldsymbol{W}_\star) & \to & \mathbf{Hom}^\bullet(\mathcal{P}_\natural(\boldsymbol{V}), \mathcal{I}_\star \boldsymbol{W}) & \leftarrow & \mathbf{Hom}^\bullet(\boldsymbol{V}_\natural, \mathcal{I}_\star \boldsymbol{W}) \\ \mathcal{P}_\natural(\boldsymbol{V}) \otimes \boldsymbol{V}'_\star & \to & \mathcal{P}_\natural(\boldsymbol{V}) \otimes \mathcal{P}_\star(\boldsymbol{V}') & \leftarrow & \boldsymbol{V}_\natural \otimes \mathcal{P}_\star(\boldsymbol{V}') \end{array} \tag{5.31}$$

where '$\boldsymbol{V}_\natural$' is a shortcut for '$\boldsymbol{V}[0]_\natural$', and similarly for '$\boldsymbol{V}'_\star$' and '$\boldsymbol{W}_\star$'.

Applying $\mathbf{Tot}_\oplus$ for $(\natural, \star)$, we obtain canonical morphisms in $C(\mathrm{GM}(\Omega_G))$

$$\begin{array}{ccccc} \mathbf{Hom}^\bullet(\mathcal{P}_\natural(\boldsymbol{V}), \boldsymbol{W}) & \to & \mathbf{Tot}_\oplus \mathbf{Hom}^\bullet(\mathcal{P}_\natural(\boldsymbol{V}), \mathcal{I}_\star \boldsymbol{W}) & \leftarrow & \mathbf{Hom}^\bullet(\boldsymbol{V}, \mathcal{I}_\star \boldsymbol{W}) \\ \mathcal{P}_\natural(\boldsymbol{V}) \otimes \boldsymbol{V}' & \to & \mathbf{Tot}_\oplus \big(\mathcal{P}_\natural(\boldsymbol{V}) \otimes \mathcal{P}_\star(\boldsymbol{V}')\big) & \leftarrow & \boldsymbol{V} \otimes \mathcal{P}_\star(\boldsymbol{V}') \end{array} \tag{5.32}$$

The following proposition is a classical result (Godement [46] §4.8).

Proposition 5.4.3.1 *The morphisms* (5.32) *are quasi-isomorphisms.*

Proof Since two projective (resp. injective) resolutions of a given object are homotopic, and since Ω_G is of finite dimensional cohomology, we can choose the resolutions to be of finite length. In that case the diagonals of constant $(\natural + \star)$ degree in the corresponding bicomplexes are of finite length, the associated spectral sequences (5.4.2.1) converge, and we can conclude.

For example, in the first line of (5.32), we note that the morphisms of complexes are compatible with row and column filtrations of bicomplexes. In the case of

$$\mathbf{Hom}^\bullet(\mathcal{P}_\natural(\boldsymbol{V}), \boldsymbol{W}) \to \mathbf{Tot}^\circ_\oplus \mathbf{Hom}^\bullet(\mathcal{P}_\natural(\boldsymbol{V}), \mathcal{I}_\star \boldsymbol{W}), \tag{5.33}$$

for each $i \in \mathbb{Z}$, the morphism $\mathbf{Hom}^\bullet(\mathcal{P}_i(\boldsymbol{V}), \boldsymbol{W}) \to \mathbf{Tot}^\circ_\oplus \mathbf{Hom}^\bullet(\mathcal{P}_i(\boldsymbol{V}), \mathcal{I}_\star \boldsymbol{W})$ is a quasi-isomorphism since $\mathcal{P}_i(\boldsymbol{V})$ is projective. Therefore, the induced map on the $''\mathbb{E}$ terms of the associated spectral sequences (5.4.2.1) is an isomorphism, which implies that (5.33) is a quasi-isomorphism. In the case of

$$\mathbf{Tot}^\circ_\oplus \mathbf{Hom}^\bullet(\mathcal{P}_\natural(\boldsymbol{V}), \mathcal{I}_\star \boldsymbol{W}) \longleftarrow \mathbf{Hom}^\bullet(\boldsymbol{V}, \mathcal{I}_\star \boldsymbol{W}),$$

the same idea works, except that now we must consider the row filtration and use the $'\mathbb{E}$ spectral sequence.

We deal with the second line in (5.32) in the same way. □

As a consequence of this proposition, in each line of (5.32), the three complexes represent the same object in the derived category $\mathcal{D}(\mathrm{GM}(\Omega_G))$. These are classically denoted by the following notations:

$$\boldsymbol{V},\ \boldsymbol{V}',\boldsymbol{W} \in \mathrm{GM}(\Omega_G) \rightsquigarrow \left\{ \begin{matrix} \mathbb{R}\,\mathbf{Hom}^\bullet_{\Omega_G}(\boldsymbol{V},\boldsymbol{W})) \\ \\ \boldsymbol{V} \otimes^{\mathbb{L}}_{\Omega_G} \boldsymbol{V}' \end{matrix} \right\} \in \mathcal{D}(\mathrm{GM}(\Omega_G))\,.$$

These constructions are natural w.r.t. each entry and, if we fix one entry, they can be extended naturally to bounded above complexes of Ω_G-gm in the free entries to be replaced by projective resolutions, and symmetrically on the other entry.[12] We therefore have two well-defined bifunctors

$$\begin{aligned} \mathbb{R}\,\mathbf{Hom}^\bullet_{\Omega_G}(\_,\_) &: \mathcal{D}(\mathrm{GM}(\Omega_G)) \times \mathcal{D}^+(\mathrm{GM}(\Omega_G)) \rightsquigarrow \mathcal{D}(\mathrm{GM}(\Omega_G)) \\ (\_ \otimes^{\mathbb{L}}_{\Omega_G} \_)^\bullet &: \mathcal{D}(\mathrm{GM}(\Omega_G)) \times \mathcal{D}^-(\mathrm{GM}(\Omega_G)) \rightsquigarrow \mathcal{D}(\mathrm{GM}(\Omega_G)) \end{aligned} \tag{5.34}$$

which are biadditive and have the usual variances and exactnesses.

Comment 5.4.3.2 In (5.34), there are other possible domains of definition, for example, for $\mathbb{R}\,\mathbf{Hom}^\bullet_{\Omega_G}(\_,\_)$, we can also choose $\mathcal{D}^-(\mathrm{GM}(\Omega_G)) \times \mathcal{D}(\mathrm{GM}(\Omega_G))$. The compatibility with (5.34) in the intersection $\mathcal{D}^-(\mathrm{GM}(\Omega_G)) \times \mathcal{D}^+(\mathrm{GM}(\Omega_G))$ is then warranted by Proposition 5.4.3.1.

5.4.4 *The* Ext• *and* Tor• *Bifunctors*

These are obtained by composing functors (5.34) with the cohomology functors

$$h^i : \mathcal{D}(\mathrm{GM}(\Omega_G)) \rightsquigarrow \mathrm{GM}(\Omega_G)\,,$$

where we must take care that the cohomology degree is relative to the *subscript index* in the complex and not to the degree of Ω_G-gm's.

The definitions of **Ext** and **Tor**, are then given by

$$\left\{ \begin{aligned} \mathbf{Ext}^{i,*}_{\Omega_G}(\_,\_) &:= h^i\big(\mathbb{R}\,\mathbf{Hom}^\bullet_{\Omega_G}(\_,\_)\big) \\ \mathbf{Tor}^*_{\Omega_G,i}(\_,\_) &:= h^{-i}\big(\_ \otimes^{\mathbb{L}}_{\Omega_G} \_\big) \end{aligned} \right\} \in \mathrm{GM}^*(\Omega_G)\,, \tag{5.35}$$

where 'i' is the subscript index in the complex and '$*$' is the Ω_G-gm grading.

[12] In a category with enough projective objects a bounded above complex is always quasi-isomorphic to a bounded above complex of projective objects, and symmetrically for injective objects, see Weibel [95] §10.5 Derived Functors (p. 390) and §10.7 Ext and RHom (p. 394).

5.4.5 The Duality Functor on $\mathcal{D}(\mathbf{GM}(\boldsymbol{\Omega}_G))$

By restriction of (5.34) we get the (derived) duality functor

$$\mathbb{R}\,\mathbf{Hom}^\bullet_{\Omega_G}(\_, \Omega_G) : \mathcal{D}(\mathrm{GM}(\Omega_G)) \rightsquigarrow \mathcal{D}(\mathrm{GM}(\Omega_G)), \tag{5.36}$$

defined by

$$\mathbb{R}\,\mathbf{Hom}^\bullet_{\Omega_G}(\_, \Omega_G) := \mathbf{Hom}^\bullet_{\Omega_G}(\_, \mathcal{I}_\star\Omega_G), \tag{5.37}$$

where $\Omega_G \rightarrowtail \mathcal{I}_\star\Omega_G$ is an injective resolution.

Comment 5.4.5.1 After Proposition 5.4.3.1, the restriction of this functor to $\mathcal{D}^-(\Omega_G)$ is canonically isomorphic to

$$\mathbb{R}\,\mathbf{Hom}^\bullet_{\Omega_G}(\_, \Omega_G) := \mathbf{Hom}^\bullet_{\Omega_G}(\mathcal{P}_\natural(\_), \Omega_G). \tag{5.38}$$

5.4.6 The Duality Functor on $\mathcal{D}(\mathbf{DGM}(\boldsymbol{\Omega}_G))$

Refer to Sect. A.2.5 for definitions of the homotopy category $\mathcal{K}(\mathrm{DGM}(\Omega_G))$ and the derived category $\mathcal{D}(\mathrm{DGM}(\Omega_G))$ associated with $\mathrm{DGM}(\Omega_G)$.

As stated in Theorem A.2.5.3, the derived duality functor $\mathbb{R}\,\mathbf{Hom}^\bullet_{\Omega_G}(\_, \Omega_G)$ acting on $\mathcal{D}(\mathrm{DGM}(\Omega_G))$ (5.2.3), is defined as in $\mathcal{D}(\mathrm{GM}(\Omega_G))$ (5.4.3), considering an injective resolution $\Omega_G \rightarrowtail \mathcal{I}_\star\Omega_G$. Hence, we set

$$\mathbb{R}\,\mathbf{Hom}^\bullet_{\Omega_G}(\_, \Omega_G) := \mathbf{Hom}^\bullet_{\Omega_G}(\_, \mathbf{Tot}_\Pi\, \mathcal{I}_\star\Omega_G), \tag{5.39}$$

which defines the *derived duality functor*

$$\mathbb{R}\,\mathbf{Hom}^\bullet_{\Omega_G}(\_, \Omega_G) : \mathcal{D}(\mathrm{DGM}(\Omega_G)) \rightsquigarrow \mathcal{D}(\mathrm{DGM}(\Omega_G)), \tag{5.40}$$

which, composed with the cohomology functor

$$\boldsymbol{h}^i : \mathcal{D}(\mathrm{DGM}(\Omega_G)) \rightsquigarrow \mathrm{GM}(\Omega_G),$$

give the *i'th extension functor*

$$\mathbf{Ext}^{i,*}_{\Omega_G}(\_, \Omega_G) := \mathbb{R}^i\,\mathbf{Hom}^\bullet_{\Omega_G}(\_, \Omega_G) : \mathcal{D}(\mathrm{DGM}(\Omega_G)) \rightsquigarrow \mathrm{GM}^*(\Omega_G). \tag{5.41}$$

The family $\{\mathbf{Ext}^{i,*}_{\Omega_G}(\_, \Omega_G)\}_{i\in\mathbb{N}}$ is a ∂-functor in $\mathcal{D}(\mathrm{DGM}(\Omega_G))$.[13]

[13]Which means that, applied to exact triangles in $\mathcal{D}(\mathrm{DGM}(\Omega_G))$, they give rise to long exact sequences of cohomology. We will use this fact in the proof of Proposition 5.4.7.2-(2). See Weibel [95] §10.5 Derived Functors (p. 390).

The relationship between the actions of $I\!R\,\mathbf{Hom}^\bullet_{\Omega_G}(\_,\_)$ on $C(\mathrm{GM}(\Omega_G))$ and on $\mathrm{DGM}(\Omega_G)$ is given in Proposition 5.3.2.1, which we restate on a restricted category where, modulo $\mathbf{Tot}_\oplus$, the two $I\!R\,\mathbf{Hom}^\bullet$ coincide.

Proposition 5.4.6.1 *On* $\mathcal{D}^b(\mathrm{GM}(\Omega_G)) \times \mathcal{D}^b(\mathrm{GM}(\Omega_G))$, *one has*

$$I\!R\,\mathbf{Hom}^\bullet_{\Omega_G}\big(\mathbf{Tot}_\oplus(\_), \mathbf{Tot}_\oplus(\_)\big) = \mathbf{Tot}_\oplus\, I\!R\,\mathbf{Hom}^\bullet_{\Omega_G}(\_,\_)$$

Proof Since Ω_G is of finite homological dimension (Proposition 5.1.2.2), bounded complexes admit projective and injective resolutions of finite lengths, in which case the second part of Proposition 5.3.2.1 establishes the equality. □

5.4.7 Spectral Sequences Associated with $I\!R\,\mathbf{Hom}^\bullet_{\Omega_G}(\_, \Omega_G)$

Let $\epsilon : \Omega_G \rightarrowtail \mathcal{I}_\star\Omega_G$ be an injective resolution of finite length in $\mathrm{GM}(\Omega_G)$.

For $(\boldsymbol{V}, \boldsymbol{d}) \in \mathrm{DGM}(\Omega_G)$, we can see the right-hand side in the definition

$$I\!R\,\mathbf{Hom}^\bullet_{\Omega_G}\big((\boldsymbol{V}, \boldsymbol{d}), \Omega_G\big) := \mathbf{Hom}^\bullet_{\Omega_G}\big((\boldsymbol{V}, \boldsymbol{d}), \mathbf{Tot}_\oplus(\mathcal{I}_\star\Omega_G, \delta_\star)\big)\,, \tag{5.42}$$

a bicomplex with rows indexed by the degree '$\natural$' of $\boldsymbol{V}$ and columns by the index '$\star$' of the injective resolution. The differentials $\boldsymbol{d}$ and $\delta_\star$ commute, $\boldsymbol{d}$ increases the $\natural$-degree and leaves the $\star$-degree unchanged, while $\delta_\star$ does the opposite.

Proposition 5.4.7.1 *Let* $(\boldsymbol{V}, \boldsymbol{d}) \in \mathrm{DGM}(\Omega_G)$.

1. *The filtration by increasing the $\star$-degree in $\mathcal{I}_\star(\Omega_G)$ is regular, giving rise to a convergent spectral sequence*

$$''I\!E_2^{p,q} := \mathbf{Ext}^{p,q}_{\Omega_G}(\boldsymbol{h}\boldsymbol{V}, \Omega_G)) \Rightarrow I\!R^{p+q}\,\mathbf{Hom}^\bullet_{\Omega_G}(\boldsymbol{V}, \Omega_G)\,. \tag{5.43}$$

2. *Assume $(\boldsymbol{V}, \boldsymbol{d})$ endowed with a regular filtration such that the associated spectral sequence $(I\!E_r(\boldsymbol{V}), d_r)$ is degenerated, i.e. there exists $N \in \mathbb{N}$, such that $d_r = 0$, for all $r > N$. Then, endowing $I\!R\,\mathbf{Hom}^\bullet_{\Omega_G}((\boldsymbol{V}, \boldsymbol{d}), \Omega_G)$ with the induced filtration, we get a degenerated, hence convergent, spectral sequence*

$$I\!R\,\mathbf{Hom}^\bullet_{\Omega_G}(I\!E_r(\boldsymbol{V}), \Omega_G) \Rightarrow I\!R\,\mathbf{Hom}^\bullet_{\Omega_G}(\boldsymbol{h}\boldsymbol{V}, \Omega_G)\,. \tag{5.44}$$

3. *If $\boldsymbol{V}$ is projective as Ω_G-gm,*[14] *then the following morphism of Ω_G-dgm's induced by the augmentation $\epsilon : \Omega_G \to \mathcal{I}_\star\Omega_G$, is a quasi-isomorphism:*

$$\mathbf{Hom}^\bullet_{\Omega_G}((\boldsymbol{V}, \boldsymbol{d}), \Omega_G) \xrightarrow{(\epsilon)} I\!R\,\mathbf{Hom}^\bullet_{\Omega_G}((\boldsymbol{V}, \boldsymbol{d}), \Omega_G)\,.$$

[14] A projective Ω_G-gm is always free, *cf*. Jacobson [59] corollary of Theorem 6.21, p. 386.

Proof

(1) The $\star$-filtration is regular simply because $(\mathcal{I}_\star\Omega_G, \delta_\star)$ is of finite length. The first page of the associated spectral sequence, is then

$$({}''\mathbb{E}_1^{p,\star}, d_1) := \mathbb{R}\,\mathbf{Hom}^\bullet_{\Omega_G}(\mathbf{h}^p\,\mathbf{V}, (\mathcal{I}_\star\Omega_G, d_\star)),$$

and the statement follows (*cf*. 5.4.4).

(2) The proof uses the exactness of the functor $\mathbb{R}\,\mathbf{Hom}^\bullet_{\Omega_G}(\_, \Omega_G)$ on $\mathrm{DGM}(\Omega_G)$. It entails that, when the functor is applied to a degenerate spectral sequence, it gives a degenerate spectral sequence, the terms of which are those in (5.44).

(3) Since $\mathcal{I}_\star\Omega_G$ is of finite length, we need only, by the mapping cone construction, show that if $\mathbf{W}_\star := (0 \to \mathbf{W}_0 \to \cdots \to \mathbf{W}_{r-1} \to \mathbf{W}_r \to 0)$ is a finite exact sequence of $\mathrm{GM}(\Omega_G)$-modules, then the Ω_G-dgm

$$\mathbf{Hom}^\bullet_{\Omega_G}((\mathbf{V}, \mathbf{d}), \mathbf{Tot}_\oplus\,\mathbf{W}_\star), \tag{5.45}$$

is acyclic.

The following proof is dual to that of the Theorem A.2.5.3.

By definition, a p-cocycle in (5.45) is a morphism $\boldsymbol{\alpha} : \mathbf{V} \to \mathbf{Tot}_\oplus\,\mathbf{W}_\star[p]$ of Ω_G-dgm. Now, recalling that $\mathbf{Tot}_\oplus\,\mathbf{W}_\star = \bigoplus_{k=0}^r \mathbf{W}_k[\star - k]$ (5.3.1), let us look at the component α_r of $\boldsymbol{\alpha}$ in the last Ω_G-gm $\mathbf{W}_r$, which is necessarily a sub-Ω_G-module of cocycles. We then have the diagram

$$\begin{array}{ccccc}
\mathbf{V}[-1] & \xrightarrow{\ d[-1]\ } & \mathbf{V} & \xrightarrow{\ d\ } & \mathbf{V}[1] \\
& \swarrow{\scriptstyle h_r} & \downarrow{\scriptstyle \alpha_r} & & \\
\mathbf{W}_{r-1}[p-1] & \xrightarrow{\ d_{r-1}[p]\ } & \mathbf{W}_r[p] & \xrightarrow{\ d_r[p]\ } & 0
\end{array}$$

with $d_{r-1}[p]$ surjective since $\mathbf{Tot}_\oplus\,\mathbf{W}_\star$ is acyclic. We then factor α_r through a morphism of Ω_G-gm's $h_r : \mathbf{V} \to \mathbf{W}_{r-1}[p-1]$, which is possible since $\mathbf{V}$ is projective, and we extend h_r in a morphism of Ω_G-gm's $\mathbf{h} = \mathbf{V} \to \mathbf{Tot}_\oplus\,\mathbf{W}_\star[p-1]$, by zero on the coordinates $j \neq r$.

The morphism $\boldsymbol{\alpha}$ is now homotopic to the morphism of Ω_G-dgm's

$$\boldsymbol{\alpha} - (\mathbf{h}[1] \circ \mathbf{d} + d_\star \circ \mathbf{h}) : \mathbf{V} \to \tau_{<r}\,\mathbf{W}_\star[p], \tag{5.46}$$

where

$$\tau_{<r}\,\mathbf{W}_\star := \mathbf{Tot}_\oplus\big(0 \to \mathbf{W}_0 \to \cdots \to \mathbf{W}_{r-2} \twoheadrightarrow \ker(d_{r-1}) \to 0\big).$$

Since $\tau_{<r}\,\mathbf{W}_\star$ is also acyclic with strictly fewer terms than $\mathbf{W}_\star$, we can conclude, by induction, that the morphism (5.46), hence $\boldsymbol{\alpha}$, is homotopic to zero. This ends the proof of the acyclicity of (5.45), which proves (3) □

Corollary 5.4.7.2 *Let* $\mathbf{V} := (\mathbf{V}, \mathbf{d})$ *and* $\mathbf{V}' := (\mathbf{V}', \mathbf{d}')$ *be* Ω_G*-dgm's.*

1. *There exists a canonical morphism of* Ω_G*-modules*

$$\bar{\xi}(\mathbf{V}, \mathbf{V}') : \boldsymbol{h}\big(\mathbf{Hom}^{\bullet}_{\Omega_G}\big((\mathbf{V}, \mathbf{d}), (\mathbf{V}', \mathbf{d})\big)\big) \to \mathbf{Hom}^{\bullet}_{\Omega_G}\big(\boldsymbol{h}(\mathbf{V}, \mathbf{d}), \boldsymbol{h}(\mathbf{V}', \mathbf{d})\big),$$

functorial on $\mathbf{V}, \mathbf{V}' \in \mathrm{DGM}(\Omega_G)$.

2. *If* $\mathbf{V}$ *and* $\mathbf{hV}$ *are projective* Ω_G*-gm, then* $\bar{\xi}(\mathbf{V}, \Omega_G)$ *is an isomorphism.*
3. *Assume that* $\mathbf{V}$ *and* $\mathbf{V}'$ *are projective* Ω_G*-gm's. If* $\boldsymbol{\alpha} : (\mathbf{V}, \mathbf{d}) \to (\mathbf{V}', \mathbf{d}')$ *is a quasi-isomorphism of* Ω_G*-dgm's, then the following induced morphism of* Ω_G*-dgm's is a quasi-isomorphism:*

$$\mathbf{Hom}^{\bullet}_{\Omega_G}(\boldsymbol{\alpha}, \Omega_G) : \mathbf{Hom}^{\bullet}_{\Omega_G}\big((\mathbf{V}', \mathbf{d}'), \Omega_G\big) \to \mathbf{Hom}^{\bullet}_{\Omega_G}\big((\mathbf{V}, \mathbf{d}), \Omega_G\big).$$

Proof In order to simplify notations we write '$\mathbf{Hom}^{\bullet}$' for '$\mathbf{Hom}^{\bullet}_{\Omega_G}$'.

Let $\boldsymbol{Z} \subseteq \mathbf{V}$, resp. $\boldsymbol{B} \subseteq \mathbf{V}$, denote the Ω_G-graded submodules respectively of cocycles and coboundaries of $(\mathbf{V}, \mathbf{d})$. Let $\boldsymbol{N} := (\boldsymbol{N}, \boldsymbol{d})$ be a Ω_G-dgm.

(1) Since the statement is invariant under degree shifts, we need only consider the 0-degree case. An element $\boldsymbol{\alpha} \in \mathbf{Hom}^{0}_{\Omega_G}(\mathbf{V}, \mathbf{V}')$ is a cocycle if we have $0 = D(\boldsymbol{\alpha}) = \boldsymbol{d} \circ \boldsymbol{\alpha} - \boldsymbol{\alpha} \circ \boldsymbol{d}$ (*cf*. 5.2.2–(5.5)), hence if $\boldsymbol{\alpha}$ is a morphism of Ω_G-dgm's, in which case it induces a morphism of Ω_G-gm in cohomology $\boldsymbol{h}(\boldsymbol{\alpha}) : \boldsymbol{h}\mathbf{V} \to \boldsymbol{h}\mathbf{V}'$. If $\boldsymbol{\alpha}$ is 0-coboundary, we have $\boldsymbol{\alpha} = \boldsymbol{d} \circ \boldsymbol{\beta} + \boldsymbol{\beta} \circ \boldsymbol{d}$, for some $\boldsymbol{\beta} : \mathrm{Homgr}^{-1}_{\Omega_G}(\mathbf{V}, \mathbf{V}')$, i.e. $\boldsymbol{\alpha}$ is homotopic to zero and $\boldsymbol{h}(\boldsymbol{\alpha}) = 0$. The fact that $\bar{\xi}(\mathbf{V}, \mathbf{V}')$ is Ω_G-linear is obvious. The functoriality of $\bar{\xi}(\_, \_)$ is clear.

(2) Applying the functor $\mathbf{Hom}^{\bullet}(\_, \Omega_G)$ to the short exact sequence:

$$0 \to \boldsymbol{Z} \longrightarrow \mathbf{V} \xrightarrow{\ \mathbf{d}\ } \boldsymbol{B}[1] \to 0, \tag{5.47}$$

we get a left exact sequence of Ω_G-dgm's

$$0 \to \mathbf{Hom}^{\bullet}(\boldsymbol{B}, \Omega_G)[-1] \xrightarrow{\ \mathrm{a}\ } \mathbf{Hom}^{\bullet}(\mathbf{V}, \Omega_G) \xrightarrow{\ \mathrm{b}\ } \mathbf{Hom}^{\bullet}(\boldsymbol{Z}, \Omega_G), \tag{5.48}$$

from which, the short exact sequence of Ω_G-dgm's

$$0 \to \mathbf{Hom}^{\bullet}(\boldsymbol{B}, \Omega_G)[-1] \xrightarrow{\ \mathrm{a}\ } \mathbf{Hom}^{\bullet}(\mathbf{V}, \Omega_G) \xrightarrow{\ \mathrm{b}\ } \boldsymbol{Q}^{\bullet}(\boldsymbol{Z}, \Omega_G) \to 0, \tag{5.49}$$

where $\boldsymbol{Q}^{\bullet}(\boldsymbol{Z}, \Omega_G)$ denotes the image of β in $\mathbf{Hom}^{\bullet}(\boldsymbol{Z}, \Omega_G)$. Note that the left and right complexes in (5.49) have null differentials so that they coincide with their cohomology.

The cohomology sequence associated with (5.49) is then the long exact sequence

$$\begin{array}{c} \boldsymbol{Q}^i(\boldsymbol{Z},\Omega_G) \longrightarrow \mathbf{Hom}^i(\boldsymbol{Z},\Omega_G) \longrightarrow \mathbf{Hom}^i(\boldsymbol{B},\Omega_G) \\ \underbrace{\hspace{6cm}}_{c_i} \end{array}$$

In this way, we obtain the long exact sequence Ω_G-graded modules

$$\xrightarrow{a_i} h^i\,\mathbf{Hom}^\bullet(\boldsymbol{V},\Omega_G) \xrightarrow{b_i} \boldsymbol{Q}^i(\boldsymbol{Z},\Omega_G) \xrightarrow{c_i} \mathbf{Hom}^i(\boldsymbol{B},\Omega_G) \xrightarrow{a_{i+1}} . \tag{5.50}$$

On the other hand, applying $\mathbf{Hom}^\bullet(\_,\Omega_G)$ to the short exact sequence

$$0 \to \boldsymbol{B} \subseteq \boldsymbol{Z} \to \boldsymbol{hV} \to 0\,, \tag{5.51}$$

we obtain the left exact sequence

$$0 \to \mathbf{Hom}^\bullet(\boldsymbol{hV},\Omega_G) \xrightarrow{b'} \mathbf{Hom}^\bullet(\boldsymbol{Z},\Omega_G) \xrightarrow{c'} \mathbf{Hom}^\bullet(\boldsymbol{B},\Omega_G)\,,$$

which, joined to (5.50), gives rise to the commutative diagram with exact lines:

$$\begin{array}{ccccccc} & \boldsymbol{h}\mathbf{Hom}^\bullet(\boldsymbol{V},\Omega_G) & \xrightarrow{b} & \boldsymbol{Q}^\bullet(\boldsymbol{V},\Omega_G) & \xrightarrow{c} & \mathbf{Hom}^\bullet(\boldsymbol{B},\Omega_G) & \\ & {\scriptstyle \bar{\xi}(\boldsymbol{V},\Omega_G)}\downarrow & \oplus & \downarrow{\scriptstyle \subseteq} & \oplus & \downarrow{\scriptstyle =} & \\ 0 \longrightarrow & \mathbf{Hom}^\bullet(\boldsymbol{hV},\Omega_G) & \xrightarrow{b'} & \mathbf{Hom}^\bullet(\boldsymbol{Z},\Omega_G) & \xrightarrow{c'} & \mathbf{Hom}^\bullet(\boldsymbol{B},\Omega_G) & \dashrightarrow 0 \end{array} \tag{5.52}$$

When $\boldsymbol{hV}$ is projective, the sequence (5.51) is a split sequence of Ω_G-gm's, and $\boldsymbol{c}'$ in (5.52) is surjective, and $\bar{\xi}(\boldsymbol{V},\Omega_G)$ is an isomorphism if and only if the inclusion $\boldsymbol{i} : \boldsymbol{Q}^\bullet(\boldsymbol{V},\Omega_G) \subseteq \mathbf{Hom}^\bullet(\boldsymbol{Z},\Omega_G)$ is an equality. Indeed, by diagram chase, if $\bar{\xi}(\boldsymbol{V},\Omega_G)$ is injective, then $\boldsymbol{b}$ is injective, hence $\boldsymbol{c}$ is surjective and, together, this implies that $\boldsymbol{i}$ is an isomorphism by the five lemma. Conversely, if $\boldsymbol{i}$ is an isomorphism, then $\boldsymbol{c}$ is surjective, $\boldsymbol{b}$ is injective, and $\bar{\xi}(\boldsymbol{V},\Omega_G)$ is bijective.

This leads us to proving that $\boldsymbol{Q}^\bullet(\boldsymbol{V},\Omega_G) = \mathbf{Hom}^\bullet(\boldsymbol{Z},\Omega_G)$, or, equivalently, to showing that $\boldsymbol{b}$ in (5.48) is surjective.

The long exact sequences of $\mathbf{Ext}^\bullet$'s associated with the short sequences (5.51) and (5.47) in conjunction with the projectivity of $\boldsymbol{hV}$ and $\boldsymbol{V}$, give us the equalities

$$\begin{cases} \text{(i) } \mathbf{Ext}^i_{\Omega_G}(\boldsymbol{Z},\Omega_G) = \mathbf{Ext}^i_{\Omega_G}(\boldsymbol{B},\Omega_G)\,, \\ \text{(ii) } \mathbf{Ext}^i_{\Omega_G}(\boldsymbol{Z},\Omega_G) = \mathbf{Ext}^{i+1}_{\Omega_G}(\boldsymbol{B}[1],\Omega_G)\,, \end{cases} \quad \forall i \geq 1\,, \tag{5.53}$$

from which: $\mathbf{Ext}^i_{\Omega_G}(\boldsymbol{B}, \Omega_G) = \mathbf{Ext}^{i+1}_{\Omega_G}(\boldsymbol{B}[1], \Omega_G)$, for all $i \geq 1$. But then, for every fixed $i \geq 1$, we have

$$\mathbf{Ext}^i_{\Omega_G}(\boldsymbol{B}, \Omega_G) = \mathbf{Ext}^{i+\ell}_{\Omega_G}(\boldsymbol{B}[\ell], \Omega_G)\,, \quad \forall \ell \geq 0\,,$$

where $\mathbf{Ext}^m_{\Omega_G}(\_, \_) = 0$ for $m \gg 0$, since Ω_G is of finite homological dimension (*cf*. 5.1.2.2-(2)). Therefore, returning to (5.53), we get

$$\mathbf{Ext}^i_{\Omega_G}(\boldsymbol{Z}, \Omega_G) = \mathbf{Ext}^i_{\Omega_G}(\boldsymbol{B}, \Omega_G) = 0\,, \quad \forall i \geq 1\,, \tag{5.54}$$

and $\boldsymbol{b}$ in (5.48) is surjective as expected.

(3) The statement is equivalent to the fact that $\mathbf{Hom}^\bullet_{\Omega_G}(\hat{c}(\boldsymbol{\alpha}), \Omega_G)$ is acyclic, where $\hat{c}(\boldsymbol{\alpha})$ is the *cone* of $\boldsymbol{\alpha}$ (see A.1.3). To see this, note that $\hat{c}(\boldsymbol{\alpha})$ is a projective Ω_G-gm since, by definition, $\hat{c}(\boldsymbol{\alpha}) = \boldsymbol{V}' \oplus \boldsymbol{V}$. But then, by Proposition 5.4.7.1-(3), the natural map

$$\mathbf{Hom}^\bullet_{\Omega_G}(\hat{c}(\boldsymbol{\alpha}), \Omega_G) \xrightarrow[\text{q.i.}]{} I\!R\,\mathbf{Hom}^\bullet_{\Omega_G}(\hat{c}(\boldsymbol{\alpha}), \Omega_G)\,.$$

is a quasi-isomorphism, and we can conclude using the fact that derived functors preserve acyclicity (*cf*. A.1.4.2).

□

Remarks 5.4.7.3

1. Given $\boldsymbol{N} \in \mathrm{GM}(\Omega_G)$ and an injective resolution of finite length $\epsilon : \boldsymbol{N} \rightarrowtail \mathcal{I}_\star\boldsymbol{N}$ in $\mathrm{GM}(\Omega_G)$, all the considerations for $I\!R\,\mathbf{Hom}^\bullet_{\Omega_G}(\_, \Omega_G)$, notably Proposition 5.4.7.1 and Corollary 5.4.7.2 and their proofs, also apply to the functor

$$I\!R\,\mathbf{Hom}^\bullet_{\Omega_G}(\_, \boldsymbol{N}) := \mathbf{Hom}^\bullet_{\Omega_G}(\_, \mathcal{I}_\star\boldsymbol{N}) : \mathrm{DGM}(\Omega_G) \rightsquigarrow \mathrm{DGM}(\Omega_G)\,,$$

which respects quasi-isomorphisms and defines a functor in derived category

$$I\!R\,\mathbf{Hom}^\bullet_{\Omega_G}(\_, \boldsymbol{N}) : \mathcal{D}(\mathrm{DGM}(\Omega_G)) \rightsquigarrow \mathcal{D}(\mathrm{DGM}(\Omega_G))\,. \tag{5.55}$$

2. In 5.4.7.2 the projectivity assumption on the Ω_G-gm $\boldsymbol{W} \in \{\boldsymbol{V}, \boldsymbol{hV}\}$ can be weakened to assume only that $I\!R^i\,\mathbf{Hom}^\bullet_{\Omega_G}(\boldsymbol{W}, \Omega_G) = 0$, for all $i > 0$.
Exercise: Check that 5.4.7.2-(2,3) *remain valid under this weaker assumption.*

5.5 Equivariant Integration

5.5.1 Definition

A G-manifold M is said to be an *orientable G-manifold* if M is orientable and that every $g \in G$ acts on M as a preserving orientation diffeomorphism of M(*cf*. 2.3.1).

Given an oriented G-manifold M of dimension M, the $\mathbb{R}$-linear integration map introduced in 2.3.2

$$\int_M : \Omega_c(M)[d_M] \to \mathbb{R},$$

is a G-equivariant morphism of complexes (by Proposition 2.3.2.2), as is its $S(\mathfrak{g}^\vee)$-linear extension

$$\int_M : S(\mathfrak{g}^\vee) \otimes \Omega_c(M)[d_M] \to S(\mathfrak{g}^\vee), \quad P \otimes \omega \mapsto P \int_M \omega. \tag{5.56}$$

where G acts on $S(\mathfrak{g}^\vee) \otimes \Omega_c(M)$ by $g \cdot (P \otimes \omega) := g \cdot P \otimes g \cdot \omega$. Therefore, the restriction of (5.56) to the subspace of G-equivariant differential forms compactly supported:

$$\Omega_{G,c}(M) := \big(S(\mathfrak{g}^\vee) \otimes \Omega_c(M)\big)^G,$$

takes its values in $\Omega_G := S(\mathfrak{g}^\vee)^G$ (Sect. 4.4.4). We denote this restriction by

$$\int_M : \Omega_{G,c}(M)[d_M] \to \Omega_G,$$

and call it *the G-equivariant integration map*, which is clearly a morphism of Ω_G-graded modules of degree $-d_M$.

The graded algebra $\Omega_{G,c}(M)$ was endowed in Sect. 4.4.4, with Cartan's differential

$$d_G(P \otimes \omega) = P \otimes d\,\omega + \sum\nolimits_i Pe^i \otimes \iota(e_i)\,\omega,$$

and if $\zeta \in \Omega_{G,c}(M)$ is homogeneous of total degree d, it can be written in a unique way as a sum

$$\zeta = \sum\nolimits_{0 \le i \le d/2} \Big(\sum\nolimits_{Q \in \mathscr{B}(i)} Q \otimes \omega_Q\Big),$$

where $\mathscr{B}(i)$ denotes a vector space basis of $S^i(\mathfrak{g}^\vee)$ and where $\omega_Q \in \Omega^{d-2\deg Q}(M)$. Consequently, if ζ is an equivariant coboundary, the terms $Q \otimes \omega_Q$ in the above decomposition such that $\omega_Q \in \Omega_c(M)$ is of top degree d_M are necessary coboundaries in $\Omega_c(M)$, in which case $\int_M \zeta = 0$. We have thus proved the following proposition.

Proposition 5.5.1.1 *Let G be a compact Lie group and let M be an oriented G-manifold. Then, $\int_M d_G(\Omega_{G,c}(M)) = 0$. In other words, the G-equivariant integration map*

$$\boxed{\int_M : (\Omega_{G,c}(M), d_G)[d_M] \to (\Omega_G, 0)} \tag{5.57}$$

is a morphism of Ω_G-dg-modules, inducing in cohomology a morphism of Ω_G-graded modules denoted by

$$\int_M : H_{G,c}(M)[d_M] \to H_G\,. \tag{5.58}$$

5.5.2 *Equivariant Integration vs. Integration Along Fibers*

Let G be a compact Lie group. Denote by G_0 the connected component of the identity element $e \in G$ and set $W := G/G_0$, which is a finite group. Given a G-manifold M, we have already considered (*cf*. 4.7.3) the Cartesian diagram of fiber bundles

$$\begin{array}{ccccc}
\nu_M : M_{G_0} & \xrightarrow[{[W]}]{} & M_{G_0}/W & \xrightarrow[\simeq]{} & M_G \\
{\scriptstyle [M]}\big\downarrow{\scriptstyle \pi_M} & \square & {\scriptstyle [M]}\big\downarrow{\scriptstyle \pi_M} & \square & {\scriptstyle [M]}\big\downarrow{\scriptstyle \pi_M} \\
\nu_{\mathbb{B}} : \mathbb{B}G_0 & \xrightarrow[{[W]}]{} & \mathbb{B}G_0/W & \xrightarrow[\simeq]{} & \mathbb{B}G
\end{array}$$

where the fibers are shown in brackets. When M is an oriented manifold, the fiber bundle in the first column $M_{G_0} := (M_{G_0}, \mathbb{B}G_0, M, \pi_M)$, is oriented since $\mathbb{B}G_0$ is simply connected (3.1.7.2-(3c)), and when, in addition, M is an oriented G-manifold, the elements $w \in W$ preserve the orientation of the fibers of M_{G_0} and, therefore, the quotient fiber bundle $M_{G_0}/W = M_G$ canonically inherits an orientation making the maps ν_M and $\nu_{\mathbb{B}}$ to preserve orientation.

We then have a commutative diagram of integrations along fibers

$$\begin{array}{ccccc}
\Omega_{\mathrm{cv}}(M_G;\mathbb{R})[d_M] & = & \Omega_{\mathrm{cv}}(M_{G_0};\mathbb{R})[d_M]^W & \xrightarrow{\subseteq} & \Omega_{\mathrm{cv}}(M_{G_0};\mathbb{R})[d_M] \\
\int_M\big\downarrow & \oplus & \int_M\big\downarrow & \oplus & \big\downarrow\int_M \\
\Omega(\mathbb{B}G;\mathbb{R}) & = & \Omega(\mathbb{B}G_0;\mathbb{R})^W & \xrightarrow{\subseteq} & \Omega(\mathbb{B}G_0;\mathbb{R})
\end{array}$$

where $\Omega_\bullet(\_;\mathbb{R})$ denotes the complexes of Alexander-Spanier cochains.

We now have two operations:

- equivariant integration $\int_M : \Omega_{G,c}(M)[d_M] \to \Omega_G$,
- integration along fibers $\int_M : \Omega_{\mathrm{cv}}(M_G;\mathbb{R}), \to \Omega(\mathbb{B}G;\mathbb{R})$.

Besides, the Equivariant de Rham comparison Theorem 4.8.3.1 tells us that there exists a canonical quasi-isomorphism $\Omega_{G,c}(M) \simeq \Omega_{\mathrm{cv}}(M_G;\mathbb{R})$, so we are led to ask whether these operations coincide.

Proposition 5.5.2.1 *Let G be a compact Lie group and let M be an oriented G-manifold. Then, integration along fibers and G-equivariant integration coincide* *modulo the equivariant de Rham comparison theorem 4.8.3.1, more precisely, we have a canonical commutative diagram in* $\mathcal{D}\,\mathrm{DGM}(\Omega_G)$

$$\begin{array}{ccc} \Omega_{\mathrm{cv}}(M_G;\mathbb{R}) & \xleftrightarrow[\text{q.i.}]{} & \Omega_{\mathrm{cv}}(M) \\ {\scriptstyle\int_M}\downarrow & \oplus & \downarrow{\scriptstyle\int_M} \\ \Omega_{\mathrm{cv}}(\{\bullet\}_G;\mathbb{R}) & \xleftrightarrow[\text{q.i.}]{} & \Omega_{\mathrm{cv}}(\{\bullet\}) \end{array}$$

which is functorial on $M \in G\text{-}\mathrm{Man}^{\mathrm{or}}_{\mathrm{pr}}$.

Sketch of Proof Starting from the G-equivariant Cartesian diagram of fiber bundles of same fiber M:

$$\begin{array}{ccccc} M_G(n) & \xleftarrow{\nu_M} & \mathbb{E}G(n)\times M & \xrightarrow{p_2} & M \\ {\scriptstyle\pi_M}\downarrow & \square & \downarrow{\scriptstyle \mathrm{id}\times c_M} & \square & \downarrow{\scriptstyle c_M} \\ \mathbb{B}G(n) & \xleftarrow{\nu_{\{\bullet\}}} & \mathbb{E}G(n)\times\{\bullet\} & \xrightarrow{p_2} & \{\bullet\} \end{array} \tag{5.59}$$

where G acts diagonally on $\mathbb{E}G(n)\times M$, we get for each $n \in \mathbb{N}$, the diagram

$$\begin{array}{ccccccc} \Omega_{G,\mathrm{c}}(M) & \xrightarrow{p_2^*} & \Omega_{G,\mathrm{c}}(\mathbb{E}G(n)\times M) & \xleftrightarrow{\text{q.i.}} & \Omega_{\mathrm{cv}}(M_G(n)) & \xleftarrow{p_2^*} & \Omega_{\mathrm{cv}}(M) \\ {\scriptstyle\int_M}\downarrow & \text{(I)} & \downarrow{\scriptstyle\int_M} & \text{(II)} & {\scriptstyle\int_M}\downarrow & \text{(III)} & \downarrow{\scriptstyle\int_M} \\ \Omega_G & \xrightarrow{p_2^*} & \Omega_{G,\mathrm{c}}(\mathbb{E}G(n)) & \xleftrightarrow{\text{q.i.}} & \Omega_{\mathrm{cv}}(\mathbb{B}G(n)) & \xleftarrow{p_2^*} & \Omega_{\mathrm{cv}}(\mathbb{B}G) \end{array}$$

where $\Omega_{\mathrm{cv}}(\_)$ stands for the Alexander-Spanier cochain complex $\Omega_{\mathrm{cv}}(\_;\mathbb{R})$.

It is clear that subdiagrams (I) and (III) are commutative since (5.59) is Cartesian, and it only remains to explain commutativity of (II). But there, the spaces under consideration are manifolds, and we can replace Alexander-Spanier complexes $\Omega_{\mathrm{cv}}(\_;\mathbb{R})$ by de Rham complexes $\Omega_{\mathrm{cv}}(\_)$, identify $\Omega_{G,\mathrm{c}}(\mathbb{E}G(n)\times\_)$ with $\Omega_{\mathrm{c}}(\mathbb{E}G(n)\times\_)^{\mathrm{Bas}}$, by Proposition 4.4.5.2, and $\Omega_{\mathrm{c}}(\mathbb{E}G(n)\times\_)^{\mathrm{Bas}}$ with $\Omega_{\mathrm{c}}((\_)_G(n))$, by the pullbacks ν^*. After what, we are done, since it is clear that each of these identifications is compatible with integrations along fibers.

The fact that the quasi-isomorphism Theorem 4.8.2.1 and 4.7.2.2-(4-iii), respectively

$$\begin{cases} \Omega_{G,\mathrm{c}}(M) \underset{\simeq}{\to} \varprojlim_n \Omega_{G,\mathrm{c}}(\mathbb{E}G(n)\times M) \\ \Omega_{\mathrm{cv}}(M_G) \underset{\simeq}{\to} \varprojlim_n \Omega_{\mathrm{c}}(M_G(n))\,, \end{cases}$$

are compatible with the integration operations, ends the proof. $\square$

5.6 Equivariant Poincaré Duality

Equivariant integration leads us to mimic the Poincaré pairing (2.4) in the equivariant framework. Let G be a compact Lie group and let M be an oriented G-manifold of dimension d_M.

5.6.1 The Ω_G-Poincaré Pairing

The composition of the Ω_G-bilinear map $\Omega_G(M) \times \Omega_{G,\mathrm{c}}(M) \to \Omega_{G,\mathrm{c}}(M)$, $(\alpha, \beta) \mapsto \alpha \wedge \beta$, with equivariant integration $\int_M : \Omega_{G,\mathrm{c}}(M) \to \Omega_G$, gives rise to the Ω_G-*Poincaré pairing*

$$\langle \cdot, \cdot \rangle_{M,G} : \Omega_G(M) \times \Omega_{G,\mathrm{c}}(M) \to \Omega_G\,, \qquad (\alpha, \beta) \mapsto \int_M \alpha \wedge \beta\,, \tag{5.60}$$

inducing the *Poincaré pairing in equivariant cohomology*

$$\langle \cdot, \cdot \rangle_{G,M} : H_G(M) \times H_{G,\mathrm{c}}(M) \to H_G\,, \qquad ([\alpha], [\beta]) \mapsto \int_M [\alpha] \cup [\beta] \tag{5.61}$$

5.6.1.1 The *Poincaré left adjunction* associated with $\langle \cdot, \cdot \rangle_{G,M}$ (see 2.4) is the map

$$\begin{aligned} I\!D_{G,M} : \Omega_G(M)[d_M] &\longrightarrow \mathbf{Hom}^{\bullet}_{\Omega_G}\big(\Omega_{G,\mathrm{c}}(M), \Omega_G\big) \\ \alpha &\longmapsto I\!D_{G,M}(\alpha) := \Big(\beta \mapsto \int_M \alpha \wedge \beta\Big) \end{aligned} \tag{5.62}$$

and one has, following Lemma 5.5.1.1, for α homogeneous

$$\begin{aligned} I\!D_{G,M}\big((-1)^{d_M} d_G\alpha\big)(\beta) &= \int_M (-1)^{d_M} d_G\alpha \wedge \beta \\ &= \int_M (-1)^{d_M+[\alpha]+1} \alpha \wedge d_G\beta = (-1)^{[\beta]} I\!D_G(\alpha)(d_G\beta)\,, \end{aligned}$$

Hence, by 2.1.7 and 5.2.2.1, the adjunction $I\!D_{G,M}$ in (5.62) is a morphism of Ω_G-graded complexes from $\Omega_G(M)[d_M]$ to $\mathbf{Hom}^{\bullet}_{\Omega_G}(\Omega_{G,\mathrm{c}}(M), \Omega_G)$.

5.6.1.2 The *Poincaré right adjunction* associated with $\langle \cdot, \cdot \rangle_G$ (see 2.6.1) is

$$\begin{aligned} I\!D'_{G,M} : \Omega_{G,\mathrm{c}}(M)[d_M] &\longrightarrow \mathbf{Hom}^{\bullet}_{\Omega_G}\big(\Omega_G(M), \Omega_G\big) \\ \beta &\longmapsto I\!D'_{G,M}(\beta) := \Big(\alpha \mapsto \int_M \alpha \wedge \beta\Big) \end{aligned} \tag{5.63}$$

which is also a morphism of Ω_G-graded complexes.

Proposition 5.6.1.1 *The equivariant Poincaré pairing* (5.60) *is nondegenerate.*

Proof Given $\alpha \in (S(\mathfrak{g}^\vee) \otimes \Omega(M))^G$, let $\alpha = \alpha_1 + \cdots + \alpha_{d_M}$ be its decomposition with $\alpha_i \in (S(\mathfrak{g}^\vee) \otimes \Omega^i(M))^G$. Assume $\alpha_i \neq 0$.

There exist $Y_0 \in \mathfrak{g}$ and $x_0 \in M$ such that $0 \neq \alpha_i(Y_0)(x_0) \in \bigwedge^i T_{x_0}(M)$. Then, proceeding as in the proof of Proposition 2.4.1.1, we can use nonnegative bump functions with sufficiently small support to construct a differential form $\beta_i \in \Omega_c^{d_M - i}(M)$ such that

$$\alpha_i(Y_0)(x_0) \wedge \beta_i(x_0) = \tilde{\omega}(x_0)\,, \tag{5.64}$$

where $\tilde{\omega} \in \Omega^{d_M}(M)$ denotes a nowhere vanishing G-invariant differential form defining the orientation of M (*cf*. 2.3.1.1-(1)).[15]

We identify β_i with $1 \otimes \beta_i \in S(\mathfrak{g}^\vee) \otimes \Omega(M)$, and similarly for $\tilde{\omega}$, and we extend equality (5.64) to all of $\mathfrak{g}$ thanks to the nowhere vanishing property of $\tilde{\omega}$, after which for every $Y \in \mathfrak{g}$ there exists a unique $f(Y) \in \Omega^0(M)$, such that

$$\alpha_i(Y) \wedge \beta_i = f(Y)\,\tilde{\omega}. \tag{5.65}$$

Furthermore, $f(Y)$ is compactly supported since $|f(Y)| \subseteq |\beta_i|$.

Lemma A *The map* $f : \mathfrak{g} \to \Omega_c^0(M)$ *is a polynomial map, i.e.* $f \in S(\mathfrak{g}^\vee) \otimes M$.

Proof of Lemma A Since $\alpha_i \in S(\mathfrak{g}^\vee) \otimes \Omega^i(M)$, we can write $\alpha_i = \sum_j P_j \otimes \alpha_{i,j}$ in which case

$$\begin{aligned} \alpha_i \wedge \beta_i &= \sum\nolimits_j P_j \otimes \alpha_{i,j} \wedge \beta_i \\ &= \sum\nolimits_j P_j \otimes (\phi_j\,\tilde{\omega}) = \big(\sum\nolimits_j P_j \otimes \phi_j\big)\,\tilde{\omega} \end{aligned}$$

where $\phi_j \in \Omega_c^0(M)$ is determined by the the equality $\alpha_{i,j} \wedge \beta_i = \phi_j\,\tilde{\omega}$, thanks to the nowhere vanishing property of $\tilde{\omega}$. Hence, we have

$$f(Y) = \sum\nolimits_j P_j(Y) \otimes \phi_j\,,$$

and Lemma A is proved. ⊟

We can therefore consider $f(Y)\,\beta_i$ as member of the algebra $S(\mathfrak{g}^\vee) \otimes \Omega(M)$, where we therefore have the equality

$$\alpha_i(Y)(f(Y)\,\beta_i) = f(Y)^2\,\tilde{\omega}. \tag{5.66}$$

[15] Notice that a G-invariant nowhere vanishing $\tilde{\omega}$ exists since M is an orientable G-manifold. Indeed, given $\omega \in \Omega^{d_M}(M)$ nowhere vanishing, we have $g \cdot \omega = \phi_g\,\omega$, with $\phi_g : M \to \mathbb{R}_{>0}$, since g preserves orientation. Consequently, the G-average $\tilde{\omega} := \int_G g \cdot \omega\, dg$ is G-invariant and is still nowhere vanishing.

Now, if dg denotes a Haar measure on G such that $1 = \int_M dg$, then the G-average operator

$$\int_G g\cdot(\_) = S(\mathfrak{g}^\vee)\otimes\Omega(M) \to (S(\mathfrak{g}^\vee)\otimes\Omega(M))^G = \Omega_G(M)$$

is a well-defined projector[16] and applied to (5.66) gives

$$\int_G g\cdot\big(\alpha_i(Y)(f(Y)\,\beta_i)\big) =_1 \alpha_i(Y)\int_G g\cdot\big(f(Y)\,\beta_i\big) = \Big(\int_G g\cdot f(Y)^2\Big)\,\tilde\omega.$$

where:

- $(=_1)$ is justified by the fact that α_i is G-equivariant;
- the element $\tilde\beta_i(Y) := \int_G g\cdot(f(Y)\,\beta_i)$ belongs to $\Omega^{d_M-i}_{G,c}(M)$;
- the element $\tilde f(Y) := \int_G g\cdot f(Y)^2$ belongs to $\Omega^0_{G,c}(M)$ and is nonzero since for all $Y\in\mathfrak{g}$ and $x\in M$ we have $f(Y)(x)^2\geq 0$, and $f(Y_0)(x_0)^2\neq 0$ after (5.64) and (5.65).

We can then conclude that

$$\int_M \alpha(Y)\,\tilde\beta_i(Y) = \int_M \alpha_i(Y)\,\tilde\beta_i(Y) = \int_M \tilde f(Y)^2\,\tilde\omega > 0\,,$$

and, this conclusion being independent of the support of α, the nondegeneracy of the equivariant Poincaré pairing is thus proved. □

5.6.2 G-Equivariant Poincaré Duality Theorem

Theorem 5.6.2.1 *Let G be a compact Lie group, and let M an oriented G-manifold of dimension d_M.*

1. The Ω_G-graded morphism of complexes[17]

$$\boxed{\mathbb{D}_{G,M} : \Omega_G(M)[d_M] \longrightarrow \mathbf{Hom}^{\bullet}_{\Omega_G}\big(\Omega_{G,c}(M), \Omega_G\big)} \qquad (5.67)$$

is an injective quasi-isomorphism.

[16] See Tu [91] §13.2 Integrating over a Compact connected Lie Group, p. 105.

[17] Compare with Allday-Franz-Puppe [3] §4. The Main Result. Lemma 4.10, p. 6579.

2. *The morphism* $I\!D_{G,M}$ *induces in* $\mathrm{GM}(\Omega_G)$ the Poincaré duality morphism in G-equivariant cohomology (see 5.4.7.2-(1))

$$\boxed{D_{G,M} : H_G(M)[d_M] \longrightarrow \mathbf{Hom}^{\bullet}_{H_G}\big(H_{G,\mathrm{c}}(M), H_G\big)} \tag{5.68}$$

which is an isomorphism if $\mathrm{Ext}^i_{H_G}(H_{G,\mathrm{c}}(M), H_G) = 0$ *for all* $i > 0$, *for example if* $H_{G,\mathrm{c}}(M)$ *is a free* H_G*-module.*[18]

3. *If G is connected, then there exist spectral sequences converging to* $H_G(M)[d_M]$

$$\begin{cases} I\!E_2^{p,q}(M) = \big(\mathrm{Ext}^p_{H_G}\big(H_{G,\mathrm{c}}(M), H_G\big)\big)^q \Rightarrow H_G^{d_M+p+q}(M) \\ I\!F_2^{p,q}(M) = H_G^p \otimes_{\mathbb{R}} \mathbf{Hom}^{\bullet}_{\mathbb{R}}(H_\mathrm{c}^{-q}(M), \mathbb{R}) \Rightarrow H_G^{d_M+p+q}(M) \end{cases}$$

where, in the first, q refers to the graded vector space grading.

4. *If, in addition, M is of finite type, the right Poincaré adjunction*

$$\boxed{I\!D'_{G,M} : \Omega_{G,\mathrm{c}}(M)[d_M] \longrightarrow \mathbf{Hom}^{\bullet}_{\Omega_G}\big(\Omega_G(M), \Omega_G\big)}$$

is an injective quasi-isomorphism, and mutatis mutandis *for (2) and (3).*

Proof The injectivity of adjunction morphisms $I\!D_{G,M}$ and $I\!D'_{G,M}$ is consequence of the nondegeneracy of the equivariant Poincaré pairing of Proposition 5.6.1.1.

For the rest, since the action of $W := G/G_0$ is exact, we will assume, as usual, that the group G connected.

(1) We recall the filtration of the Cartan complex that we used earlier in the proof of 4.3.3.1-(2): Let $m \in \mathbb{N}$. An equivariant form in $(\Omega_G(M), d_G)$ is said to be *of index m* if it belongs to the subspace

$$\Omega_G(M)_m := \big(S^{\geq m}(\mathfrak{g}^\vee) \otimes \Omega(M)\big)^G .$$

We easily check that each $\Omega_G(M)_m$ is stable under the Cartan differential d_G, that $\Omega_G(M) = \Omega_G(M)_m$ for all $m \leqslant 0$ and that we have a decreasing filtration

$$\Omega_G(M) = \Omega_G(M)_0 \supseteq \Omega_G(M)_1 \supseteq \Omega_G(M)_2 \supseteq \cdots \tag{5.69}$$

Furthermore, $\Omega^i_G(M) \cap \Omega_G(M)_m = 0$ whenever $m > i$, so that (5.69) is a *regular filtration* (see [46] §4 pp. 76-).

Similarly, $\lambda \in \mathbf{Hom}^{\bullet}_{\Omega_G}(\Omega_G(M), \Omega_G)$ is said to be *of index m* whenever

$$\lambda\big(\big(S^a(\mathfrak{g}^\vee) \otimes \Omega_\mathrm{c}(M)\big)^G\big) \subseteq \Omega_G^{\geq a+m}, \qquad \forall a \in \mathbb{N},$$

[18] See Ginzburg [44] Corollary 3.9, Brion [23] Proposition 1, Franz [42] Corollary 1.5.

and we denote $\mathbf{Hom}^{\bullet}_{\Omega_G}(\Omega_{G,c}(M), \Omega_G)_m$ the vector subspace of such maps. As before, each of these is a subcomplex of $(\mathbf{Hom}^{\bullet}_{\Omega_G}(\Omega_G(M)), \boldsymbol{D})$ and the resulting decreasing filtration

$$\cdots \supseteq \mathbf{Hom}^{\bullet}_{\Omega_G}(\Omega_{G,c}, \Omega_G)_m \supseteq \mathbf{Hom}^{\bullet}_{\Omega_G}(\Omega_{G,c}, \Omega_G)_{m+1} \supseteq \cdots \tag{5.70}$$

verifies, for each λ homogeneous of degree i,

$$a + \dim M + i \geq \deg \lambda\big((S^a(\mathfrak{g}^\vee) \otimes \Omega_c(M))^G\big) \geq a + i\,, \qquad \forall a \in \mathbb{N}\,,$$

so that the filtration (5.70) is also regular. An immediate verification shows that $I\!\!D_{G,M}$ is a morphism of graded filtered modules, i.e.

$$I\!\!D_{G,M}\big(\Omega_G(M)[d_M]_m\big) \subseteq \mathbf{Hom}^{\bullet}_{\Omega_G}\big(\Omega_{G,c}(M), \Omega_G\big)_m\,, \qquad \forall m \in \mathbb{Z}\,,$$

therefore giving rise to a morphism between the associated spectral sequences[19] whose $I\!\!F_0$ terms are

$$\begin{cases} \text{(i) } {}'I\!\!F_0^p := \big((S^p(\mathfrak{g}^\vee) \otimes \Omega(M))^G, 1 \otimes d\big)[d_M] \\ \text{(ii) } {}''I\!\!F_0^p := \mathbf{Hom}^{\bullet}_{\Omega_G}\big((S(\mathfrak{g}^\vee) \otimes \Omega_c(M))^G, 1 \otimes d\big), \Omega_G^p\big)\,, \end{cases} \tag{5.71}$$

where $\Omega_G^p := \Omega_G^{\geq p} / \Omega_G^{>p}$ as Ω_G-gm.

Lemma *The complexes* (5.71) *are respectively quasi-isomorphic to*

$$\begin{cases} \text{(i') } \Omega_G^p \otimes \big(\Omega(M), d\big)[d_M]\,, \\ \text{(ii') } \mathbf{Hom}^{\bullet}_{\Omega_G}\big(\Omega_G \otimes (\Omega_c(M), d), \Omega_G^p\big)\,. \end{cases}$$

Proof of Lemma 1

(i') is a straightforward consequence of 4.4.6.1 applied to the following inclusions where the differentials come from $(\Omega(M), d)$:

$$\Omega_G^p \otimes \Omega(M) \supseteq \Omega_G^p \otimes \Omega(M)^G \subseteq (S^p(\mathfrak{g}^\vee) \otimes \Omega(M))^G\,.$$

[19] See Godement [46], §4 Thm. 4.3.1, p. 80.

(ii') Consider the following commutative diagram of complexes where the differentials are those coming from $(\Omega_c(M), d)$ only:

$$
\begin{array}{ccc}
\mathbf{Hom}^\bullet_{\Omega_G}\big((S(\mathfrak{g}^\vee)\otimes\Omega_c(M))^G, \Omega_G^p\big) & \xrightarrow[\text{q.i.}]{(\epsilon)} & I\!R\,\mathbf{Hom}^\bullet_{\Omega_G}\big((S(\mathfrak{g}^\vee)\otimes\Omega_c(M))^G, \Omega_G^p\big) \\
\xi\downarrow & & \text{q.i.}\downarrow\xi \\
\mathbf{Hom}^\bullet_{\Omega_G}\big(\Omega_G\otimes\Omega_c(M)^G, \Omega_G^p\big) & \xrightarrow[\text{q.i.}]{(\epsilon)} & I\!R\,\mathbf{Hom}^\bullet_{\Omega_G}\big(\Omega_G\otimes\Omega_c(M)^G, \Omega_G^p\big) \\
\xi\uparrow & & \text{q.i.}\uparrow\xi \\
\mathbf{Hom}^\bullet_{\Omega_G}\big(\Omega_G\otimes\Omega_c(M), \Omega_G^p\big) & \xrightarrow[\text{q.i.}]{(\epsilon)} & I\!R\,\mathbf{Hom}^\bullet_{\Omega_G}\big(\Omega_G\otimes\Omega_c(M), \Omega_G^p\big)
\end{array}
\tag{5.72}
$$

– The vertical arrows ξ are those induced by the inclusions

$$\Omega_G\otimes\Omega_c(M)\supseteq\Omega_G\otimes\Omega_c(M)^G\subseteq(S^p(\mathfrak{g}^\vee)\otimes\Omega_c(M))^G .$$

which we know to be quasi-isomorphisms after (4.4.6.1). We can then immediately claim that the ξ's in the right-hand side of the diagram, concerned by the derived duality functor, are quasi-isomorphisms also.

– The horizontal arrows (ϵ), induced by the augmentation morphisms (*cf.* 5.4.1), are also quasi-isomorphisms. Indeed, since G is compact, we have a decomposition $S(\mathfrak{g}^\vee)=\Omega_G\otimes_{\mathbb{R}}\mathcal{H}$, where $\mathcal{H}$ denotes the (graded) subspace of G-harmonic polynomials of $S(\mathfrak{g}^\vee)$.[20]

Hence, we have

$$\big(S(\mathfrak{g}^\vee)\otimes_{\mathbb{R}}\Omega_c(M)\big)^G\simeq\Omega_G\otimes_{\mathbb{R}}\big(\mathcal{H}\otimes\Omega_c(M)\big)^G . \tag{5.73}$$

We thus see that in the diagram (5.72), the first entries in the duality functors are all *free* Ω_G-gm's, in which case, we can apply 5.4.7.1-(3) and conclude that the morphisms (ϵ) are all quasi-isomorphisms as expected.

⊟

By the lemma, the action of the morphism of spectral sequences $I\!F(I\!D_{G,M})$ induced by $I\!D_{G,M}$ at the $I\!F_0$ pages, i.e. the morphism

$$\big(S^p(\mathfrak{g}^\vee)\otimes\Omega(M)\big)^G[d_M]\xrightarrow{I\!F_0(I\!D_{G,M})}\mathbf{Hom}^\bullet_{\Omega_G}\big((S(\mathfrak{g}^\vee)\otimes\Omega_c(M))^G, \Omega_G^p\big),$$

[20] This is Kostant Theorem 0.2, in Kostant [67], p. 521, and in Kostant [68], p. 330.

for the differential $1 \otimes d$, is equivalent to the action of $1 \otimes \mathbb{D}_M$ in

$$\Omega_G^p \otimes \Omega(M)[d_M] \xrightarrow{1 \otimes \mathbb{D}_M} \mathbf{Hom}^\bullet_{\Omega_G}\left(\Omega_G \otimes \Omega_c(M), \Omega_G^p\right) = \mathbf{Hom}^\bullet_{\mathbb{R}}\left(\Omega_c(M), \Omega_G^p\right)$$

From this we see the action of $\mathbb{F}(\mathbb{D}_{G,M})$ in the $\mathbb{F}_1$ pages:

$$\begin{array}{ccc} {}'\mathbb{F}_1^{p,q} & \xrightarrow[\simeq]{\mathbb{F}_1(\mathbb{D}_{G,M})} & {}''\mathbb{F}_1^{p,q} \\ \simeq\updownarrow & & \updownarrow\simeq \\ \Omega_G^p \otimes H^{q+d_M}(M) & \xrightarrow[\simeq]{1 \otimes D_M} & \mathbf{Hom}^\bullet_{\mathbb{R}}(H_c^{-q}(M), \Omega_G^p) \end{array}$$

where we recognize in the second row the classical Poincaré duality 2.4.1.3. The morphism $\mathbb{F}_1(\mathbb{D}_{G,M})$ is therefore an isomorphism, which implies that $\mathbb{D}_{G,M}$ is a quasi-isomorphism of complexes as announced in (1).

We can go a little further and easily determine the $\mathbb{F}_2$ pages. Indeed, since the differential $d_1 : \mathbb{F}_1^{p,q} \to \mathbb{F}_1^{p+1,q}$ is zero because Ω_G is evenly graded with zero differential, we have an equality of page $\mathbb{F}_2 = \mathbb{F}_1$, in which case

$$\mathbb{F}_2^{p,q}(M) = H_G^p \otimes_{\mathbb{R}} \mathbf{Hom}^\bullet_{\mathbb{R}}(H_c^{-q}(M), \mathbb{R}) \Rightarrow H_G^{d_M+p+q}(M),$$

as stated in (3).

(2) Thanks to (1) $D_{G,M}$ factors through the natural morphism

$$\boldsymbol{h}\,\mathbf{Hom}^\bullet_{\Omega_G}(\Omega_{G,c}(M), \Omega_G) \to \mathbf{Hom}^\bullet_{H_G}(\boldsymbol{h}(\Omega_{G,c}(M)), H_G)$$

and we can apply Proposition 5.4.7.2, since $\Omega_G(M)$ is a free Ω_G-gm, as was noted in the previous paragraphs.

(3) According to (1), $\mathbb{D}_{G,M}$ stablishes a canonical quasi-isomorphism

$$\Omega_G(M)[d_M] \underset{\text{q.i.}}{\longrightarrow} \mathbf{Hom}^\bullet_{\Omega_G}\left(\Omega_{G,c}(M), \Omega_G\right), \tag{5.74}$$

and the spectral sequence $\mathbb{E}(M)$ is simply the $''\mathbb{E}$ spectral sequence of Proposition 5.4.7.1 converging to the cohomology of the right-hand side of (5.74). The existence of the spectral sequence $\mathbb{F}(M)$ was justified at the end of the proof of (1). Statement (4) follows by the same arguments. □

5.6.3 Torsion-Freeness, Freeness and Reflexivity

Proposition 5.6.2.1-(2,4) shows that freeness of equivariant cohomology as H_G-gm is sufficient for equivariant Poincaré duality to hold. The question then arises whether some weaker condition could be equivalent to duality.

The following two properties are related to this question and have been thoroughly studied in Allday et al. [3].

AFP-1. **Torsion-freeness.** An H_G-gm $\boldsymbol{V}$ is said to be *torsion-free* if

$$\mathrm{Ann}(v) := \{P \in H_G \mid P \cdot v = 0\} = 0\,, \qquad \text{for all } v \in \boldsymbol{V}.$$

The torsion-freeness of equivariant cohomology, which is clearly a necessary condition for duality, since the modules $\mathbf{Hom}^{\bullet}_{H_G}(\_, H_G)$ are torsion-free, is also a sufficient condition for the injectivity of the Poincaré duality morphism (Prop. 5.9 [3], see Exercise 7.2.2), but not for duality, as examples of Franz-Puppe [43] show.

AFP-2. **Reflexivity.** An H_G-gm $\boldsymbol{V}$ is said to be *reflexive* if the natural map

$$\boldsymbol{V} \to \mathbf{Hom}^{\bullet}_{H_G}\big(\mathbf{Hom}^{\bullet}_{H_G}(\boldsymbol{V}, H_G), H_G\big)$$

is an isomorphism.
For a finite type manifold M, while the reflexivity of $H_G(M)$ and $H_{G,\mathrm{c}}(M)$ are clearly necessary conditions to H_G-duality, the converse, which is also true, is much more subtle. The equivalence between duality and reflexivity has been established in [3] (Prop. 5.10) for G abelian, and in Franz [42] (Cor. 5.1) for general G and real coefficients.

The following diagram illustrates the relationship between the different kinds of nontorsions in equivariant cohomology and significant properties of the equivariant Poincaré pairing.

$$\begin{array}{ccccc}
\{\texttt{free}\} \subseteq & \{\texttt{reflexive}\} & \subseteq & \{\texttt{torsion-free}\} \\
& \Updownarrow & & \Updownarrow \\
& \left\{\begin{array}{c}\text{Perfect}\\ \text{Poincaré pairing}\end{array}\right\} & \subseteq & \left\{\begin{array}{c}\text{Nondegenerate}\\ \text{Poincaré pairing}\end{array}\right\}
\end{array}$$

It is worth noting that in [41], Franz gives the first known examples of compact manifolds having reflexive but nonfree equivariant cohomology.

5.6.4 T-Equivariant Poincaré Duality Theorem

When G is a **compact connected torus** $T = \mathbb{S}^1 \times \cdots \times \mathbb{S}^1$, we have:

$$\begin{cases} \Omega_T = S(\mathfrak{t}^\vee) \\ \Omega_T(M) = S(\mathfrak{t}^\vee) \otimes_{\mathbb{R}} \Omega(M)^T \\ \Omega_{T,\mathrm{c}}(M) = S(\mathfrak{t}^\vee) \otimes_{\mathbb{R}} \Omega_{\mathrm{c}}(M)^T \end{cases}$$

so that

$$\begin{aligned} \mathbf{Hom}^{\bullet}_{\Omega_T}(\Omega_{T,\mathrm{c}}, \Omega_T) &= \mathbf{Hom}_{S(\mathfrak{t}^\vee)}\big(S(\mathfrak{t}^\vee) \otimes \Omega_{\mathrm{c}}(M)^T, S(\mathfrak{t}^\vee)\big) \\ &= \mathbf{Hom}^{\bullet}_{\mathbb{R}}\big(\Omega_{\mathrm{c}}(M)^T, S(\mathfrak{t}^\vee)\big) \\ &= S(\mathfrak{t}^\vee) \otimes_{\mathbb{R}} \mathbf{Hom}^{\bullet}_{\mathbb{R}}\big(\Omega_{\mathrm{c}}(M)^T, \mathbb{R}\big) \end{aligned}$$

The left adjunction $I\!D_T(M)$ associated with the T-equivariant Poincaré pairing $\langle \cdot, \cdot \rangle_T$ (see 5.6.1.1) identifies naturally to $\mathbf{1} \otimes I\!D_M$,

$$\begin{array}{ccc} S(\mathfrak{t}^\vee) \otimes \Omega(M)^T[d_M] & \xrightarrow[\mathbf{1} \otimes I\!D_M]{I\!D_T(M)} & S(\mathfrak{t}^\vee) \otimes \mathbf{Hom}^{\bullet}_{\mathbb{R}}\big(\Omega_{\mathrm{c}}(M)^T, \mathbb{R}\big) \\ P \otimes \alpha & \longmapsto & P \otimes \Big(\beta \mapsto \int_M \alpha \wedge \beta\Big) \end{array}$$

and the right adjunction (see 5.6.1.2) to

$$\begin{array}{ccc} S(\mathfrak{t}^\vee) \otimes \Omega_{\mathrm{c}}(M)^T[d_M] & \xrightarrow[\mathbf{1} \otimes I\!D'_M]{I\!D'_T(M)} & S(\mathfrak{t}^\vee) \otimes \mathbf{Hom}^{\bullet}_{\mathbb{R}}\big(\Omega(M)^T, \mathbb{R}\big) \\ P \otimes \beta & \longmapsto & P \otimes \Big(\alpha \mapsto \int_M \alpha \wedge \beta\Big) \end{array}$$

Theorem 5.6.4.1 *Let T be a compact connected torus, and M an oriented T-manifold of dimension d_M.*

1. The H_T-graded morphism of complexes

$$\boxed{I\!D_T(M) : \Omega_T(M)[d_M] \longrightarrow \mathbf{Hom}^{\bullet}_{\Omega_T}\big(\Omega_{T,\mathrm{c}}(M), \Omega_T\big)}$$

is an injective quasi-isomorphism.

2. *The morphism* $I\!D_T(M)$ *induces the* 'Poincaré duality morphism in T-equivariant cohomology' (see 5.4.7.2-(1))

$$\boxed{D_{T,M} : H_T(M)[d_M] \longrightarrow \mathbf{Hom}^\bullet_{H_T}\big(H_{T,c}(M), H_T\big)}$$

If $H_{T,c}(M)$ *is a free* H_T*-module,* $D_{T,M}$ *is an isomorphism.*

3. *There are natural spectral sequences converging to* $H_T(M)[d_M]$

$$\begin{cases} I\!E_2^{p,q}(M) = \operatorname{Ext}^{p,q}_{H_T}\big(H_{T,c}(M), H_T\big) \Rightarrow H_T^{p+q}(M)[d_M] \\ I\!F_2^{p,q}(M) = H_T^p \otimes_{\mathbb{R}} \operatorname{Hom}_{\mathbb{R}}(H_c^q(M), \mathbb{R}) \Rightarrow H_T^{p+q}(M)[d_M], \end{cases}$$

where, in the first, q *denotes the* H_T*-graded vector space degree.*

4. *If, in addition,* M *is of finite type, then the* Ω_G*-graded morphism of complexes*

$$\boxed{I\!D'_T(M) : \Omega_{T,c}(M)[d_M] \longrightarrow \mathbf{Hom}^\bullet_{H_T}\big(\Omega_T(M), H_T\big)}$$

is a quasi-isomorphism, and mutatis mutandis *for (2) and (3).*

Proof The theorem is a particular case of 5.6.2.1 taking $G := T$. Notice however that since we have the identification $I\!D_T(M) = \mathbf{1} \otimes I\!D_M$, we may also conclude using 4.3.3.1-(4). □

Comment 5.6.4.2 Recall that $H_{T,c}(M)$ is a free H_T-module whenever M has no odd (or no even) degree compactly supported cohomology (4.3.3.1-(4c)). Clearly, though not very interesting, this is true also when T acts trivially on M, since then $\iota(Y) = \theta(Y) = 0$, $\forall\, Y \in \mathfrak{t}$, and $H_{T,c}(M) = H_T \otimes_{\mathbb{R}} H_c(M)$.

Chapter 6
Equivariant Gysin Morphism and Euler Classes

We define the Gysin morphisms within the equivariant framework following the approach described in Sect. 2.8. Note that thanks to the equivariant de Rham comparison Theorem 4.8, we know a priori that equivariant Gysin morphisms correspond to relative Gysin morphisms over $H(\mathbb{B}G;\mathbb{R})$ (3.5). In Sect. 6.5, we recall the definition of equivariant Euler classes as an application of Gysin morphisms and state their basic properties.

6.1 G-Equivariant Gysin Morphism

6.1.1 Equivariant Finite de Rham Type Coverings

Recall that by 2.5.3.1, when G is a compact Lie Group, a G-manifold M is the union of a countable ascending chain $\mathscr{U} := \{U_0 \subseteq U_1 \subseteq \cdots\}$ of G-stable open subsets of finite type. The following theorem, a simple corollary of the G-equivariant Poincaré duality Theorem 5.6.2.1, is the equivariant analogue to Proposition 2.6.3.1. Its proof is the same, so details are left to the reader.

Theorem 6.1.1.1 *Let G be a compact Lie group, and let M be an oriented G-manifold. Then, for every filtrant cover $\mathscr{U}$ of M by G-stable open subsets, the natural morphism $\varinjlim_{U\in\mathscr{U}} \Omega_{G,c}(U) \to \Omega_{G,c}(M)$ is an isomorphism, and the analogue to 2.6.3.1–(2) is a well-defined morphism of complexes:*

$$\mathbb{D}'_{G,\mathscr{U}} : (\Omega_{G,c}(M), d_G)[d_M] \longrightarrow \varinjlim_{U\in\mathscr{U}} \big(\mathbf{Hom}^{\bullet}_{\Omega_G}(\Omega_G(U), \Omega_G), -D\big),$$

and, if the open sets in $\mathscr{U}$ are of finite type, then $\mathbb{D}'_{G,\mathscr{U}}$ is a quasi-isomorphism.

A. Arabia, *Equivariant Poincaré Duality on G-Manifolds*, Lecture Notes in Mathematics 2288, https://doi.org/10.1007/978-3-030-70440-7_6

6.1.2 G-Equivariant Gysin Morphism for General Maps

We now follow closely the steps of Sect. 2.8.2. Let $f : M \to N$ be a G-equivariant map between oriented G-manifolds.

To $\beta \in \Omega_{G,\mathrm{c}}(M)$ we assign by equivariant integration the Ω_G-linear form

$$\mathbb{D}'_{G,f}(\beta) : \Omega_G(N) \to \Omega_G\,, \quad \alpha \mapsto \int_M f^*\alpha \wedge \beta\,.$$

We thus have the diagram

$$\begin{array}{ccc} \Omega_{G,\mathrm{c}}(M)[d_M] & \overset{f_!}{\dashrightarrow} & \Omega_{G,\mathrm{c}}(N)[d_N] \\ & \searrow^{\mathbb{D}'_{G,f}} \quad \oplus & \downarrow \mathbb{D}'_{G,N} \\ & & \mathbf{Hom}^\bullet_{\Omega_G}(\Omega_G(N), \Omega_G) \end{array} \quad \left(\begin{array}{c}\text{quasi-iso if } N \text{ is of} \\ \text{finite type}\end{array}\right)$$

which can be commutatively closed in derived category $\mathcal{D}(\mathrm{DGM}(\Omega_G))$ when N is of **finite type**, since, in that case, $\mathbb{D}'_{G,N}$ is a quasi-isomorphism (5.6.2.1-(4)).

When N is not of finite type, we fix an equivariant cover $\mathcal{U}$ of N of open finite type subspaces (6.1.1), and replace $\mathbb{D}'_{G,N}$ with the morphism $\mathbb{D}'_{G,\mathcal{U}}$ of Theorem 6.1.1.1. Then, the following diagram, defined as in 2.8.2,

$$\begin{array}{ccc} \Omega_{G,\mathrm{c}}(M)[d_M] & \overset{f_!}{\dashrightarrow} & \Omega_{G,\mathrm{c}}(N)[d_N] \\ & \searrow^{\mathbb{D}'_{G,f,\mathcal{U}}} \quad \oplus & \downarrow \mathbb{D}'_{G,\mathcal{U}} \quad \text{(quasi-iso)} \\ & & \varinjlim_{U \in \mathcal{U}} \mathbf{Hom}^\bullet_{\Omega_G}(\Omega_G(U), \Omega_G) \end{array}$$

can be commutatively closed in $\mathcal{D}(\mathrm{DGM}(\Omega_G))$ since $\mathbb{D}'_{G,\mathcal{U}}$ is a quasi-isomorphism.

The closing arrow in $\mathcal{D}(\mathrm{DGM}(\Omega_G))$

$$f_! := (\mathbb{D}'_{G,\mathcal{U}})^{-1} \circ \mathbb{D}'_{G,f,\mathcal{U}} : \Omega_{G,\mathrm{c}}(M)[d_M] \to \Omega_{G,\mathrm{c}}(N)[d_N]\,. \tag{6.1}$$

is *the equivariant Gysin morphism associated with* f. It defines in cohomology the morphism of Ω_G-gm's

$$\boxed{f_! : H_{G,\mathrm{c}}(M)[d_M] \to H_{G,\mathrm{c}}(N)[d_N]} \tag{6.2}$$

Theorem 6.1.2.1 (And Definitions) *Let G be a compact Lie group. With the above notations, the following statements are verified.*

1. The equality

$$\int_M f^*[\alpha] \cup [\beta] = \int_N [\alpha] \cup f_![\beta] \tag{6.3}$$

holds for all $[\alpha] \in H_G(N)$ *and* $[\beta] \in H_{G,\mathrm{c}}(M)$.

2. *Furthermore, $f_!$ is a morphism of $H_G(N)$-modules, i.e. the equality, called the* equivariant projection formula,

$$f_!\big(f^*[\alpha] \cup [\beta]\big) = [\alpha] \cup f_!([\beta]) \qquad (6.4)$$

holds for all $[\alpha] \in H_G(N)$ and $[\beta] \in H_{G,c}(M)$.

3. *The correspondence*

$$(\_)_! : G\text{-Man}^{\text{or}} \rightsquigarrow \mathcal{D}\text{DGM}(\Omega_G) \quad \textit{with} \quad \begin{cases} M \rightsquigarrow M_! := \Omega_{G,c}(M)[d_M] \\ f \rightsquigarrow f_! \end{cases}$$

is a covariant functor, which we refer to as the equivariant Gysin functor for general maps. *And,* mutatis mutandis *replacing $\mathcal{D}$DGM(Ω_G) by* GM(Ω_G) *and $\Omega_{G,c}(\_)$ by $H_{G,c}(\_)$.*

4. *Suppose that M and N are manifolds of finite type. If the pullback morphism $f^* : H_G(N) \to H_G(M)$ is an isomorphism, then the Gysin morphism $f_! : H_{G,c}(M)[d_M] \to H_{G,c}(N)[d_N]$ is an isomorphism too.*

Proof (1) Immediate from the definition of the Gysin morphism.

(2) Unlike the proof of the nonequivariant statement 2.6.2.1–(2), this claim is no longer a formal consequence of (1) because equivariant cohomology may have torsion elements, which are killed by equivariant integration. Instead, when N is of finite type and since then $I\!D'_N$ is a quasi-isomorphism, we can check that the following equality holds at the *cochain* level,

$$I\!D'_{G,f}(f^*(\alpha) \cup \beta) = \text{‘}I\!D'_N(\alpha \cup f_!(\beta))\text{’} = I\!D'_{G,f}(\beta) \circ \mu_{\mathrm{r}}(\alpha)\,, \qquad (6.5)$$

where the central term is included for purely heuristic reasons and where we denote $\mu_{\mathrm{r}}(\alpha) : \Omega_G(N) \to \Omega_G(N)$ the right multiplication by α, i.e. $\mu_{\mathrm{r}}(\alpha)(\_) = (\_) \cup \alpha$. The identification of the left and right-hand terms in 6.5 is then a straightforward verification from the definition of $I\!D'_{G,f}$. When N is not of finite type, we use the same arguments with $I\!D'_{G,f,\mathcal{U}}$ instead of $I\!D'_{G,f}$.

(3) is clear. (4) As $f^* : \Omega_G(N) \to \Omega_G(M)$ is a quasi-isomorphism, the induced map $\mathbf{Hom}^\bullet_{H_G}(\Omega_G(N), H_G) \to \mathbf{Hom}^\bullet_{H_G}(\Omega_G(M), H_G)$ is also a quasi-isomorphism, after 5.4.7.2–(3). We can thus conclude, since $I\!D'_{G,M}$ and $I\!D'_{G,N}$ are both quasi-isomorphisms. □

Exercise 6.1.2.2 Prove the following enhancement of the statement 6.1.2.1–(4). If $\pi : V \to B$ is a vector bundle over an oriented manifold B, then π is of finite type (2.5.2.1), and $\pi^* : H_G(B) \to H_G(V)$ and $\pi_! : H_{G,c}(V)[d_V] \to H_{G,c}(B)[d_B]$ and both isomorphisms (Exercise 2.5.2.4–(1)).

6.1.3 G-Equivariant Gysin Morphism for Proper Maps

Following 2.8.1, let $f : M \to N$ be a **proper** G-equivariant map between oriented G-manifolds. To $\alpha \in \Omega_G(M)$ we assign the H_G-linear form on $\Omega_{G,\mathrm{c}}(N)$ defined by $I\!D'_{G,f}(\alpha) : \beta \mapsto \int_M f^*\beta \wedge \alpha$. In this way we obtain the diagram

$$\begin{array}{ccc} \Omega_G(M)[d_M] & \overset{f_*}{\dashrightarrow} & \Omega_G(N)[d_N] \\ & \searrow^{I\!D'_{G,f}} \quad \oplus & \downarrow {\scriptstyle I\!D'_{G,N}\ \text{(quasi-iso)}} \\ & & \Omega_{G,\mathrm{c}}(N)^\vee \end{array}$$

which may be commutatively closed in $\mathcal{D}(\mathrm{DGM}(\Omega_G))$ because $I\!D'_{G,N}$ is a quasi-isomorphism after 5.6.2.1-(1). The closing arrow in $\mathcal{D}(\mathrm{DGM}(\Omega_G))$

$$f_* := (I\!D'_{G,N})^{-1} \circ I\!D'_{G,f} : \Omega_G(M)[d_M] \to \Omega_G(N)[d_N], \tag{6.6}$$

is *the equivariant Gysin morphism associated with the proper map* f. It defines in cohomology the morphism of Ω_G-gm's

$$\boxed{f_* : H_G(M)[d_M] \to H_G(N)[d_N]} \tag{6.7}$$

Theorem 6.1.3.1 (And Definitions) *Let G be a compact Lie group. With the above notations, the following statements are verified.*

1. *The equality*

$$\int_M f^*[\beta] \cup [\alpha] = \int_N [\beta] \cup f_*[\alpha] \tag{6.8}$$

   holds for all $[\alpha] \in H_G(M)$ *and* $[\beta] \in H_{G,\mathrm{c}}(N)$.
2. *Furthermore,* f_* *is a morphism of* $H_{G,\mathrm{c}}(N)$*-modules, the equality, called the* equivariant projection formula for proper maps,

$$f_*\big(f^*[\beta] \cup [\alpha]\big) = [\beta] \cup f_*[\alpha] \tag{6.9}$$

   holds for all $[\beta] \in H_{G,\mathrm{c}}(N)$ *and* $[\alpha] \in H_G(M)$.
3. *The correspondence*

$$f_* : G\text{-}\mathrm{Man}^{\mathrm{or}}_{\mathrm{pr}} \rightsquigarrow \mathcal{D}\mathrm{DGM}(\Omega_G) \qquad \textit{with} \qquad \begin{cases} M \rightsquigarrow M_* := \Omega_G(M)[d_M] \\ f \rightsquigarrow f_* \end{cases}$$

   is a covariant functor, which we will refer to as the equivariant Gysin functor for proper maps. *And,* mutatis mutandis *replacing* $\mathcal{D}\mathrm{DGM}(\Omega_G)$ *by* $\mathrm{GM}(\Omega_G)$ *and* $\Omega_G(\_)$ *by* $H_G(\_)$.

4. *If the pullback morphism* $f^* : H_{G,c}(N) \to H_{G,c}(M)$ *is an isomorphism, then the Gysin morphism* $f_* : H_G(M)[d_M] \to H_G(N)[d_N]$ *is also an isomorphism.*
5. *The natural map* $\phi(\_) : H_{G,c}(\_)[d\_] \to H_G(\_)[d\_]$ *(2.2.3) is a homomorphism between the two equivariant Gysin functors* $(\_)_! \to (\_)_*$ *over the category* $G\text{-Man}^{\rm or}_{\rm pr}$*, i.e. we have natural commutative diagrams*

$$\begin{array}{ccc} H_{G,c}(M) & \xrightarrow{\phi(M)} & H_G(M) \\ f_! \downarrow & \oplus & \downarrow f_* \\ H_{G,c}(N) & \xrightarrow[\phi(N)]{} & H_G(N) \end{array}$$

Proof (1, 2, 3, and 4) Same as the proof of Theorem 6.1.2.1. (5) Immediate after definitions. □

6.1.4 Gysin Morphisms through Spectral Sequences

The next theorem establishes a close connection between the nonequivariant and the equivariant Gysin morphisms. It is a basic tool for the generalization of known properties of classical Gysin morphisms within the equivariant framework.

Theorem 6.1.4.1 *Let G be a compact connected Lie group and* $f : M \to N$ *a G-equivariant map between oriented G-manifolds. There exists a natural morphism of the spectral sequences* $\mathbb{F}$ *of Theorem 5.6.2.1–*(3) *converging to the Gysin morphism* $f_! : H_{G,c}(M)[d_M] \to H_{G,c}(N)[d_N]$,

$$\begin{array}{ccc} \mathbb{F}_{c,2}(M) = H_G \otimes H_c(M)[d_M] & \Rightarrow & H_{G,c}(M)[d_M] \\ {\scriptstyle 1\otimes f_!}\downarrow & & \downarrow {\scriptstyle f_!} \\ \mathbb{F}_{c,2}(N) = H_G \otimes H_c(N)[d_N] & \Rightarrow & H_{G,c}(M)[d_N] \end{array}$$

and in the proper case to $f_* : H_G(M)[d_M] \to H_G(N)[d_N]$,

$$\begin{array}{ccc} \mathbb{F}_2(M) = H_G \otimes H(M)[d_M] & \Rightarrow & H_G(M)[d_M] \\ {\scriptstyle 1\otimes f_*}\downarrow & & \downarrow {\scriptstyle f_*} \\ \mathbb{F}_2(N) = H_G \otimes H(N)[d_N] & \Rightarrow & H_G(M)[d_N] \end{array}$$

In particular, an equivariant Gysin morphism is an isomorphism if the corresponding nonequivariant Gysin morphism is so.

Proof Clear from the proof of Theorem 5.6.2.1 and the definition of Gysin morphisms. □

6.2 Group Restriction and Equivariant Gysin Morphisms

6.2.1 Group Restriction and Equivariant Cohomology

Since a closed subgroup H of a compact Lie group G is a compact Lie group, we have projection maps, functorial on $M \in G$-Man, between Borel constructions

$$\begin{array}{ccc} & \mathbb{E}G \times M & \\ {}^{\nu_{H,M}}\swarrow & & \searrow^{\nu_{G,M}} \\ M_H := \mathbb{E}G \times_H M & \xrightarrow{\ \nu^G_{H,M}\ } & \mathbb{E}G \times_G M =: M_G \\ (x,m) \bmod H & \longmapsto & (x,m) \bmod G \end{array}$$

where the reader will have observed that we are using the same space $\mathbb{E}G$ as universal fiber bundle for G and H simultaneously. In particular, the classifying space for H is $\mathbb{B}H := \mathbb{E}G/H$ (Sect. 4.9).

Proposition 6.2.1.1 *Let $H \subseteq G$ be an inclusion of compact Lie groups, and let M be a G-manifold.*

1. *We have a Cartesian diagram of fiber bundles*

$$\begin{array}{ccc} M_H := \mathbb{E}G \times_H M & \xrightarrow[{[G/H]}]{\ \nu^G_{H,M}\ } & M_G := \mathbb{E}G \times_G M \\ {\scriptstyle \pi_{H,M}}\big\downarrow{\scriptstyle [M]} & \square & {\scriptstyle [M]}\big\downarrow{\scriptstyle \pi_{G,M}} \\ \mathbb{B}H := \mathbb{E}G/H & \xrightarrow[{[G/H]}]{\ \nu^G_{H,\mathbb{B}}\ } & \mathbb{B}G := \mathbb{E}G/G \end{array} \tag{6.10}$$

where the vertical arrows are induced by the constant map $c_M : M \to \{\bullet\}$, and where fiber spaces are shown in brackets.

2. *The map $\nu^G_{H,M} : M_H \to M_G$ is a morphism of fiber bundles*

$$\begin{array}{ccc} M_H & \xrightarrow[{[G/H]}]{\ \nu^G_{H,M}\ } & M_G \\ {\scriptstyle \nu^G_{H,\mathbb{B}} \circ \pi_{H,M}}\searrow{\scriptstyle [G \times_H M]} & & {\scriptstyle [M]}\swarrow{\scriptstyle \pi_{G,M}} \\ & \mathbb{B}G & \end{array} \tag{6.11}$$

with base $\mathbb{B}G$ and fibers respectively $G \times_H M$ and M, as shown in brackets.

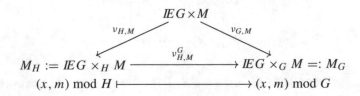

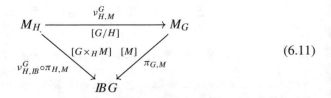

Proof (1) After 3.1.4.2–(2), the diagram is Cartesian if and only if it is so locally relative to $\mathbb{B}G$. For every open subspace $V \subseteq \mathbb{B}G$ which is trivializing for the canonical projection $\nu : \mathbb{E}G \twoheadrightarrow \mathbb{B}G$, let $\sigma : V \to \mathbb{E}G$ be a section of ν defining the slice $S := \sigma(V) \subseteq \mathbb{E}G$ (4.5.2). We then have $\nu^{-1}(V) \simeq S{\times}G$, and the local view of (6.10) is given by the diagram

$$\begin{array}{ccc} S{\times}G{\times}_H M & \twoheadrightarrow & S{\times}G{\times}_G M \\ \downarrow & \square & \downarrow \\ S{\times}G/H & \twoheadrightarrow & S \times G/G \end{array} \tag{6.12}$$

which is Cartesian after Exercise 3.1.4.3.

(2) The same diagram (6.12) shows that the map $\nu^G_{H,M} \circ \pi_{H,M} : M_H \to \mathbb{B}G$ is a locally trivial fibration of fiber $G{\times}_H M$. □

Definition 6.2.1.2 Let $H \subseteq G$ be an inclusion of compact Lie groups, and let M be a G-manifold. We call *group restriction morphisms* and we denote

$$\mathrm{Res}^G_H : \Omega(M_G; \Bbbk) \to \Omega(M_H; \Bbbk) \quad \text{and} \quad \mathrm{Res}^G_H : \Omega_G(M) \to \Omega_H(M)\,,$$

the pullback morphisms induced by the map $\nu^G_{H,M} : M_H \to M_G$.

Thanks to 3.1.10.2–(2), these restrictions respect compact supports and therefore induce the group restriction morphisms of complexes

$$\mathrm{Res}^G_H : \Omega_{\mathrm{cv}}(M_G; \Bbbk) \to \Omega_{\mathrm{cv}}(M_H; \Bbbk) \quad \text{and} \quad \mathrm{Res}^G_H : \Omega_{G,\mathrm{c}}(M) \to \Omega_{H,\mathrm{c}}(M)\,,$$

functorial on $M \in G\text{-}\mathrm{Man}_{\mathrm{pr}}$.

6.2.2 Group Restriction and Integration

An oriented G-manifold M, is automatically an oriented H-manifold, and, because of 6.2.1.1–(1) and 3.2.1.3–(1), we get commutative diagrams combining group restriction and integration morphisms

$$\begin{array}{ccc} \Omega_{\mathrm{cv}}(M_G; \mathbb{R}) & \xrightarrow{\mathrm{Res}^G_H} & \Omega_{\mathrm{cv}}(M_H; \mathbb{R}) \\ {\scriptstyle\int_M}\downarrow & \oplus & \downarrow{\scriptstyle\int_M} \\ \Omega(\mathbb{B}G; \mathbb{R}) & \xrightarrow{\mathrm{Res}^G_H} & \Omega(\mathbb{B}G; \mathbb{R}) \end{array} \qquad \begin{array}{ccc} \Omega_{G,\mathrm{c}}(M) & \xrightarrow{\mathrm{Res}^G_H} & \Omega_{H,\mathrm{c}}(M) \\ {\scriptstyle\int_M}\downarrow & \oplus & \downarrow{\scriptstyle\int_M} \\ \Omega_G & \xrightarrow{\mathrm{Res}^G_H} & \Omega_H \end{array}$$

It is then natural to expect some kind of compatibility between G and H-equivariant Gysin morphism. The following theorem addresses this question.

Theorem 6.2.2.1 *Let $H \subseteq G$ be an inclusion of compact Lie groups. For every G-equivariant map $f : M \to N$ between oriented G-manifolds, the following* *diagrams of Gysin morphisms are commutative.*

$$\begin{array}{ccccccc} H_G(M) & \xrightarrow{f_*} & H_G(N) & \quad & H_{G,c}(M) & \xrightarrow{f_!} & H_{G,c}(N) \\ {\scriptstyle \mathrm{Res}^G_H}\downarrow & \text{(if } f \text{ is proper)} & \downarrow{\scriptstyle \mathrm{Res}^G_H} & & {\scriptstyle \mathrm{Res}^G_H}\downarrow & & \downarrow{\scriptstyle \mathrm{Res}^G_H} \\ H_H(M) & \xrightarrow{f_*} & H_H(N) & & H_{H,c}(M) & \xrightarrow{f_!} & H_{H,c}(N) \end{array} \tag{6.13}$$

Proof By the equivariant de Rham comparison Theorem 4.8, this is a particular case of the commutativity between Gysin functors and base change established in Proposition 3.5.2.1. □

6.3 Adjointness of Equivariant Gysin Morphisms

This is the equivariant analogue to Sect. 3.6.1 where we discussed the concept of adjoint Poincaré morphisms at the level of complexes of differential forms. There, the concept was based on the property of nondegeneracy of the relative Poincaré pairing, a property which holds also for the equivariant Poincaré pairing, as established in Proposition 5.6.1.1, which allows us to introduce the same concept in the equivariant context.

6.3.1 Adjointness Property

The following proposition, analogue to Proposition 3.6.1.1, can help find explicit Gysin morphisms at the level of Ω_G-dg-modules. The proof, is word for word the proof of Proposition 3.6.1.1.

Proposition 6.3.1.1 *Let $f : M \to N$ be an equivariant map between oriented G-manifolds.*

1. *A set-theoretic map $f_\natural : \Omega_{G,c}(M)[d_M] \to \Omega_{G,c}(N)[d_N]$ verifying,*

$$\int_{M'} f^*(\alpha) \wedge \beta = \int_M \alpha \wedge f_\natural(\beta)\,, \tag{6.14}$$

for all $\alpha \in \Omega_G(N)$ and $\beta \in \Omega_{G,c}(M)$, is automatically a morphism of Ω_G-dg-modules inducing the Gysin morphism $f_!$ in $\mathcal{D}(\mathrm{DGM}(\Omega_G, d))$. Furthermore,

$$f_\natural\big(f^*(\alpha) \wedge \beta\big) = \alpha \wedge f_\natural(\beta)\,, \tag{6.15}$$

for all $\alpha \in \Omega_G(N)$ and all $\beta \in \Omega_{G,c}(M)$.

2. *Let f be proper. A set-theoretic map $f_\natural : \Omega_G(M)[d_M] \to \Omega_G(N)[d_N]$ verifying,*

$$\int_{M'} f^*(\beta) \wedge \alpha = \int_M \beta \wedge f_\natural(\alpha)\,, \tag{6.16}$$

for all $\alpha \in \Omega_G(M)$ and $\beta \in \Omega_{G,\mathrm{c}}(N)$, is automatically a morphism of Ω_G-dg-modules inducing the Gysin morphism f_* in $\mathcal{D}(\mathrm{DGM}(\Omega_G, d))$. Furthermore,

$$f_\natural\big(f^*(\beta) \wedge \alpha\big) = \beta \wedge f_\natural(\alpha)\,, \tag{6.17}$$

for all $\alpha \in \Omega_G(M)$ and all $\beta \in \Omega_{G,\mathrm{c}}(N)$.

Remarks 6.3.1.2

1. Take care that the converse of the statements in 6.3.1.1 are not true. Indeed, since the equivariant Poincaré pairing at the level of complexes is nondegenerated (5.6.1.1), expressions 6.14 and 6.16, where they exist, characterize unique morphisms of complexes $\int_M$ and φ_*, while in general many induce the same Gysin morphisms in cohomology.[1]
2. Unlike the nonequivariant case (3.6.1.2), the equivariant Gysin morphism *in cohomology* is generally not characterized by the equality:

$$\int_M f^*[\alpha] \cup [\beta] = \int_N [\alpha] \cup f_![\beta]\,, \quad \forall [\alpha] \in H_G(N)\,,\ \forall [\beta] \in H_{G,\mathrm{c}}(M)\,, \tag{6.18}$$

due to the presence of torsion elements.

For example, the uniqueness of $f_!$ satisfying the relation 6.18, results only from the injectivity of the map:

$$\begin{aligned} I\!D_{G,N} : H_{G,\mathrm{c}}(N) &\longrightarrow \mathbf{Hom}^\bullet_{H_G}\big(H_G(N), H_G\big) \\ [\beta] &\longrightarrow \Big([\alpha] \to \int_N [\alpha] \wedge [\beta]\Big) \end{aligned}$$

a property which is not always satisfied.

Indeed, let T be a torus and let N a compact oriented T-manifold without fixed points. We will see from the localization theorem (Sect. 7.5.1.1) that $H_T(N)$ is a torsion H_T-module (7.4.1). This implies that $\mathbf{Hom}^\bullet_{H_T}(H_G(N), H_T) = 0$, in which case $I\!D_{G,N}$ is null, although $H_T(N) \neq 0$.

By contrast, when $H_G(N)$ and $H_{G,\mathrm{c}}(N)$ are nontorsion (free for example) conditions (6.18) et (6.8) characterize well Gysin morphisms in cohomology.

[1] I thank the referee for this clarification.

6.4 Explicit Constructions of Equivariant Gysin Morphisms

Based in the adjunction existence criterion 6.3.1.1, we now review well-known explicit constructions of Gysin morphisms associated with familiar maps, just as in the nonequivariant case discussed in Sect. 3.6.

6.4.1 *Equivariant Open Embedding*

Let M be an oriented G-manifold, and let U be a G-stable open set in M. Denote by $j : U \subseteq M$ the injection and endow U with the induced orientation.

We have the natural *extension by zero* $j_! : \Omega_{G,c}(U) \hookrightarrow \Omega_{G,c}(M)$ inclusion of Cartan complexes, and the obvious equality

$$\int_U j^*(\alpha) \wedge \beta = \int_M \alpha \wedge j_!(\beta) , \quad \forall \alpha \in \Omega_G(M) , \forall \beta \in \Omega_{G,c}(U) .$$

The pair $(j^*, j_!)$ is thus an equivariant Poincaré adjoint pair in $\mathrm{DGM}(\Omega_G, d)$, and this gives immediately, by Proposition 6.3.1.1–(2), the identification of the equivariant Gysin morphism associated with the open restriction $j^* : \Omega_G(M) \to \Omega_G(U)$ and the extension by zero morphism

$$j_! : \Omega_{G,c}(U)[d_U] \hookrightarrow \Omega_{G,c}(M)[d_M] .$$

6.4.2 *Equivariant Constant Map*

Let M be an oriented G-manifold. The constant map $c_M : M \to \{\bullet\}$ is G-equivariant, and $\Omega_G(\{\bullet\}) = \Omega_G$. The pullback $c_M^* : \Omega_G \to \Omega_G(M)$ is the Ω_G-module structure morphism and the equivariant integration map $\int_M : \Omega_{G,c}(M)[d_M] \to \Omega_G$ is Ω_G-linear (*cf*. Sect. 5.5). Hence, we have

$$\int_M c_M^*(\alpha) \wedge \beta = \int_M \alpha \wedge \beta = \alpha \wedge \int_M \beta = \int_{\{\bullet\}} \alpha \wedge \textstyle\int_M \beta .$$

The pair $(c_M^*, \int_M)$ is thus an equivariant Poincaré adjoint pair in $\mathrm{DGM}(\Omega_G, d)$, and this gives, by Proposition 6.3.1.1–(2), the identification of the equivariant Gysin morphism associated with the pullback $c_M^* : \Omega_G(\{\bullet\}) \to \Omega_G(M)$ and the equivariant integration morphism:

$$c_{M!} = \int_M : \Omega_{G,c}(M)[d_M] \to \Omega_{G,c}(\{\bullet\}) = \Omega_G .$$

6.4.3 Equivariant Projection

Given two oriented G-manifolds M and N, let $\pi : N \times M \twoheadrightarrow N$, denote the projection map $(x, y) \mapsto x$. The morphisms of integration along fibers

$$\textstyle\int_M : \Omega_{\mathrm{cv}}(N\times M)[d_M] \to \Omega(N) \quad \text{and} \quad \int_M : \Omega_{\mathrm{c}}(N\times M)[d_M] \to \Omega_{\mathrm{c}}(N) \tag{6.19}$$

which verify after Proposition 3.2.1.3–(1)

$$\textstyle\int_M \pi^*\alpha \wedge \beta = \alpha \wedge \int_M \beta\,, \quad \forall \alpha \in \Omega(N)\,,\ \forall \beta \in \Omega_{\mathrm{cv}}(N\times M)\,, \tag{6.20}$$

commute with the action of G and with $\mathfrak{g}$-interior products since, locally on $N\times M$, we have

$$\begin{aligned}\textstyle\int_M (g\cdot(\nu\otimes\mu)) &= \textstyle\int_M (g\cdot\nu)\otimes(g\cdot\mu) = (g\cdot\nu)\otimes\int_M (g\cdot\mu)\\ &= \textstyle(g\cdot\nu)\otimes\int_M \mu = g\cdot\int_M \nu\otimes\mu\\ \textstyle\int_M \big(\iota(X)(\nu\otimes\mu)\big) &= \textstyle\int_M \big(\iota(X)(\nu)\otimes\mu\big) + (-1)^{\deg\nu}\int_M \big(\nu\otimes\iota(X)(\mu)\big)\\ &=_1 \textstyle\int_M \iota(X)(\nu)\otimes\mu = \iota(X)\big(\nu\otimes\int_M\mu\big)\,,\end{aligned} \tag{6.21}$$

where $(=_1)$ is justified since $\int_M \iota(X)(\mu) = 0$.

The morphisms (6.19) can then be naturally extended to Cartan complexes

$$\textstyle\int_M : \Omega_{G,\mathrm{cv}}(N\times M)[d_M] \to \Omega_G(N) \quad \text{and} \quad \int_M : \Omega_{G,\mathrm{c}}(N\times M)[d_M] \to \Omega_{G,\mathrm{c}}(N)\,,$$

where $\Omega_{G,\mathrm{cv}}(N\times M) := \big(S(\mathfrak{g}^\vee)\otimes\Omega_{\mathrm{cv}}(N\times M)\big)^G$.

In addition, by Fubini's theorem, we have:

$$\int_{N\times M} \pi^*(\alpha)\wedge\beta = \int_N\int_M \pi^*(\alpha)\wedge\beta =_1 \int_N \alpha\wedge\textstyle\int_M\beta\,, \tag{6.22}$$

for all $\alpha \in \Omega(N)$ and $\beta \in \Omega_{\mathrm{cv}}(N\times M)$, and where $(=_1)$ is justified by (6.20), which also extends naturally to the equivariant context, showing that the pair $(\pi^*, \int_M)$ is an equivariant Poincaré adjoint pair in $\mathrm{DGM}(\Omega_G, d)$. We can thus again conclude, by Proposition 6.3.1.1–(2), to the identification

$$\pi_! = \int_M : \Omega_{G,\mathrm{c}}(N\times M)[d_M] \to \Omega_{G,\mathrm{c}}(N)\,,$$

thus extending the case of the constant map discussed in Sect. 6.4.2.

6.4.4 Equivariant Fiber Bundle

Equivariant projections are particular cases of equivariant fiber bundles. Let (E, B, π, M) be an oriented G-equivariant fiber bundle with fiber M. Integration along fibers gives a morphism of complexes $\int_M : \Omega_{cv}(E) \to \Omega_c(B)$ such that if $\psi : E \to E$ is an isomorphism exchanging fibers, then $\int_M \circ \psi^* = \psi^* \circ \int_M$, so that $\int_M$ is G-equivariant. On the other hand, $\int_M$ commutes with the interior products $\iota(X)$. Indeed, since these are local operators, it suffices (modulo partitions of unity if necessary) to verify the claim over a trivializing open subset of E, i.e. over $\pi^{-1}(U)$ for U s.t. $\pi^{-1}U \sim U \times M$, where we are in the case of a projection already discussed in Sect. 6.4.3. (Beware that although $U \times M$ is not the product of two G-manifolds, the infinitesimal action of G induces '*vertical*' and '*horizontal*' interior products $\iota(X)$, and equalities (6.21) remain valid.)

The study of equivariant projections of Sect. 6.4.3 extends then naturally to the case of equivariant fiber bundles, and the pair $(\pi^*, \int_M)$ is an equivariant Poincaré adjoint pair in $\mathrm{DGM}(\Omega_G, d)$. Hence the identification

$$\pi_! = \int_M : \Omega_{G,c}(E)[d_M] \to \Omega_{G,c}(B)\,.$$

6.4.5 Zero Section of an Equivariant Vector Bundle

Let $(V, B, \pi, \mathbb{R}^n)$ be a G-equivariant oriented vector bundle, and let $\sigma : B \to V$ be the zero section map, which is also a G-equivariant map.

6.4.5.1 The Equivariant Thom Class

The underlying formalism of the nonequivariant Thom class (Sects. 3.2.3 and 3.6.5) remains valid in the equivariant framework.

In Proposition 3.2.3.1, we proved that integration along fibers

$$\int_M : \Omega_{cv}(V)[n] \to \Omega(B) \tag{6.23}$$

is a quasi-isomorphism, and we defined the Thom class of (V, B, π, M) as the unique cohomology class $\Phi_\pi \in H_{cv}^{d_M}$ such that

$$\int_M \Phi_\pi = 1\,. \tag{6.24}$$

The following proposition generalizes this fact in the equivariant setting.

Proposition 6.4.5.1 (and Definition)

Let G be a compact Lie group and let $(V, B, \pi, \mathbb{R}^n)$ be a G-equivariant oriented vector bundle.

1. *The submodule $\big(S(\mathfrak{g}^\vee) \otimes \Omega_{\mathrm{cv}}(V)\big)^G$ of $\Omega_G(V)$ is stable under the action of Cartan's differential $d_G := d + \iota(X)$ and defines the $\Omega_G(B)$-dg-module of G-equivariant differential forms with π-proper supports*

$$\Omega_{G,\mathrm{cv}}(V) := \big(S(\mathfrak{g}^\vee) \otimes \Omega_{\mathrm{cv}}(V), d_G\big)^G . \tag{6.25}$$

The $S(\mathfrak{g}^\vee)$-linear extension of the integration along fibers (6.23) *defines morphisms of $\Omega_G(B)$-dg-modules*

$$\begin{cases} \text{(i) } \int_{\mathbb{R}^n} : \Omega_{G,\mathrm{cv}}(V)[n] \to \Omega_G(B) \\ \text{(ii) } \int_{\mathbb{R}^n} : \Omega_{G,\mathrm{c}}(V)[n] \to \Omega_{G,\mathrm{c}}(B) , \end{cases} \tag{6.26}$$

where (i) *is a quasi-isomorphism.*

2. *There exists a homogeneous G-equivariant cocycle in $\Omega_{\mathrm{cv}}(V)$ of total degree n*

$$\Phi_G = \Phi^{[n]} + \Phi^{[n-2]} + \Phi^{[n-4]} + \cdots$$

with $\Phi^{[n-2k]} \in \big(S^k(\mathfrak{g}^\vee) \otimes \Omega^{n-2k}_{\mathrm{cv}}(V)\big)^G$, and where $\Phi^{[n]} \in \Omega^n_{\mathrm{cv}}(V)^G$ represents the Thom class of $(V, B, \pi, \mathbb{R}^n)$ (see Proposition 3.2.3.1). *Two such cocycles are cohomologous, and the maps*

$$\begin{array}{rcl|rcl} \sigma_! : \Omega_G(B) & \longrightarrow & \Omega_{G,\mathrm{cv}}(V)[n] & \sigma_! : \Omega_{G,\mathrm{c}}(B) & \longrightarrow & \Omega_{G,\mathrm{c}}(V)[n] \\ \nu & \longmapsto & \pi^*\nu \wedge \Phi_G & \nu & \longmapsto & \pi^*\nu \wedge \Phi_G \end{array}$$

are morphisms of Cartan complexes which are right quasi-inverses to (6.26).

3. *The* zero section *$\sigma : B \to V$ of the vector bundle $(V, B, \pi, \mathbb{R}^n)$ is a proper G-equivariant map and we have*

$$\sigma^*(\alpha) = \int_{\mathbb{R}^n} \alpha \wedge \Phi_G , \qquad \forall \alpha \in \Omega_G(V) .$$

Proof (1) As recalled in Sect. 4.4, the natural action of G on the de Rham complex $(\Omega(V), d)$ induces its structure of $\mathfrak{g}$-dg-algebra. The relations $g \cdot \alpha = |\alpha|$, $|d\alpha| \subseteq |\alpha|$ and $|\iota(X)(\alpha)| \subseteq |\alpha|$, for all $\alpha \in \Omega(V)$, then show that the subcomplexes $(\Omega_{\mathrm{c}}(V), d)$ and $(\Omega_{\mathrm{cv}}(V), d)$ are $(G, \mathfrak{g})$-dg-submodules of $(\Omega(V), d)$.

Furthermore, for reasons similar to those in Sects. 6.4.3 and 6.4.4, the maps

$$\int_{\mathbb{R}^n} : \Omega_{\mathrm{cv}}(V)[n] \to \Omega(B) \quad \text{and} \quad \int_{\mathbb{R}^n} : \Omega_{\mathrm{c}}(V)[n] \to \Omega_{\mathrm{c}}(B) , \tag{6.27}$$

are morphisms of $(\Omega(B), \mathfrak{g})$-dg-modules. Consequently, and by functoriality of Cartan complexes (*cf*. Theorem 4.3.3.1), these morphisms (6.27) induce the morphisms of Cartan complexes in (6.26), i.e.

$$\begin{cases} \text{(i) } \int_{\mathbb{R}^n} : \Omega_{G,\mathrm{cv}}(V)[n] \to \Omega_G(B) \\ \text{(ii) } \int_{\mathbb{R}^n} : \Omega_{G,\mathrm{c}}(V)[n] \to \Omega_{G,\mathrm{c}}(B)\,, \end{cases} \tag{6.28}$$

which are clearly also morphisms of $\Omega_G(B)$-dg-modules.

To see that (i) is a quasi-isomorphism, it will suffice, as usual, to assume G connected, in which case we can consider the induced action of $\int_{\mathbb{R}^n}$ on the second page of the spectral sequences associated with the complexes $\Omega_{G,\mathrm{cv}}(V)$ and $\Omega_G(B)$ filtered by polynomial degrees, which is simply the map:

$$\mathrm{id}_{S(\mathfrak{g}^\vee)^\mathfrak{g}} \otimes \pi_! : S(\mathfrak{g}^\vee)^\mathfrak{g} \otimes H_{\mathrm{cv}}(V) \to S(\mathfrak{g}^\vee)^\mathfrak{g} \otimes H(B)\,,$$

where $\pi_! : H_{\mathrm{cv}}(V) \to H(B)$ is an isomorphism as it is the relative Gysin morphism over $H(B)$ associated with the pullback $\pi^* : H(B) \to H(V)$, which is an isomorphism because π is a homotopy equivalence.

(2) The fact that (i) in (6.26) is a quasi-isomorphism proves the existence of the equivariant cocycle Φ_G as well as its uniqueness modulo coboundaries. There is however a constructive proof of its existence, which we now recall.

Since G is compact, every Thom form in $\Omega^n_{\mathrm{cv}}(V)$ can be G-averaged, allowing us to assume that there exists a Thom form $\Phi^{[n]} \in \mathcal{Z}^n_{\mathrm{cv}}(V)^G$. We then have

$$d_G(\Phi^{[n]}) = d\,\Phi^{[n]} + \textstyle\sum_{i=1}^r e^i \otimes \iota(e_i)(\Phi^{[n]}) = \sum_{i=1}^r e^i \otimes \iota(e_i)(\Phi^{[n]})\,,$$

where $\{e_i\}_{i=1}^r$ and $\{e^i\}_{i=1}^r$ are dual bases of $\mathfrak{g}$, and where

$$d(\iota(e_i)(\Phi^{[n]})) = \theta(e_i)(\Phi^{[n]}) = 0\,, \qquad \forall i = 1, \ldots, r\,,$$

since $\Phi^{[n]}$ is G-invariant.

It follows that $\iota(e_i)(\Phi^{[n]}) \in \mathcal{Z}^{n-1}_{\mathrm{cv}}(V)$, and since after (6.23), we have

$$H^j_{\mathrm{cv}}(V) \simeq H^{j-n}(B) = 0\,, \qquad \forall j < n\,, \tag{6.29}$$

there exists $\varphi_i \in \Omega^{n-2}_{\mathrm{cv}}(V)$ such that $d\,\varphi_i = \iota(e_i)(\Phi^{[n]})$, in which case

$$d_G\big(\Phi^{[n]} - \textstyle\sum_{i=1}^r e^i \otimes \varphi_i\big) \in S^2(\mathfrak{g}^\vee) \otimes \Omega^{n-3}_{\mathrm{cv}}(V)\,.$$

We then define

$$\Phi^{[n-2]} := -\int_G g \cdot \big(\textstyle\sum_{i=1}^r e^i \otimes \varphi_i\big)\, dg \in \big(S^1(\mathfrak{g}^\vee) \otimes \Omega^{n-2}_{\mathrm{cv}}(V)\big)^G \subseteq \Omega^n_{G,\mathrm{cv}}(V)\,,$$

so that, by construction, $d_G\big(\Phi^{[n]} + \Phi^{[n-2]}\big) = \iota(X)(\Phi^{[n-2]})\,.$

The same procedure on $\iota(X)(\Phi^{[n-2]})$, possible because of the vanishing condition (6.29), leads to an element $\Phi^{[n-4]} \in \big(S^2(\mathfrak{g}^\vee) \otimes \Omega_{\mathrm{cv}}^{n-4}(V)\big)^G$ such that

$$d_G\big(\Phi^{[n]} + \Phi^{[n-2]} + \Phi^{[n-4]}\big) = \iota(X)(\Phi^{[n-4]}) \in S^3(\mathfrak{g}^\vee) \otimes \Omega_{\mathrm{cv}}^{n-5}(V)\,.$$

The iteration of this procedure ends with a cocycle $\Phi_G := \sum_{k\in\mathbb{N}} \Phi^{[n-2k]}$, which verifies, by construction, $\int_{\mathbb{R}_n} \Phi_G = \int_{\mathbb{R}_n} \Phi^{[n]} = 1$, hence the equality after (6.27)

$$\int_{\mathbb{R}^n} \pi^*(\alpha) \wedge \Phi_G = \alpha, \qquad \forall \alpha \in \Omega_G(B)\,, \tag{6.30}$$

ending the effective construction of an equivariant Thom form.

The maps $\sigma_!$ are morphisms of complexes since Φ_G is a d_G-cocycle, and the fact that they are right inverses to $\int_{\mathbb{R}^n}$ is clear by (6.30).

(2) Since $\pi^* : H_G(B) \to H_G(V)$ and $\sigma^* : H_G(V) \to H_G(B)$ are isomorphisms inverse to each other (6.1.2.2), we have $\alpha \sim \pi^*(\sigma^*(\alpha))$ for all $\alpha \in \Omega_G(V)$, in which case, by (6.30),

$$\int_{\mathbb{R}^n} \alpha \wedge \Phi_G = \int_{\mathbb{R}^n} \pi^*(\sigma^*(\alpha)) \wedge \Phi_G = \sigma^*(\alpha)\,, \qquad \forall \alpha \in \Omega_G(V)\,,$$

which ends the proof of the proposition. □

Proposition 6.4.5.1 can be summarized in the following equivariant analogue to diagram (3.84), p. 101,

$$\begin{array}{ccccc}
\Omega_G(B) & \xleftarrow[\text{q.i.}]{\sigma^*} & \Omega_G(V) & \xleftarrow[\text{q.i.}]{\pi^*} & \Omega_G(B) \\
 & & \text{id} & & \\[1em]
\Omega_G(B) & \xrightarrow[\text{q.i.}]{\sigma_!} & \Omega_{G,\mathrm{cv}}(V)[n] & \xrightarrow[\text{q.i.}]{\pi_!} & \Omega_G(B) \\
 & & \text{id} & & \\[1em]
\Omega_{G,\mathrm{c}}(B) & \xrightarrow{\sigma_!} & \Omega_{G,\mathrm{c}}(V)[n] & \xrightarrow{\pi_!} & \Omega_{G,\mathrm{c}}(B) \\
 & & \text{id} & &
\end{array} \tag{6.31}$$

where the middle row represents the *relative* Gysin morphisms over $\Omega_G(B)$. (Take care that the Gysin morphisms in the bottom row are generally not quasi-isomorphisms, as already observed in 3.6.5.)

Corollary 6.4.5.2 *The equivariant Gysin morphisms associated with the zero section $\sigma : B \hookrightarrow V$ of the vector bundle $(V, B, \pi, \mathbb{R}^n)$:*

$$\begin{cases} \sigma_! : H_{G,\mathrm{c}}(B))[d_B] \to H_{G,\mathrm{c}}(V)[d_V] \\ \sigma_* : H_G(B))[d_B] \to H_G(V)[d_V]\,, \end{cases}$$

are both induced by the morphism of complexes $\pi^*(\_) \wedge \Phi_G$*, where* Φ_G *is a representative of the equivariant Thom class of* $(V, B, \pi, \mathbb{R}^n)$ *(cf. Proposition 6.4.5.1–(2)).*

Proof By the adjointness of the equivariant Gysin morphisms 6.3.1.1, it suffices to check the equality

$$\int_B \sigma^*(\alpha) \wedge \beta = \int_V \alpha \wedge \pi^*(\beta) \wedge \Phi_G \,.$$

But since $\int_V = \int_B \circ \int_{\mathbb{R}^n}$, it is enough to check

$$\sigma^*(\alpha) \wedge \beta = \int_{\mathbb{R}^n} \alpha \wedge \pi^*(\beta) \wedge \Phi_G \,, \quad \forall \alpha \in \Omega_G(V)\,, \forall \beta \in \Omega_G(B)\,,$$

which is obvious since, by 6.4.5.1–(3), we have

$$\int_{\mathbb{R}^n} \alpha \wedge \pi^*(\beta) \wedge \Phi_G = \sigma^*(\alpha \wedge \pi^*(\beta)) = \sigma^*(\alpha) \wedge \beta \,. \qquad \square$$

6.4.6 Equivariant Gysin Long Exact Sequence

Let $i : N \hookrightarrow M$ be a *closed equivariant embedding* of oriented G-manifolds, and denote by $j : U := M \smallsetminus N \hookrightarrow M$ the complementary open embedding.

The sequence of Ω_G-dg-modules

$$0 \longrightarrow \Omega_{G,\mathrm{c}}(U) \xrightarrow{j_!} \Omega_{G,\mathrm{c}}(M) \xrightarrow{i^*} \Omega_{G,\mathrm{c}}(N) \longrightarrow 0\,,$$

where $j_!$ denotes the extension by zero morphism (6.4.1), clearly left exact, is also right exact since every compactly supported differential form of N can be extended (eventually with the help of partitions of unit) to a compactly supported differential form on M.

It is a standard fact that the morphism

$$\begin{array}{ccccc}
\Omega_{G,\mathrm{c}}(U) & \xrightarrow{j_!} & \Omega_{G,\mathrm{c}}(M) & \xrightarrow{\iota} & \hat{c}(j_!) := \Omega_{G,\mathrm{c}}(M) \oplus \Omega_{G,\mathrm{c}}(U)[1] \longrightarrow \\
\| & & \| & & \text{q.i.}\big\downarrow q \\
\Omega_{G,\mathrm{c}}(U) & \xrightarrow{j_!} & \Omega_{G,\mathrm{c}}(M) & \xrightarrow{i^*} & \Omega_{G,\mathrm{c}}(N) \longrightarrow
\end{array} \tag{6.32}$$

where $q(x, y) = i^*(x)$, is a quasi-isomorphism of complexes (see A.1.5.7–(1)), hence an isomorphism in the derived category $\mathcal{D}\mathrm{DGM}(\Omega_G)$, which allows embedding the bottom row in the top row exact triangle (see Comment A.1.6.5).

The duality functor $I\!R\,\mathbf{Hom}^\bullet_{\Omega_G}(\_, \Omega_G)$ applied to (6.32) and the equivariant Poincaré duality theorem 5.6.2.1, then give rise to an isomorphism of exact triangles in $\mathcal{D}\mathrm{DGM}(\Omega_G)$:

$$\begin{array}{ccccccc}
\Omega_G(U)[d_M] & \xleftarrow{j^*} & \Omega_G(M)[d_M] & \longleftarrow & I\!R\,\mathbf{Hom}^\bullet_{\Omega_G}(\hat{c}(j_!), \Omega_G) & \longleftarrow \\
\| & & \| & & \simeq\uparrow & \\
\Omega_G(U)[d_M] & \xleftarrow{j^*} & \Omega_G(M)[d_M] & \xleftarrow{i_!} & \Omega_G(N)[d_N] & \longleftarrow
\end{array} \tag{6.33}$$

where $i_! : \Omega_G(N)[d_N] \to \Omega_G(M)[d_M]$ is the Gysin morphism associated with restriction $i^* : \Omega_G(M) \to \Omega_G(N)$.

It is worth noticing that local equivariant cohomology (Definition 4.10.1.1) makes its appearance in the right term of the first row as the abutment of the quasi-isomorphism:

$$\Omega_G(N)[d_N] \to I\!R\,\mathbf{Hom}^\bullet_{\Omega_G}(\hat{c}(j_!), \Omega_G) = \hat{c}(j^*)[d_M - 1] = \Omega_{G,N}(M)[d_M].$$

We therefore have a canonical isomorphism

$$\boxed{H_G(N)[d_N - d_M] \simeq H_{G,N}(M)} \tag{6.34}$$

The following proposition extends 3.7.1–(2) to the equivariant context.

Proposition 6.4.6.1 *The equivariant Gysin long exact sequence associated with a closed embedding $i : N \hookrightarrow M$ of oriented G-manifolds is the sequence*

$$\to H_G(N)[d_N] \xrightarrow{i_!} H_G(M)[d_M] \xrightarrow{j^*} H_G(U)[d_U] \xrightarrow{c} H_G(N)[d_N + 1],$$

where $j : U := M \smallsetminus N \hookrightarrow M$ is the open inclusion map, j^* is the restriction, $i_!$ is the Gysin morphism (equivariant analogue to Proposition 3.6.6.1), and c is the Gysin morphism associated with the (proper) projection map of the fiber bundle $(\mathbb{S}_\epsilon(N), N, \pi, \mathbb{R}^{d_M - d_N - 1})$, where $\mathbb{S}_\epsilon(N)$ denotes the sphere bundle around N defined in 3.7.1–(3.86).

Proof Same as for 3.7.1–(2). □

6.4.6.1 Exercises

1. The equivariant version of the Lefschetz fixed point theorem (*cf.* Sect. 3.7.2) defines the *G-equivariant Lefschetz class of f* by the expression

$$L_G(f) := \mathrm{Gr}(f)^*(\delta_*(1)) \in H_G^{d_M}(M),$$

where $\delta : M \to M \times M$ denotes the diagonal embedding. The analog of the *equivariant Lefschetz number* is now *a priori* an equivariant cohomology class

$$\Lambda_{G,f} := \int_M L_G(f) \in H_G\,.$$

Prove that

$$\begin{cases} \mathrm{Res}^G_{\{e\}} L_G(f) = L(f) \in H^{d_M}(M) \\ \Lambda_{G,f} = \Lambda_f\,, \end{cases}$$

where $\mathrm{Res}^G_{\{e\}} : H_G(M) \to H(M)$ is the group restriction (Definition 6.2.1.2). Conclude that the *equivariant Lefschetz number* coincides with the nonequivariant one.

Deduce that if $H_G(M)$ is a torsion module (Sect. 7.4.1), then the Euler characteristic of M is zero (*cf*. Sect. 3.7.2–(4)).

2. Show that if $f : N \to M$ is an equivariant map between oriented G-manifolds, then the projective limit of nonequivariant Gysin morphisms (*cf*. Theorem 4.7.2.2–(4))

$$\varprojlim_n \big(f(n)_! : H_c(N_G(n))[d_N] \to H_c(M_G(n))[d_M]\big)$$

is well-defined and coincides with the equivariant Gysin morphism

$$f_! : H_{G,c}(N)[d_N] \to H_{G,c}(M)[d_M]\,.$$

And *mutatis mutandis* for f proper and the equivariant Gysin morphism

$$f_* : H_G(N)[d_N] \to H_G(M)[d_M]\,.$$

6.5 Equivariant Euler Classes

The reference for this section is Atiyah-Bott's paper [7], notably §2 and §3.

6.5.1 The Nonequivariant Euler Class

Let $i : N \subseteq M$ be a closed inclusion of oriented manifolds.

Denote by N_ϵ a tubular neighborhood of N in M (see Sect. 3.2.3.1). As the inclusion $N \subseteq N_\epsilon$ is of the same nature as the inclusion of the zero section of a vector bundle $\sigma : B \subseteq V$ (*cf*. Sect. 3.6.5), we can define the *Thom class* $\Phi(N, M)$

of the pair (N, M) using the same principle, that is, by means of Gysin morphisms. We thus set:

$$\boxed{\Phi(N, M) := i_*(1) \in H^{d_M - d_N}(M)}$$

Definition 6.5.1.1 The *Euler class* $\mathrm{Eu}(N, M)$ *is the restriction to* N *of the Thom class of the pair* (N, M), [2] i.e. :

$$\mathrm{Eu}(N, M) := i^* i_*(1) = \Phi(N, M)|_N \in H^{d_M - d_N}(N)\,. \tag{6.35}$$

6.5.2 G-Equivariant Euler Class

Generalizing the concept of Euler class to the equivariant framework is straightforward thanks to the equivariant Gysin morphism formalism.

Let G be a compact Lie group and let $i : N \subseteq M$ be a closed inclusion of oriented G-manifolds.

Definition 6.5.2.1 The *G-equivariant Euler class* $\mathrm{Eu}_G(N, M)$ is defined by the same formula (6.35):

$$\boxed{\mathrm{Eu}_G(N, M) := i^* i_*(1) = \Phi_G(N, M)|_N \in H_G^{d_M - d_N}(N)} \tag{6.36}$$

where now $i_* : H_G(N)[d_N] \to H_G(M)[d_M]$ is the equivariant Gysin morphism.

Proposition 6.5.2.2 *Let* H *be a closed subgroup of* G, *then*

$$\mathrm{Eu}_H(N, M) = \mathrm{Res}_H^G(\mathrm{Eu}_G(N, M))\,.$$

Proof Immediate after the compatibility between Gysin morphisms and group restrictions established in Theorem 6.2.2.1. □

Exercise 6.5.2.3 Given oriented G-manifolds $L \subseteq N \subseteq M$, prove the following formula for nested equivariant Euler classes (🕮, p. 349)

$$\mathrm{Eu}_G(L, M) = \mathrm{Eu}_G(L, N) \cup \mathrm{Eu}_G(N, M)|_L\,. \tag{6.37}$$

[2] *Cf.* formula (2.19), p. 5, in Atiyah-Bott [7].

6.5.3 G-Equivariant Euler Class of Fixed Points

In the sequel, we denote by M^G the subspace of G-fixed points of M.

For a discrete subspace N of M^G, we have

$$\mathrm{Eu}_G(N, M) \in H_G^{d_M}(N) = \prod\nolimits_{b \in N} S^{d_M/2}(\mathfrak{g}^\vee)^G\,, \tag{6.38}$$

and $\mathrm{Eu}_G(N, M)$ is simply the family of invariant polynomials

$$\mathrm{Eu}_G(N, M) = \big\{\mathrm{Eu}_G(b, M)\big\}_{b \in N} \subseteq S^{d_M/2}(\mathfrak{g}^\vee)^G\,.$$

Comment 6.5.3.1 Note that the exponent $d_M/2$ in (6.38) refers to the polynomial degree, in which case $\mathrm{Eu}_G(N, M) = 0$ when M is odd dimensional. We will see in Proposition 6.5.4.2 that this elementary observation is closely related to the fact that in odd dimensional manifolds there are no isolated T-fixed points, where T denotes a maximal torus in G.

Proposition 6.5.3.2 *Let G be a compact Lie group and let M be an oriented G-manifold. If N is a finite subset of M^G, then*

$$\sum\nolimits_{b \in N} \mathrm{Eu}_G(b, M) = \int_M \Phi_G(N, M) \cup \Phi_G(N, M)$$

and

$$|N| = \int_M \Phi_G(N, M)\,.$$

Proof The constant function $\mathbf{1}_N$ and *a fortiori* the Thom class $\Phi_G(N, M)$, both have compact supports in which case $i_* = i_! : H_{G,\mathrm{c}}(N) \to H_{G,\mathrm{c}}(M)$. The adjointness property of Gysin morphisms then gives:

$$\sum\nolimits_{b \in N} \alpha|_b = \int_M i_!(\mathbf{1}_N) \cup \alpha\,, \quad \forall \alpha \in H_G(M)\,.$$

The proposition results by taking $\alpha := \Phi_G(N, M)$ and $\alpha := 1_M$ respectively. □

6.5.4 T-Equivariant Euler Class of Fixed Points

Let T be a maximal torus of the compact connected Lie group G and denote by $T' := N_G(T)$ the *normalizer* of T in G. We have $T \subseteq T' \subseteq G$.

The universal fiber bundle $\mathbb{E}G$ (*cf*. Sect. 4.6.2) will be used as universal fibre bundle for every closed subgroup $H \subseteq G$. We therefore set for every G-manifold M

$$M_H := \mathbb{E}G \times_H M\,.$$

We then have a natural commutative diagram of Borel constructions:

$$
\begin{array}{ccccc}
M_T := \mathbb{E}G \times_T M & \overset{p}{\twoheadrightarrow} & M_{T'} := \mathbb{E}G \times_{T'} M & \overset{q}{\twoheadrightarrow} & M_G := \mathbb{E}G \times_G M \\
\downarrow & & \downarrow & & \downarrow \\
\mathbb{B}T & \overset{p}{\twoheadrightarrow} & \mathbb{B}T' & \overset{q}{\twoheadrightarrow} & \mathbb{B}G
\end{array}
\tag{6.39}
$$

where p and q denote the obvious quotient maps.

The *Weyl group of the pair* (T, G), i.e. the finite group $W := N_G(T)/T$, acts on the *right* of M_T by $\overline{(x, m)} \cdot w := \overline{(x \cdot \tilde{w}, \tilde{w}^{-1}m)}$, where $\tilde{w}$ is a lift of w in T'. We have therefore an identification $\mathbb{E}G \times_{T'} M \simeq (\mathbb{E}G \times_T M)/W$, and p is the orbit map for this action.

Consequently, the pullback

$$p^* : H(M_{T'}; \mathbb{R}) \to H(M_T; \mathbb{R})^W$$

is an isomorphism, as is also the morphism

$$p^* : H_{T'}(M) \underset{\simeq}{\to} H_T(M)^W,$$

after the equivariant de Rham Theorem 4.8.3.1. [3]

The fibers of $q : M_{T'} \twoheadrightarrow M_G$ are isomorphic to G/T' which is a rational acyclic space. Indeed, this space is the orbit space of G/T for the right action of W and we know from Leray[4] that, under this action, $H(G/T)$ is isomorphic to the regular representation of W. In particular $H(G/T') = H(G/T)^W = \mathbb{R}$ (sec Sect. 8.4.1). This implies that

$$q^* : H(M_G; \mathbb{R}) \to H(M_{T'}; \mathbb{R})$$

is an isomorphism, as is, again after the equivariant de Rham Theorem, the pullback morphism

$$q^* : H_G(M) \underset{\simeq}{\to} H_{T'}(M).$$

Summing up, we have the following canonical isomorphisms

$$H_G(M) \xrightarrow{q^*} H_{T'}(M) \xrightarrow{p^*} H_T(M)^W. \tag{6.40}$$

[3]Recall that the fact that over a field of characteristic 0 and for finite groups, the cohomology of the orbit space is the invariant subspace of the cohomology of the space, is a general result, see Grothendieck [49] Théorème 5.3.1 and Corollaire de la Proposition 5.2.3. But here the context is much simpler since the action of W is free.

[4]Leray [73] Théorème 2.1, p. 103, and Lemma 27.1 in the Ph.D. thesis of Borel [13], p. 193.

Comment 6.5.4.1 When $M = \{\bullet\}$, we obtain a commutative diagram of Chern-Weil homomorphisms

$$\begin{array}{ccccc} S(\mathfrak{g}^\vee)^G & \xrightarrow{\text{Chv}} & S(\mathfrak{t}^\vee)^W & \xrightarrow[\subseteq]{} & S(\mathfrak{t}^\vee) \\ \wr\!\!\downarrow & & \wr\!\!\downarrow & & \wr\!\!\downarrow \\ H_G & \xrightarrow[\simeq]{q^*} & H_{T'} & \overset{p^*}{\rightarrowtail} & H_T \end{array} \tag{6.41}$$

where $\text{Chv} : S(\mathfrak{g}^\vee)^G \to S(\mathfrak{t}^\vee)^W$ associates with a symmetric polynomial function on $\mathfrak{g}$, its restriction to the subspace $\mathfrak{t}$. The diagram shows that Chv is an isomorphism, a claim known in the theory of Lie algebra representations as the *Chevalley isomorphism* or the *Chevalley's restriction theorem.* [5]

It is worth noting that for each $b \in M^G$ the group G acts naturally on the tangent space $T_b(M)$ through a *linear representation*. Now, if we endow M with a G-invariant Riemannian metric, then the exponential map $\exp : T_b(M) \to M$ (*cf*. Sect. 4.5.3) is a G-equivariant diffeomorphism between $T_b(M)$ and an open neighborhood of b in M, and the computation of equivariant Euler classes on fixed points may be greatly simplified by linearizing the data through the exponential map. The following proposition deals with the linear case.

Proposition 6.5.4.2 *Let V be a linear representation of a compact connected Lie group G. Let T be a maximal torus of G.*

1. *The equivariant Euler class* $\mathrm{Eu}_T(0, V)$ *belongs to* $S(\mathfrak{t}^\vee)^W$ *and the Chevalley isomorphism* $\text{Chv} : S(\mathfrak{g}^\vee)^G \to S(\mathfrak{t}^\vee)^W$ *exchanges* $\mathrm{Eu}_G(0, V)$ *and* $\mathrm{Eu}_T(0, V)$.
2. *If* $V := V_1 \oplus V_2$ *as G-module, then* $\mathrm{Eu}_G(0, V) = \mathrm{Eu}_G(0, V_1)\,\mathrm{Eu}_G(0, V_2)$.
3. *Denote by $\mathbb{C}(\alpha)$ the complex vector space $\mathbb{C}$ endowed with the representation of T corresponding to the* weight $\alpha \in \mathfrak{t}^\vee$, *i.e.*

$$\exp(tx)(z) = e^{2\pi\, it\alpha(X)}z\,, \quad \forall(t \in \mathbb{R}),\ \forall(X \in \mathrm{Lie}(T)),\ \forall(z \in \mathbb{C})\,.$$

If $V = \mathbb{R}^{\mu_0} \oplus \bigoplus_\alpha \mathbb{C}(\alpha)^{\mu(\alpha)}$, is the decomposition of V in irreducible representations of T, where μ_0 and $\mu(\alpha)$ denote respectively the multiplicities of $\mathbb{R}$ and $\mathbb{C}(\alpha)$ in V, then the T-equivariant Euler class at $0 \in V$ is given by the formula

$$\mathrm{Eu}_T(0, V) = 0^{\mu_0} \textstyle\prod_\alpha \alpha^{\mu(\alpha)}\,.$$

In particular, $\mathrm{Eu}_G(0, V) \neq 0$ *if and only if* $V^T = 0$.

Proof (1) Results from Proposition 6.5.2.2. (2) Left to the reader. (3) Following (2), it suffices to show that $\mathrm{Eu}_T(0, \mathbb{R}) = 0$ and $\mathrm{Eu}_T(0, \mathbb{C}(\alpha)) = \alpha$.

[5] See Chriss-Ginzburg [31] §3.1.37 Chevalley Restriction Theorem, p. 140.

$\mathrm{Eu}_T(0, \mathbb{R})$. Since the action of T on $\mathbb{R}$ is trivial, we have $H_T(\mathbb{R}) = H_T \otimes H(\mathbb{R})$ and $\mathrm{Eu}_T(0, \mathbb{R}) = 1 \otimes \mathrm{Eu}(0, \mathbb{R}) = 0$, simply because $H^1(\mathbb{R}) = 0$.
$\mathrm{Eu}_T(0, \mathbb{C}(\alpha))$. Taking polar coordinates $(\rho, \theta) \in \mathbb{R}_+ \times [0, 2\pi]$ in $\mathbb{C}$, the nonequivariant Thom form $\Phi(0, \mathbb{C})$ is of the form

$$\Phi^{[2]} = \lambda(\rho)\ \rho\, d\rho \wedge d\theta\,,$$

where $\lambda : \mathbb{R} \to \mathbb{R}$ is a nonnegative bump function with compact support equal to 1 in a neighborhood of 0 and such that $\int_0^\infty \lambda(\rho)\ \rho\, d\rho = 1/2\pi$. By definition of the action of $g \in T$ in $\mathbb{C}(\alpha)$, we have $g \cdot \Phi^{[2]} = \Phi^{[2]}$ so that $\Phi^{[2]}$ is T-invariant. We can thus construct an equivariant Thom class following the procedure described in the proof of 6.4.5.1–(2). We have

$$\begin{aligned}(d_T\, \Phi^{[2]})(X) &= \iota(X)\ \big(\lambda(\rho)\ \rho\, d\rho \wedge d\theta\big)\\ &= -\lambda(\rho)\rho \wedge \iota(X) d\theta = -\lambda(\rho)\rho \wedge \mathcal{L}(\vec{X})(\theta)\\ &= -\lambda(\rho)\rho \wedge \frac{d}{dt}(2\pi\ t\alpha(X)) = -2\pi\ \alpha(X)\ \lambda(\rho)\ \rho\, d\rho\,,\end{aligned}$$

so that $\Phi^{[0]}(X)$ must satisfy the following two conditions

$$\Phi^{[0]}(X) \in S^1(\mathfrak{t}^\vee) \otimes \Omega^0_{\mathrm{c}}(\mathbb{C}(\alpha))^T \quad \text{and} \quad d\,\Phi^{[0]}(X) = -2\pi\ \alpha(X)\ \lambda(\rho)\ \rho\, d\rho\,.$$

But then, the only possibility is

$$\Phi^{[0]}(X) = -2\pi\ \alpha(X)\Big(\int_0^\rho \lambda(\rho)\ \rho d\rho - \int_0^{+\infty} \lambda(\rho)\ \rho d\rho\Big)\,,$$

since $\Phi^{[0]}(X)$ is of compact support. Consequently,

$$\mathrm{Eu}_T(0, \mathbb{C}(\alpha)) = \Phi_T(0, \mathbb{C}(\alpha))|_0 = \Phi^{[0]}(0) = \alpha\,.$$

as stated. □

Exercise 6.5.4.3 If $G = \mathrm{SO}(3)$, show that $\mathrm{Eu}_G(0, \mathbb{R}^3) = 0$. Conclude that isolated G-fixed points may have a null equivariant Euler class when G is nonabelian, contrary to the abelian case. (p. 349)

Chapter 7
Localization

We describe the behavior of de Rham Equivariant Poincaré Duality and Gysin Morphisms under the Localization Functor. In Sect. 8.6 the same topic will be addressed for equivariant cohomologie in positive characteristic.

7.1 The Localization Functor

The *graded ring of fractions* of $\Omega_G = H_G$ (Appendix D) will be denoted by Q_G (simply Q when G is understood). The *localization functor* is then defined as the base change functor

$$Q_G \otimes_{\Omega_G} (\_) : \mathrm{GM}(\Omega_G) \rightsquigarrow \mathrm{Vec}(Q_G)\,.$$

Following Proposition D.1, for any Ω_G-module N, the H_G-module $Q_G \otimes_{H_G} N$ is flat and injective. Consequently, the localization functor is exact and when applied to Cartan complexes, we obtain the *localized Cartan complexes*

$$\begin{cases} Q \circledast \Omega_G(M) := \big(Q_G \otimes_{\Omega_G} \Omega_G(M), \mathrm{id} \otimes d_G\big) \\ Q \circledast \Omega_{G,\mathrm{c}}(M) := \big(Q_G \otimes_{\Omega_G} \Omega_{G,\mathrm{c}}(M), \mathrm{id} \otimes d_G\big) \end{cases}$$

whose cohomologies, the *localized equivariant cohomologies*, respectively denoted by $Q \circledast H_G(M)$ and $Q \circledast H_{G,\mathrm{c}}(M)$, satisfy:

$$Q \circledast H_G(M) = Q_G \otimes_{H_G} H_G(M) \quad \text{and} \quad Q \circledast H_{G,\mathrm{c}}(M) = Q_G \otimes_{H_G} H_{G,\mathrm{c}}(M)\,.$$

A. Arabia, *Equivariant Poincaré Duality on G-Manifolds*, Lecture Notes in Mathematics 2288, https://doi.org/10.1007/978-3-030-70440-7_7

7.2 Localized Equivariant Poincaré Duality

The localized equivariant cohomology is very close to the non equivariant cohomology in that the Poincaré duality pairings are perfect. The following, analogue to the equivariant Poincaré duality Theorem 5.6.2.1, results straight forwardly from the fact that Q_G is a flat and injective Ω_G-module in which case:

$$\mathbb{R}\,\mathbf{Hom}^\bullet_{Q_G}(Q_G \otimes (\_), Q_G) = \mathbb{R}\,\mathbf{Hom}^\bullet_{\Omega_G}((\_), Q_G) = \mathbf{Hom}^\bullet_{\Omega_G}((\_), Q_G)\,,$$

since the functor $\mathbf{Hom}^\bullet_{\Omega_G}((\_), Q_G)$ is exact in $\mathrm{DGM}(\Omega_G)$.

Theorem 7.2.1 *Let G be a compact Lie group, and let M be an oriented G-manifold of dimension d_M.*

1. *The morphism of (nongraded) complexes*

$$\mathbb{D}_{G,M} : Q \otimes \Omega_G(M)[d_M] \to \mathbf{Hom}^\bullet_{Q_G}\big(Q \otimes \Omega_{G,\mathrm{c}}(M), Q_G\big)$$

induces an isomorphism

$$D_{G,M} : Q \otimes H_G(M)[d_M] \to \mathbf{Hom}^\bullet_{Q_G}\big(Q \otimes H_{G,\mathrm{c}}(M), Q_G\big)$$

2. *If, in addition, M is of finite type, then the morphism of complexes*

$$\mathbb{D}'_{G,M} : Q \otimes \Omega_{G,\mathrm{c}}(M)[d_M] \to \mathbf{Hom}^\bullet_{Q_G}\big(Q \otimes \Omega_G(M), Q_G\big)$$

induces an isomorphism

$$D'_{G,M} : Q \otimes H_{G,\mathrm{c}}(M)[d_M] \to \mathbf{Hom}^\bullet_{Q_G}\big(Q \otimes H_G(M), Q_G\big)$$

Exercise 7.2.2 Let M be of finite type. Prove that the torsion-freeness (*cf*. Sect. 5.6.3) of $H_G(M)$ (resp. $H_{G,\mathrm{c}}(M)$) is a necessary and sufficient condition for

$$D_{G,M} : H_G(M)[d_M] \to \mathbf{Hom}^\bullet_{H_G}\big(H_{G,\mathrm{c}}(M), H_G\big)$$

(resp. $D'_{G,M}$) to be injective.

Discuss the case where M is not of finite type. (☛, p. 350)

7.3 Localized Equivariant Gysin Morphisms

As a consequence of Theorem 7.2.1, if $f : M \to N$ is a map between oriented G-manifolds, then the localized Gysin morphisms

$$\begin{cases} f_! : Q \otimes H_{G,\mathrm{c}}(M) \to Q \otimes H_{G,\mathrm{c}}(N) \\ f_* : Q \otimes H_G(M) \to Q \otimes H_G(N)\,, \quad \text{if } f \text{ is proper,} \end{cases}$$

are uniquely determined by the adjointness equalities,

$$\begin{cases} \int_M f^*[\beta] \cup [\alpha] = \int_N [\beta] \cup f_![\alpha] \\ \int_M f^*[\beta] \cup [\alpha] = \int_N [\beta] \cup f_*[\alpha], \quad \text{if } f \text{ is proper.} \end{cases}$$

as in the nonequivariant framework.

7.4 Torsion in Equivariant Cohomology Modules

7.4.1 Torsion

In Sect. 5.6.3 we defined the *annihilator of an element* v of an H_G-gm V as the ideal

$$\mathrm{Ann}(v) := \{P \in H_G \mid P \cdot v = 0\}.$$

We say that an element $v \in V$ is *torsion* if $\mathrm{Ann}(v) \neq 0$, otherwise we say that v is *nontorsion*. An H_G-gm V is a *torsion module* if all its elements are torsion, it is called *nontorsion* or *torsion-free* if 0 is its only torsion element.

Exercise 7.4.1.1 Given an H_G-gm V, let $\tau(V)$ be the subset of its torsion elements. Show that

1. $\tau(V)$ is a torsion module and the quotient $\varphi(V) := V/\tau(V)$ is torsion-free. The natural map: $Q_G \otimes_{H_G} V \to Q_G \otimes_{H_G} \varphi(V)$ is an isomorphism.
2. $Q_G \otimes_{H_G} V = 0$ if and only if V is a torsion module.
3. $\mathrm{Homgr}^0_{H_G}(V, Q_G) = 0$ if and only if V is torsion. (🕮, p. 350)
4. An inductive limit of torsion modules is a torsion module.
5. A projective limit of torsion modules may be a nontorsion module.

Exercise 7.4.1.2 The *annihilator of an H_G-module* is the ideal

$$\mathrm{Ann}(V) := \{P \in H_G \mid P \cdot V = 0\} = \bigcap\nolimits_{v \in V} \mathrm{Ann}(v).$$

1. Show that if $\mathrm{Ann}(V) \neq 0$, then V is a torsion module, but the converse statement may fail to be true. (🕮, p. 350)
2. Show that if V is a unital H_T-algebra, then $\mathrm{Ann}(V) = \mathrm{Ann}(1)$.
3. Let $\{U_1 \subseteq U_2 \subseteq \cdots U_n \cdots\}$ be an increasing family of G-stable open subsets of a G-manifold M such that $M = \bigcup_n U_n$. Suppose that $H_G(U_n)$ and $H_{G,\mathrm{c}}(U_n)$ are torsion for all $n \in \mathbb{N}$. Show that $H_{G,\mathrm{c}}(M)$ is torsion, whereas $H_G(M)$ may fail to be torsion.

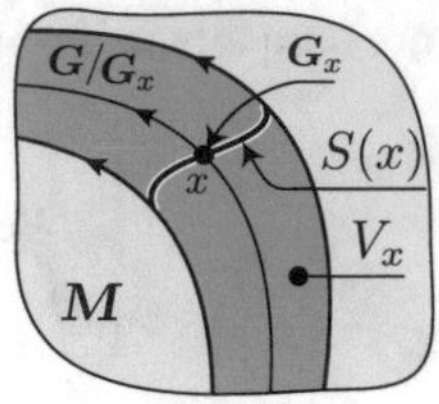

Fig. 7.1 The slice $S(x)$

4. In (3) show that $\{\mathrm{Ann}(H_G(U_n))\}_n$ is a decreasing sequence of ideals and that

$$\mathrm{Ann}(H_G(M)) = \bigcap\nolimits_{n\in\mathbb{N}} \mathrm{Ann}(H_{G,\mathrm{c}}(U_n)) .$$

In particular, if the set $\{\mathrm{Ann}(H_G(U_n))\}$ is finite, then $H_G(M)$ is torsion.

7.5 Slices and Orbit Types

7.5.1 The General Slice Theorem

In Sect. 4.5.2 we recalled the theorem of slices for free G-manifolds, but the same proof works for any G-manifold M, whether the action of G is free or not. The *general slice theorem*[1] then claims that, for every $x \in M$ there exists a (locally closed) submanifold $S(x) \ni x$, stable under the action of *isotropy subgroup group* $G_x \subseteq G$ *of* x, such that the map

$$G \times_{G_x} S(x) \to M\,, \quad [(g, x)] \mapsto g \cdot x\,,$$

is a diffeomorphism onto a G-stable neighborhood V_x of x (Fig. 7.1)

Such submanifold is called *a slice*. We then have

$$H_G(V_x) = H(\mathbb{E}G \times_G G \times_{G_x} S(x)) = H_{G_x}(S(x))\,,$$

and $H_{G,\mathrm{c}}(V_x) = H_{G_x,\mathrm{c}}(S(x))$ (exercise) (p. 350). As a consequence, the H_G-module structures of $H_G(V_x)$ and $H_{G,\mathrm{c}}(V_x)$ factors through the natural ring homomorphism $\rho_x : H_G \to H_{G_x}$.

Proposition 7.5.1.1 *Let T be a torus. For every point x in a T-manifold M, the following equivalences hold.*

1. $\rho_x : H_T \to H_{T_x}$ *is injective if and only if* $x \in M^T$. *The* H_T*-modules* $H_G(V_x)$ *and* $H_{G,\mathrm{c}}(V_x)$ *are torsion if and only if* $x \notin M^T$
2. *If* $x \in M^T$, *then* $\mathrm{Eu}_T(x, M) \neq 0$ *if and only if x is an isolated point of* M^T.

[1] See Hsiang [55] §I.3 Theorem (I.5'), pp. 11–12.

Proof (1) If $T_x \neq T$, there exist closed subtorus $H \subseteq T$ such that $T = H \times T_x$ and $\dim(H) > 0$, in which case $\ker(\rho_x) = H_H^+ \otimes H_{T_x} \neq 0$. (2) is 6.5.4.2–(3). □

Remarks 7.5.1.2

1. Proposition 7.5.1.1 is interesting in that it translates topological properties of a point in a T-manifold into algebraic properties of H_T-modules, opening the way to the algebraic study of the equivariant cohomology of T-spaces.
2. Take care that in 7.5.1.1, we cannot replace T by a non abelian Lie group G, since both claims may fail. For 7.5.1.1–(1), if T is a maximal torus in G and if $M = G/T$, then the isotropy group of $x = g[T] \in M$ is the maximal torus $G_x = gTg^{-1}$, and ρ_x is the inclusion $H_G = (H_T)^W \subseteq H_{G_x}$. Thus, ρ_x is injective although x is not a G-fixed point. A counterexample for 7.5.1.1–(2) is given in Exercise 6.5.4.3.

7.5.2 *Orbit Type of T-Manifolds*

The torsions of the H_T-modules $H_{T,\mathrm{c}}(M)$ and $H_T(M)$ play a central role in the *fixed point theorem.*

When $M^T = \emptyset$, the slice theorem and 7.5.1.1–(1) show that M can be covered by a family of T-stable open subspaces V_x where $H_T(V_x)$ is killed by the elements of the nontrivial kernel $\rho_x : H_T \to H_{T_x}$. Any finite union of those subspaces will also have torsion equivariant cohomology thanks to Mayer-Vietoris sequences. We therefore conclude that if M is compact without fixed points, then $H_T(M)$ is a torsion H_T-module. On the contrary, when M is not compact, the same conclusion can fail to be true (Exercise 7.4.1.2–(3)) unless we have a better control on the set of kernels $\mathcal{K}_T(M) := \{\ker(\rho_x) \mid x \in M\}$. For example, a sufficient condition, as shown in Exercises 7.4.1.2–(4), is the finiteness of $\mathcal{K}_T(M)$, or, which amounts to the same, the finiteness of the set of the isotropy groups $\mathcal{O}_T(M) := \{T_x \mid x \in X\}$, usually referred to as the *orbit type of the T-space M*.[2]

Definition 7.5.2.1 A T-manifold M is said *of finite orbit type* if $\mathcal{O}_T(M)$ is finite.

Exercise 7.5.2.2 Show that a T-manifold M is always locally of finite orbit type. In particular, if M is compact, then it is of finite orbit type. (🕮, p. 351)

Proposition 7.5.2.3 *If $M^T = \emptyset$ and M is of finite orbit type, then*

$$H_{T,\mathrm{c}}(M) \otimes_{H_T} \mathcal{Q}_T = H_T(M) \otimes_{H_T} \mathcal{Q}_T = 0\,.$$

[2]See Hsiang [55] ch. IV §2, p. 54, for the general definition especially for non abelian groups.

Proof

- *Torsion of $H_{T,c}(M)$*. Let $(\mathscr{U}, \subseteq)$ be the set of G-stable open subspaces $U \subseteq M$, such that $H_{T,c}(U)$ is torsion, partially ordered by set inclusion. The set $\mathscr{U}$ is non empty as it contains every slice neighborhood V_x (7.5.1) and is an inductive poset by Exercise 7.4.1.2–(3), so that the Zorn Lemma can be applied. Let U be a maximal element in $\mathscr{U}$. For any $y \in M$, let V_y be a slice neighborhood of y. By the exactness of the Mayer-Vietoris sequence for compactly supported cohomology:

$$\cdots \to H^0_{G,c}(U\cap V_y) \to H^0_{G,c}(U)\oplus H^0_{G,c}(V_y) \to H^0_{G,c}(U\cup V_y) \to H^0_{G,c}(U\cap V_y)[1] \to ,$$

  we easily conclude that $H^0_{G,c}(U\cup V_y)$ is torsion. Then $U \supseteq V_y$, by the maximality of U, hence $U = M$.
- *Torsion of $H_T(M)$*. We cannot use the same argument as in the compactly supported case because a projective limit of torsion modules is not necessarily torsion. The finiteness assumption on the set of orbit types will now be crucial.

  Let I be the intersection of all the ideals $\ker(\rho_x : H_T \to H_{T_x})$ for $x \in M$. The finiteness of the orbit type of M ensures that $I \neq 0$. Let $(\mathscr{U}, \subseteq)$ be the set of G-stable open subspaces $U \subseteq M$, such that $I \subseteq \mathrm{Ann}(H_T(U))$, partially ordered by set inclusion. The set $\mathscr{U}$ is non empty as it contains every slice neighborhood V_x (*cf*. Sect. 7.5.1) and it is an inductive poset by Exercise 7.4.1.2–(4), so that the Zorn Lemma can be applied. Let U be a maximal element in $\mathscr{U}$. For any $y \in M$, let V_y be a slice neighborhood of y. Thanks to the exactness of the first terms of the Mayer-Vietoris sequence: $0 \to H^0_G(U \cup V_y) \to H^0_G(U) \oplus H^0_G(V_y) \to H^0_G(U \cap V_y) \to$, we easily see that $1 \in H^0_G(U \cup V_y)$ is killed by I. Then $I \subseteq \mathrm{Ann}(H_G(U \cup V_y))$ by 7.4.1.2–(2) and $U \supseteq V_y$, by the maximality of U, hence $U = M$. □

7.6 Localized Gysin Morphisms

Given a T-manifold M and a *closed* subgroup $H \subseteq T$, the fixed point set $M^H := \{x \in M \mid h \cdot x = x\ \forall h \in H\}$ is a submanifold whose connected components (not necessarily of equal dimensions) are stable under the action of T, and, in addition, orientable if M is so.[3]

Terminology A homomorphism of H_T-modules $\alpha : L \to L'$ is called an *isomorphism modulo torsion* if its kernel and cokernel are both torsion H_T-modules,

[3]We recall the reason: under the action of H, the tangent spaces $T_x(M)$ for $x \in M^H$ split as the direct sum of $T_x(M^H)$ and a sum of H-irreducible two dimensional representations $\mathbb{C}(\alpha)$ (Propositions 6.5.4.2–(3)), canonically oriented by their character. Therefore, the orientation of $T_x(M^H)$ determines that of $T_x(M)$ and vice versa.

i.e. if the following induced homomorphism of Q_T-modules is an isomorphism

$$\alpha \otimes_{H_T} \mathrm{id} : Q_T \otimes_{H_T} L \to Q_T \otimes_{H_T} L' .$$

Proposition 7.6.1 *Let M be an oriented T-manifold of finite orbit type. For any H closed subgroup of T, denote by $\iota_H : M^H \hookrightarrow M$ the set inclusion. The following morphisms of H_T-gm* [4] *are isomorphisms modulo torsion.*

$$\text{Gysin morphisms} \begin{cases} \iota_{H*} : \; H_T(M^H)[d_{M^H}] \;\to\; H_T(M)[d_M] \\ \iota_{H!} : H_{T,c}(M^H)[d_{M^H}] \to H_{T,c}(M)[d_M] \end{cases}$$

$$\text{Restriction morphisms} \begin{cases} \iota_H^* : H_{T,c}(M) \to H_{T,c}(M^H) \\ \iota_H^* : \; H_T(M) \;\to\; H_T(M^H) \end{cases}$$

Proof The kernel and cokernel of the restriction $\iota_H^* : H_{T,c}(M) \to H_{T,c}(M^H)$ lay within $H_{T,c}(U)$, where $U := M \setminus M^H$. Now, as the isotropy groups of the points of U are *strict* subgroups of T, there are no T-fixed points, i.e. $U^T = \emptyset$, and we can conclude that $H_{T,c}(U)$ is an H_T-torsion module by Proposition 7.5.2.3. In particular, any submodule of $H_{T,c}(U)$, viz. the kernel and the cokernel of ι_H^*, is a torsion H_T-module. By duality the same is true for $\iota_{H*} : H_T(M^H) \to H_T(M)$.

The other restriction $\iota_H^* : H_T(M) \to H_T(M^H)$ is a little more tricky as its kernel and cokernel lay within $H_{T,U}(X)$ which we have not yet proved to be an H_T-torsion module. For that, recall that since we have short exact sequences of local section functors over open subspaces

$$0 \to \Gamma_{U_1 \cap U_1}(\_) \longrightarrow \Gamma_{U_1}(\_) \oplus \Gamma_{U_2}(\_) \longrightarrow \Gamma_{U_1 \cup U_2}(\_) \to 0$$

where $\Gamma_U(\_)$ denotes the kernel of restriction $\Gamma(M, \_) \to \Gamma(M \setminus U, \_)$, we may follow Mayer-Vietoris procedure to approach $H_{T,U}(X)$ by successively adding slice open sets $V_x \subseteq U$ (*cf*. Sect. 7.5.1). In this way, to show that $H_{T,U}(M)$ is a torsion module, we need only show that each $H_{T,V_x}(M)$ is so. Now, this H_T-module occurs in the long exact sequence $\to H_{T,V_x}(M) \to H(M) \to H_T(M \setminus V_x) \to$ where $M \setminus V_x$ is T-equivariantly homotopic to $M \setminus T{\cdot}x$ since the slice $S(x)$ is a submanifold of M. Hence $H_{T,V_x}(M) \simeq H_{T,T\cdot x}(M) \simeq H_T(T{\cdot}x) = H_{T_x}$, proving that $H_{T,V_x}(M)$ is a torsion H_T-module. □

[4] As the submanifold M^H need be neither connected nor equidimensional the shift indication in a notation as $H_T(M^H)[d_{M^H}]$ must be understood component-wise.

7.7 The Localization Formula

7.7.1 *Inversibility of Euler Classes*

Let M be an oriented T-manifold of dimension d_M and of finite orbit type.

As already mentioned, a connected component F of M^T is an oriented submanifold of M of some dimension d_F. Let $(F_\epsilon, F, \pi, \mathbb{R}^{d_M-d_F})$ be a tubular neighborhood of F in M (Sect. 3.2.3.1) and denote by $\iota : F \hookrightarrow F_\epsilon$ its zero section. Since F is a retract of F_ϵ, the Gysin morphism $\iota_! : Q \otimes H_{T,c}(F)[d_F] \to Q \otimes H_{T,c}(F_\epsilon)[d_M]$ is an isomorphism. On the other hand, Proposition 7.5.2.3 applied to $F_\epsilon \setminus F$ in the long exact sequence $\to Q \otimes H_{T,c}(F_\epsilon \setminus F) \to Q \otimes H_{T,c}(F_\epsilon) \to Q \otimes H_{T,c}(F) \to$ shows that the pullback

$$Q \otimes H_{T,c}(F_\epsilon) \xrightarrow[\simeq]{\iota^*} Q \otimes H_{T,c}(F) \tag{7.1}$$

is an isomorphism (note that this is almost never the case before localization). We therefore have isomorphisms

$$\underbrace{Q \otimes H_{T,c}(F)[d_F] \xrightarrow[\simeq]{\iota_!} Q \otimes H_{T,c}(F_\epsilon)[d_M] \xrightarrow[\simeq]{\iota^*} Q \otimes H_{T,c}(F)[d_M]}_{\mathrm{Eu}_T(F, F_\epsilon)}\ , \tag{7.2}$$

whose composition is the multiplication by the equivariant Euler class $\mathrm{Eu}_T(F, F_\epsilon)$ (Corollary 6.4.5.2 and Sect. 6.5.2.1). This class is in fact invertible in the ring $Q \otimes H_T(F)$. Indeed, applying equivariant Poincaré duality to (7.1), the corresponding equivariant Gysin morphism

$$i_* : Q \otimes H_T(F)[d_F] \xrightarrow[\simeq]{} Q \otimes H_T(F_\epsilon)[d_M], \quad \iota_*(\alpha) = \pi^*(\alpha) \cup [\Phi_T(F, F_\epsilon)]$$

with $\Phi_T(F, F_\epsilon)$ the Thom class (Sects. 6.4.5.1 and 6.4.6.1), is an isomorphism too. If we then compose ι_* with $\iota^* : Q \otimes H_T(F_\epsilon)[d_M] \to Q \otimes H_T(F)[d_M]$, yet an isomorphism, we get the isomorphism $\iota^* \circ \iota_* : Q \otimes H_T(F)[d_F] \xrightarrow[\simeq]{} Q \otimes H_T(F)[d_M]$, given by the multiplication by the equivariant Euler class $\mathrm{Eu}_T(F, F_\epsilon)$ (Definition 6.5.2.1). Therefore, $\mathrm{Eu}_T(F, F_\epsilon)$ is invertible in $Q \otimes H_T(F)$, where we have:

$$\frac{1}{\mathrm{Eu}_T(F, F_\epsilon)} = \frac{1}{\mathrm{Eu}_T(F, M)} \in Q \otimes H_T(F)\,.$$

Then, applying $\int_{F_\epsilon}$ to the middle term in (7.2), and thanks to the fundamental property of Thom class (Proposition 6.4.5.1–(3)), we get the equality

$$\int_{F_\epsilon} \beta = \int_F \frac{\beta|_F}{\mathrm{Eu}_T(F, M)}\,, \quad \forall \beta \in Q \otimes H_{T,c}(F_\epsilon)\,, \tag{7.3}$$

which is the essential ingredient in the proof of the following statement.

Proposition 7.7.1 (Localization Formula)
Let M be an oriented T-manifold of dimension d_M and of finite orbit type. Then, denoting by $\mathcal{F}$ the set of connected components F of M^T, we have:

1. *The following equality known as* the localization formula
$$\int_M \beta = \sum\nolimits_{F\in\mathcal{F}} \int_F \frac{\beta|_F}{\mathrm{Eu}_T(F,M)}\,, \tag{7.4}$$
holds for all $\beta \in Q \otimes H_{T,\mathrm{c}}(M)$.
2. *The localized equivariant Poincaré pairing*
$$\langle\cdot,\cdot\rangle_{T,M} : Q\otimes H_T(M)\times Q\otimes H_{T,\mathrm{c}}(M)\to Q_T\,,$$
is perfect and the formula
$$\langle\alpha,\beta\rangle_{T,M} = \sum_{F\in\mathcal{F}} \int_F \frac{\alpha|_F\,\beta|_F}{\mathrm{Eu}_T(F,M)}\,,$$
holds for all $\alpha\in Q\otimes H_T(M)$ and $\beta\in Q\otimes H_{T,\mathrm{c}}(M)$.
3. *If M is compact of positive dimension, then*
$$0=\sum_{F\in\mathcal{F}}\int_F \frac{1}{\mathrm{Eu}_T(F,M)}\,.$$

Proof (1) Let $\iota : M^T \hookrightarrow M^T_\epsilon$ be a tubular neighborhood of M^T in M. By applying Proposition 7.5.2.3 to the left-hand term in the long exact sequence
$$\to Q\otimes H_{T,\mathrm{c}}(X\smallsetminus M^T)\to Q\otimes H_{T,\mathrm{c}}(X)\to Q\otimes H_{T,\mathrm{c}}(M^T)\to\,,$$
for $X = M,\, M^T_\epsilon$, we immediately deduce that the following *extension by zero* morphism, is an isomorphism,
$$\bigoplus\nolimits_F Q\otimes H_{T,\mathrm{c}}(F_\epsilon) = Q\otimes H_{T,\mathrm{c}}(M^T_\epsilon)\xrightarrow[\simeq]{\iota_!} Q\otimes H_{T,\mathrm{c}}(M)\,.$$
Therefore, $\int_M = \sum_F \int_{F_\epsilon}$, and statement (1) follows from (7.3).
(2) is obvious after (1). (3) results from (7.4) with $\beta := 1 \in H^0_{T,\mathrm{c}}(M)$. □

The following corollary was originally proved independently by Atiyah-Bott [7] and Berline-Vergne [9].

Corollary 7.7.2 *Let M be an oriented T-manifold of dimension d_M, of finite orbit type and such that M^T is a discrete subspace of M. Then*

1. *For all $\beta \in H_{T,\mathrm{c}}(M)$ the following* 'localization formula' *is satisfied:*

$$\int_M \beta = \sum_{x \in M^T} \frac{\beta|_x}{\mathrm{Eu}_T(x, M)}\,.$$

2. *The localized equivariant Poincaré pairing*

$$\langle \cdot, \cdot \rangle_{T,M} : Q \otimes H_T(M) \times Q \otimes H_{T,\mathrm{c}}(M) \to Q_T\,,$$

is perfect and given by the formula

$$\langle \alpha, \beta \rangle_{T,M} = \sum_{x \in M^T} \frac{\alpha|_x \, \beta|_x}{\mathrm{Eu}_T(x, M)}\,.$$

3. *If M is compact of positive dimension*

$$0 = \sum_{x \in M^T} \frac{1}{\mathrm{Eu}_T(x, M)}\,.$$

Proof Particular case of Proposition 7.7.1. □

Chapter 8
Changing the Coefficients Field

The constructions and results described in the previous chapters are available with coefficients in a field $\Bbbk$ of positive characteristic prime to the cardinality of G/G_0.[1] In this section, we prove equivariant Poincaré duality and define Gysin morphisms over an arbitrary field $\Bbbk$ as an application of Grothendieck-Verdier duality.

Interested readers can learn the basis of this theory by consulting Iversen [58], Borel [16] or Kashiwara-Schapira [61], and obviously the article of Verdier [92], on the topics of sheaf theory, derived categories and Grothendieck-Verdier duality. We also recommend recent articles by Allday-Franz-Puppe [3–5] which lead to new results and conjectures regarding Equivariant Cohomology.

8.1 Comments about Notations

Given a map $f : M \to N$ between manifolds, the notations f_* and $f_!$ for Gysin morphisms:

$$\begin{cases} f_* : \Omega(M)[d_M] \to \Omega(N)[d_N], & (f \text{ proper}) \\ f_! : \Omega_c(M)[d_M] \to \Omega_c(N)[d_N], \end{cases} \tag{8.1}$$

can be confused with identical notations in sheaf theory, which traditionally denote the functors of *direct image* and *direct image with proper supports*

$$\begin{cases} f_* : \mathrm{Sh}(M; \Bbbk) \to \mathrm{Sh}(N; \Bbbk), \\ f_! : \mathrm{Sh}(M; \Bbbk) \to \mathrm{Sh}(N; \Bbbk). \end{cases}$$

where $\mathrm{Sh}(\_; \Bbbk)$ denotes the category of sheaves of $\Bbbk$-vector spaces on $(\_)$.

[1] In fact, on any ring, but the increase of technicalities that would impose is not warranted by the advantages of the resulting generality, so we prefer to limit ourselves to fields.

A. Arabia, *Equivariant Poincaré Duality on G-Manifolds*, Lecture Notes in Mathematics 2288, https://doi.org/10.1007/978-3-030-70440-7_8

The notations are nevertheless consistent and their meaning should be clear from the context. Indeed, a central chapter in Grothendieck-Verdier's duality is concerned with the extension of the familiar concept of *integration along fibers* (Sect. 3.6.2) beyond locally trivial fibrations. The result, for manifolds, is a canonical morphism in derived category of sheaves[2]

$$f_!(\underline{\underline{or}}_M)[d_M] \to (\underline{\underline{or}}_N)[d_N]\,,$$

where $\underline{\underline{or}}$ denotes the orientation sheaf. Then, if c denotes the constant map, applying the derived functors $I\!R\,c_*$ and $I\!R\,c_!$ gives natural morphisms

$$\begin{cases} f_* : \; I\!R\,c_{M*}(\underline{\underline{or}}_M)[d_M] \to \; I\!R\,c_{N*}(\underline{\underline{or}}_N)[d_N]\,, & (f \text{ proper}) \\ f_! : \; I\!R\,c_{M!}(\underline{\underline{or}}_M)[d_M] \to \; I\!R\,c_{N!}(\underline{\underline{or}}_N)[d_N]\,, \end{cases} \tag{8.2}$$

which happen to be the Gysin morphisms (8.1).

Note the consistency in (8.2) in how the signs $\{*, !\}$ are used in Gysin morphisms and in direct images of sheaves.

8.1.1 Preliminaries

By *(topological) space* we mean a Hausdorff, local contractible and paracompact space[3] (see also Sect. B.1), for example manifolds, open subspaces of CW-complexes, and in particular, the universal fiver bundle $I\!E G$ and classifying space $I\!B G := I\!E G/G$ of a compact Lie group G (Sect. 4.6).

The category of spaces and continuous maps will be denoted by Top.

For a space X, we denote by

$$\mathrm{Sh}(X) := \mathrm{Sh}(X; \Bbbk), \qquad \mathcal{C}(X) := \mathcal{C}(\mathrm{Sh}(X; \Bbbk),$$
$$\mathcal{K}(X) := \mathcal{K}(\mathrm{Sh}(X; \Bbbk), \quad \mathcal{D}(X) := \mathcal{D}(\mathrm{Sh}(X; \Bbbk),$$

respectively the categories of sheaves of $\Bbbk$-vector spaces on X, its category of complexes, its homotopy category and its derived category.

---

[2]For a locally compact space M, the theory defines the *dualizing complex of sheaves* $\underline{\underline{\omega}}^{\bullet}_M(\Bbbk)$, an object in the derived category of sheaves $\mathcal{D}^+(M; \Bbbk)$. The analogue to integration along fibers then appears as a canonical morphism $I\!R\,f_! : \underline{\underline{\omega}}^{\bullet}_M(\Bbbk) \to \underline{\underline{\omega}}^{\bullet}_N(\Bbbk)$ in $\mathcal{D}^+(N; \Bbbk)$. In the case of a manifold, the complex $\underline{\underline{\omega}}^{\bullet}_M(\Bbbk)$ coincides in $\mathcal{D}^+(M; \Bbbk)$ with the orientation sheaf $\underline{\underline{or}}_M[d_M]$. Details on these subjects can be found in Kashiwara-Schapira [62] chap. III Poincaré-Verdier duality, p. 139, and in Iversen [58] chap. VI. Poincaré duality with general coefficients, p. 289.

[3]Recall that for these spaces Alexander-Spanier, Singular, Čech, and Sheaf cohomologies are isomorphic, see Bredon [20], ch. III, Comparison with other cohomology theories.

The word '*cohomology*' stands for *cohomology of sheaves* and the notation $H(X;\mathbb{k})$ is a shortcut for $H(X;\underline{\mathbb{k}}_X)$, where $\underline{\mathbb{k}}_X$ denotes the constant sheaf on X with fiber $\mathbb{k}$.

8.2 Sheafification of Cartan Models over Arbitrary Fields

We denote by G a compact Lie group[4] and we shrink notations

$$\mathbb{E} := \mathbb{E}G\,, \quad \mathbb{B} := \mathbb{B}G \quad \text{and} \quad H_G := H(\mathbb{B};\mathbb{k})\,.$$

As recalled in Sect. 4.7, the Borel construction is a functor from the category of G-spaces to the category of spaces based on $\mathbb{B}$,

$$\left(\begin{array}{ccc} X & \xrightarrow[G\text{-equiv.}]{f} & Y \\ & \searrow \quad \swarrow & \\ & \bullet & \end{array}\right) \rightsquigarrow \left(\begin{array}{ccc} X_G & \xrightarrow{f_G} & Y_G \\ & {\scriptstyle\pi}\searrow \quad \swarrow{\scriptstyle\pi} & \\ & \mathbb{B} & \end{array}\right)$$

The analogues of the ordinary and the compactly supported *equivariant* cohomology are the *ordinary* and the π*-properly supported* cohomologies over $\mathbb{k}$, of the Borel construction,[5] i.e. we set:

$$H_G(X;\mathbb{k}) := H(X_G;\mathbb{k}) \quad \text{and} \quad H_{G,\mathrm{c}}(X;\mathbb{k}) := H_{\pi\text{-cv}}(X_G;\mathbb{k})\,,$$

both endowed with the structure of H_G-modules induced by the projections π. In the language of sheaf cohomology, this amounts to setting:[6]

$$H_G(X;\mathbb{k}) := \mathbb{R}\Gamma(\mathbb{B}, \mathbb{R}\pi_*\,\underline{\mathbb{k}}_{X_G}) \quad \text{and} \quad H_{G,\mathrm{c}}(X;\mathbb{k}) := \mathbb{R}\Gamma(\mathbb{B}, \mathbb{R}\pi_!\,\underline{\mathbb{k}}_{X_G})\,.$$

where $\underline{\mathbb{k}}_{X_G}$ denotes the constant sheaf on X_G.

Convention When no confusion is likely to arise, we will omit the field $\mathbb{k}$ *in the notations. For example,* $H_G(X)$, $H_{G,\mathrm{c}}(X)$ *will stand for* $H_G(X;\mathbb{k})$, $H_{G,\mathrm{c}}(X;\mathbb{k})$.

These rewritings provide clues to the replacements needed in order to transpose the work on equivariant Poincaré duality in the framework of sheaf cohomology over the coefficients field $\mathbb{k}$.

---

[4]Later we will require that the cardinality of G/G_0 be prime to char($\mathbb{k}$).

[5] $H_{\mathrm{cv}}(\_)$ for *cohomology with compact vertical supports*, after Bott-Tu [18] p. 61.

[6]For a continuous map $f: X \to X'$, the notations $f_*, f_! : \mathrm{Sh}(X;\mathbb{k}) \rightsquigarrow \mathrm{Sh}(X';\mathbb{k})$ denote the functors of *direct image* and of *direct image with proper supports*, respectively. One then denotes by $\mathbb{R}f_*, \mathbb{R}f_! : \mathcal{D}^+(X;\mathbb{k}) \rightsquigarrow \mathcal{D}^+(X';\mathbb{k})$ the corresponding derived functors.

P-1. The complexes $\Omega_G(X)$ and $\Omega_{G,c}(X)$ in $\mathrm{DGM}(\Omega_G)$ should respectively be replaced by the complexes $I\!R\,\pi_*\underline{\underline{\Bbbk}}_{X_G}$ and $I\!R\,\pi_!\underline{\underline{\Bbbk}}_{X_G}$ in $\mathcal{D}^+(I\!B)$.

P-2. The graded algebra $\Omega_G := S(\mathfrak{g}^\vee)^G$ should be replaced by some $\Gamma(I\!B, \_)$-acyclic resolution of the constant sheaf $\underline{\underline{\Bbbk}}_{I\!B}$.

For a space X, we choose the complex of sheaves $\underline{\underline{\Omega}}(X_G) := \underline{\underline{\Omega}}(X_G; \Bbbk)$ *of Alexander-Spanier cochains of* X as resolution of its constant sheaf $\underline{\underline{\Bbbk}}_X$,[7]

$$0 \to \underline{\underline{\Bbbk}}_{X_G} \xrightarrow{\epsilon} \underline{\underline{\Omega}}^0(X_G; \Bbbk) \xrightarrow{d_0} \underline{\underline{\Omega}}^1(X_G; \Bbbk) \xrightarrow{d_1} \cdots$$

Although there are many other possible choices, this one has the advantage of being familiar to topologists and useful in the theory of sheaves. Among its properties, we note the following.

a. The sheaves $\underline{\underline{\Omega}}^i(X)$ are $\Gamma_\Phi(X; \_)$-acyclic (B.6.3.4–(1)), for every family of supports Φ (see B.6.2). This has convenient implications.
   – The complex of *global Alexander-Spanier cochains supported in* Φ:

$$(\Omega_\Phi(X), d_*) := \Gamma_\Phi(X; (\underline{\underline{\Omega}}^\bullet(X), d_*)), \tag{8.3}$$

   computes the cohomology of X with supports in Φ.
   We will be concerned mainly with Φ being the families of closed subspaces of X, of compact subspaces of X, of closed subspaces of $Y \subseteq X$, which respectively compute $H(X)$, $H_c(X)$ and $H_Y(X)$.
   – For any continuous map $f : X \to X'$, the natural morphisms

$$\begin{cases} \pi_*(\underline{\underline{\Omega}}^i(X)) \to & I\!R\,\pi_*(\underline{\underline{\Omega}}^i(X)) \\ \pi_!(\underline{\underline{\Omega}}^i(X)) \to & I\!R\,\pi_!(\underline{\underline{\Omega}}^i(X)), \end{cases} \tag{8.4}$$

   are isomorphisms in $\mathcal{D}^+(X')$.[8]
   – The sheaves $\pi_*(\underline{\underline{\Omega}}^i(X))$ and $\pi_!(\underline{\underline{\Omega}}^i(X))$ are $\Gamma(X', \_)$-acyclic, which implies that the natural morphism of complexes

$$\begin{cases} \Omega(X) \ := \Gamma(X'; \pi_*\,\underline{\underline{\Omega}}(X)) \to I\!R\,\Gamma(X'; \pi_*\,\underline{\underline{\Omega}}(X)) \\ \Omega_{\mathrm{cv}}(X) := \Gamma(X'; \pi_!\,\underline{\underline{\Omega}}(X)) \to I\!R\,\Gamma(X'; \pi_!\,\underline{\underline{\Omega}}(X)). \end{cases} \tag{8.5}$$

   are isomorphisms in $\mathcal{D}^+(\mathrm{Vec}(\Bbbk))$.

b. The *cup product* of Alexander-Spanier cochains endows $(\underline{\underline{\Omega}}^\bullet(X), d_*)$ of a structure of sheaves of differential graded algebras.

[7] Godement [46] §2.5, example 2.5.2, p. 134, and in §3.7, example 3.7.1, p. 157.
[8] Kashiwara-Schapira [61], ch. II.2.5 Sheaves on locally compact spaces, p. 102.

c. The correspondence $X \rightsquigarrow \underline{\underline{\Omega}}(X)$ has a functorial behavior on the category Top, in the sense that, given a continuous map $f : X \to X'$, the maps

$$f_i^\# : \underline{\underline{\Omega}}^i(X') \to f_* \underline{\underline{\Omega}}^i(X)\,, \quad \zeta \mapsto \zeta \circ f\,, \quad \forall i \in \mathbb{N}\,,$$

where ζ denotes a section of $\underline{\underline{\Omega}}^i(X')$, define a morphism of differential graded algebras $f^\# : \underline{\underline{\Omega}}(X') \to f_* \underline{\underline{\Omega}}(X)$. For every continuous map $g : X' \to X''$, we then have

$$(g \circ f)^\# = g_*(f^\#) \circ g^\#\,.$$

The usual functoriality behavior of the correspondence $X \rightsquigarrow H(X)$, respectively of $X \rightsquigarrow H_c(X)$ for proper maps, then follows.

d. In the specific case of a G-space X, the map

$$\pi^\# : \underline{\underline{\Omega}}(\mathbb{B}) \to \pi_* \underline{\underline{\Omega}}(X_G)$$

is compatible with the underlying differential graded algebra structures, and $\pi_* \underline{\underline{\Omega}}(X_G)$ appears as an $\underline{\underline{\Omega}}(\mathbb{B})$-dg-algebra.
Similarly, the subcomplex $\pi_! \underline{\underline{\Omega}}(X_G) \subseteq \pi_* \underline{\underline{\Omega}}(X_G)$ is also stabilized by the action of $\pi^\#$ and is therefore an $\underline{\underline{\Omega}}(\mathbb{B})$-sub-dg-module of $\pi_* \underline{\underline{\Omega}}(X_G)$.
At this point, the reader will have noticed the parallel with the preliminaries of Chap. 5 for the category $\mathrm{DGM}(\Omega_G)$ (Sect. 5.1.2). To emphasize the similarities further, we introduce the notations:

$$\underline{\underline{\Omega}}_G := \underline{\underline{\Omega}}(\mathbb{B})\,, \quad \underline{\underline{\Omega}}_G(X) := \pi_* \underline{\underline{\Omega}}(X_G)\,, \quad \underline{\underline{\Omega}}_{G,c}(X) := \pi_! \underline{\underline{\Omega}}(X_G)\,,$$

and denote by $\mathrm{DGM}(\underline{\underline{\Omega}}_G)$ the *category of sheaves of differential graded modules over* $\underline{\underline{\Omega}}_G$.
This allows us to rephrase things in a more convenient language, which we set out in more detail in Sect. 8.5, formulas (8.34) and (8.35).
– The correspondence $X \rightsquigarrow \underline{\underline{\Omega}}_G(X)$ is a contravariant functor of the category G-Top of G-spaces and equivariant maps to the category $\mathrm{DGM}(\underline{\underline{\Omega}}_G)$

$$\left(\begin{array}{c} X \xrightarrow[G\text{-equiv.}]{f} Y \\ \searrow \qquad \swarrow \\ \{\bullet\} \end{array}\right) \rightsquigarrow \left(\begin{array}{c} \underline{\underline{\Omega}}_G(Y) \xrightarrow{\pi_*(f_G^\#)} \underline{\underline{\Omega}}_G(X) \\ {\scriptstyle \pi^\#}\nwarrow \qquad \nearrow{\scriptstyle \pi^\#} \\ \underline{\underline{\Omega}}_G \end{array}\right)$$

– The correspondence $X \rightsquigarrow \big(\underline{\underline{\Omega}}_{G,c}(X) \subseteq \underline{\underline{\Omega}}_G(X)\big)$ is a contravariant functorial inclusion from the category G-$\mathrm{Top}_{\mathrm{pr}}$ of G-spaces and equivariant *proper* maps to the category DGM($\underline{\underline{\Omega}}_G$)

$$
\left(\begin{array}{ccc} X & \xrightarrow[G\text{-equiv.}]{\text{proper } f} & Y \\ & \searrow \quad \swarrow & \\ & \{\bullet\} & \end{array}\right)
\rightsquigarrow
\left(\begin{array}{ccccc} \underline{\underline{\Omega}}_{G,\mathrm{c}}(Y) & & \xrightarrow{\pi_!(f^\#_G)} & & \underline{\underline{\Omega}}_{G,\mathrm{c}}(X) \\ \downarrow & & & & \downarrow \\ \underline{\underline{\Omega}}_G(Y) & & \xrightarrow{\pi_*(f^\#_G)} & & \underline{\underline{\Omega}}_G(X) \\ & \nwarrow^{\pi^\#} & & {}^{\pi^\#}\nearrow & \\ & & \underline{\underline{\Omega}}_G & & \end{array}\right)
$$

e. After P-2-(a), the sheaves $\underline{\underline{\Omega}}^i_{G,\mathrm{c}}(X)$ and $\underline{\underline{\Omega}}^i_G(X)$ are acyclic for the functors $\Gamma(X, \_)$, $\Gamma_\mathrm{c}(X; \_)$ and $\Gamma_Z(X, \_)$, in which case,

$$
\text{if} \quad \left\{\begin{array}{l} \Omega_G(X) := \Gamma(X; \underline{\underline{\Omega}}^\bullet_G(X)) \\ \Omega^i_{G,\mathrm{c}}(X) := \Gamma_\mathrm{c}(X; \underline{\underline{\Omega}}^\bullet_G(X)) \\ \Omega^i_{G,Z}(X) := \Gamma_Z(X; \underline{\underline{\Omega}}^\bullet_G(X)) \end{array}\right\} \quad \text{then} \quad \left\{\begin{array}{l} H^i_G(X) = \boldsymbol{h}^i(\Omega^\bullet_G(X)) \\ H^i_{G,\mathrm{c}}(X) = \boldsymbol{h}^i(\Omega^\bullet_{G,\mathrm{c}}(X)) \\ H^i_{G,Z}(X) = \boldsymbol{h}^i(\Omega^\bullet_{G,Z}(X)) \end{array}\right\}
$$

P-3. The obvious candidate to replace the duality functor $I\!R\,\mathbf{Hom}^\bullet_{\Omega_G}(\_, \Omega_G)$ (*cf.* Sect. 5.2.3) should be the functor $(\_) \rightsquigarrow I\!R\,\underline{\underline{\boldsymbol{Hom}}}^\bullet_{\underline{\underline{\Omega}}_G}(\_, \underline{\underline{\Omega}}_G)$ but since we are considering $\mathcal{D}\,\mathrm{DGM}(\underline{\underline{\Omega}}_G)$ within $\mathcal{D}^+(I\!B)$, where $\underline{\underline{\Bbbk}}_{I\!B}$ is isomorphic to $\underline{\underline{\Omega}}_G$, we can achieve the same functor by instead considering

$$
(\_) \rightsquigarrow I\!R\,\underline{\underline{\boldsymbol{Hom}}}^\bullet_{\Bbbk}(\_, \underline{\underline{\Bbbk}}_{I\!B})\,.
$$

Indeed, for $\mathcal{M}, \mathcal{N} \in \mathrm{DGM}(\underline{\underline{\Omega}}_G)$, the natural morphisms of sheaves

$$
\underline{\underline{Hom}}_{\underline{\underline{\Omega}}_G}(\mathcal{M}, \mathcal{N}) \to \underline{\underline{Hom}}_{\underline{\underline{\Omega}}_G}(\underline{\underline{\Omega}}_G \otimes_\Bbbk \mathcal{M}, \mathcal{N}) \to \underline{\underline{Hom}}_\Bbbk(\mathcal{M}, \mathcal{N})
$$

are isomorphisms in $\mathcal{D}^+(I\!B)$.

Proof The first, induced by the morphism $\underline{\underline{\Omega}}_G \otimes_\Bbbk \mathcal{M} \to \mathcal{M}$, $\omega \otimes m \mapsto \omega m$ is a quasi-isomorphism in DGM($\underline{\underline{\Omega}}_G$), as germ analysis shows, while the second is classical (almost algebraic), already isomorphism at sheaves level. □

8.2.1 Dictionary

To summarize, we are exchanging:

Real de Rham cohomology	$\leftrightarrow$	Sheaf cohomology over $\Bbbk$
$\Omega_G := S(\mathfrak{g}^\vee)^G$	$\leftrightarrow$	$\underline{\underline{\Omega}}_G \underset{\text{q.i.}}{\leftarrow} \underline{\underline{\Bbbk}}_{I\!B}$
$\mathcal{D}(\mathrm{DGM}(\Omega_G))$	$\leftrightarrow$	$\mathcal{D}^+(I\!B)$
$\Omega_G(X)$	$\leftrightarrow$	$\underline{\underline{\Omega}}_G(X) := \pi_* \underline{\underline{\Omega}}(X_G) \underset{\text{q.i.}}{\leftarrow} I\!R\,\pi_*\underline{\underline{\Bbbk}}_{X_G}$
$\Omega_{G,c}(X)$	$\leftrightarrow$	$\underline{\underline{\Omega}}_{G,c}(X) := \pi_! \underline{\underline{\Omega}}(X_G) \underset{\text{q.i.}}{\leftarrow} I\!R\,\pi_!\underline{\underline{\Bbbk}}_{X_G}$
$I\!R\,\mathbf{Hom}^\bullet_{\Omega_G}(\_,\Omega_G)$	$\leftrightarrow$	$I\!R\,\underline{\underline{\mathit{Hom}}}^\bullet_{\Bbbk}(\_,\underline{\underline{\Bbbk}}_{I\!B})$

Note however that, unlike Ω_G, the differentials in $\underline{\underline{\Omega}}_G$ and in $\mathit{\Omega}_G := \Gamma(I\!B;\underline{\underline{\Omega}}_G)$ are generally not zero.

8.2.2 Reformulation of The Poincaré Adjunctions

Under the current substitutions, the sheafification of the equivariant Poincaré adjunctions (Sect. 5.6) for an oriented G-manifold M of dimension d_M:

$$\begin{cases} \text{(i)} \quad I\!D_{G,M} : \Omega_G(M)[d_M] \to I\!R\,\mathbf{Hom}^\bullet_{\Omega_G}\big(\Omega_{G,c}(M),\Omega_G\big) & \text{(Sect. 5.6.1.1)} \\ \text{(ii)} \quad I\!D'_{G,M} : \Omega_{G,c}(M)[d_M] \to I\!R\,\mathbf{Hom}^\bullet_{\Omega_G}\big(\Omega_G(M),\Omega_G\big), & \text{(Sect. 5.6.1.2)} \end{cases}$$

become the morphisms in the derived category of sheaves $\mathcal{D}^+(I\!B)$

$$\begin{cases} \text{(I)} \quad \underline{\underline{I\!D}}_{G,M} : \underline{\underline{\Omega}}_G(M)[d_M] \to I\!R\,\underline{\underline{\mathit{Hom}}}^\bullet_{\Bbbk}\big(\underline{\underline{\Omega}}_{G,c}(M),\underline{\underline{\Bbbk}}_{I\!B}\big) \\ \text{(II)} \quad \underline{\underline{I\!D}}'_{G,M} : \underline{\underline{\Omega}}_{G,c}(M)[d_M] \to I\!R\,\underline{\underline{\mathit{Hom}}}^\bullet_{\Bbbk}\big(\underline{\underline{\Omega}}_G(M),\underline{\underline{\Bbbk}}_{I\!B}\big). \end{cases} \tag{8.6}$$

Notice that, by Grothendieck-Verdier duality, the last term in (I) verifies

$$I\!R\,\underline{\underline{\mathit{Hom}}}^\bullet_{\Bbbk}(I\!R\,\pi_!\,\underline{\underline{\Bbbk}}_{X_G},\underline{\underline{\Bbbk}}_{I\!B}) = I\!R\,\pi_* I\!R\,\underline{\underline{\mathit{Hom}}}^\bullet_{\Bbbk}(\underline{\underline{\Bbbk}}_X,\pi^!\underline{\underline{\Bbbk}}_{I\!B}) = I\!R\,\pi_*(\pi^!\underline{\underline{\Bbbk}}_{I\!B}),$$

so that (I) is simply a morphism in $\mathcal{D}^+(I\!B)$:

$$\text{(I)} \quad \underline{\underline{I\!D}}_{G,M} : I\!R\,\pi_*(\pi^{-1}\underline{\underline{\Bbbk}}_{I\!B})[d_M] \to I\!R\,\pi_*(\pi^!\underline{\underline{\Bbbk}}_{I\!B}), \tag{8.7}$$

On the other hand, (II) is the dual of (I) composed with the natural morphism

$$\text{(II)} \quad \underline{\underline{\Omega}}_{G,c}(M) \to I\!R\,\underline{\underline{\mathit{Hom}}}^\bullet_{\Bbbk}\big(I\!R\,\underline{\underline{\mathit{Hom}}}^\bullet_{\Bbbk}(\underline{\underline{\Omega}}_{G,c}(M),\underline{\underline{\Bbbk}}_{I\!B}),\underline{\underline{\Bbbk}}_{I\!B}\big) \tag{8.8}$$

so that proving that (I) and (II) are isomorphisms in $\mathcal{D}^+(\mathbb{B})$ is equivalent to proving that $(\underline{\mathrm{I}})$ and $(\underline{\mathrm{II}})$ are too. The next proposition establishes that this is the case in a slightly more general situation.

Proposition 8.2.2.1 *Let $(E, \mathbb{B}, \pi, M)$ be an oriented fiber bundle where M is an equidimensional manifold of dimension d_M.*[9]

1. The Poincaré adjunction *applied to the morphism $\mathbb{R}\pi_!(\underline{\underline{\Bbbk}}_E) \to \mathbb{R}\pi_*(\underline{\underline{\Bbbk}}_E)$, defines a natural morphism of complexes*

$$\pi^{-1}\underline{\underline{\Bbbk}}_{\mathbb{B}}[d_M] \to \pi^!\underline{\underline{\Bbbk}}_{\mathbb{B}}\,, \tag{8.9}$$

which is an isomorphism in $\mathcal{D}^+(E;\Bbbk)$.

2. If $\dim_{\Bbbk} H_c(M;\Bbbk) < +\infty$, the natural morphism

$$\mathbb{R}\pi_!\underline{\underline{\Bbbk}}_E \to \mathbb{R}\,\underline{\underline{\mathbf{Hom}}}^\bullet_{\Bbbk}\big(\mathbb{R}\,\underline{\underline{\mathbf{Hom}}}^\bullet_{\Bbbk}(\mathbb{R}\pi_!\underline{\underline{\Bbbk}}_E, \underline{\underline{\Bbbk}}_{\mathbb{B}}), \underline{\underline{\Bbbk}}_{\mathbb{B}}\big) \tag{8.10}$$

is an isomorphism in $\mathcal{D}^+(\mathbb{B})$.

Proof (1) We must first construct the morphism.
Grothendieck-Verdier duality canonically identifies

$$\begin{aligned}\mathbb{R}\pi_*\,\mathbb{R}\,\underline{\underline{\mathbf{Hom}}}^\bullet_{\Bbbk}\big((\pi^{-1}\underline{\underline{\Bbbk}}_{\mathbb{B}})[d_M], \pi^!\underline{\underline{\Bbbk}}_{\mathbb{B}}\big) &= \\ &= \mathbb{R}\,\underline{\underline{\mathbf{Hom}}}^\bullet_{\Bbbk}\big(\mathbb{R}\pi_!(\pi^{-1}\underline{\underline{\Bbbk}}_{\mathbb{B}})[d_M], \underline{\underline{\Bbbk}}_{\mathbb{B}}\big)\,,\end{aligned} \tag{8.11}$$

where $\pi^{-1}\underline{\underline{\Bbbk}}_{\mathbb{B}}$ is the constant sheaf $\underline{\underline{\Bbbk}}_E$, and where the complex of sheaves $\mathbb{R}\pi_!(\underline{\underline{\Bbbk}}_E)$ admits a simple description over a trivializing open subset $U \subseteq \mathbb{B}$ since $\pi : E \to \mathbb{B}$ is a locally trivial fibration. Indeed, the open subset $\tilde{U} := \pi^{-1}(U)$ is simply $U \times M$ and since $\underline{\underline{\Bbbk}}_{\tilde{U}} = \underline{\underline{\Bbbk}}_U \boxtimes \underline{\underline{\Bbbk}}_M$,[10] we get:

$$\mathbb{R}\pi_!(\underline{\underline{\Bbbk}}_{\tilde{U}}) = (\mathrm{id} \boxtimes \mathbb{R}c_{M!})(\underline{\underline{\Bbbk}}_U \boxtimes \underline{\underline{\Bbbk}}_M) = \underline{\underline{\Bbbk}}_U \otimes \mathbb{R}c_{M!}(\underline{\underline{\Bbbk}}_M)$$

[9]As shown in Exercise 3.1.1.1, the total space E of a locally trivial fibration $\pi : E \to \mathbb{B}$ with non-connected fiber manifold M, is a disjoint union of open subspaces E_i on which the restrictions $\pi_i := \pi|_{E_i}$ are locally trivial fibrations with equidimensional fibers. The hypothesis on equidimensionality in the proposition is therefore not really restrictive.

The hypothesis regarding orientability is related to the comment 3.1.7.3 and means that we choose an atlas of E made of trivializations $U \times M$, where U is open in $\mathbb{B}$, and such that the transition maps induce *oriented* isomorphisms on M. Recall that $\mathbb{B} = \text{lim-ind}\,\mathbb{B}G(n)$ (4.6), and that when G is connected the spaces $\mathbb{B}G(n)$ are *simply connected*, in which case such *fiber-oriented atlases* always exist for M orientable (Corollary 3.1.7.2–(3c)).

[10]Recall that if $p_i : Z_1 \times Z_2 \to Z_i$ is the canonical projection, and if $\mathscr{F}_i$ is a sheaf in Z_i, then, the sheaf $\mathscr{F}_1 \boxtimes \mathscr{F}_2$ is, by definition, $\mathscr{F}_1 \boxtimes \mathscr{F}_2 := p_1^{-1}\mathscr{F}_1 \otimes p_2^{-1}\mathscr{F}_2$.

where $c_M : M \to \{\bullet\}$ is the constant map. As a consequence, the restriction of complex of sheaves, at the right-hand side of (8.11), to U verifies

$$\begin{aligned} I\!R\,\underline{\underline{\mathbf{Hom}}}^\bullet_{\Bbbk}\left(I\!R\,\pi_!(\pi^{-1}\underline{\underline{\Bbbk}}_{I\!B})[d_M],\underline{\underline{\Bbbk}}_{I\!B}\right)\Big|_U &= \\ &= I\!R\,\underline{\underline{\mathbf{Hom}}}^\bullet_{\Bbbk}\left(\underline{\underline{\Bbbk}}_{I\!B}\boxtimes I\!R\,c_{M!}(\Bbbk_M)[d_M],\underline{\underline{\Bbbk}}_{I\!B}\right) \qquad (8.12)\\ &= \underline{\underline{\Bbbk}}_{I\!B}\otimes \mathbf{Hom}^\bullet_{\Bbbk}\left(I\!R\,c_{M!}(\underline{\underline{\Bbbk}}_M)[d_M];\Bbbk\right), \end{aligned}$$

where the reader will have noticed that, in the third line, instead of $I\!R\,\mathbf{Hom}^\bullet_{\Bbbk}$, we wrote $\mathbf{Hom}^\bullet_{\Bbbk}$, the reason being that the functor $\mathbf{Hom}^\bullet_{\Bbbk}(\_;\Bbbk)$ is already exact on the category $\mathrm{Vec}(\Bbbk)$ of $\Bbbk$-vector spaces.

Now, if $0 \to \underline{\underline{\Bbbk}}_E \to \underline{\underline{\Omega}}^*(E)$ is the resolution by Alexander-Spanier cochains (see Sect. 8.2-(P-2)), the complex of $\Bbbk$-vector spaces

$$I\!R\,c_{M!}(\underline{\underline{\Bbbk}}_E) = \Gamma_c(M,\underline{\underline{\Omega}}^*_E),$$

computes the compactly supported cohomology $H_c(M;\Bbbk)$. On the other hand, an orientation of M is a choice of a generator $\zeta_i \in H_c^{d_M}(M_i)$ on each connected component M_i of M, and, as such, it determines a unique linear form taking the value 1 on each ζ_i, which is the familiar integration operator over M,

$$\int_M : H_c(M;\Bbbk)[d_M] \to \Bbbk. \qquad (8.13)$$

At this point, recall that since the category $\mathcal{D}^+(\mathrm{Vec}(\Bbbk))$ is *split*, i.e. a complex is isomorphic to its cohomology, we have

$$\mathbf{Hom}^\bullet_{\Bbbk}(I\!R\,c_{M!}(\underline{\underline{\Bbbk}}_M)[d_M];\Bbbk) = \mathbf{Hom}^\bullet_{\Bbbk}(H_c(M)[d_M];\Bbbk). \qquad (8.14)$$

Therefore, (8.13) determines a 0-cycle of the left-hand term in (8.14), hence in (8.12) and in (8.11). In this way, taking global sections over $I\!B$ and 0-cohomology, we get a well-defined element in

$$H^0\Gamma\big(I\!B; I\!R\,\pi_*\,I\!R\,\underline{\underline{\mathbf{Hom}}}^\bullet_{\Bbbk}\big(\pi^{-1}\underline{\underline{\Bbbk}}_{I\!B}[d_M],\pi^!\underline{\underline{\Bbbk}}_{I\!B}\big)\big) = \mathrm{Mor}_{\mathcal{D}^+(I\!B)}\big(\pi^{-1}\underline{\underline{\Bbbk}}_{I\!B}[d_M],\pi^!\underline{\underline{\Bbbk}}_{I\!B}\big),$$

which is the morphism $\pi^{-1}\underline{\underline{\Bbbk}}_{I\!B}[d_M] \to \pi^!\underline{\underline{\Bbbk}}_{I\!B}$ (8.9) stated in (1).[11]

To show that (8.9) is an isomorphism in $\mathcal{D}^+(E;\Bbbk)$, we need only show that the induced morphism at the germs of those complexes at each $x \in E$ are quasi-isomorphisms, and, for convenience, to restrict ourselves to the open subsets $\tilde{U}$, as

[11] The procedure glues together the individual Poincaré adjunctions on the fibers $\pi^{-1}(x)$, for all $x \in I\!B$, in a coherent way over the whole base space $I\!B$, which is possible by having a coherent choice of orientations for these fibers. This is what we meant, in 8.2.2.1-(1), by *Poincaré adjunction applied to the morphism* $I\!R\,\pi_!(\underline{\underline{\Bbbk}}_E) \to I\!R\,\pi_*(\underline{\underline{\Bbbk}}_E)$.

they cover E. In that case, for $x := (b, m) \in \tilde{U} := U \times M$, we have the following identification at the level of germs:

$$\big\{\big(\pi^{-1}\underline{\underline{\Bbbk}}_{I\!B}[d_M]\big)_x \to \big(\pi^{!}\underline{\underline{\Bbbk}}_{I\!B}\big)_x\big\} = \big\{\big(\underline{\underline{\Bbbk}}_M[d_M]\big)_m \to \big(c_M^{!}\Bbbk\big)_m = \underline{\underline{\omega}}_{M,m}\big\} \tag{8.15}$$

where $\underline{\underline{\omega}}_M := \underline{\underline{\omega}}_M(\Bbbk)$ denotes the *dualizing complex* on M (fn. ([2]), p. 246), i.e. the complex of sheaves defined by the complex of presheaf on M

$$\underline{\underline{\omega}}_M : V \mapsto \mathbf{Hom}^\bullet_{\Bbbk}(I\!R\, c_{V!}\underline{\underline{\Bbbk}}_V; \Bbbk) = \mathbf{Hom}^\bullet_{\Bbbk}(H_c(V); \Bbbk)\,.$$

Likewise, the morphism in the right-hand side of (8.15) is induced by the morphisms of presheaves $\underline{\underline{\Bbbk}}_M \to \underline{\underline{\omega}}_M[-d_M]$:

$$\begin{array}{ccc} \Gamma(V;\underline{\underline{\Bbbk}}) & \longrightarrow & \Gamma(V;\underline{\underline{\omega}}_M)[-d_M] \\ \| & & \| \\ \Bbbk & \longrightarrow & \mathbf{Hom}^\bullet_{\Bbbk}(H_c(V;\Bbbk);\Bbbk)[-d_M] \end{array} \tag{8.16}$$

which assigns to 1 the integration $\int_V$ relative to the orientation of V, induced by the orientation of the fiber bundle E.

Now, as M is a manifold, a point $m \in M$ has a basis of open neighborhoods V homeomorphic to $\mathbb{R}^{d_M}$, in which case $H_c(V;\Bbbk)$ is concentrated in degree d_M, with $H_c^{d_M}(V;\Bbbk) = \Bbbk$. As a consequence, the second row in (8.16) is an isomorphism for such V's, which implies that the morphisms in (8.15) are quasi-isomorphisms for all $x \in E$, ending the proof of (1).

(2) A complex of sheaves of $\Bbbk$-vector spaces $\mathscr{F}^* := (\mathscr{F}^*, d_*)$ on a topological space $I\!B$ is said to be *cohomologically bounded* if its cohomology sheaves $\mathcal{H}^i(\mathscr{F}_*, d_*) = 0$ vanish for $|i|$ big. It is said to be *perfect* if its cohomology sheaves $\mathcal{H}^i(\mathscr{F}_*, d_*)$ are locally trivial sheaves of finite rank, i.e. if for all $i \in \mathbb{Z}$, the space $I\!B$ can be covered by open subsets V such that $\mathcal{H}^i(\mathscr{F}_*, d_*)|_V \sim \underline{\underline{\Bbbk}}_V^{n(V)}$, for some $n(V) \in \mathbb{N}$.

Lemma *If $\mathscr{F}^*$ is cohomologically bounded and perfect, the natural morphism*

$$\iota : \mathscr{F}^* \to I\!R\ \underline{\underline{\boldsymbol{Hom}}}^\bullet_{\Bbbk}\big(I\!R\ \underline{\underline{\boldsymbol{Hom}}}^\bullet_{\Bbbk}(\mathscr{F}^*, \underline{\underline{\Bbbk}}_{I\!B}), \underline{\underline{\Bbbk}}_{I\!B}\big)\,,$$

is an isomorphism in $\mathcal{D}^+(I\!B)$.

Proof of the Lemma First, notice that since

$$\underline{\underline{Hom}}_{\Bbbk}(\Bbbk_V^n, \underline{\underline{\Bbbk}}_V) \sim \underline{\underline{Hom}}_{\Bbbk}(\Bbbk_V, \underline{\underline{\Bbbk}}_V)^n \sim \underline{\underline{\Bbbk}}_V^n\,,$$

we immediately see that if $\mathcal{L}$ is a locally trivial sheaf of finite local rank, then the functor $\underline{\underline{Hom}}_{\Bbbk}(\mathcal{L}, \_)$ is exact on the category of sheaves. As a consequence,

$$I\!R\ \underline{\underline{Hom}}_{\Bbbk}(\mathcal{L}, \underline{\underline{\Bbbk}}_{I\!B}) = \underline{\underline{Hom}}_{\Bbbk}(\mathcal{L}, \underline{\underline{\Bbbk}}_{I\!B})\,,$$

is again a locally trivial sheaf of finite ranks, so that

$$\iota : \mathcal{L} \to I\!R\,\underline{\underline{Hom}}_{\Bbbk}(I\!R\,\underline{\underline{Hom}}_{\Bbbk}(\mathcal{L}, \Bbbk_{I\!B}), \underline{\underline{\Bbbk}}_{I\!B}) = \underline{\underline{Hom}}_{\Bbbk}(\underline{\underline{Hom}}_{\Bbbk}(\mathcal{L}, \Bbbk_{I\!B}), \underline{\underline{\Bbbk}}_{I\!B}) \tag{8.17}$$

is an isomorphism in $\mathcal{D}^{+}(I\!B)$.

Next, if $0 \to \mathscr{F}^0 \to \mathscr{F}^1 \to \cdots \to \mathscr{F}^{\ell} \to 0$ is a perfect complex of sheaves, the *truncated complex* $\tau_{<\ell}(\mathscr{F}^*)$ is still perfect and we have a small exact sequence of perfect complexes of sheaves $0 \to \tau_{<\ell}(\mathscr{F}^*) \to \mathscr{F}^* \to \mathcal{H}^{\ell}(\mathscr{F}) \to 0$, giving rise in $\mathcal{D}^{+}(I\!B, \Bbbk)$ to the morphism of exact triangles[12]

$$\begin{array}{ccccc} \tau_{<\ell}(\mathscr{F}^*) & \longrightarrow & \mathscr{F}^* & \longrightarrow & \mathcal{H}^{\ell}(\mathscr{F}^*) \longrightarrow \\ {\scriptstyle\iota}\downarrow & & {\scriptstyle\iota}\downarrow & & {\scriptstyle\iota}\downarrow \\ (\tau_{<\ell}(\mathscr{F}^*))^{\vee\vee} & \longrightarrow & (\mathscr{F}^*)^{\vee\vee} & \longrightarrow & (\mathcal{H}^{\ell}(\mathscr{F}^*))^{\vee\vee} \longrightarrow \end{array} \tag{8.18}$$

where $(\_)^{\vee\vee} := I\!R\,\underline{\underline{Hom}}_{\Bbbk}(I\!R\,\underline{\underline{Hom}}_{\Bbbk}(\_, \Bbbk_{I\!B}), \underline{\underline{\Bbbk}}_{I\!B})$.

In (8.18) the right-hand vertical arrow is an isomorphism, since (8.17) is too. Hence, we can conclude that the central arrow is an isomorphism if and only if the left-hand arrow is also. But since the truncated complex $\tau_{<\ell}(\mathscr{F}^*)$ has less than ℓ terms, we can assume by induction on this number, that this is indeed the case. Therefore, the central arrow in (8.18) is an isomorphism in $\mathcal{D}^{+}(I\!B)$.

To finish, we need only recall that in $\mathcal{D}^{+}(I\!B)$, a cohomologically bounded complex is always isomorphic, after a shift of subscripts, to a bounded complex as in the previous case, which ends the proof of the lemma. ⊟

The proof of (2) is now reduced to proving that the complex $I\!R\,\pi_!\underline{\underline{\Bbbk}}_E$ is cohomologically bounded and perfect.

This is already the case over a trivializing open subset $U \subseteq I\!B$. Indeed, we already explained in the poof of (1) that one has in $\mathcal{D}^{+}(U; \Bbbk)$

$$I\!R\,\pi_!\underline{\underline{\Bbbk}}_E|_U \simeq \underline{\underline{\Bbbk}}_U \otimes H_c(M; \Bbbk)\,,$$

which gives the perfection property as it is a local property. It also shows that the sheaves $\mathcal{H}^i(I\!R\,\pi_!\underline{\underline{\Bbbk}}_E)$ vanish for $i \notin [\![0, d_M]\!]$, which is also a local property. But, it is because bounds in $[\![0, d_M]\!]$ are independent of the trivializing open subset U, that in the end explains that $I\!R\,\pi_!\underline{\underline{\Bbbk}}_E$ is *globally* cohomologically bounded and that we can apply the lemma and finish the proposition's proof. □

Now we have all we need to extend the validity of the equivariant duality to any field of coefficients $\Bbbk$.

---

[12]This is the subcomplex $\mathscr{G}^* \subseteq \mathscr{F}^*$, such that $\mathscr{G}^i := \mathscr{F}^i$ for $i < \ell$, $\mathscr{G}^{\ell} := \mathrm{im}(d_{\ell-1})$ and $\mathscr{G}^i := 0$ for all $i > \ell$. Notice that, by construction, $\mathcal{H}^i(\mathscr{G}^*) = \mathcal{H}^i(\mathscr{F}^*)$ if $i < \ell$ and $\mathcal{H}^i(\mathscr{G}^*) = 0$ otherwise.

8.3 Equivariant Poincaré Duality over Arbitrary Fields

8.3.1 The Equivariant Duality Theorem

The following theorem is the analogue to the equivariant Poincaré duality Theorem 5.6.2.1. Notice its similarity with the relative Poincaré duality Theorem 3.3.3.1, which is unsurprising since both theorems use the same underlying formalism of Grothendieck-Verdier Duality. For the same reason it is worth emphasizing that all the results in Chap. 3 apply, with the obvious changes to the current framework of sheaf cohomology.

Recall the notations:

$$\underline{\underline{\Omega}}_G(M) := \pi_* \underline{\underline{\Omega}}(M_G)\,, \quad \underline{\underline{\Omega}}_{G,c}(M) := \pi_! \underline{\underline{\Omega}}(M_G) \quad \text{and} \quad H_G := H(\mathbb{B})\,.$$

Theorem 8.3.1.1 (Equivariant Poincaré Duality) *Let G be a compact Lie group such that $|G/G_0|$ is prime to* char($\mathbb{k}$), *and let M be an oriented G-manifold of dimension d_M. Then,*

1. *The sheafification of the left Poincaré adjunction (8.6)-(I)*

$$\underline{\underline{\mathbb{D}}}_{G,M} : \underline{\underline{\Omega}}_G(M)[d_M] \longrightarrow \mathbb{R}\,\underline{\underline{\mathbf{Hom}}}^{\bullet}_{(\underline{\underline{\Omega}}_G,d)}\big(\underline{\underline{\Omega}}_{G,c}(M), \underline{\underline{\Omega}}_G\big) \tag{8.19}$$

   is an isomorphism in $\mathcal{D}\,\mathrm{DGM}(\underline{\underline{\Omega}}_G)$.
2. *Applying $\mathbb{R}\,\Gamma(\mathbb{B};\,\_)$ to (8.19) we get a quasi-isomorphism*

$$D_{G,M} : \Omega_G(M)[d_M] \longrightarrow \mathbb{R}\,\mathbf{Hom}^{\bullet}_{(\Omega_G,d)}\big(\Omega_{G,c}(M), \Omega_G\big) \tag{8.20}$$

   which is the left Poincaré adjunction isomorphism in $\mathcal{D}(\mathrm{DGM}(\Omega_G))$.
3. *If, in addition, M is of finite type, then the sheafification of the right Poincaré adjunction (8.6)-(II)*

$$\underline{\underline{\mathbb{D}}}'_{G,M} : \underline{\underline{\Omega}}_{G,c}(M)[d_M] \to \mathbb{R}\,\underline{\underline{\mathbf{Hom}}}^{\bullet}_{(\underline{\underline{\Omega}}_G,d)}\big(\underline{\underline{\Omega}}_G(M), \underline{\underline{\Omega}}_G\big) \tag{8.21}$$

   is an isomorphism in $\mathcal{D}\,\mathrm{DGM}(\underline{\underline{\Omega}}_G)$, *and the analogue to* (2) *also holds.*

Proof (1) Results, after the reformulation of the Poincaré adjunction in Sect. 8.2.2, from proposition 8.2.2.1-(1) applied to the fiber bundle $(M_G, \mathbb{B}, \pi, M)$.
(2) Follows from (1), since by Proposition B.9.1.1-(1), we have

$$\mathbb{R}\,\mathbf{Hom}^{\bullet}_{(\Omega_G,d)}(\_,\_) = \mathbb{R}\,\Gamma\big(B, \mathbb{R}\,\underline{\underline{\mathbf{Hom}}}^{\bullet}_{(\underline{\underline{\Omega}}_G,d)}(\_,\_)\big)\,.$$

(3) Results from Proposition 8.2.2.1-(2) applied to the fiber bundle $(M_G, \mathbb{B}, \pi, M)$, and the same arguments as for (1,2). □

Comment 8.3.1.2 Notice an important difference between Theorems 5.6.2.1-(1) and 8.3.1.1-(2). While the first states the quasi-isomorphy for the Ω_G-dual:

$$\Omega_G(M)[d_M] \longrightarrow \mathbf{Hom}^{\bullet}_{\Omega_G}\big(\Omega_{G,c}(M), \Omega_G\big), \tag{8.22}$$

the second states the quasi-isomorphy only for the derived Ω_G-dual:

$$\Omega_G(M)[d_M] \longrightarrow \mathbb{R}\,\mathbf{Hom}^{\bullet}_{(\Omega_G,d)}\big(\Omega_{G,c}(M), \Omega_G\big). \tag{8.23}$$

The reason being that (8.22) concerns equivariant de Rham complexes, and because $\Omega_{G,c}(M)$ is a free Ω_G-graded module (Eq. (5.73), p. 206), while (8.23), concerns Alexander-Spanier cochains complexes and we do not know if $\Omega_G(M)$ and $\Omega_{G,c}(M)$ are K-projective (Ω_G, d)-graded modules (*cf*. fn. ([24]), p. 295), especially in positive characteristic, which prevents us from making the same conclusion.

8.4 Formality of $\mathbb{B}G$ over Arbitrary Fields

As recalled in the review on Cartan's 1950 Seminar (4.1.1), formality of the classifying space $\mathbb{B}G$ over the real numbers was highlighted by Cartan (4.1.1.3) as a by product of the Cartan-Weil's methods. However, the same methods do not apply over fields of positive characteristic, and the question arises if $\mathbb{B}G$ can be formal in those cases. The question was approached by Borel in a different setting in his Ph.D. thesis[13] although the result is not explicitly stated.

8.4.1 The Integral Cohomology of G/T

The key ingredient in the description of $H(\mathbb{B}G;\Bbbk)$ is the fact that $H(G/T;\mathbb{Z})$ is the regular representation of the Weyl group $W(G)$, which we now recall. Notice that we already used this result in Sect. 6.5.4 when comparing the cohomologies of $\mathbb{B}G$ and $\mathbb{B}T$.

Given a compact Lie group G denote by G_0 the connected component of $e \in G$, and let T be a maximal torus of G (hence of G_0).

Lemma 8.4.1.1 *The Weyl groups $W(G) := N_G(T)/T$, $W(G_0) := N_{G_0}(T)/T$ are finite groups.*[14] *We have $W(G_0) \lhd W(G)$, a an exact sequence of groups:*

$$\{1\} \to W(G_0) \hookrightarrow W(G) \twoheadrightarrow G/G_0 \to \{1\}. \tag{8.24}$$

[13] Borel [13] Chapitre VII. Cohomologie entière et mod p de quelques espaces homogènes, more precisely §29. Le quotient d'un groupe compact par un tore maximal, p. 197.

[14] $N_G(T)$ denotes the normalizer in G of T, i.e. the group of $g \in G$ such that $gTg^{-1} = T$.

Proof We have $W(G_0) \lhd W(G)$ because $G_0 \lhd G$ and $N_{G_0}(T) = G_0 \cap N_G(T)$. We therefore get a natural injection $W(G)/W(G_0) \subseteq G/G_0$, and claim that it is in fact an equality. To see this, we use the well-know result that for compact connected Lie groups the Weyl group $W(G_0)$ acts transitively on the set of maximal tori of G_0. Hence, given $g \in G$, there exists $h \in G_0$ such that $gTg^{-1} = hTh^{-1}$, in which case $h^{-1}g \in N_G(T)$, and $g \in N_G(T)\, G_0$, or, equivalently, $W(G)/W(G_0) = G/G_0$. The sequence (8.24) is therefore exact. The Weyl group $W(G_0)$ is well-known to be finite and G/G_0 is finite because G is compact. □

Proposition 8.4.1.2 *If the cardinality of the Weyl group $W(G)$ is nonzero in $\Bbbk$, then the space $G/N_G(T)$ is $\Bbbk$-weakly contractible.*[15]

Proof The group $W(G)$ acts naturally on G_0/T and on G/G_0 and the obvious exact sequence of G_0-spaces $G_0/T \hookrightarrow G/T \twoheadrightarrow G/G_0$ is compatible with this action. Consequently, $(G_0/T)/W(G) = (G/T)/W(G)$, which implies that

$$G/N_G(T) = \big(G_0/T\big)/W(G) = \big(G_0/N_{G_0}(T)\big)/(G/G_0)\,,$$

after Lemma 8.4.1.1. But then, to prove the proposition we need only establish the $\Bbbk$-weakly contractibility of $G_0/N_{G_0}(T) = (G_0/T)/W(G_0)$.

To see how, we recall that G_0, being connected, the *Bruhat decomposition* decomposes the manifold G_0/T as union of *even* dimensional Schubert cells indexed by the Weyl group $W(G_0)$, which immediately shows that the integral cohomology of G_0/T satisfies

$$H(G_0/T;\mathbb{Z}) = \mathbb{Z}[W(G_0)]\,,$$

in which case Leray's theorem stating that the $W(G_0)$-module $H(G_0/T;\mathbb{R})$ is the regular representation of G_0 (*cf*. fn. ([4]), p. 231) extends to integral coefficients. Therefore

$$H(G_0/N_{G_0}(T);\Bbbk) =_1 H(G_0/T;\Bbbk)^{W(G_0)} = \Bbbk[W(G_0)]^{W(G_0)} = \Bbbk\,,$$

where $(=_1)$ is justified by the fact that $|W(G_0)|$ is invertible in $\Bbbk$. □

Theorem 8.4.1.3 *Let G be a compact Lie group. If the cardinality of the Weyl group $W(G)$ is nonzero in $\Bbbk$, then the classifying space $\mathbb{B}G$ is a $\Bbbk$-formal space. In particular, the classifying space $\mathbb{B}T$ of torus T is a $\Bbbk$-formal space for every $\Bbbk$.*

Proof Let T be a maximal torus in G_0 and let $T' := N_{G_0}(T)$. The locally trivial fibration $p : \mathbb{B}T \twoheadrightarrow \mathbb{B}G$ factors as the composition of three locally trivial

[15] Recall that a topological space is said to be $\Bbbk$-*weakly contractible* is its $\Bbbk$-Betti numbers are those of the singleton $\{\bullet\}$ (*cf*. fn. ([31]), p. 146).

fibrations

$$\mathbb{B}T \xrightarrow[{[W(G_0)]}]{q} \mathbb{B}T' \xrightarrow[{[G_0/T']}]{r} \mathbb{B}G_0 \xrightarrow[{[G/G_0]}]{s} \mathbb{B}G\,,$$

where fibers are shown in brackets.
– *The fibration* $s : \mathbb{B}G_0 \to \mathbb{B}G$. This is the orbit map of the action of the finite group G/G_0 on $\mathbb{B}G_0$, and since $|G/G_0|/|W(G)|$ after Lemma 8.4.1.1, the pullback

$$s^* : \Omega(\mathbb{B}G; \Bbbk) \xrightarrow[\text{q.i.}]{} \Omega(\mathbb{B}G_0; \Bbbk)^{G/G_0} \tag{8.25}$$

is a morphism of dg-algebras and a quasi-isomorphism.
– *The fibration* $r : \mathbb{B}T' \to \mathbb{B}G_0$. Since G_0 is connected, the space $\mathbb{B}G_0$ is simply connected and the sheaves $\mathcal{H}^i(\mathbb{R}r_* \underline{\Bbbk}_{\mathbb{B}T'})$ are constant and equal to $\underline{\Bbbk}_{\mathbb{B}G_0}$, for $i = 0$, and 0, for $i > 0$. Whence, $\underline{\Bbbk}_{\mathbb{B}G_0} \simeq \mathbb{R}p_* \underline{\Bbbk}_{\mathbb{B}T'}$ in $\mathcal{D}(\mathbb{B}G_0)$, and the pullback

$$r^* : \Omega(\mathbb{B}G_0; \Bbbk) \xrightarrow[\text{q.i.}]{} \Omega(\mathbb{B}T'; \Bbbk) \tag{8.26}$$

is a morphism of dg-algebras and a quasi-isomorphism.
– *The fibration* $q : \mathbb{B}T \to \mathbb{B}T'$. This is the orbit map of the action of $W(G_0)$ on $\mathbb{B}T$, hence the pullback

$$q^* : \Omega(\mathbb{B}T'; \Bbbk) \xrightarrow[\text{q.i.}]{} \Omega(\mathbb{B}T; \Bbbk)^{W(G_0)}\,, \tag{8.27}$$

which is also is a morphism of dg-algebras and a quasi-isomorphism.

Summing up (8.25), (8.26), and (8.27), the pullback associated with the fibration $p : \mathbb{B}T \twoheadrightarrow \mathbb{B}G$ is a morphism of dg-algebras and a quasi-isomorphism:

$$p^* : \Omega(\mathbb{B}G; \Bbbk) \xrightarrow[\text{q.i.}]{} \Omega(\mathbb{B}T; \Bbbk)^{W(G)}\,. \tag{8.28}$$

We now recall how to prove that $\mathbb{B}T$ is $\Bbbk$-formal for every field $\Bbbk$, in a way that keeps track of the action of the Weyl group $W(G)$.

Let $X(T)$ denote the group of irreducible characters of T, i.e. the set of group (continuous) homomorphisms $\mathrm{Hom}(T, \mathbb{C}_u)$, where $\mathbb{C}_u$ denotes the unit circle in $\mathbb{C}$. We recall that $X(T)$ is isomorphic to the free $\mathbb{Z}$-module $\mathbb{Z}^{\dim T}$.

For every non trivial character $\alpha \in X(T)$, denote by $\mathbb{C}(\alpha)$ the corresponding representation of T, i.e. $t \cdot z := \alpha(t)\, z$, for all $t \in T$ and $z \in \mathbb{C}(\alpha)$. Then, for every $n \in \mathbb{N}$, the group T acts on the unit sphere $\mathbb{S}^{2n-1}(\alpha) \subseteq \mathbb{C}(\alpha)^n$, well-known to be $(2n-1)$-connected, and we define

$$\mathbb{S}^\infty(\alpha) := \varinjlim_{n\in\mathbb{N}} \mathbb{S}^{2n-1}(\alpha)\,,$$

where the inductive limit is relative to the usual inclusions $\mathbb{S}^{2n-1}(\alpha) \subseteq \mathbb{S}^{2n+1}(\alpha)$. It is easily seen that $\mathbb{S}^\infty(\alpha)$ is a CW-complex and a contractible space.

The Weyl group $W(G)$ acts by group automorphisms on T, inducing and right action on $X(T)$ by $\alpha \cdot w := \alpha \circ w$. Let $\mathcal{B} \subseteq X(T)$ denote a $W(G)$ stable finite subset of generators of the group $X(T)$. The representation of T on

$$\mathbb{C}(\mathcal{B}) := \prod\nolimits_{\alpha \in \mathcal{B}} \mathbb{C}(\alpha)\,,$$

is then faithful, and the induced action on the spaces

$$\mathbb{E}T(\mathcal{B}, n) := \prod\nolimits_{\alpha \in \mathcal{B}} \mathbb{S}(\alpha, n) \quad \text{and} \quad \mathbb{E}T(\mathcal{B}) := \prod\nolimits_{\alpha \in \mathcal{B}} \mathbb{S}(\alpha, \infty)$$

is free. Since $\mathbb{E}T(\mathcal{B})$ is a CW-complex and a contractible space, it can be used as universal fiber bundle for T. We then define

$$\mathbb{B}T(\mathcal{B}, n) := \prod\nolimits_{\alpha \in \mathcal{B}} \mathbb{P}(\alpha, n) \quad \text{and} \quad \mathbb{B}T(\mathcal{B}) := \prod\nolimits_{\alpha \in \mathcal{B}} \mathbb{P}(\alpha, \infty)\,,$$

where $\mathbb{P}(\alpha, n) := \mathbb{S}^{2n-1}(\alpha)/T$ is the projective space $\mathbb{P}_{n-1}(\mathbb{C})$ of complex dimension $(n-1)$. But the projective spaces $\mathbb{P}_n(\mathbb{C})$ have a canonical well-known structure of CW-complex with even dimensional cells, compatible with the inclusions $\mathbb{P}_{n-1}(\mathbb{C}) \subseteq \mathbb{P}_n(\mathbb{C})$, hence inducing a CW-complex structure on $\mathbb{B}T(\mathcal{B}) :=$ lim-ind $\mathbb{B}T(\mathcal{B}, n)$, always with even dimensional cells. The dg-algebra associated with the CW-structure of $\mathbb{B}T(\mathcal{B})$ is then canonically isomorphic to its cohomology $H(\mathbb{B}T; k)$, which gives a morphism of dg-algebras

$$\epsilon : H(\mathbb{B}T(\mathcal{B}); \Bbbk) \xrightarrow[\text{q.i.}]{} \Omega(\mathbb{B}T(\mathcal{B}); \Bbbk)\,, \tag{8.29}$$

that is a quasi-isomorphism and establishes the formality of $\mathbb{B}T$.

Now, since $\mathcal{B}$ is stable under the action of $W(G)$, this group acts naturally on $\mathbb{B}T(\mathcal{B})$ and the morphism ϵ is $W(G)$-equivariant and hence induce a morphism of dg-algebras and quasi-isomorphism

$$\epsilon : H(\mathbb{B}T(\mathcal{B}); \Bbbk)^{W(G)} \xrightarrow[\text{q.i.}]{} \Omega(\mathbb{B}T(\mathcal{B}); \Bbbk)^{W(G)}\,, \tag{8.30}$$

where

$$H(\mathbb{B}T(\mathcal{B}); \Bbbk)^{W(G)} = H(\mathbb{B}T(\mathcal{B})/W(G); \Bbbk) = H(\mathbb{B}G; \Bbbk)\,,$$

since $|W(G)|$ is invertible in $\Bbbk$, and also after (8.28).

The composition of (8.28) and (8.30) gives the zig-zag

$$\Omega(\mathbb{B}G; \Bbbk) \xrightarrow[\text{q.i.}]{p^*} \Omega(\mathbb{B}T; \Bbbk)^{W(G)} \underset{\text{q.i.}}{\longleftrightarrow} \Omega(\mathbb{B}T(\mathcal{B}); \Bbbk)^{W(G)} \xleftarrow[\text{q.i.}]{p^*} H(\mathbb{B}G; \Bbbk)\,,$$

which establishes the $\Bbbk$-formality of $\mathbb{B}G$. □

Corollary 8.4.1.4 (Enhanced Equivariant Poincaré Duality) *Let G be a compact Lie group such that the cardinality of the Weyl group $W(G)$ is invertible in $\Bbbk$. Let M be an orientable G-manifold of dimension d_M. Then*

1. *The left Poincaré adjunction isomorphism 8.3.1.1-(2) in $\mathcal{D}\,\mathrm{DGM}(\Omega_G)$:*

$$D_{B,M} : \Omega_G(M)[d_M] \simeq I\!R\,\mathbf{Hom}^{\bullet}_{(\Omega_G,d)}(\Omega_{G,c}(M), \Omega_G)\,, \tag{8.31}$$

induces a convergent spectral sequence:

$$I\!E_2^{p,q} := \mathbf{h}^p\, I\!R\,\mathbf{Hom}^{\bullet}_{H_G}(H^q_{G,c}(M), H_G) \Rightarrow H_G^{p+q+d_M}(M)\,.$$

2. *If $\dim_{\mathrm{proj}}(H_{G,c}(M)) \leq 1$ as H_G-graded module, then the Poincaré adjunction induces an isomorphism in $\mathcal{D}\,\mathrm{DGM}(H_G)$:*

$$D_{B,M} : H_G(M)[d_M] \simeq I\!R\,\mathbf{Hom}^{\bullet}_{H_G}(H_{G,c}(M), H_G)\,, \tag{8.32}$$

and the spectral sequence $(I\!E_r, d_r)$ in (1) degenerates, i.e. $d_r = 0$ for $r \geq 2$.
Furthermore, if $H_{G,c}(M)$ is a projective H_G-graded module, then (8.32) induces an isomorphism of H_G-graded modules

$$H_G(M)[d_M] \simeq \mathbf{Hom}^{\bullet}_{H_G}\left(H_{G,c}(M), H_G\right).$$

3. *If M is of finite type, the statements (1) and (2) remain true even if we swap the terms $\Omega_G(M) \leftrightarrow \Omega_{G,c}(M)$ and $H_G(M) \leftrightarrow H_{G,c}(M)$.*

Proof Since $I\!BG$ is a $\Bbbk$-formal space, the statements are immediate applications of Corollary A.3.3.2. □

Comment 8.4.1.5 When G is a torus T, we have $W(G) = \{1\}$ and Theorem 8.4.1.4 is valid over any field $\Bbbk$. Moreover, when G is the circle group $\mathbb{S}^1$, $I\!BG = \mathbb{P}_\infty(\mathbb{C})$ and the ring $H(I\!BG; \Bbbk)$ is isomorphic to the polynomial algebra $\Bbbk[X]$ which is of homological dimension 1. In that case, the condition $\dim_{\mathrm{proj}}(H_{G,c}(M)) \leq 1$ in 8.4.1.4-(2) is automatically verified and can be omitted.

8.5 Equivariant Gysin Morphisms over Arbitrary Fields

Now that we have proved Poincaré duality over a field $\Bbbk$ of characteristic prime to $|W(G)|$, we can mimic the method used to introduce Gysin morphisms in de Rham cohomology and achieve the same result over $\Bbbk$.

In 8.2-(P-2d), we explained how an equivariant map $f : X \to Y$ between G-spaces defines a morphism in $\mathcal{D}^+(Y_G)$

$$f^\# : \underline{\underline{\Omega}}(Y_G) \to f_*\underline{\underline{\Omega}}(X_G) \tag{8.33}$$

inducing the usual pullback $f^* : H_G(Y) \to H_G(X)$.

Applying the functor $\pi_{Y!}$ to (8.33), we get a morphism in $\mathcal{D}^+(I\!B)$

$$\pi_{Y!}(f^\#) : \pi_{Y!}(\underline{\underline{\Omega}}(Y_G)) \to \pi_{Y!}(f_* \underline{\underline{\Omega}}(X_G)) = \pi_{X!}(\underline{\underline{\Omega}}(X_G))$$

where one recognizes the complexes $\underline{\underline{\Omega}}_G(X)$ and $\underline{\underline{\Omega}}_G(Y)$. We can thus write:

$$\pi_{Y!}(f^\#) : \underline{\underline{\Omega}}_G(Y) \to \underline{\underline{\Omega}}_G(X) . \tag{8.34}$$

If f is, in addition, a proper map, we have $f_! = f_*$ and the same ideas that apply to π_{Y*}, now lead to a morphism in $\mathcal{D}^+(Y_G)$

$$\pi_{Y*}(f^\#) : \underline{\underline{\Omega}}_{G,c}(Y) \to \underline{\underline{\Omega}}_{G,c}(X) , \tag{8.35}$$

inducing the usual pullback $f^* : H_{G,c}(Y) \to H_{G,c}(X)$.

In the next sections, M and N are equidimensional manifolds of dimensions d_M and d_N, orientable over a field $\Bbbk$ of characteristic prime to $|W(G)|$.

8.5.1 *Gysin Morphism for General Maps*

We apply the duality functor $(\_)^\vee := I\!R\,\underline{\underline{\boldsymbol{Hom}}}^\bullet_{\Bbbk}(\_, \underline{\Bbbk}_{I\!B})$ to (8.34), and consider the diagram in $\mathcal{D}^+(I\!B)$

$$\begin{array}{ccc} \underline{\underline{\Omega}}_G(M)^\vee & \xrightarrow{(\pi_{Y!}(f^\#))^\vee} & \underline{\underline{\Omega}}_G(N)^\vee \\ {\scriptstyle \underline{\underline{I\!D}}'_{G,M}}\uparrow & & \uparrow{\scriptstyle \simeq\ \underline{\underline{I\!D}}'_{G,N}} \\ \underline{\underline{\Omega}}_{G,c}(M)[d_M] & \overset{f_!}{\dashrightarrow} & \underline{\underline{\Omega}}_{G,c}(N)[d_N] \end{array} \tag{8.36}$$

where the right-hand vertical arrow is an isomorphism when N is of finite type, after Theorem 8.3.1.1–(3). Following Sects. 2.8.2 and 6.1.2, we can proceed to defining *the Gysin morphism for a general map in* $\mathcal{D}^+(I\!B)$ as the unique morphism $f_!$ closing the diagram (8.36). The cohomology of global sections then gives *the Gysin morphism for a general map in equivariant cohomology over* $\Bbbk$:

$$f_! : H_{G,c}(M; \Bbbk)[d_M] \to H_{G,c}(N; \Bbbk)[d_N] .$$

The case where N is not of finite type is dealt with as in Sect. 6.1.2, taking limits over filtrant covers of finite type open subsets of N.

8.5.2 *Gysin Morphism for Proper Maps*

The underlying idea is the same as in the previous case, but we now apply the duality functor $(\_)^\vee$ to (8.35). We then obtain the diagram

$$\begin{array}{ccc} \underline{\underline{\Omega}}_{G,c}(M)^\vee & \xrightarrow{(\pi_{Y*}(f^\#))^\vee} & \underline{\underline{\Omega}}_{G,c}(N)^\vee \\ {\scriptstyle \underline{\underline{ID}}_{G,M}}\uparrow{\scriptstyle \simeq} & & {\scriptstyle \simeq}\uparrow{\scriptstyle \underline{\underline{ID}}_{G,N}} \\ \underline{\underline{\Omega}}_G(M)[d_M] & \overset{f_*}{\dashrightarrow} & \underline{\underline{\Omega}}_G(N)[d_N] \end{array}$$

where both vertical arrows are isomorphisms in $\mathcal{D}^+(I\!B)$, after 8.3.1.1-(1). *The Gysin morphism for proper maps in* $\mathcal{D}^+(I\!B)$ is then defined, as in Sects. 2.8.2 and 6.1.2, as the only morphism f_* that closes the diagram. The cohomology of global sections then gives *the Gysin morphism for proper maps in equivariant cohomology over* $\Bbbk$:

$$f_* : H_G(M; \Bbbk)[d_M] \to H_G(N; \Bbbk)[d_N] .$$

8.6 The Localization Formula over Arbitrary Fields

Following Sect. 8.4, we know that when $|W(G)|$ is invertible in $\Bbbk$, the classifying space $I\!BG$ is $\Bbbk$-formal, and $H(I\!BG; \Bbbk) = H(I\!BT; \Bbbk)^{W(G)} \simeq \Bbbk[X_1, \ldots, X_r]^{W(G)}$, where $r := \dim_{\mathbb{R}}(T)$. We can then set Q_G and Q_T the respective rings of graded fractions of H_G and H_T and follow word-by-word Chap. 7 on Localization, in which case all the results up to the Localization Formula 7.7.1 remain true under the only additional hypothesis that the cardinality of $W(G)$ is invertible in $\Bbbk$.

Appendix A
Basics on Derived Categories

This compendium of basic ideas aims at a quick overview of the theory of derived categories and derived functors. For more thorough study of the subject, we encourage the reader to consult the profuse literature on the subject, for example:

- R. Hartshorne, Residues and Duality (chap. I) . . . . [51]
- B. Keller, Derived categories and their uses . . . . [65]
- B. Iversen. Cohomology of Sheaves (chap. XI) . . . . [58]
- M. Kashiwara, P. Schapira, Categories and sheaves (chap. 13) . . . . [62]
- Stacks Project, Differential Graded Algebra . . . . [87]
- J.-L. Verdier, Des catégories dérivées des catégories abéliennes . . . . [93]
- C. Weibel, An introduction to homological algebra (chap. 10) . . . . [95]

A.1 Categories of Complexes

The aim of Algebraic Topology is, as the name suggests, the algebraization of topology. This entails, among other ideas, attaching to a topological space complexes whose cohomology encodes topological properties. Although the main interest of this discipline lies in the interaction between algebra and topology, the theme of complexes or, more precisely, of differential graded algebras and modules has undergone significant development, to the point of becoming a subject of study in its own right. We recall now some very basic definitions and results on the subject that so far in this book we have given little attention. We will confine to complexes of cochains.

A. Arabia, *Equivariant Poincaré Duality on G-Manifolds*, Lecture Notes in Mathematics 2288, https://doi.org/10.1007/978-3-030-70440-7

A.1.1 The Category of Complexes of an Abelian Category

Given an abelian category Ab, we denote by $C(\mathrm{Ab})$ the *category of complexes (of cochains) of an abelian category* Ab.

C-1. *A complex of cochains of* Ab, is a graded object $C \in \mathrm{Ob}(\mathrm{Ab}^{\mathbb{Z}})$ together with a *differential*, i.e. with $d \in \mathrm{Mor}_{\mathrm{Ab}}(C, C[1])$, such that $d^2 = 0$:

$$(C, d) := \Big(\cdots \xrightarrow{d_{-3}} C_{-2} \xrightarrow{d_{-2}} C_{-1} \xrightarrow{d_{-1}} C_0 \xrightarrow{d_0} C_1 \xrightarrow{d_1} C_2 \xrightarrow{d_2} \cdots\Big).$$

C-2. A *morphism of complexes* $f \in \mathrm{Mor}_{C(\mathrm{Ab})}((C, d), (C', d'))$ is a morphism of graded objects $f \in \mathrm{Mor}_{\mathrm{Ab}^{\mathbb{Z}}}(C, C')$ such that $f \circ d = d' \circ f$.

C-3. The *cohomology functor* is the covariant (additive) functor

$$\boldsymbol{h} : C(A) \rightsquigarrow \mathrm{Ab}^{\mathbb{Z}}, \tag{A.1}$$

which associates with (C, d), its *cohomology*, i.e. the *graded object*

$$\boldsymbol{h}(C, d) := \big\{\ker(d_i)/\mathrm{im}(d_{i-1})\big\}_{i\in\mathbb{Z}} \in \mathrm{Ab}^{\mathbb{Z}}.$$

C-4. A complex (C, d) is said to be *acyclic* if $\boldsymbol{h}(C, d) = 0$.

C-5. For $m \in \mathbb{Z}$, the category Ab is embedded in $C(\mathrm{Ab})$ as the full subcategory of complexes *concentrated in (degree)* m, i.e. complexes (C, d) with $C^i = 0$ for all $i \neq m$ and (hence) $d = 0$. The embedding is given by the functor

$$[m] : \mathrm{Ab} \rightsquigarrow C(\mathrm{Ab}),$$

which associates with $X \in \mathrm{Ob}(\mathrm{Ab})$, the complex $X[0]$ concentrated in m, where $X[m]_m := X$, and likewise for $f \in \mathrm{Mor}_{\mathrm{Ab}}(X, X')$, where $f[m]_m := f$.

C-6. For every $m \in \mathbb{Z}$, the *shift functor* '$[m]$'

$$[m] : C(\mathrm{Ab}) \to C(\mathrm{Ab}), \tag{A.2}$$

associates with (C, d) the complex $(C, d)[m] := (V, D)$, where $V_i := C_{i+m}$ and $D_i := (-1)^m d_{i+m}$, and it associates with $\alpha : (C, d) \to (C', d')$, the morphism $\alpha[m] : (C, d)[m] \to (C', d')[m]$, where $\alpha[m]_i = \alpha_{i+m}$.

A.1.2 *Extending Additive Functors from* Ab *to* C(Ab)

Let $F : \text{Ab} \to \text{Ab}'$ be an additive functor between abelian categories.[1]

Applying F to a complex $(C, d) \in C(\text{Ab})$, term to term, results in a graded object $F(C, d) := (F(C), F(d))$, with $F(C)_i = F(C_i)$ and $F(d)_i := F(d_i)$:

$$(C,d) := \big(C_i \xrightarrow{d_i} C_{i+1}\big) \overset{F}{\rightsquigarrow} F(C,d) := \big(F(C_i) \xrightarrow{F(d_i)} F(C_{i+1})\big),$$

The fact that F is additive implies that $F(C, d)$ is also a complex since

$$F(d_{i+1}) \circ F(d_i) = F(d_{i+1} \circ d_i) = F(0) = 0\,.$$

If $f \in \text{Mor}_{C(\text{Ab})}((C, d), (C', d'))$, we then denote by $F(f) : F(C, d) \to F(C', d')$ the morphism of graded objects $F(f)_i := F(f_i)$, which is clearly a morphism of complexes of $C(\text{Ab}')$.

We thus extend the definition of F from Ab to C(Ab), which we denote by the same letter $F : C(\text{Ab}) \rightsquigarrow C(\text{Ab}')$. The extension is clearly compatible with the shift functors $[m]$, making commutative the diagram of functors

$$\begin{array}{ccc} \text{Ab} & \overset{[m]}{\rightsquigarrow} & C(\text{Ab}) \\ {\scriptstyle F}\big\downarrow & & \big\downarrow{\scriptstyle F} \\ \text{Ab}' & \overset{[m]}{\rightsquigarrow} & C(\text{Ab}')\,. \end{array} \tag{A.3}$$

A.1.3 *The Mapping Cone*

A fundamental operation in categories of complexes, is *the mapping cone associated with a morphism of complexes*.[2]

Given a morphism $\alpha : (C, d) \to (C', d')$ in C(Ab), consider, for all $i \in \mathbb{Z}$, the morphism Δ_i in Ab defined by

[1] We recall that in an abelian category Ab the sets of morphisms $\text{Mor}_{\text{Ab}}(X, Y)$ are abelian groups for all $X, Y \in \text{Ob(Ab)}$. A covariant functor between abelian categories $F : \text{Ab} \to \text{Ab}'$ is said to be *additive* if the maps $F_{X,Y} : \text{Mor}_{\text{Ab}}(X, Y) \to \text{Mor}_{\text{Ab}'}(F(X), F(Y))$ are homomorphisms of groups for all $X, Y \in \text{Ob(Ab)}$; and likewise for contravariant functors. The functor is then said to be exact, left exact or right exact, if it transforms every short exact sequence of Ab in respectively an exact, left exact or right exact sequence of Ab$'$, and this regardless of whether F is covariant or contravariant.

[2] Useful references to consult are Weibel [95] §1.5 Mapping Cones and Cylinders, p. 18, and Kashiwara-Schapira [62] ch. 11.1 Differential Objects and Mapping Cones, p. 270.

$$\begin{aligned} C'_i \oplus C_{i+1} &\xrightarrow{\Delta_i} C'_{i+1} \oplus C_{i+2} \\ (x', x) &\longmapsto (d'_i x' + \alpha_{i+1}(x), -d_{i+1}(x)) \end{aligned} \tag{A.4}$$

One has $\Delta_{i+1} \circ \Delta_i = 0$, whence a complex, called the *cone of* α,

$$\hat{c}(\alpha) := (C' \oplus C[1], \Delta) \tag{A.5}$$

We then consider the morphisms of complexes

$$\begin{array}{ccc} (C', d') & \xrightarrow{\iota_1} & (C' \oplus C[1], \Delta) \\ x' & \longmapsto & (x', 0) \end{array} \qquad \begin{array}{ccc} (C' \oplus C[1], \Delta) & \xrightarrow{p_2} & (C, d)[1] \\ (x', x) & \longmapsto & x \end{array} \tag{A.6}$$

giving rise to a triangle (Definition A.1.5.4) of morphisms of complexes:

$$\begin{array}{ccc} (C, d) & \xrightarrow{\quad \alpha \quad} & (C', d') \\ & {\scriptstyle p_2}\nwarrow^{[1]} \quad \swarrow{\scriptstyle \iota_1} & \\ & (\hat{c}(\alpha), \Delta) & \end{array}$$

equivalently denoted by

$$(C, d) \xrightarrow{\alpha} (C', d') \xrightarrow{\iota_1} \hat{c}(\alpha) \xrightarrow[{[1]}]{p_2} \tag{A.7}$$

where the notation '[1]' recalls that degrees are increased by 1.[3]

A.1.4 Homotopies

A morphism of complexes $f : (C, d) \to (C', d')$ is said to be *homotopic to zero* if there is a family of morphisms $h_i \in \mathrm{Mor}_{\mathrm{Ab}}(C_i, {C'}_{i-1})$, such that

$$f_i = d'_{i-1} \circ h_i + h_{i+1} \circ d_i \tag{A.8}$$

$$\begin{array}{ccccccccc} \cdots \longrightarrow & C_{i-1} & \longrightarrow & C_i & \xrightarrow{d_i} & C_{i+1} & \longrightarrow \cdots \\ & \downarrow {\scriptstyle f_{i-1}} & \swarrow {\scriptstyle h_i} & \downarrow {\scriptstyle f_i} & \swarrow {\scriptstyle h_{i+1}} & \downarrow {\scriptstyle f_{i+1}} & \\ \cdots \longrightarrow & C'^{i-1} & \xrightarrow{d'_{i-1}} & C'^{i} & \longrightarrow & C'^{i+1} & \longrightarrow \cdots \end{array}$$

[3]Beware that the compositions $\iota_1 \circ \alpha$ and $\alpha \circ p_2$ are not equal to 0.

A.1.4.1 Terminology and Notations

$\mathcal{K}$-1. Two morphisms of complexes $f, g : C \to C'$ are said to be *homotopic*, and we write $f \sim g$, if the difference $f - g$ is homotopic to zero.

$\mathcal{K}$-2. A complex C is said to be *homotopic to zero*, and we write $C \sim 0$, if the identity morphism id_C is homotopic to zero.

$\mathcal{K}$-3. A morphism $f : C \to C'$ is said to be *an isomorphism up to homotopy*, if there exists a morphism $g : C' \to C$ such that $g \circ f \sim \mathrm{id}_C$ and $f \circ g \sim \mathrm{id}_{C'}$. We then say that C and C' are *homotopy-equivalent*, and write $C \sim C'$.

$\mathcal{K}$-4. The relation '$\sim$', is an equivalence relation compatible with the additive structure of $C(\mathrm{Ab})$, i.e. if $f \sim f'$ and $g \sim g'$ then $(f + g) \sim (f' + g')$. As a consequence, the sets of homotopy-equivalence classes:

$$\mathbf{Hot}^{\bullet}_{\mathrm{Ab}}(C, C') := \mathbf{Hom}^{\bullet}_{\mathrm{Ab}}(C, C')/_{\sim} . \tag{A.9}$$

is an abelian group.

$\mathcal{K}$-5. The relation '$\sim$' is also compatible with the composition of morphisms of complexes, i.e. if $f \sim f'$ et $g \sim g'$ then $(f \circ g) \sim (f' \circ g')$, whenever the compositions make sense.

The following proposition summarizes well-known properties of the definitions and constructions that we have introduced so far.

Proposition A.1.4.1 *Let* Ab *be an abelian category.*

1. *The category of complexes* $C(\mathrm{Ab})$ *is abelian.*
2. *Let* $(0 \to A \xrightarrow{\alpha} B \xrightarrow{\beta} C \to 0)$ *be an exact sequence in* $C(\mathrm{Ab})$. *There exists a connecting morphism* $c : \mathbf{h}(C) \to \mathbf{h}(A)[1]$ *such that the following long sequence of cohomology is exact:*

$$\cdots \longrightarrow \mathbf{h}(A) \xrightarrow{\mathbf{h}(\alpha)} \mathbf{h}(B) \xrightarrow{\mathbf{h}(\beta)} \mathbf{h}(C) \xrightarrow{c} \mathbf{h}(A)[1] \longrightarrow \cdots$$

3. *Let* $A \xrightarrow{\alpha} B \xrightarrow{\iota} \hat{c}(\alpha) \xrightarrow{p} A[1]$ *be a mapping cone in* $C(\mathrm{Ab})$.
   a. *The following long sequence of cohomology is exact:*

$$\cdots \longrightarrow \mathbf{h}(A) \xrightarrow{\mathbf{h}(\alpha)} \mathbf{h}(B) \xrightarrow{\mathbf{h}(\iota)} \mathbf{h}(\hat{c}(\alpha)) \xrightarrow{\mathbf{h}(p)} \mathbf{h}(A)[1] \longrightarrow \cdots$$

   b. α *is a quasi-isomorphism if and only if* $\mathbf{h}(\hat{c}(\alpha)) = 0$.
   c. α *is an isomorphism modulo homotopy if and only if* $\hat{c}(\alpha) \sim 0$.

Sketch of Proof (1) and (2) are well-known standard facts. (3a) Results, for example, from Lemma 1.5.3 in Weibel [95] (p. 19). Rather than giving details, we encourage the reader to prove the claim himself by straightforward verification after the definitions of ι and p in formula (A.6). (3b) is immediate after (3a) and (3c) demands manual verifications which are also left to the reader. $\square$

Exercise A.1.4.2 Let $F : \mathrm{Ab} \rightsquigarrow \mathrm{Ab}'$ be an additive functor between abelian categories. Show the equivalence of the following statements. (p. 351)

1. $F : \mathrm{Ab} \rightsquigarrow \mathrm{Ab}'$ is exact.
2. $F : C(\mathrm{Ab}) \rightsquigarrow C(\mathrm{Ab}')$ preserves acyclicity.
3. $F : C(\mathrm{Ab}) \rightsquigarrow C(\mathrm{Ab}')$ preserves quasi-isomorphisms.

A.1.5 The Homotopy Category $\mathcal{K}$**(Ab)**

When interested in the homotopy of topological spaces, continuous maps which coincide up to homotopy induce homotopic morphisms of associated complexes. One is thus led to defining the *homotopy category* $\mathcal{K}(\mathrm{Ab})$ whose objects are the same as in $C(\mathrm{Ab})$ but whose morphisms are the homotopy classes of morphisms of complexes A.1.4.1-($\mathcal{K}$-4), i.e. we set

$$\mathrm{Mor}_{\mathcal{K}(\mathrm{Ab})}(C, C') := \mathbf{Hot}^0_{\mathrm{Ab}}(C, C') \,. \tag{A.10}$$

The category $\mathcal{K}(\mathrm{Ab})$ is additive. Furthermore, the complexes in $C(\mathrm{Ab})$ which are homotopy-equivalent become isomorphic, and the morphisms in $C(\mathrm{Ab})$ which are isomorphisms up to homotopy are invertible. Furthermore, the well-known fact that homotopic morphisms of complexes induce the same morphism in cohomology makes the cohomology functor $\boldsymbol{h}$ in (A.1) to factor through $\mathcal{K}(\mathrm{Ab})$.

We thus have the factorization

$$C(\mathrm{Ab}) \overset{\iota}{\rightsquigarrow} \mathcal{K}(\mathrm{Ab}) \overset{\boldsymbol{h}}{\rightsquigarrow} \mathrm{Ab}^{\mathbb{Z}} \,, \qquad \boldsymbol{h} : C(\mathrm{Ab}) \rightsquigarrow \mathrm{Ab}^{\mathbb{Z}} \tag{A.11}$$

where $\iota : C(\mathrm{Ab}) \rightsquigarrow \mathcal{K}(\mathrm{Ab})$ denotes the functor which is the identity on objects, and which associates its homotopy class with a morphism.

Exercise A.1.5.1 Show that the mapping cones of homotopic morphisms in $C(\mathrm{Ab})$ are homotopy-equivalent complexes. Conclude that the mapping cone of a morphism in $\mathcal{K}(\mathrm{Ab})$ is canonically defined (p. 351)

The following Proposition is an important foundational result.

Proposition A.1.5.2

1. *Given* C , $C' \in \mathrm{Ob}(C(\mathrm{Ab}))$, *there is a canonical identification*

$$h^i\big(\mathbf{Hom}^{\bullet}_{\mathrm{Ab}}\big(C, C'\big)\big) \simeq \mathrm{Mor}_{\mathcal{K}(\mathrm{Ab})}\big(C, C'[i]\big) \,, \quad \forall i \in \mathbb{Z} \,. \tag{A.12}$$

2. *For* $C \in \mathrm{Ob}(C(\mathrm{Ab}))$ *and every mapping cone* $A \xrightarrow{\alpha} B \xrightarrow{\iota} \hat{c}(\alpha) \xrightarrow{p} A[1]$ *in* $C(\mathrm{Ab})$, *the following induced long sequences* ($\mathcal{K}$ *stands for* $\mathcal{K}(\mathrm{Ab})$) *are exact.*

$$\to \mathrm{Mor}_{\mathcal{K}}(C, A) \longrightarrow \mathrm{Mor}_{\mathcal{K}}(C, B) \longrightarrow \mathrm{Mor}_{\mathcal{K}}(C, \hat{c}(\alpha)) \longrightarrow \mathrm{Mor}_{\mathcal{K}}(C, A[1]) \longrightarrow$$

$$\to \mathrm{Mor}_{\mathcal{K}}(\hat{c}(\alpha), C) \longrightarrow \mathrm{Mor}_{\mathcal{K}}(B, C) \longrightarrow \mathrm{Mor}_{\mathcal{K}}(A, C) \longrightarrow \mathrm{Mor}_{\mathcal{K}}(A, C[1]) \longrightarrow$$

in particular the morphisms $\iota \circ \alpha$, $p \circ \iota$, $\alpha[1] \circ p$ *are all homotopic to* 0.

Sketch of Proof (1) The isomorphism (A.12) is immediate after the definition of the complex $(\mathbf{Hom}^\bullet(\_, \_), \boldsymbol{D})$ (Sect. 2.1.8). (2) The exactness of the long sequences induced by the functors $\mathrm{Mor}_{\mathcal{K}}(C, \_)$ and $\mathrm{Mor}_{\mathcal{K}}(\_, C)$ can be proven by hand using only the definition of homotopy. This is a great exercise to do at least once in your life. The result is however generally shown as a corollary of the fact that $\mathcal{K}(\mathrm{Ab})$ is a triangulated category (Sect. A.1.5.2), which has the advantage of proving (2) in greater generality.[4] ⊡

A main difference between $C(\mathrm{Ab})$ and $\mathcal{K}(\mathrm{Ab})$ is that the homotopy category is generally not abelian as shown in the following exercise.

Exercises A.1.5.3 The aim is to show that $\mathcal{K}(\mathrm{Ab})$ is an abelian category if and only if Ab is a split category, in which case the fully faithful embedding $\mathrm{Ab}^{\mathbb{Z}} \subseteq \mathcal{K}(\mathrm{Ab})$ is an equivalence of categories, and $\mathcal{K}(\mathrm{Ab})$ too is a split category.[5]

1. Given a *monomorphism* $\alpha : A \to B$ in $\mathcal{K}(\mathrm{Ab})$ and its mapping cone: $A \xrightarrow{\alpha} B \xrightarrow{\iota} \hat{c}(\alpha) \xrightarrow{p} A[1]$, show that $p = 0$, that α admits a *retraction* $\rho : B \to A$, and that ι is an *epimorphism* admitting a *section* $\sigma : \hat{c}(\alpha) \to B$.[6]

   If, in addition, $\mathcal{K}(\mathrm{Ab})$ is an abelian category, then the following sequence is a split sequence in $\mathcal{K}(\mathrm{Ab})$,

$$0 \longrightarrow A \underset{\rho}{\overset{\alpha}{\rightleftarrows}} B \underset{\sigma}{\overset{\iota}{\rightleftarrows}} \hat{c}(\alpha) \longrightarrow 0,$$

[4]See Hartshorne [51] Proposition 1.1 (b), p. 23, or Weibel [95] Example 10.2.8, p. 377.

[5]The result is valid for any triangulated category (Sect. A.1.5.2) in lieu of $\mathcal{K}(\mathrm{Ab})$.

[6]In any category C a morphism $\alpha : X \to X'$ is said to be a *monomorphism*, if it is *left cancellable*, i.e. $(\alpha \circ \beta = \alpha \circ \beta') \Rightarrow (\beta = \beta')$, in other words, if for every $Y \in \mathrm{Ob}(C)$, the pushforward $\alpha_* : \mathrm{Mor}_C(Y, X) \to \mathrm{Mor}_C(Y, X')$, $\alpha_*(\beta) := \alpha \circ \beta$, is injective. Dually, $\alpha : X \to X'$ is said to be an *epimorphism*, if it is *right cancellable*, i.e. $(\beta \circ \alpha = \beta' \circ \alpha) \Rightarrow (\beta = \beta')$, in other words, if for every $Y \in \mathrm{Ob}(C)$, the pullback $\alpha^* : \mathrm{Mor}_C(X', Y) \to \mathrm{Mor}_C(X, Y)$, $\alpha^*(\beta) := \beta \circ \alpha$, is injective. If in addition C is an additive category, a *kernel* for $\alpha : X \to X'$, is any monomorphism $\iota : K \to X$ such that $\alpha \circ \iota = 0$ and such that that every morphism $\beta : Y \to X$ verifying $\alpha \circ \beta = 0$ factors through ι (in a unique way). Dually, a *cokernel* for $\alpha : X \to X'$ is any epimorphism $\nu : X' \to C$ such that $\nu \circ \alpha = 0$ and such that every morphism $\beta : X' \to Y$ verifying $\beta \circ \alpha = 0$ factors through ν (in a unique way).

and there are isomorphisms inverse of each other,

$$B \underset{\alpha+\sigma}{\overset{\rho\oplus\iota}{\rightleftarrows}} (A \oplus \hat{c}(\alpha)) \quad \text{and} \quad \boldsymbol{h}(B) \underset{\boldsymbol{h}(\alpha)+\boldsymbol{h}(\sigma))}{\overset{\boldsymbol{h}(\rho)\oplus\boldsymbol{h}(\iota)}{\rightleftarrows}} \boldsymbol{h}(A) \oplus \boldsymbol{h}(\hat{c}(\alpha))\,,$$

respectively in $\mathcal{K}(\mathrm{Ab})$ and $\mathrm{Ab}^{\mathbb{Z}}$.

The categories $\mathcal{K}(\mathrm{Ab})$ and Ab are therefore split categories. (♟, p. 351)

2. Show that if Ab is split, then the embedding $\mathrm{Ab}^{\mathbb{Z}} \subseteq \mathcal{K}(\mathrm{Ab})$ is an equivalence of categories and $\mathcal{K}(\mathrm{Ab})$ is a split abelian category. (♟, p. 352)

A.1.5.1 Triangles and Exact Triangles

Since the category $\mathcal{K}(\mathrm{Ab})$ is generally not abelian, we are led to relegate short exact sequences in favor of mapping cones, or, more generally, sequences isomorphic *in* $\mathcal{K}(\mathrm{Ab})$ to mapping cones, which bring us to recall the definitions of *triangles* and *exact triangles* in the category $\mathcal{K}(\mathrm{Ab})$.

Definition A.1.5.4 A *triangle* in $\mathcal{K}(\mathrm{Ab})$ is a sextuple (A, B, C, u, v, w) where A, B, C are complexes and $u : A \to B$, $v : B \to C$ and $w : C \to A[1]$ are morphisms of complexes. These data can be represented by a triangular diagram, hence the name '*triangle*',

$$\begin{array}{ccc} A & \xrightarrow{\quad u \quad} & B \\ {\scriptstyle [1]}\nwarrow^{w} & & \swarrow^{v} \\ & C & \end{array} \qquad \text{equivalently denoted by} \qquad \begin{cases} A \xrightarrow{u} B \xrightarrow{v} C \xrightarrow{w} A[1]\,, \text{ or} \\ A \xrightarrow{u} B \xrightarrow{v} C \underset{[1]}{\xrightarrow{w}}\,. \end{cases}$$

A *morphism of triangles* from (A, B, C, u, v, w) to (A', B', C', u', v', w') is any triple (α, β, γ) of morphisms in $\mathcal{K}(\mathrm{Ab})$ making a commutative diagram in $\mathcal{K}(\mathrm{Ab})$

$$\begin{array}{ccccccc} A & \xrightarrow{u} & B & \xrightarrow{v} & C & \xrightarrow{w} & A[1] \\ \downarrow{\scriptstyle \alpha} & & \downarrow{\scriptstyle \beta} & & \downarrow{\scriptstyle \gamma} & & \downarrow{\scriptstyle \alpha[1]} \\ A' & \xrightarrow{u'} & B' & \xrightarrow{v'} & C' & \xrightarrow{w'} & A[1] \end{array} \tag{A.13}$$

The composition of morphisms of triangles is defined component-wise. The morphism (u, v, w) is an isomorphism if its components u, v, w are isomorphisms.

The triangles of most interest to us are those giving rise to long exact sequences when applying the cohomology functor $\boldsymbol{h}$ and the functors $\mathrm{Mor}_{\mathcal{K}}(C, \_)$ and $\mathrm{Mor}_{\mathcal{K}}(\_, C)$ (Proposition A.1.5.2). These are exactly those triangles which are isomorphic, *in* $\mathcal{K}(\mathrm{Ab})$, to mapping cones. One is thus led to the following definition.

Definition A.1.5.5 A triangle in $\mathcal{K}(\mathrm{Ab})$ is called *exact* (or *distinguished*) if it is isomorphic to the mapping cone of a morphism.

Comment A.1.5.6 Beware that if $(0 \to A \xrightarrow{\alpha} B \xrightarrow{\beta} C \to 0)$ is a short exact sequence in $C(\mathrm{Ab})$, none of the triangles

$$T := (A, B, C, \alpha, \beta, 0) \quad \text{and} \quad \mathbf{h}(T) := (\mathbf{h}(A), \mathbf{h}(B), \mathbf{h}(C), \mathbf{h}(\alpha), \mathbf{h}(\beta), c)$$

where c is the connection morphism, is necessarily an exact triangle.

However the triangles T remain important for us as they generate long exact sequences in cohomology. This pinpoints the fact that exact triangles do not completely replace short exact sequences, which is an issue of the category $\mathcal{K}(\mathrm{Ab})$, since we still need the category $C(\mathrm{Ab})$ in order not to lose the information carried by short exact sequences. We will see that this is a lack of the category $\mathcal{K}(\mathrm{Ab})$ which the derived category $\mathcal{D}(\mathrm{Ab})$ is free from (Comment A.1.6.5), which is another reason to prefer the latter.

Exercise A.1.5.7 We justify in two different ways the usually claimed fact that not every short exact sequence in $C(\mathrm{Ab})$ can be embedded in an exact triangle of $\mathcal{K}(\mathrm{Ab})$.

1. Given an exact sequence $E := (0 \to A \xrightarrow{\alpha} B \xrightarrow{\beta} C \to 0)$ in $C(\mathrm{Ab})$, show that the map $\epsilon : \hat{c}(\alpha) \to C$, defined by $\epsilon(x, y) := \beta(x)$ is a quasi-isomorphism of complexes.

$$\begin{array}{ccccccccc}
 & & A & \xrightarrow{\alpha} & B & \xrightarrow{\iota} & \hat{c}(\alpha) & \xrightarrow{p} & A[1] \\
 & & \| & & \| & & {\scriptstyle \epsilon}\downarrow{\scriptstyle \text{q.i.}} & & \\
0 & \longrightarrow & A & \overset{\alpha}{\rightarrowtail} & B & \overset{\beta}{\twoheadrightarrow} & C & \longrightarrow & 0
\end{array}$$

Give an example in $C(\mathrm{Mod}(\mathbb{Z}))$ where ϵ has no homotopy inverse. (♜, p. 352)

2. Given an exact sequence $E := (0 \to A \xrightarrow{p} B \xrightarrow{q} C \to 0)$ in Ab, show that an exact triangle of the form

$$A[0] \xrightarrow{p[0]} B[0] \xrightarrow{q[0]} C[0] \xrightarrow{\gamma} A[1]$$

exists in $\mathcal{K}(\mathrm{Ab})$, if and only if E is split in Ab, in which case $\gamma = 0$.
Hint. Use Exercise A.1.5.3-(1). (♜, p. 353)

A.1.5.2 Triangulated Categories[7]

We owe to Verdier[8] the study of the properties of the family of exact triangles in the homotopy category $\mathcal{K}(\mathrm{Ab})$ allowing the axiomatization of the underlying abstract structure, today referred to as a *triangulated category*. This is an additive category $\mathcal{C}$ together with an automorphism $[1] : \mathcal{C} \rightsquigarrow \mathcal{C}$, the *translation functor*, and a collection $\mathcal{T}$ of triangles (defined as in A.1.5.4), called *exact* (or *distinguished*) triangles, which is required to satisfy four axioms[9]

TR-1. Any triangle isomorphic to a triangle in $\mathcal{T}$ belongs to $\mathcal{T}$. Every morphism $\alpha : A \to B$ can be imbedded in a triangle $(A, B, C, \alpha, \beta, \gamma) \in \mathcal{T}$. For every $A \in \mathrm{Ob}(\mathcal{C})$, the triangle $(A, A, 0, \mathrm{id}_A, 0, 0)$ belongs to $\mathcal{T}$.

TR-2. $(A, B, C, \alpha, \beta, \gamma) \in \mathcal{T}$ if and only if $(B, C, A[1], \beta, \gamma, \alpha[1]) \in \mathcal{T}$.

TR-3. Given $T := (A, B, C, u, v, w)$, $T' := (A', B', C', u', v', w')$ in $\mathcal{T}$, and given morphisms $\alpha : A \to A'$ and $\beta : B \to B'$ such that the diagram

$$\begin{array}{ccccccc} A & \xrightarrow{u} & B & \xrightarrow{v} & C & \xrightarrow{w} & A[1] \\ \downarrow{\scriptstyle \alpha} & \oplus & \downarrow{\scriptstyle \beta} & & \vdots\, {\scriptstyle \exists \gamma} & & \downarrow{\scriptstyle \alpha[1]} \\ A' & \xrightarrow{u'} & B' & \xrightarrow{v'} & C' & \xrightarrow{w'} & A[1] \end{array} \tag{A.14}$$

is commutative, there exists $\gamma : C \to C'$ (not necessarily unique) such that $(\alpha, \beta, \gamma) = T \to T'$ is a morphism of triangles.

TR-4. The Octahedral Axiom. Given exact triangles

$$(A, B, C, u, j, \_), \quad (B, C, A', v, \_, i), \quad (A, C, B', v \circ u, \_, \_)$$

there exist morphisms $f : C' \to B'$ and $g : B' \to A'$ such that the triangle

$$(C', B', A', f, g, j[1] \circ i)$$

is exact and the two other faces of the octahedron with f and g as edges, are commutative (view Fig. A.1).

Readers discovering these axioms for the first time, especially TR-3 and TR-4, should bear in mind that that they are natural extensions to triangles of familiar and useful elementary properties of short exact sequences in abelian categories. For example, TR-3 is obvious if T and T' are exact sequences, in which case v and v' are cokernels.

---

[7]The reader will find a thorough presentation of Triangulated Categories in Hartshorne [51], Chapter I §1, p. 20, Kashiwara-Schapira [62], Chapter 10, p. 241, Weibel [95], Chapter 10.2, p. 373, and obviously in Verdier [93].

[8]As reported by Illusie in the preface to Verdier [93].

[9]As presented in Hartshorne [51], Chapter I-§1, Triangulated Categories, p. 20.

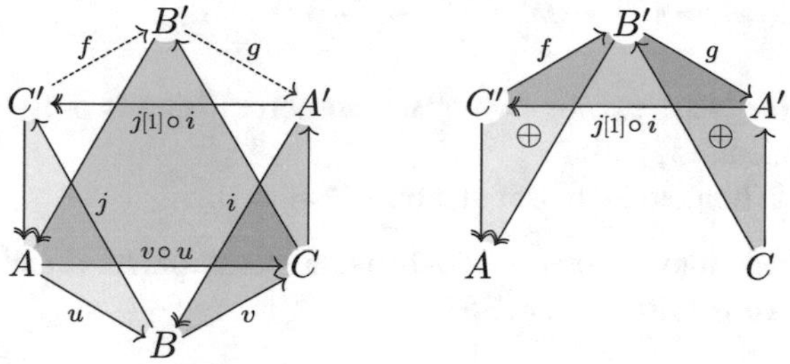

(A.15)

Fig. A.1 The Octahedral Axiom

The verification of the four axioms in the homotopy category $\mathcal{K}(\mathrm{Ab})$ is not difficult and leads to the following basic fact.[10]

Proposition A.1.5.8

1. *The category* $\mathcal{K}(\mathrm{Ab})$ *together with the collection of exact triangles (A.1.5.5) is a triangulated category.*
2. *A triangle* $(A, B, C, ., ., .)$ *in* $\mathcal{K}(\mathrm{Ab})$ *is exact if and only if the induced sequences*

$$\longrightarrow \mathrm{Mor}_{\mathcal{K}}(\_, A) \longrightarrow \mathrm{Mor}_{\mathcal{K}}(\_, B) \longrightarrow \mathrm{Mor}_{\mathcal{K}}(\_, C) \longrightarrow \mathrm{Mor}_{\mathcal{K}}(\_, A[1]) \longrightarrow$$

$$\longrightarrow \mathrm{Mor}_{\mathcal{K}}(C, \_) \longrightarrow \mathrm{Mor}_{\mathcal{K}}(B, \_) \longrightarrow \mathrm{Mor}_{\mathcal{K}}(A, \_) \longrightarrow \mathrm{Mor}_{\mathcal{K}}(A, \_[1]) \longrightarrow$$

(where $\mathcal{K}$ *stands for* $\mathcal{K}(\mathrm{Ab})$*) are long exact sequences.*

A.1.6 The Derived Category $\mathcal{D}(\mathrm{Ab})$

When interested in homology, one quickly realizes that homology is much more flexible than complexes for essentially two reasons: quasi-isomorphic complexes are not always isomorphic in $\mathcal{K}(\mathrm{Ab})$ and acyclic complexes[11] are not always homotopy-equivalent to zero (Exercise A.1.6.1). We owe Grothendieck for the idea of filling this gap by formally inverting the quasi-isomorphisms of $\mathcal{K}(\mathrm{Ab})$ following a procedure of *localization* analogue to the construction of noncommutative rings of quotients, which we now describe.[12],[13]

Exercise A.1.6.1 (Counterexamples)

1. Consider in $\mathrm{Mod}(\mathbb{Z})$, the acyclic complexes, for $m \geq 0$,

[10] See Hartshorne [51] Chapter I-§2, p. 25, or Weibel [95] Proposition 10.2.4, p. 376.

[11] A complex (C, d) is said to be *acyclic* if it is exact, i.e. if it has zero cohomology.

[12] Hartshorne [51] Chapter I. The Derived Category, §3 Localization of Categories p. 28.

[13] Lam [71] Chapter 4. Rings of Quotients, §10A. Ore Localizations, p. 299.

$$C(m) := \big(0 \to (m) \to \mathbb{Z} \to \mathbb{Z}/(m) \to 0\big)\,.$$

a. Show that the $C(m)$'s are two-by-two quasi-isomorphic and two-by-two non homotopy-equivalent. (, p. 353)

b. Show that $C(m)$ is homotopic to 0 if and only if $m = 0$.

2. Show that every quasi-isomorphism in $\mathcal{K}(\mathrm{Ab})$ is an isomorphism if and only if every morphism in Ab is *split*.[14] (, p. 354)

A.1.6.1 Multiplicative Collection of Morphisms

A collection $\mathcal{S}$ of morphisms of a category $\mathcal{C}$ is said to be *multiplicative* if it contains the identity morphisms id_X for all $X \in \mathrm{Ob}(C)$, if it is stable under composition of morphisms in $\mathcal{C}$, and if it satisfies the following conditions:

$\mathcal{S}$-1. Given $s \in \mathcal{S}$, any diagram in $\mathcal{C}$ of shape $\left(\begin{smallmatrix} & \bullet \\ & \downarrow s \\ \bullet \xrightarrow{t} & \bullet \end{smallmatrix}\right)$ can be completed to a commutative diagram $\left(\begin{smallmatrix} \bullet & \xrightarrow{t'} & \bullet \\ s' \downarrow & & \downarrow s \\ \bullet & \xrightarrow{t} & \bullet \end{smallmatrix}\right)$, for some $s' \in \mathcal{S}$. And the same, reversing arrows.

$\mathcal{S}$-2. Given $t_1, t_2 \in \mathrm{Mor}_{\mathcal{C}}(X, Y)$, the following conditions are equivalent.

a. There exists $s \in \mathcal{S}$ such that $t_1 \circ s = t_2 \circ s$.

b. There exists $s \in \mathcal{S}$ such that $s \circ t_1 = s \circ t_2$.

A.1.6.2 Universal Property of Localized Categories

Let $\mathcal{C}$ be a category, and let $\mathcal{S}$ be a multiplicative collection *of morphisms of $\mathcal{C}$.* The localization of $\mathcal{C}$ with respect to $\mathcal{S}$ *is a category $\mathcal{C}[\mathcal{S}^{-1}]$, together with a functor $\mathcal{Q} : \mathcal{C} \rightsquigarrow \mathcal{C}[\mathcal{S}^{-1}]$ such that the following conditions are satisfied.*

a) $\mathcal{Q}(s)$ is an isomorphism for every $s \in \mathcal{S}$.

b) Any functor $F : \mathcal{C} \rightsquigarrow \mathcal{C}'$ such that $F(s)$ is an isomorphism for every $s \in \mathcal{S}$ factors uniquely through $\mathcal{Q}$. (A.16)

The reader will find a constructive proof of the following existence theorem of localized categories in Hartshorne [51] (Proposition 3.1, p. 29.).

Theorem A.1.6.2 *If $\mathcal{S}$ is a multiplicative collection of morphisms in a category $\mathcal{C}$, then the localized category $\mathcal{C}[\mathcal{S}^{-1}]$ exists and is unique up to isomorphism. In addition, if $\mathcal{C}$ is additive, then so is $\mathcal{C}[\mathcal{S}^{-1}]$.*

[14] A morphism $f : X \to Y$ in an abelian category is said to be *split* if there are isomorphisms $X \sim K \oplus Z$ and $Z \oplus L \sim Y$ through which f reads as $(k, z) \mapsto (z, 0)$.

A.1.6.3 The Derived Category $\mathcal{D}(\mathrm{Ab})$ as Localization of $\mathcal{K}(\mathrm{Ab})$

A remarkable property of the category $\mathcal{K}(\mathrm{Ab})$, which is not verified by $C(\mathrm{Ab})$ (Exercise A.1.6.4), is that in $\mathcal{K}(\mathrm{Ab})$ the collection $\mathcal{S}$ of quasi-isomorphisms, that we want to invert, is indeed a multiplicative system. Furthermore, $\mathcal{S}$ is *compatible* with the triangulated category structure of $\mathcal{K}(\mathrm{Ab})$, which means that the following two additional conditions, obvious in $\mathcal{K}(\mathrm{Ab})$, are satisfied:

$\mathcal{S}$-3 if $s \in \mathcal{S}$ then $s[1] \in \mathcal{S}$;
$\mathcal{S}$-4 in axiom A.1.5.2-TR-3, if $\alpha, \beta \in \mathcal{S}$ then $\gamma \in \mathcal{S}$.

The category $\mathcal{K}(\mathrm{Ab})[\mathcal{S}^{-1}]$ therefore exists, it is the *derived category* of Ab.

Proposition A.1.6.3 (And Definition) *Let* Ab *be an abelian category. The family* $\mathcal{S}$ *of quasi-isomorphisms in* $\mathcal{K}(\mathrm{Ab})$ *is a multiplicative collection of morphisms compatible with the triangulated category of* $\mathcal{K}(\mathrm{Ab})$*. The* derived category of Ab *is then, by definition, the localized category*

$$\mathcal{D}(\mathrm{Ab}) := \mathcal{K}(\mathrm{Ab})[\mathcal{S}^{-1}]\,. \tag{A.17}$$

Furthermore, the category $\mathcal{D}(\mathrm{Ab})$ *is a triangulated category relative to the collection of triangles isomorphic (in* $\mathcal{D}(\mathrm{Ab})$*) to cones of morphisms.*

Sketch of Proof See Hartshorne [51] §I, Proposition 3.2, p. 32, and Proposition 4.1, p. 35. The proof that $\mathcal{S}$ is multiplicative is an application of the mapping cone construction (Sect. A.1.3). This is one of the first examples where it is necessary to soften the category of complexes $C(\mathrm{Ab})$ by identifying homotopic morphisms. We encourage the reader to provide his own detailed proof of this proposition without looking at ours (p. 354). ⊡

Exercise A.1.6.4 (A Counterexample) It is instructive to see why the collection $\mathcal{S}$ of quasi-isomorphisms is not multiplicative in the category $C(\mathrm{Ab})$.

Let $\big(X \xrightarrow{\mathrm{id}_X} X \xrightarrow{\iota} \hat{c}(\mathrm{id}_X) \xrightarrow{p_2}\big)$ be the mapping cone (A.1.3).

1. Show that the diagram $D := \left(\begin{array}{ccc} & & 0 \\ & & \downarrow \\ X & \xrightarrow{\iota} & \hat{c}(\mathrm{id}_X)\end{array}\right)$ cannot always be completed in the category $C(\mathrm{Ab})$ to a commutative diagram $\left(\begin{array}{ccc} \bullet & \longrightarrow & 0 \\ s\downarrow & & \downarrow \\ X & \xrightarrow{\iota} & \hat{c}(\mathrm{id}_X)\end{array}\right)$ where s is a quasi-isomorphism. (p. 356)
2. Complete the diagram D in $\mathcal{K}(\mathrm{Ab})$.

Comment A.1.6.5 Coming back to Exercise A.1.5.7-(1), given the exact sequence $E := (0 \to A \xrightarrow{\alpha} B \xrightarrow{\beta} C \to 0)$ in $C(\mathrm{Ab})$, the quasi-isomorphism $\epsilon(x, y) := \beta(x)$

$$\begin{array}{ccccccccc}
 & & A & \xrightarrow{\alpha} & B & \xrightarrow{\iota} & \hat{c}(\alpha) & \xrightarrow{p} & A[1] \\
 & & \| & & \| & & \epsilon\downarrow \text{q.i.} & & \\
0 & \longrightarrow & A & \xrightarrow{\alpha} & B & \xrightarrow{\beta} & C & \longrightarrow & 0
\end{array}$$

is invertible in $\mathcal{D}(\mathrm{Ad})$, so that we can now state that every short exact sequence in $C(\mathrm{Ab})$ can be embedded in an exact triangle of $\mathcal{D}(\mathrm{Ab})$ (*cf*. Comment A.1.5.6). Notice that the embedding associates E with the triangle $(A, B, C, \alpha, \beta, \gamma)$, where $\gamma = p \circ \epsilon^{-1}$ is generally different from 0.

A.1.6.4 The Morphisms in the Derived Category $\mathcal{D}(\mathrm{Ab})$

The morphisms in $\mathcal{D}(\mathrm{Ab})$ from a complex C to another C' are represented by zig-zag paths of morphisms in $\mathcal{K}(\mathrm{Ab})$ of the form

$$C \xleftarrow[q.i.]{s_1} \bullet \xrightarrow{t_1} \bullet \xleftarrow[q.i.]{s_2} \bullet \cdots \bullet \xleftarrow[q.i.]{s_{r-1}} \bullet \xrightarrow{t_r} \bullet \xleftarrow[q.i.]{s_r} C'. \tag{A.18}$$

denoted by $s_r^{-1} \circ t_r \circ s_{r-1}^{-1} \circ \cdots \circ s_2^{-1} \circ t_1 \circ s_1^{-1}$, where the arrows in the '*wrong*' direction are quasi-isomorphisms.

Two paths are said to be *equivalent* if applying the property A.1.6.1-($\mathcal{S}$-2) of multiplicative systems we can transform one into the other by means of a finite number of exchanging steps of the form

$$\begin{array}{c} t_1 \circ s_1^{-1} \leftrightarrow s_2^{-1} \circ t_2 \\ \Updownarrow \\ \big(s_2 \circ t_1 = t_2 \circ s_1\big) \text{ in } \mathcal{K}(\mathrm{Ab}) \end{array} \qquad \begin{array}{ccccc} & & \bullet & & \\ & {}^{s_1}\swarrow & & \searrow^{t_1} & \\ \bullet & & \oplus & & \bullet \\ & {}_{t_2}\searrow & & \swarrow_{s_2} & \\ & & \bullet & & \end{array} \tag{A.19}$$

where $s_i \in \mathcal{S}$ et $t_i \in \mathrm{Mor}(\mathcal{C})$.

A morphism in $\mathcal{D}(\mathrm{Ab})$ from C to C') is then an equivalence class of paths of the form (A.18). We denote by $\mathrm{Mor}_{\mathcal{D}(\mathrm{Ab})}(C, C')$ the set of such morphisms. The composition law is the concatenation of zig-zag paths.

A.1.6.5 Factorization of the Cohomology Functor

Thanks to the universal property of localizations A.1.6.2-(b), we can now add a new step to the factorization of the cohomology functor in (A.1):

$$C(\mathrm{Ab}) \overset{\iota}{\rightsquigarrow} \mathcal{K}(\mathrm{Ab}) \overset{\mathcal{Q}}{\rightsquigarrow} \mathcal{D}(\mathrm{Ab}) \overset{h}{\rightsquigarrow} \mathrm{Ab}^{\mathbb{Z}}, \qquad C(\mathrm{Ab}) \overset{h}{\rightsquigarrow} \mathrm{Ab}^{\mathbb{Z}} \tag{A.20}$$

The category $\mathcal{D}(\mathrm{Ab})$ thus appears to be the closest possible to cohomology while preserving complexes, which is where the strength of the derived category lies.

Exercise A.1.6.6 Show that a morphism in $\mathcal{D}(\mathrm{Ab})$ can always be lifted in $\mathcal{K}(\mathrm{Ab})$ as two-step paths $\bullet \xleftarrow[q.i.]{} \bullet \longrightarrow \bullet$ and $\bullet \longrightarrow \bullet \xleftarrow[q.i.]{} \bullet$. (p. 357)

A.1.7 *The Subcategories* $C^*(\mathbf{Ab})$, $\mathcal{K}^*(\mathbf{Ab})$ *and* $\mathcal{D}^*(\mathbf{Ab})$

A complex $C \in C(\mathrm{Ab})$ is said to be *bounded* (resp. *bounded below, bounded above*) if $C^m = 0$ for $|m| \gg 0$ (resp. $m \ll 0$, $m \gg 0$). We denote by $C^b(\mathrm{Ab})$, $C^+(\mathrm{Ab})$ and $C^-(\mathrm{Ab})$ the full subcategories of $C(\mathrm{Ab})$ whose objects are the complexes respectively bounded, bounded below and bounded above.

The categories $\mathcal{K}^*(\mathrm{Ab})$ and $\mathcal{D}^*(\mathrm{Ab})$ for $* \in \{b, +, -\}$ are defined in the same way.

The shift functors $[m]$ (A.1.1-C-6) extend to $\mathcal{K}^*(\mathrm{Ab})$ and $\mathcal{D}^*(\mathrm{Ab})$, where they clearly preserve exactness of triangles, quasi-isomorphisms and acyclicity.

Everything we have said since Sect. A.1.1 applies *mutatis mutandis* to the categories of bounded complexes.

A.2 Deriving Functors

A.2.1 *Extending Functors from* Ab *to* $\mathcal{D}(\mathbf{Ab})$

Given an abelian category Ab, we have introduced the sequence of '*extensions*'

$$\mathrm{Ab} \xrightarrow{[m]} C(\mathrm{Ab}) \xrightarrow{\iota} \mathcal{K}(\mathrm{Ab}) \xrightarrow{\mathcal{Q}} \mathcal{D}(\mathrm{Ab}), \tag{A.21}$$

and now ask whether it is possible to '*extend*' a functor $F : \mathrm{Ab} \rightsquigarrow \mathrm{Ab}'$. More precisely, whether it is possible to construct a commutative diagram of functors

$$\begin{array}{ccccccc} \mathrm{Ab} & \xrightarrow{[m]} & C(\mathrm{Ab}) & \xrightarrow{\iota} & \mathcal{K}(\mathrm{Ab}) & \xrightarrow{\mathcal{Q}} & \mathcal{D}(\mathrm{Ab}) \\ {\scriptstyle F}\downarrow & & \downarrow{\scriptstyle \exists ?} & & \downarrow{\scriptstyle \exists ?} & & \downarrow{\scriptstyle \exists ?} \\ \mathrm{Ab}' & \xrightarrow{[m]} & C(\mathrm{Ab}') & \xrightarrow{\iota} & \mathcal{K}(\mathrm{Ab}') & \xrightarrow{\mathcal{Q}} & \mathcal{D}(\mathrm{Ab}') \end{array} \tag{A.22}$$

We will consider this question only when $F : \mathrm{Ab} \to \mathrm{Ab}'$ is an additive functor.

A.2.2 *Extending Functors from* $C(\mathbf{Ab})$ *to* $\mathcal{K}(\mathbf{Ab})$

Consider the functor

$$C(\mathrm{Ab}) \xrightarrow{\iota} \mathcal{K}(\mathrm{Ab}), \tag{A.23}$$

which is the identity on complexes and which associates with a morphism of complexes its homotopy class.

The map

$$\iota_{\_,\_} : \mathrm{Mor}_{\mathcal{C}(\mathrm{Ab})}(\_,\_) \twoheadrightarrow \mathrm{Mor}_{\mathcal{K}(\mathrm{Ab})}(\_,\_)\,, \quad f \rightsquigarrow (f \bmod \sim)\,, \tag{A.24}$$

is a *surjective* homomorphism of groups, so that the diagram

$$\begin{array}{ccc} \mathcal{C}(\mathrm{Ab}) & \overset{\iota}{\rightsquigarrow} & \mathcal{K}(\mathrm{Ab}) \\ {\scriptstyle G}\wr\downarrow & & \wr\downarrow{\scriptstyle \exists ?\, G} \\ \mathcal{C}(\mathrm{Ab}') & \overset{\iota}{\rightsquigarrow} & \mathcal{K}(\mathrm{Ab}') \end{array} \tag{A.25}$$

can be closed in a *unique* way to a commutative diagram of functors, if and only if the functor $G : \mathcal{C}(\mathrm{Ab}) \rightsquigarrow \mathcal{C}(\mathrm{Ab}')$ is *compatible with homotopies*, in other words such that if $(f \sim f')$, then $(G(f) \sim G(f'))$, in which case, we will denote by the same letter the induced functor $G : \mathcal{K}(\mathrm{Ab}) \to \mathcal{K}(\mathrm{Ab}')$.

Notice that when $G : \mathcal{C}(\mathrm{Ab}) \to \mathcal{C}(\mathrm{Ab}')$ is induced by an *additive* functor $F : \mathrm{Ab} \rightsquigarrow \mathrm{Ab}'$, as in (A.3), the homotopy condition (A.8) is automatically respected:

$$F(f_i) = F(d'_{i-1} \circ h_i + h_{i+1} \circ d_i) = F(d'_{i-1}) \circ F(h_i) + F(h_{i+1}) \circ F(d_i)\,,$$

and the (unique) extension (A.25) exists. We have thus all the arguments proving the following proposition.

Proposition A.2.2.1 *An additive functor of abelian categories $F : \mathrm{Ab} \rightsquigarrow \mathrm{Ab}'$ admits unique extensions $F : \mathcal{C}(\mathrm{Ab}) \rightsquigarrow \mathcal{C}(\mathrm{Ab}')$ and $F : \mathcal{K}(\mathrm{Ab}) \rightsquigarrow \mathcal{K}(\mathrm{Ab}')$ making commutative the diagram of functors*

$$\begin{array}{ccccc} \mathrm{Ab} & \overset{[m]}{\rightsquigarrow} & \mathcal{C}(\mathrm{Ab}) & \overset{\iota}{\rightsquigarrow} & \mathcal{K}(\mathrm{Ab}) \\ {\scriptstyle F}\wr\downarrow & & {\scriptstyle F}\wr\downarrow & & {\scriptstyle F}\wr\downarrow \\ \mathrm{Ab}' & \overset{[m]}{\rightsquigarrow} & \mathcal{C}(\mathrm{Ab}') & \overset{\iota}{\rightsquigarrow} & \mathcal{K}(\mathrm{Ab}')\,. \end{array} \tag{A.26}$$

And likewise for every homomorphism of additive functors $(F \to G) : \mathrm{Ab} \to \mathrm{Ab}'$.

A.2.3 Extending Functors from $\mathcal{K}$(Ab) to $\mathcal{D}$(Ab)

As the category $\mathcal{D}(\mathrm{Ab})$ is the localization of $\mathcal{K}(\mathrm{Ab})$ relative to the collection of quasi-isomorphisms, the universal property of localized categories A.1.6.2 tells us that the necessary and sufficient condition to complete commutatively a diagram

$$\begin{array}{ccc} \mathcal{K}(\mathrm{Ab}) & \overset{Q}{\rightsquigarrow} & \mathcal{D}(\mathrm{Ab}) \\ {\scriptstyle G}\wr\downarrow & & \wr\downarrow{\scriptstyle \exists ?\, G} \\ \mathcal{K}(\mathrm{Ab}') & \overset{Q}{\rightsquigarrow} & \mathcal{D}(\mathrm{Ab}') \end{array} \tag{A.27}$$

is that the functor $G : \mathcal{K}(\mathrm{Ab}) \rightsquigarrow \mathcal{K}(\mathrm{Ab}')$ preserves quasi-isomorphisms.

When G is the extension of an additive functor $F : \mathrm{Ab} \rightsquigarrow \mathrm{Ab}'$, this condition expresses the fact that F is exact (*cf*. Exercise A.1.4.2), a property that is too strong, and in practice without real interest. The difficulty of F not being exact is that in $\mathcal{K}(\mathrm{Ab})$ classes of quasi-isomorphic complexes generally include more than one homotopy class (A.1.6.1-(2)) and it is not clear a priori how to choose one of them *in a functorial way*. Yet, this can be done if Ab is an abelian category with enough *injective* or *projective* objects, provided that we restrict ourselves to the bounded subcategories $\mathcal{D}^+(\mathrm{Ab})$ or $\mathcal{D}^-(\mathrm{Ab})$, respectively.

A.2.3.1 Projective and Injective Objects

- An object $I \in \mathrm{Ob}(\mathrm{Ab})$ is said to be *injective* if the functor

$$\mathrm{Hom}_{\mathrm{Ab}}(\_, I) : \mathrm{Ab} \to \mathrm{Mod}(\mathbb{Z}), \tag{A.28}$$

is exact, in other words,

if every diagram $\left(\begin{array}{ccc} X & \rightarrowtail & Y \\ \downarrow & & \\ I & & \end{array}\right)$ can be closed to a commutative diagram $\left(\begin{array}{ccc} X & \rightarrowtail & Y \\ \downarrow & \oplus & \swarrow \\ I & & \end{array}\right)$

The category Ab is said to *have enough injective objects* if for every $X \in \mathrm{Ob}(\mathrm{Ab})$, there exists a monomorphism $X \rightarrowtail I$ where I is an injective object. In that case, for every bounded below complex $C \in C^{\geq \ell}(\mathrm{Ab})$, there exist a complex $\boldsymbol{I} \in C^{\geq \ell}(\mathrm{Ab})$ with injective terms and a quasi-isomorphic monomorphism

$$C \rightarrowtail \boldsymbol{I}. \tag{A.29}$$

Such data constitute *an injective resolution of* C. By the universal property of injective objects, two injective resolutions of C are homotopy-equivalent.

- An object $P \in \mathrm{Ob}(\mathrm{Ab})$ is said to be *projective* if the functor

$$\mathrm{Hom}_{\mathrm{Ab}}(P, \_) : \mathrm{Ab} \to \mathrm{Mod}(\mathbb{Z}), \tag{A.30}$$

is exact, in other words,

if every diagram $\left(\begin{array}{ccc} Y & \twoheadrightarrow & X \\ & & \uparrow \\ & & P \end{array}\right)$ can be closed to a commutative diagram $\left(\begin{array}{ccc} Y & \twoheadrightarrow & X \\ & \nwarrow \; \oplus & \uparrow \\ & & P \end{array}\right)$

The category Ab is said to *have enough projective objects* if for every $X \in \mathrm{Ob}(\mathrm{Ab})$, there exists an epimorphism $P \twoheadrightarrow X$ where P is a projective object. In that case, for every bounded above complex $C \in \mathcal{C}^{\leq \ell}(\mathrm{Ab})$, there exist a complex $\boldsymbol{P} \in \mathcal{C}^{\leq \ell}(\mathrm{Ab})$ with projective terms and a quasi-isomorphic epimorphism

$$\boldsymbol{P} \twoheadrightarrow C\,. \tag{A.31}$$

These data constitute *a projective resolution of* C. By the universal property of projective objects, two projective resolutions of C are homotopy-equivalent.

The next proposition states that in a wide diversity of abelian categories there are indeed enough projective and injective objects.[15]

Proposition A.2.3.1 ([16]) *Let A be a ring (resp. a graded unital algebra). The category* $\mathrm{Mod}(A)$ *of A-modules (resp. the category* $\mathrm{GM}(A)$ *of A-graded modules (5.1.1.1)) has enough projective and injective objects.*

The main interest of these resolutions lies in the following theorem.

Theorem A.2.3.2 ([17]) *Let* Ab *be an abelian category.*

1. *Denote by $\mathcal{K}_I^+(\mathrm{Ab})$ the full subcategory of $\mathcal{K}(\mathrm{Ab})$ of bounded below complexes of injective objects in* Ab.
   a. *Complexes in $\mathcal{K}_I^+(\mathrm{Ab})$ are isomorphic if and only if they are quasi-isomorphic.*
   b. *If* Ab *has enough injective objects, the localization functor*

$$\mathcal{Q} : \mathcal{K}_I^+(\mathrm{Ab}) \rightsquigarrow \mathcal{D}^+(\mathrm{Ab}) \tag{A.32}$$

   is an equivalence of categories.
2. *Denote by $\mathcal{K}_P^-(\mathrm{Ab})$ the full subcategory of $\mathcal{K}(\mathrm{Ab})$ of bounded above complexes of projective objects in* Ab.
   a. *Two complexes in $\mathcal{K}_P^-(\mathrm{Ab})$ are isomorphic if and only if they are quasi-isomorphic.*
   b. *If* Ab *has enough projective objects, the localization functor*

$$\mathcal{Q} : \mathcal{K}_P^-(\mathrm{Ab}) \rightsquigarrow \mathcal{D}^-(\mathrm{Ab}) \tag{A.33}$$

   is an equivalence of categories.

We can now enhance proposition A.2.2.1 with a new extension.

---

[15]For a general overview on the problem of the existence of projective and injective objects, see Grothendieck [49], chapter **I**, §1.10, p. 135.

[16]Godement [46] §1.3-4, pp. 4-7, and Weibel [95] §2.2-2, pp. 33-43, for nongraded rings. The case of graded unital algebras follows by the same arguments as for proposition 5.1.2.2-(2).

[17]Hartshorne [51]. §I.4 Prop. 4.7, p. 46.

Theorem A.2.3.3

Let $F : \text{Ab} \to \text{Ab}'$ be a <u>*covariant*</u> *additive functor of abelian categories.*

1. *If* Ab *has enough injective objects, then there exists a unique functor*

$$\mathbb{R}F : \mathcal{D}^{+}(\text{Ab}) \rightsquigarrow \mathcal{D}^{+}(\text{Ab}') , \tag{A.34}$$

called the right derived functor of F*, such that, for all $m \in \mathbb{Z}$, the following diagram is commutative.*

$$\begin{array}{ccccccc}
\text{Ab} & \overset{[m]}{\rightsquigarrow} & \mathcal{C}^{+}(\text{Ab}) & \overset{\iota}{\rightsquigarrow} & \mathcal{K}^{+}(\text{Ab}) & \overset{Q}{\rightsquigarrow} & \mathcal{D}^{+}(\text{Ab}) \\
F\big\downarrow & & F\big\downarrow & & F\big\downarrow & & \big\downarrow \mathbb{R}F \\
\text{Ab}' & \overset{[m]}{\rightsquigarrow} & \mathcal{C}^{+}(\text{Ab}') & \overset{\iota}{\rightsquigarrow} & \mathcal{K}^{+}(\text{Ab}') & \overset{Q}{\rightsquigarrow} & \mathcal{D}^{+}(\text{Ab}')
\end{array} \tag{A.35}$$

Moreover, if $\epsilon_{\mathbf{I}} : C \rightarrowtail \mathbf{I}$ is an injective resolution of $C \in \text{Ob}(\mathcal{D}^{+})$ (A.29)*, we have $F(\mathbf{I}) = \mathbb{R}F(C)$ and a natural transformation of functors*

$$\epsilon : F \to \mathbb{R}F . \tag{A.36}$$

2. *If* Ab *has enough projective objects, then there exists a unique functor*

$$\mathbb{L}F : \mathcal{D}^{-}(\text{Ab}) \rightsquigarrow \mathcal{D}^{-}(\text{Ab}') , \tag{A.37}$$

called the left derived functor of F*, such that, for all $m \in \mathbb{Z}$, the following diagram is commutative.*

$$\begin{array}{ccccccc}
\text{Ab} & \overset{[m]}{\rightsquigarrow} & \mathcal{C}^{-}(\text{Ab}) & \overset{\iota}{\rightsquigarrow} & \mathcal{K}^{-}(\text{Ab}) & \overset{Q}{\rightsquigarrow} & \mathcal{D}^{-}(\text{Ab}) \\
F\big\downarrow & & F\big\downarrow & & F\big\downarrow & & \big\downarrow \mathbb{L}F \\
\text{Ab}' & \overset{[m]}{\rightsquigarrow} & \mathcal{C}^{-}(\text{Ab}') & \overset{\iota}{\rightsquigarrow} & \mathcal{K}^{-}(\text{Ab}') & \overset{Q}{\rightsquigarrow} & \mathcal{D}^{-}(\text{Ab}')
\end{array} \tag{A.38}$$

Moreover, if $\epsilon_{\mathbf{P}} : \mathbf{P} \twoheadrightarrow C$ is a projective resolution of $C \in \text{Ob}(\mathcal{D}^{-})$ (A.31)*, we have $\mathbb{L}F(C) = F(\mathbf{P})$ and a natural transformation of functors*

$$\epsilon : \mathbb{L}F \to F . \tag{A.39}$$

Comments A.2.3.4

1. In the theorem, the word *covariant* has been underlined to emphasize that care must be taken with the variance of functors. By convention, $G : \text{Ab} \to \text{Ab}'$ is *contravariant* if $G : \text{Ab}^{\text{op}} \to \text{Ab}'$ is covariant (fn. ([4]), p. 14). At the same time, the exchange $\text{Ab} \leftrightarrow \text{Ab}'$ entails exchange of $\mathcal{C}^{\pm}(\text{Ab}) \leftrightarrow \mathcal{C}^{\mp}(\text{Ab})$ as well as that of projective $\leftrightarrow$ injective objects. Hence, the right derived functor of a

contravariant functor G is defined on $\mathcal{D}^-(\text{Ab})$ and not in $\mathcal{D}^+(\text{Ab})$. We therefore have

$$\mathbb{R}\,G : \mathcal{D}^-(\text{Ab}) \to \mathcal{D}^+(\text{Ab})\,.$$

If $\boldsymbol{\epsilon} : \boldsymbol{P} \twoheadrightarrow C$ is a projective resolution in $C^-(\text{Ab})$, then $\mathbb{R}\,G(C) = G(\boldsymbol{P})$, and we get, by contravariance, a natural morphism $G(C) \to G(\boldsymbol{P}) = \mathbb{R}\,G(C)$. We therefore have a natural transformation of functors

$$\boldsymbol{\epsilon} : G \to \mathbb{R}\,G,$$

where, again, the right derived functor appears at the right-hand side, as in the covariant case (A.36), justifying the choice of the notation $\mathbb{R}$.

The same can be said *mutatis mutandis* for the left derived functor of a contravariant functor which is hence defined as

$$\mathbb{L}G : \mathcal{D}^+(\text{Ab}) \to \mathcal{D}^-(\text{Ab})\,,$$

and for which we have a natural transformation

$$\boldsymbol{\epsilon} : \mathbb{L}G \to G\,,$$

where the left derived functor still appears at the left-hand side.

2. The different notations for the right and left derived functors are justified because they do not necessarily coincide. Even when Ab has enough injective and projective objects, in which case the two derived functors are defined on the category of bounded complexes $\mathcal{K}^b(\text{Ab})$, they still can be different.

   Let us take a look at the case of the functor

$$F := \mathbb{Z}/(2) \otimes_{\mathbb{Z}} (\_) : \text{Mod}(\mathbb{Z}) \to \text{Mod}(\mathbb{Z})\,,$$

   and for the complex $\mathbb{Z}[0] \in C^b(\text{Ab})$.

   We have $\mathbb{Z}[0] \in \mathcal{K}^b_P(\text{Ab})$ and $\mathbb{Z}[0] \sim (0 \to \mathbb{Q} \to \mathbb{Q}/\mathbb{Z} \to 0) \in \mathcal{K}^b_I(\text{Ab})$, hence

$$\mathbb{L}F(\mathbb{Z}[0]) = \mathbb{Z}/(2)[0] \neq 0 \quad \text{whereas} \quad \mathbb{R}\,F(\mathbb{Z}[0]) = 0\,.$$

Notation A.2.3.5 Derived functors take their values in categories of complexes, so their cohomology can be calculated. The traditional notations are:

$$\mathbb{R}^i\,F(\_) := H^i(\mathbb{R}\,F(\_)) \quad \text{et} \quad \mathbb{L}^i F(\_) := H^{-i}(\mathbb{L}F(\_))\,. \tag{A.40}$$

Notice the change of sign on the degree i for left derived functors.

Exercise A.2.3.6 Let Ab be an abelian category with enough injective (resp. projective) objects. According to Theorem A.2.3.3, if $F : \mathrm{Ab} \to \mathrm{Ab}'$ is an additive covariant functor, we have commutative diagrams of functors (A.35)

$$\begin{array}{ccc} \mathrm{Ab} & \rightsquigarrow[m] \rightsquigarrow & \mathcal{D}^+(\mathrm{Ab}) \\ {\scriptstyle F}\big\downarrow & \xrightarrow{\ \epsilon\ } & \big\downarrow{\scriptstyle I\!\!R F} \\ \mathrm{Ab}' & \rightsquigarrow[m] \rightsquigarrow & \mathcal{D}^+(\mathrm{Ab}') \end{array} \qquad \left(\text{resp. } \begin{array}{ccc} \mathrm{Ab} & \rightsquigarrow[m] \rightsquigarrow & \mathcal{D}^-(\mathrm{Ab}) \\ {\scriptstyle F}\big\downarrow & \xleftarrow{\ \epsilon\ } & \big\downarrow{\scriptstyle I\!\!L F} \\ \mathrm{Ab}' & \rightsquigarrow[m] \rightsquigarrow & \mathcal{D}^-(\mathrm{Ab}') \end{array} \right) \tag{A.41}$$

1. Show that if F left (resp. right) exact, then the morphisms

$$\epsilon_{\_} : F(\_) \to I\!\!R^0\, F((\_)[0]) \qquad \left(\text{resp. } \epsilon_{\_} : I\!\!L^0 F((\_)[0]) \to F(\_)\right),$$

   where ϵ is the natural transformation (A.36) (resp. (A.39)), is an isomorphism.
2. What changes do we need to do to obtain the same conclusions for contravariant F? (p. 357)

A.2.4 Acyclic Resolutions

Although the injective (resp. projective) resolutions are sufficient to derive a left (resp. right) exact functor F, their use is limited by the fact that these resolutions are constructed by abstract means generally foreign to the working context. An alternative is given by F-*acyclic* objects. We will recall their definition only for left exact functors, as the case of right exact functors follow symmetrically.

Definition A.2.4.1 Let $F : \mathrm{Ab} \to \mathrm{Ab}'$ be a left exact functor. An object $O \in \mathrm{Ob}(\mathrm{Ab})$ is said to be F-*acyclic* if

$$I\!\!R^i\, F(O) = 0\,, \quad \forall i > 0\,.$$

An F-*acyclic resolution* of $C \in C^+(\mathrm{Ab})$ is then any quasi-isomorphism $C \to \boldsymbol{O}$ where $\boldsymbol{O} \in C^+(\mathrm{Ab})$ is a complex with F-acyclic terms.

The usefulness of acyclic resolutions is justified by the following proposition.

Proposition A.2.4.2 *Let* $F : \mathrm{Ab} \to \mathrm{Ab}'$ *be a left exact functor. For every complex* $\boldsymbol{O} \in C^+(\mathrm{Ab})$ *with* F*-acyclic terms, the augmentation* (cf. *Theorem A.2.3.3-(1)*)

$$\epsilon(\boldsymbol{O}) : F(\boldsymbol{O}) \to I\!\!R\, F(\boldsymbol{O})\,.$$

is an isomorphism in $\mathcal{D}(\mathrm{Ab})$. *In particular, if* $\alpha : C \to \boldsymbol{O}$ *is an* F*-acyclic resolution, then, in* $\mathcal{D}(\mathrm{Ab})$, *we have*

$$I\!\!R\, F(C) = F(\boldsymbol{O})\,.$$

Proof The idea is to construct a bi-complex $I_\star^*$

$$\begin{array}{ccccccccc}
C & & 0 \longrightarrow & C_0 & \longrightarrow & C_1 & \longrightarrow & C_2 & \longrightarrow \\
\alpha\downarrow & & & \alpha_0\downarrow & & \alpha_1\downarrow & & \alpha_2\downarrow & \\
\boldsymbol{O} & & 0 \longrightarrow & O_0 & \longrightarrow & O_1 & \longrightarrow & O_2 & \longrightarrow \\
\beta\downarrow & & & \downarrow & & \downarrow & & \downarrow & \\
\mathbf{Tot}_\oplus I_\star^* & & 0 \longrightarrow & I_0^0 & \longrightarrow & I_0^1 & \longrightarrow & I_0^2 & \longrightarrow \\
 & & & \downarrow & & \downarrow & & \downarrow & \\
 & & 0 \longrightarrow & I_1^0 & \longrightarrow & I_1^1 & \longrightarrow & I_1^2 & \longrightarrow \\
 & & & \downarrow & & \downarrow & & \downarrow &
\end{array} \tag{A.42}$$

- the morphism $\boldsymbol{O} \to I_0^*$ in an injective resolution of $\boldsymbol{O}$;
- for each $i \in \mathbb{N} \smallsetminus \{0\}$, the i-th row I_i^* is an injective resolution of the cokernel of morphism between the previous two rows, either $\boldsymbol{O} \to I_0^*$ or $I_{i-2}^* \to I_{i-1}^*$;
- for each $j \in \mathbb{N}$, the j-th column $O_j \to I_\star^j$ is an injective resolution.

The column $*$-filtration of the complex $\mathbf{Tot}_\oplus I_\star^*$, is then clearly regular and the $I\!E_2$ page of the corresponding spectral spectral sequence is concentrated in $q = 0$, where we recover the complex $\boldsymbol{O}$, i.e. $I\!E_2^{p,0} = O_p$. The spectral sequence is degenerated ($d_r = 0$ for $r \geq 2$), and the morphisms $\beta : \boldsymbol{O} \to \mathbf{Tot}_\oplus I_\star^*$ and $\beta \circ \alpha : C \to \mathbf{Tot}_\oplus I_\star^*$ is a quasi-isomorphism. In particular,

$$I\!R\, F(C) = F(\mathbf{Tot}_\oplus I_\star^*) = I\!R\, F(\boldsymbol{O})\,.$$

Besides, the columns $F(I_\star^j)$ have their cohomology concentrated in degree zero, and $F(O_j) = \mathbf{h}^0(F(I_\star^j))$, since O_j is F-acyclic. The arguments in the previous paragraph apply again showing that $F(\boldsymbol{O}) \to F(\mathbf{Tot}_\oplus I_\star^*)$ is a quasi-isomorphism, which ends the proof of the proposition. □

Example A.2.4.3 Alexander-Spanier cochains sheaves $\underline{\underline{\Omega}}^i(B; \Bbbk)$ over a mild topological space X (*cf*. Sect. B.1) are $\Gamma_\Phi(X; \_)$-acyclic for every family of supports Φ. The same is true for every $\mathcal{O}_X$-module, where $\mathcal{O} := \underline{\underline{\Omega}}^0(X; \Bbbk)$. Therefore, for every fiber bundle (E, X, π, F), we can write

$$\begin{aligned}
I\!R\,\Gamma(X; \pi_! \underline{\underline{\Omega}}^*(E; \Bbbk)) &= \Gamma(X; \pi_! \underline{\underline{\Omega}}^*(E; \Bbbk)) = (\Omega^*_{\mathrm{cv}}(E; \Bbbk), d)\,,\\
I\!R\,\Gamma_{\mathrm{c}}(X; \pi_! \underline{\underline{\Omega}}^*(E; \Bbbk)) &= \Gamma_{\mathrm{c}}(X; \pi_! \underline{\underline{\Omega}}^*(E; \Bbbk)) = (\Omega^*_{\mathrm{c}}(E; \Bbbk), d)\,,\\
I\!R\,\Gamma(X; \pi_* \underline{\underline{\Omega}}^*(E; \Bbbk)) &= \Gamma(X; \pi_* \underline{\underline{\Omega}}^*(E; \Bbbk)) = (\Omega^*(E; \Bbbk), d)\,.
\end{aligned}$$

Mutatis mutandis for a manifold X, the sheaves of differential forms $\underline{\underline{\Omega}}_X^i$ and every $\mathcal{O}_X$-module, where $\mathcal{O}_X := \underline{\underline{\Omega}}_X^0$ (*cf*. Theorem B.6.3.4-(1)).

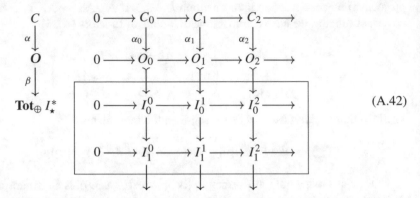

A.2.5 The Duality Functor on $\mathcal{D}(\mathbf{DGM}(\Omega_G))$

A.2.5.1 $I\!R\,\mathbf{Hom}^\bullet_{\Omega_G}(\_,\_)$ in $\mathcal{D}(\mathbf{GM}(\Omega_G))$

In the previous sections we recalled the basis of derivation of functors on the homotopy category $\mathcal{K}^*(\mathrm{Ab})$ induced by additive functors on Ab. A particular case of this is the functor of Ω_G-duality defined on Ab := $\mathrm{GM}(\Omega_G)$ (5.1.1)-(5.1):

$$\mathrm{Homgr}^*_{\Omega_G}(\_,\Omega_G) : \mathrm{GM}^*(\Omega_G) \rightsquigarrow \mathrm{GM}^*(\Omega_G)\,. \tag{A.43}$$

or, more generally, that of the bifunctor

$$\mathrm{Homgr}^*_{\Omega_G}(\_,\_) : \quad \mathrm{GM}^*(\Omega_G) \times \mathrm{GM}^*(\Omega_G) \rightsquigarrow \mathrm{GM}^*(\Omega_G)\,. \tag{A.44}$$

which was the topic of Sect. 5.3.2. There, we showed that since the category $\mathrm{GM}(\Omega_G)$ has enough injective objects (Proposition 5.1.2.2), this bifunctor has a right derived functor

$$I\!R\,\mathbf{Hom}^\bullet_{\Omega_G}(\_,\_) : \quad \mathcal{D}(\mathrm{GM}(\Omega_G)) \times \mathcal{D}^+(\mathrm{GM}(\Omega_G)) \rightsquigarrow \mathcal{D}(\mathrm{GM}(\Omega_G))\,, \tag{A.45}$$

given by the formula

$$I\!R\,\mathbf{Hom}^\bullet_{\Omega_G}(C,C') := \mathbf{Hom}^\bullet_{\Omega_G}(C,\boldsymbol{I})\,, \tag{A.46}$$

where $\boldsymbol{\epsilon}_I : C' \rightarrowtail \boldsymbol{I}$ is any injective resolution. However, since the category $\mathrm{GM}(\Omega_G)$ also has enough projective objects, we can also fix a projective resolution $\boldsymbol{\epsilon}_P : \boldsymbol{P} \twoheadrightarrow C$ and then consider the induced morphisms by $\boldsymbol{\epsilon}_I$ et $\boldsymbol{\epsilon}_P$:

$$\mathbf{Hom}^\bullet_{\Omega_G}(C,\boldsymbol{I}) \xrightarrow{\mathbf{Hom}^\bullet(\boldsymbol{\epsilon}_P,\boldsymbol{I})} \mathbf{Hom}^\bullet_{\Omega_G}(\boldsymbol{P},\boldsymbol{I}) \xleftarrow{\mathbf{Hom}^\bullet(\boldsymbol{P},\boldsymbol{\epsilon}_I)\,.} \mathbf{Hom}^\bullet_{\Omega_G}(\boldsymbol{P},C)$$

These are also quasi-isomorphisms (see Proposition 5.4.3.1), giving us an alternative method for calculating derived functors.

As a particular case of (A.45), the right derived functor of $\mathbf{Hom}^\bullet_{\Omega_G}(\_,\Omega_G)$: $\mathrm{GM}(\Omega_G) \rightsquigarrow \mathrm{GM}(\Omega_G)$ is given by the same formula (A.46), hence

$$\begin{array}{ccc} \mathcal{D}(GM(\Omega_G)) & \xrightarrow{\;I\!R\,\mathbf{Hom}^\bullet_{\Omega_G}(\_,\Omega_G)\;} & \mathcal{D}(GM(\Omega_G)) \\ C & \rightsquigarrow & \mathbf{Hom}^\bullet_{\Omega_G}(C,\boldsymbol{I})\,, \end{array} \tag{A.47}$$

where $\Omega_G \rightarrowtail \boldsymbol{I}$ is an injective resolution of Ω_G in $\mathcal{C}^{\geq 0}(\mathrm{GM}(\Omega_G))$.

A.2.5.2 The Categories $\mathcal{K}(\mathrm{DGM}(\Omega_G))$ and $\mathcal{D}(\mathrm{DGM}(\Omega_G))$

We have so far discussed the derivation of a functor between categories of complexes, $F : C(\mathrm{Ab}) \rightsquigarrow C(\mathrm{Ab}')$, which is induced by an additive functor defined at the level of the abelian categories themselves, i.e. $F : \mathrm{Ab} \to \mathrm{Ab}'$. Now we turn to the case of the category $\mathrm{DGM}(\Omega_G)$ which is particular because, while it contains the category $C(\mathrm{GM}(\Omega_G))$ (via the **Tot** functors in Sect. A.2.5.3), the categories are generally not equivalent. In simple terms, there are Ω_G-dgm's which are not complexes of Ω_G-gm's. In that case, and since we will see that the derived category $\mathcal{D}(\mathrm{DGM}(\Omega_G))$ does exist (Proposition A.2.5.2), the question arises of whether it is possible to extend the duality functor from $\mathcal{D}(\mathrm{GM}(\Omega_G))$ to $\mathcal{D}(\mathrm{DGM}(\Omega_G))$. The aim of this section is to explain that the answer is yes.

Although different, the categories $C(\mathrm{GM}(\Omega_G))$ and $\mathrm{DGM}(\Omega_G)$ are very close to each other. They are both abelian with shift functors whose objects are complexes of vector spaces endowed with a structure of Ω_G-graded module.

- In both we have the mapping cone construction. In $C(\mathrm{GM}(\Omega_G))$ it was defined in Sect. A.1.3. In $\mathrm{DGM}(\Omega_G)$, for a morphism $\alpha : X \to X'$ the definition is the same

$$\hat{c}(\alpha) := (X' \oplus X[1], \Delta) \quad \text{with} \quad \Delta(x', x) := (d'(x') + \alpha(x), -d(x))\,. \tag{A.48}$$

- In both we have homotopies. In $C(\mathrm{GM}(\Omega_G))$ it was defined in Sect. A.1.4. In $\mathrm{DGM}(\Omega_G)$, a morphism $f : X \to X'$ is said to be *homotopic to zero* if

$$\exists h \in \mathrm{Mor}_{\mathrm{GM}(\Omega_G)}(X, X'[-1]) \quad \text{such that} \quad f = h \circ d + d' \circ h\,.$$

  The corresponding relations of *homotopic morphisms* and *homotopy-equivalent Ω_G-dgm's*, both noted '$\sim$', are the same as in $C(\mathrm{GM}(\Omega_G))$. The relation '$\sim$', is an equivalence relation compatible with the additive structure of $\mathrm{DGM}(\Omega_G)$ and the set of equivalence classes:

$$\mathrm{Hot}_{\mathrm{DGM}(\Omega_G)}(X, X') := \mathrm{Mor}_{\mathrm{DGM}(\Omega_G)}(X, X')/_{\sim}\,. \tag{A.49}$$

  is an abelian group.

Definition A.2.5.1 The category $\mathcal{K}(\mathrm{DGM}(\Omega_G))$ is the category whose objects are Ω_G-dgm's, and whose morphisms are the homotopy classes of morphisms of Ω_G-dgm's (A.49). We denote by $\iota : \mathrm{DGM}(\Omega_G) \rightsquigarrow \mathcal{K}(\mathrm{DGM}(\Omega_G))$ the functor which is the identity on Ω_G-dgm's and which associates with a morphism its homotopy class.

The following proposition summarizes basic properties of these concepts.

Proposition A.2.5.2 (And Definition)

1. *The functors* $\mathbf{Tot}_* : C(\mathrm{GM}(\Omega_G)) \rightsquigarrow \mathrm{DGM}(\Omega_G)$ *introduced in Sect. 5.3.1 preserve cones, homotopies and quasi-isomorphisms.*
2. *The collection* $\mathcal{S}$ *of quasi-isomorphisms of* $\mathcal{K}(\mathrm{DGM}(\Omega_G))$ *constitute a multiplicative collection of morphisms (A.1.6.1). The derived category of* Ω_G*-differential graded modules is, by definition:*

$$\mathcal{D}(\mathrm{DGM}(\Omega_G)) := \mathcal{K}(\mathrm{DGM}(\Omega_G))[\mathcal{S}^{-1}] .$$

3. *The* $\mathbf{Tot}_*$ *functors are well-defined in the homotopy and derived categories and the following diagram of functors is commutative.*

$$\begin{array}{ccc} \mathcal{D}(GM(\Omega_G)) & \xrightarrow{\mathbb{R}\,\mathbf{Hom}^\bullet_{\Omega_G}(\_,\Omega_G)} & \mathcal{D}(GM(\Omega_G)) \\ \wr\downarrow \mathbf{Tot}_* & & \wr\downarrow \mathbf{Tot}_* \\ \mathcal{D}(\mathrm{DGM}(\Omega_G)) & \xrightarrow{\exists ?} & \mathcal{D}\mathrm{DGM}(\Omega_G) \end{array}$$

Sketch of Proof Statements (1,3) follow by straightforward verifications. For (2), the proof of the same result for $\mathcal{K}(\mathrm{Ab})$ (proof of Proposition A.1.6.3, page 354) which is based on the mapping cone construction, can be easily adapted to $\mathcal{K}(\mathrm{DGM}(\Omega_G))$. Details are left to the reader. ⊡

A.2.5.3 Extending Duality from $\mathcal{D}(\mathrm{GM}(\Omega_G))$ to $\mathcal{D}(\mathrm{DGM}(\Omega_G))$

Now that we have the $\mathbf{Tot}_*$ functors

$$\mathbf{Tot}_* : \mathcal{D}(\mathrm{GM}(\Omega_G)) \rightsquigarrow \mathcal{D}(\mathrm{DGM}(\Omega_G)) ,$$

we can ask if the duality functor $\mathbb{R}\,\mathbf{Hom}^\bullet_{\Omega_G}(\_, \Omega_G)$ can be extended to the derived category $\mathcal{D}(\mathrm{DGM}(\Omega_G))$ to complete a commutative diagram

$$\begin{array}{ccc} \mathcal{D}(GM(\Omega_G)) & \xrightarrow{\mathbb{R}\,\mathbf{Hom}^\bullet_{\Omega_G}(\_,\Omega_G)} & \mathcal{D}(GM(\Omega_G)) \\ \wr\downarrow \mathbf{Tot}_* & & \wr\downarrow \mathbf{Tot}_* \\ \mathcal{D}(\mathrm{DGM}(\Omega_G)) & \xrightarrow{\exists ?} & \mathcal{D}\mathrm{DGM}(\Omega_G) \end{array} \tag{A.50}$$

It turns out we can, as we see in the following proposition.

Theorem A.2.5.3 *For every finite*[18] *injective resolution* $\epsilon : \Omega_G \rightarrowtail \boldsymbol{I}$ *in the category* $C(\mathrm{GM}(\Omega_G))$*, the functor*

$$\mathbf{Hom}^\bullet_{\Omega_G}(\_, \mathbf{Tot}_\Pi\, \boldsymbol{I}) : \mathcal{K}(\mathrm{DGM}(\Omega_G)) \rightsquigarrow \mathcal{K}(\mathrm{DGM}(\Omega_G)), \tag{A.51}$$

preserves quasi-isomorphisms and is independent of the resolution ϵ*, inducing a canonical functor*

$$\mathbb{R}\,\mathbf{Hom}^\bullet_{\Omega_G}(\_, \Omega_G) : \mathcal{D}(\mathrm{DGM}(\Omega_G)) \rightsquigarrow \mathcal{D}(\mathrm{DGM}(\Omega_G)), \tag{A.52}$$

which makes commutative the following diagram

$$\begin{array}{ccc} \mathcal{D}(GM(\Omega_G)) & \xrightarrow{\mathbb{R}\,\mathbf{Hom}^\bullet_{\Omega_G}(\_,\Omega_G)} & \mathcal{D}(GM(\Omega_G)) \\ \wr\downarrow \mathbf{Tot}_\oplus & & \wr\downarrow \mathbf{Tot}_\Pi \\ \mathcal{D}(\mathrm{DGM}(\Omega_G)) & \xrightarrow{\mathbb{R}\,\mathbf{Hom}^\bullet_{\Omega_G}(\_,\Omega_G)} & \mathcal{D}\mathrm{DGM}(\Omega_G)\,. \end{array} \tag{A.53}$$

In addition, if we restrict the first term to $\mathcal{D}^b(\mathrm{GM}(\Omega_G))$*, then the* $\mathbf{Tot}_\Pi$ *functor on the second column can be replaced by* $\mathbf{Tot}_\oplus$.

Proof Since the functor $\mathbf{Hom}^\bullet_{\Omega_G}(\_, \mathbf{Tot}_\Pi\, \boldsymbol{I})$ is well-defined in the category $\mathrm{DGM}(\Omega_G)$ and preserves homotopies, the functor (A.51) is also well-defined. The uniqueness relative to the choice of the resolution ϵ is due the fact that two such resolutions are homotopic. To finish, the equality

$$\mathbf{Tot}_\Pi\, \mathbb{R}\,\mathbf{Hom}^\bullet_{\Omega_G}(\_, \Omega_G) \simeq \mathbb{R}\,\mathbf{Hom}^\bullet_{\Omega_G}(\mathbf{Tot}_\oplus(\_), \Omega_G)$$

is a consequence of Proposition 5.3.2.1, which gives the identification

$$\mathbf{Tot}_\Pi\, \mathbf{Hom}^\bullet_{\Omega_G}(\_, \boldsymbol{I}) \simeq \mathbf{Hom}^\bullet_{\Omega_G}(\mathbf{Tot}_\oplus(\_), \mathbf{Tot}_\Pi\, \boldsymbol{I})\,.$$

We still have to show that the functor $\mathbb{R}\,\mathbf{Hom}^\bullet_{\Omega_G}(\_, \Omega_G)$ is well-defined on the entire category $\mathcal{D}(\mathrm{DGM}(\Omega_G))$. In other words, that (A.51) preserves quasi-isomorphisms or, equivalently, preserves acyclicity.

As usual, as Ω_G is of finite homological dimension, we can choose an injective resolution of Ω_G in $\mathrm{GM}(\Omega_G)$ of *finite* length:

$$\Omega_G \rightarrowtail \boldsymbol{I} := (0 \to I_0 \to I_1 \to \cdots \to I_r \to 0)\,.$$

[18] The theorem is true without this restriction. Cf. Hartshorne [51] §I.6. Lemma 6.2, p. 64.

We have the exact sequence of complexes in $C(\Omega_G)$

$$\begin{array}{ccccccccccccc}
\boldsymbol{I}_{>0} := & (0 & \to & 0 & \to & I_1 & \to \cdots & \to & I_{r-1} & \to & I_r & \to & 0) \\
\downarrow & \downarrow & & \downarrow & & \downarrow & & & \downarrow & & \downarrow & & \downarrow \\
\boldsymbol{I} := & (0 & \to & I_0 & \to & I_1 & \to \cdots & \to & I_{r-1} & \to & I_r & \to & 0) \\
\downarrow & \downarrow & & \downarrow & & \downarrow & & & \downarrow & & \downarrow & & \downarrow \\
I_0[0] := & (0 & \to & I_0 & \to & 0 & \to \cdots & \to & 0 & \to & 0 & \to & 0)
\end{array}$$

which is split if we forget the (horizontal) differentials. This implies that the *a priori* left exact short sequence of functors

$$0 \to \mathbf{Hom}^{\bullet}_{\Omega_G}(\_, \mathbf{Tot}_{\Pi}\, \boldsymbol{I}_{>0}) \to \mathbf{Hom}^{\bullet}_{\Omega_G}(\_, \mathbf{Tot}_{\Pi}\, \boldsymbol{I}) \to \mathbf{Hom}^{\bullet}_{\Omega_G}(\_, I_0[0]) \to 0\,,$$

is also right exact. Then, an easy induction shows that if we can assume that for every injective module J in $\mathrm{GM}(\Omega_G)$, the functor $\mathbf{Hom}^{\bullet}_{\Omega_G}(\_, J[0])$ preserves acyclicity, then the same will be true for $\mathbf{Hom}^{\bullet}_{\Omega_G}(\_, \mathbf{Tot}_{\Pi}\, \boldsymbol{I})$.

Let us show that $\mathbf{Hom}^{\bullet}_{\Omega_G}(\_, J[0])$ preserves acyclicity.

Let (X, d) be an acyclic Ω_G-dgm. A p-cocycle in $\mathbf{Hom}^{\bullet}_{\Omega_G}(X, J[0])$ is a graded morphism $\alpha : X \to J[p]$ of Ω_G-dgm's, i.e. we have the commutative diagram

$$\begin{array}{ccccc}
X[-1] & \xrightarrow{\ d[-1]\ } & X & \xrightarrow{\ d\ } & X[1] \\
\downarrow & & \downarrow{\scriptstyle\alpha} & & \downarrow \\
0 & \longrightarrow & J[p] & \longrightarrow & 0
\end{array}$$

We can therefore factor d through $\dfrac{X}{\mathrm{im} d[-1]}$ as $d = \iota \circ \boldsymbol{\nu}$

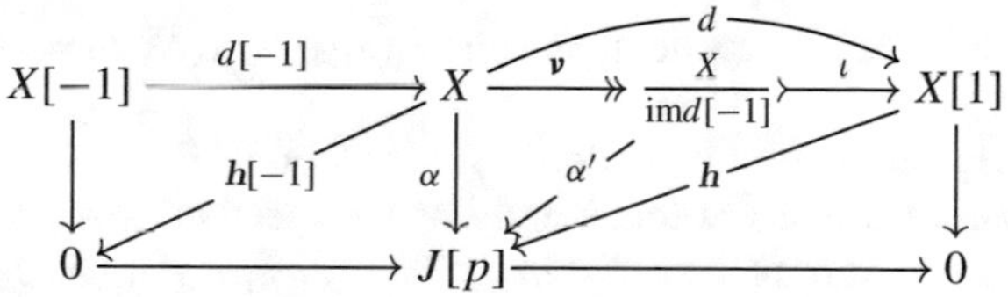

where ι is injective since X is acyclic and the extension h of α' exists since J is an injective Ω_G-gm. We can then conclude that we have

$$\alpha = h \circ d = (-1)^p \mathbf{D}(h)\,,$$

where $\mathbf{D}$ is the differential in $\mathbf{Hom}^{\bullet}_{\Omega_G}(X, J[0])$, because the differential in $J[p]$ is zero (Sect. 5.2.2-(5.5)). The Ω_G-dgm $\mathbf{Hom}^{\bullet}_{\Omega_G}(X, J[0])$ is therefore acyclic.

The possibility of substitution of $\mathbf{Tot}_{\Pi}$ by $\mathbf{Tot}_{\oplus}$, is an application of the last statement in Proposition 5.3.2.1, justified by the fact that one can always choose the injective resolution of Ω_G to belong to $C^b(\mathrm{GM}(\Omega_G))$ after 5.1.2.2-(3). □

Exercises A.2.5.4

1. Rewrite the last part of the Proof of Theorem A.2.5.3 without using the fact that Ω_G is of finite homological dimension.
2. Following the procedure of Sect. A.2.5, define the *left* derived functor of the bifunctor $(\_ \otimes_{\Omega_G} \_)^\bullet$ (5.3.3-(5.23), 5.2.2.1-(5.6)), on the derived categories $\mathcal{D}^-(\mathrm{GM}(\Omega_G))$ and $\mathcal{D}^-(\mathrm{DGM}(\Omega_G))$. Generalize Proposition 5.3.3.1 in order to construct the commutative diagram

$$\begin{array}{ccc} \mathcal{D}^-(\mathrm{GM}(\Omega_G)) \times \mathcal{D}^-(\mathrm{GM}(\Omega_G)) & \xrightarrow{(\_\otimes^{\mathbb{L}}_{\Omega_G}\_)^\bullet} & \mathcal{D}^-(\mathrm{GM}(\Omega_G)) \\ \wr\downarrow \mathbf{Tot}_{\oplus} \quad \wr\downarrow \mathbf{Tot}_{\oplus} & & \wr\downarrow \mathbf{Tot}_{\oplus} \\ \mathcal{D}^-(\mathrm{DGM}(\Omega_G)) \times \mathcal{D}^-(\mathrm{DGM}(\Omega_G)) & \xrightarrow{(\_\otimes^{\mathbb{L}}_{\Omega_G}\_)^\bullet} & \mathcal{D}^-\,\mathrm{DGM}(\Omega_G)\,. \end{array} \tag{A.54}$$

A.3 DG-Modules over DG-Algebras[19]

A.3.1 K-Injective ($\mathcal{A}$, d)-Differential Graded Modules

We begin recalling that if (A, d) is a differential graded algebra (dg-algebra in short), the category $\mathrm{DGM}(A, d)$ of (A, d)-dg-modules is such that for each $M \in \mathrm{DGM}(A, d)$ there exist a quasi-isomorphism $M \to I$ in $\mathrm{DGM}(A, d)$, where $I \in \mathrm{DGM}(A, d)$ is an injective A-graded module such that

$$\text{if } N \in \mathrm{DGM}(A, d) \text{ is acyclic, then } \mathrm{Hom}_{K(\mathrm{DGM}(A,d))}(N, I) = 0. \tag{A.55}$$

This last property being essential to derive functors on $\mathrm{DGM}(A, d)$ (Sect. A.2.3).

It is important to emphasize that simply being injective as A-graded module is not sufficient for I to warrant the vanishing of $\mathrm{Hom}_{K(\mathrm{DGM}(A,d))}(N, I)$ for all acyclic (A, d)-dg-modules N. This enhanced property entails some technicalities we will omit, instead referring interested readers to the excellent account given in the Stack Project.

---

[19]We refer to Stacks Project [87, 90] for a thorough introduction to the derived category of dg-modules over dg-algebras. In particular sections: 24.26, The derived category, and 24.30 Equivalences of derived categories.

The (A, d)-dg-modules I which are injective as A-graded modules and satisfy the vanishing condition (A.55) are called *K-injective*.[20]

In a number of parts of this book we have made use of the following theorem.

Theorem A.3.1.1 ([21]) *Let $(C, \mathcal{O})$ be a ringed site. Let $(\mathcal{A}, d)$ be a sheaf of differential graded algebras on $(C, \mathcal{O})$. For every differential graded $\mathcal{A}$-module $\mathcal{M}$ there exists a quasi-isomorphism $\mathcal{M} \to \mathcal{I}$ where $\mathcal{I}$ is an $\mathcal{A}$-graded injective module and a K-injective $(\mathcal{A}, d)$-differential graded module. Moreover, the construction is functorial in $\mathcal{M} \in \mathrm{DGM}(\mathcal{A}, d)$.*

Corollary A.3.1.2 *Let X be a topological space and let $(\underline{\underline{\Omega}}, d)$ be a sheaf of differential graded algebras over a field $\Bbbk$ (of arbitrary characteristic). Denote by $\mathrm{DGM}(\underline{\underline{\Omega}}, d)$ the category of $\underline{\underline{\Omega}}$-differential graded modules. For every $\mathcal{M} \in \mathrm{DGM}(\underline{\underline{\Omega}}, d)$ there exists a quasi-isomorphism $\mathcal{M} \to \mathcal{I}$ where*

- *$\mathcal{I}$ is an $\underline{\underline{\Omega}}$-graded injective module;*
- *$\mathcal{I}$ is a complex of injective sheaves of $\Bbbk$-vector spaces;*
- *$\mathcal{I}$ is an K-injective $\underline{\underline{\Omega}}$-differential graded module.*

Moreover, the construction is functorial in $\mathcal{M} \in \mathrm{DGM}(\underline{\underline{\Omega}}, d)$.

Proof After Theorem A.3.1.1, we need now only justify that $\mathcal{I}$ is a complex of injective sheaves, which follows easily from the usual identity

$$\mathbf{Hom}^\bullet(\_, \mathcal{I}) = \mathbf{Hom}^\bullet_{\underline{\underline{\Omega}}}\left(\underline{\underline{\Omega}} \otimes_{\underline{\underline{\Bbbk}}_X} (\_), \mathcal{I}\right)$$

and the fact that $\underline{\underline{\Bbbk}}_X \to \underline{\underline{\Omega}}$ is a flat extension of sheaves of rings, which is clear at the stalks level, since $\Bbbk$ is a field. □

A.3.2 *Formality of DGA's*

Following DGMS,[22] we recall the concept of *formality* for dga's.

Definition A.3.2.1 A dg-algebra (A, d) is said to be *formal* if there exists a zig-zag diagram

$$\begin{array}{ccccccccc} (A,d) & & (A_2,d) & & (A_2,d) & \cdots & (A_n,d) & & \\ & \searrow \quad \swarrow & & \searrow \quad \swarrow & & & & \searrow & \\ & (A_1,d) & & (A_3,d) & & & & & H(A) \end{array}$$

of quasi-isomorphic homomorphisms of dg-algebras joining (A, d) to its cohomology $H(A) := \mathbf{h}(A, d)$, considered as dg-algebra with zero differential.

[20] For the existence of K-injectives, see Stacks Project [87], ch. 22 Differential Graded Algebra. §21 *I-resolutions, Lemma 21.4, p. 27.*

[21] See Stacks Project [90] Theorem 24.25.13

[22] Deligne-Griffiths-Morgan-Sullivan [36] §4. Formality of Differential Algebras, p. 260.

The interest of formal dga's lies in the following equivalence of categories.

Proposition A.3.2.2 *Let $\rho : (A, d) \to (B, d)$ be a homomorphism of dga's. Consider the following functors:*

- $\mathrm{res}_\rho(\_) : \mathrm{DGM}(B, d) \rightsquigarrow \mathrm{DGM}(A, d)$ *the* restriction *functor which endows a (B, d)-dgm with the structure of (A, d)-dgm through the homomorphism ρ.*
- $(B, d) \otimes_{(A,d)} (\_) : \mathrm{DGM}(A, d) \rightsquigarrow \mathrm{DGM}(B, d)$ *the base change functor.*

1. *The functors $(B, d) \otimes_{(A,d)} (\_)$ and $\mathrm{res}_\rho(\_)$ are adjoints of each other, i.e. the following natural map is bijective:*

$$\mathbf{Hom}^\bullet_{(A,d)}\big((B, d) \otimes_{(A,d)} (\_), \_\big) \to \mathbf{Hom}^\bullet_{(A,d)}\big(\_, \mathrm{res}_\rho(\_)\big).$$

2. *If ρ is a quasi-isomorphism, then the derived functors*

$$\begin{cases} \mathrm{res}_\rho(\_) : \mathcal{D}\mathrm{DGM}(B, d) \rightsquigarrow \mathcal{D}\mathrm{DGM}(A, d) \\ (B, d) \otimes^{\mathbb{L}}_{(A,d)} (\_) : \mathcal{D}\mathrm{DGM}(A, d) \rightsquigarrow \mathcal{D}\mathrm{DGM}(B, d) \end{cases}$$

are equivalences of categories inverse of each other.

In particular, the adjunction map

$$(\_) \to \mathrm{res}_\rho\big((B, d) \otimes^{\mathbb{L}}_{(A,d)} (\_)\big) \tag{A.56}$$

is a quasi-isomorphism.

Sketch of Proof ([23]) (1) is the well-known adjunction for graded modules, it then suffices to check compatibility of differentials which is straightforward. (2) Since $\mathrm{res}_\rho(\_)$ does not modify the dgm's, it trivially preserves quasi-isomorphisms and acyclicity, hence tautologically defining a functor in derived categories, in which case, the derived functor $(B, d) \otimes^{\mathbb{L}}_{(A,d)} (\_)$ is its left adjoint. The conclusion then follows by proving that the adjunction morphism (A.56) is a quasi-isomorphism. But this is clear since $\rho : (A, d) \to (B, d)$ is an isomorphism in $\mathcal{D}(\mathrm{DGM}(A, d))$, so that, by definition of derived functors, the morphism

$$\rho \otimes \mathrm{id} : (A, d) \otimes^{\mathbb{L}}_{(A,d)} (\_) \to (B, d) \otimes^{\mathbb{L}}_{(A,d)} (\_)$$

is also a quasi-isomorphism, and we can conclude since the morphism

$$(\_) \to (A, d) \otimes_{(A,d)} (\_), \quad x \mapsto 1 \otimes x,$$

is an (obvious) isomorphism. □

[23] Stacks Project [90], §24.30 Equivalences of derived categories.

A.3.3 Formality of DGM's

We now consider a dga $A := (A, d)$ such that $d = 0$, in which case, the category $\mathrm{DGM}(A)$ has the particularity of containing both, an A-dg-module (M, d) and its cohomology $H(M) := \mathbf{h}(M, d)$, considered as a dg-module with zero differential. We can therefore consider the property of *formality* within the category $\mathrm{DGM}(A)$ and say that (M, d) is *formal in* $\mathrm{DGM}(A)$ if there exists a zig-zag of quasi-isomorphic morphisms in $\mathrm{DGM}(A)$ joining (M, d) to $H(M)$

$$\begin{array}{ccccccccccc} (M,d) & & & & (M_2,d) & & & & (M_2,d) & \cdots & (M_n,d) & & \\ & \searrow & & \swarrow & & \searrow & & \swarrow & & & & \searrow & \\ & & (M_1,d) & & & & (M_3,d) & & & & & & H(M) \end{array}$$

The following proposition gives a sufficient condition for formality of dgm's which is relevant in equivariant cohomology.

Proposition A.3.3.1 *Let A be an anticommutative (graded commutative) graded algebra.*

1. *If $(M, d) \in \mathrm{DGM}(A)$ is such that $\dim_{\mathrm{proj}}(H(M)) \le 1$ as A-gm, then (M, d) is formal in $\mathrm{DGM}(A)$. Moreover, there is a canonical bijection*

$$\mathrm{Iso}'_{\mathcal{D}\mathrm{DGM}(A)}\big((M,d), H(M)\big) \simeq \mathrm{Ext}^1_A\,\big(H(M), H(M)\big)\,,$$

where Iso' denotes the set of quasi-isomorphisms $\varphi : (M, d) \to H(M)$ inducing the identity in cohomology. In particular, if $H(M)$ is projective, then

$$\big|\mathrm{Iso}'_{\mathcal{D}\mathrm{DGM}(A)}\big((M,d), H(M)\big)\big| = 1\,.$$

2. *If A is* hereditary, *i.e. if $\dim_{\mathrm{coh}}(A) \le 1$, then every $(M, d) \in \mathrm{DGM}(A)$ is formal.*

Proof (1) Since K-projective resolutions exist in $\mathrm{DGM}(A)$,[24] we can set $(M, d) \to (M_1, d)$ to be such a resolution and assume in the sequel M to be a projective A-graded module.

Lemma 1. If $\dim_{\mathrm{proj}}(H(M)) \le 1$, then $B := dM$ and $Z :=$ $\ker(d)$ *are projective.*

Proof of Lemma 1. The long exact sequence of Ext^i_A functors associated with the short exact sequence of A-gm's

$$0 \to B \to Z \to H(M) \to 0\,, \tag{A.57}$$

[24]This is the dual notion of the K-injective resolutions of Sect. A.3.1. For the definition and existence of K-projective resolutions, see Stacks Project [87], ch. 22 Differential Graded Algebra. §20 *P-resolutions*, Lemma 20.2, p. 24.

where we denote $B := dM$ and $Z := \ker(d)$, gives rise to canonical maps

$$\begin{cases} \text{(i) } \mathrm{Ext}^1_A(Z,\_) \twoheadrightarrow \mathrm{Ext}^1_A(B,\_)\,, & \text{surjective,} \\ \text{(ii) } \mathrm{Ext}^i_A(Z,\_) \simeq \mathrm{Ext}^i_A(B,\_)\,, & \text{isomorphic } \forall i > 1\,. \end{cases} \tag{A.58}$$

since $\mathrm{Ext}^i(H(M),\_) = 0$ for all $i > 1$.

In the same vein, the long exact sequence of Ext^i_A functors associated with the short exact sequence of A-gm's

$$0 \to Z \longrightarrow M \xrightarrow{d} B[1] \to 0\,, \tag{A.59}$$

gives bijections

$$\mathrm{Ext}^i_A(Z,\_) \simeq \mathrm{Ext}^{i+1}_A(B[1],\_)\,, \quad \forall i \geq 1\,,$$

since $\mathrm{Ext}^i_A(M,\_) = 0$, for all $i \geq 1$. But $\mathrm{Ext}^{i+1}_A(B[1],\_) = \mathrm{Ext}^{i+1}_A(Z[1],\_)$ by (A.58)-(ii), so that we can write

$$\mathrm{Ext}^i_A(Z,\_) = \mathrm{Ext}^{i+1}_A(Z[1],\_)\,, \quad \forall i \geq 1\,,$$

whence, for fixed $i > 0$, we have

$$\mathrm{Ext}^i_A(Z,\_) = \mathrm{Ext}^{i+1}_A(Z[1],\_) = \cdots = \mathrm{Ext}^{i+n}_A(Z[n],\_), \quad \forall n > 0\,,$$

and, since A is of finite homological dimension, we can conclude that

$$\mathrm{Ext}^i_A(Z,\_) = 0, \quad \forall i \geq 1\,,$$

and also, by (A.58), that

$$\mathrm{Ext}^i_A(B,\_) = 0, \quad \forall i \geq 1\,.$$

We have thus proved that Z and B are projective A-graded modules. ⊟

By Lemma 1, the sequence (A.59) splits in $\mathrm{GM}(A)$ and $d : M \to B[1]$ admits sections. Given $\sigma : B[1] \to M$ such that $d \circ \sigma = \mathrm{id}_{B[1]}$, we consider the map:

$$\varphi_\sigma : (M, d) \to H(M) \quad \text{defined by} \quad \varphi_\sigma(x) := [x - \sigma(d(x))]\,.$$

Lemma 2. φ_σ is a quasi-isomorphism of A-dgm such that $\varphi_\sigma(z) = [z]$, $\forall z \in Z$.
Proof of Lemma 2. Indeed, φ_σ is a morphism of A-gm's verifying $\varphi_\sigma(z) = [z]$ by definition. It is also a morphism of dgm's since we have

$$d\varphi_\sigma(x) = dx - d(\sigma(dx)) = 0 = [dx] = \varphi_\sigma(dx)\,.$$

The restriction of the induced map $\boldsymbol{h}(\varphi_\sigma) : \boldsymbol{h}(M, d) \to H(M)$ to Z clearly coincides with the canonical projection $Z \twoheadrightarrow H(M)$, immediately implying that $\boldsymbol{h}(\varphi_\sigma)$ is an isomorphism. ⊟

The first part of (1) is thus proved. For its second part, we consider the map

$$\Xi : \{\sigma : B[1] \to M \mid d \circ \sigma = \mathrm{id}_{B[1]}\} \to \mathrm{Iso}'_{\mathcal{K}\mathrm{DGM}(A)}\big((M, d), H(M)\big)$$
$$\sigma \mapsto \varphi_\sigma := \big[(\mathrm{id}_M - \sigma \circ d)(\_)\big] \tag{A.60}$$

Lemma 3. The map Ξ is a surjective A-affine map.

Proof of Lemma 3. For the surjectivity of Ξ, let $\varphi : (M, d) \to H(M)$ be a quasi-isomorphic morphism of A-dgm's such that $\varphi(z) = [z]$, for all $z \in Z$. The map φ is a surjective morphism of A-gm's, and, since M is projective, can be lifted to a morphism $\tilde{\varphi} : M \to Z$, which necessarily verifies $(\tilde{\varphi}(z) - z) \in B$, for all $z \in Z$.

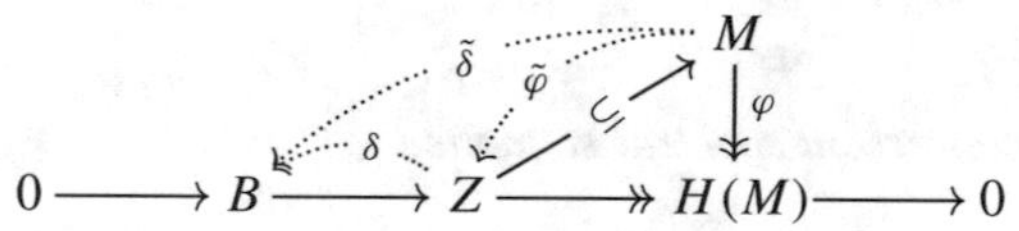

Hence, we have the morphism of A-gm's $\delta : Z \to B$, $\delta(z) := \tilde{\varphi}(z) - z$, which we can extend to a morphism of A-dgm's $\tilde{\delta} : M \to B$, since the short exact sequence (A.59) is split. As a consequence, the morphism $\tilde{\varphi}' := (\tilde{\varphi} - \tilde{\delta}) : M \to Z$ is a projector of M onto Z, and $\ker(\mathrm{id}_M - \tilde{\varphi}') = Z = \ker(d)$. Therefore,

$$(\mathrm{id}_M - \tilde{\varphi}')(m) = \sigma(dm)\,, \quad \forall m \in M\,,$$

for some unique $\sigma : B[1] \to M$, which is clearly a section of $d : M \to B[1]$, such that $\varphi_\sigma := \mathrm{id}_M - \sigma \circ d = \tilde{\varphi}'$ induces $\varphi : (M, d) \to H(M)$, as expected. ⊟

Applying the functor $\mathrm{Hom}_A(B[1], \_)$ to the split sequence (A.59), the set of sections of $d : M \to B[1]$ is identified to the vector space $\mathrm{Hom}_A(B[1], Z)$:

$$\begin{array}{ccccccc} 0 \to \mathrm{Hom}_A(B[1], Z) & \to & \mathrm{Hom}_A(B[1], M) & \to & \mathrm{Hom}_A(B[1], B[1]) & \to 0 \\ & & \sigma & \mapsto & \mathrm{id}_{B[1]} & \end{array}$$

The map Ξ is therefore defined on $\mathrm{Hom}_A(B[1], Z)$ by the A-affine formula

$$\mathrm{Hom}_A(B[1], Z) \ni \eta \quad \mapsto \quad \Xi(\eta) := \big[\big(\mathrm{id}_M - (\sigma_0 + \eta) \circ d\big)(\_)\big]\,,$$

where σ_0 denotes some fixed section of d. As a consequence, we see that

$$\Xi(\eta_1) \sim \Xi(\eta_2) \quad \Leftrightarrow \quad \big((\eta_2 - \eta_1) \circ d\big)(\_) \sim 0\,,$$

where '$\sim$' stands for '*homotopic*'. But, to ask $(\eta_2 - \eta_1) \circ d : M \to H(M)$ to be homotopic to zero is equivalent to asking that $h : M[1] \to H(M)$ exists such that $(\eta_2 - \eta_1) \circ d = h \circ d$, in other words, such that h extends $(\eta_2 - \eta_1) : B[1] \to H(M)$ to $M[1]$, or, equivalently, to $Z[1]$, since $Z[1]$ is an additive factor of $M[1]$. As a consequence, the image of Ξ is in bijection with the cokernel of the restriction map $\mathrm{Hom}_A(Z[1], H(M)) \to \mathrm{Hom}_A(B[1], H(M))$. This cokernel appears in the long exact sequence of δ-functors $\{\mathrm{Ext}^i_A(\_, H(M))\}_{i \in \mathbb{Z}}$ applied to $0 \to B \to Z \to H(M) \to 0$ (A.57), where we see the terms

$$\mathrm{Hom}_A(Z[1], H(M)) \to \mathrm{Hom}_A(B[1], H(M)) \to \mathrm{Ext}^1_A(H(M)[1], H(M)) \to 0,$$

since Z is a projective A-gm, hence proving (1).

(2) By (1), since if A is hereditary, then $\dim_{\mathrm{proj}} M \le 1$ for all $M \in \mathrm{GM}(A)$. □

Corollary A.3.3.2 *Let (A, d) be a formal dga such that the algebra $H(A) := \mathbf{h}(A, d)$ is of finite homological dimension. Let (M, d) be an (A, d)-dgm and denote $H(M) := \mathbf{h}(M, d)$.*

1. *There exists a convergent spectral sequence*

$$\begin{cases} \mathbb{E}_2^{p,q} := \mathbf{h}^p\, \mathbb{R}\,\mathbf{Hom}^\bullet_{H(A)}(H^q(M), H(A)) \\ \qquad\qquad \Downarrow \\ \qquad \mathbf{h}^{p+q}\, \mathbb{R}\,\mathbf{Hom}^\bullet_{(A,d)}((M, d), (A, d)) \end{cases}$$

2. *If $\dim_{\mathrm{proj}}(H(M)) \le 1$, then there exists an isomorphism in $\mathcal{D}\mathrm{DGM}(H(A))$:*

$$\mathbb{R}\,\mathbf{Hom}^\bullet_{(A,d)}((M, d), (A, d)) \xrightarrow[\simeq]{\xi(M,A)} \mathbb{R}\,\mathbf{Hom}^\bullet_{H(A)}(H(M), H(A))$$

functorial on $(M, d) \in \mathrm{DGM}(A, d)$, unique when $H(M)$ is projective, which makes commutative the following diagram:

$$\begin{array}{ccc} \mathbf{h}^0\, \mathbb{R}\,\mathbf{Hom}^\bullet_{(A,d)}((M, d), (A, d)) & \xrightarrow[\simeq]{\mathbf{h}^0(\xi(M,A))} & \mathbf{h}^0\, \mathbb{R}\,\mathbf{Hom}^\bullet_{H(A)}(H(M), H(A)) \\ \| & & \| \\ \mathrm{Hot}_{(A,d)}((M, d), (A, d)) & \xrightarrow[\bar{\xi}(M,A)]{} & \mathrm{Hom}_{H(A)}(H(M), H(A)) \end{array}$$

where $\bar{\xi}(M, A)$ is the morphism which associates the induced morphism in cohomology with a morphism of complexes (Corollary 5.4.7.2-(1)).

Furthermore, the spectral sequence $(\mathbb{E}_r, d_r)$ in (1) degenerates for $r \ge 2$.

Proof A sequence $(A,d) \leftrightarrow (A_1,d) \leftrightarrow \cdots \leftrightarrow (A_r,d)$ of quasi-isomorphic morphisms of dga'a gives rise, after Proposition A.3.2.2, to a sequence of equivalences of categories from $\mathcal{D}\mathrm{DGM}(A_0,d)$ to $\mathcal{D}\mathrm{DGM}(A_r,d)$ which, given a (A,d)-dgm M, generate a sequence $M \leftrightarrow M_1 \leftrightarrow \cdots \leftrightarrow M_r$ of quasi-isomorphic dgm's.

When (A,d) is formal and $(A_r,d) = (H(A),0)$, we therefore have

$$\mathbb{R}\,\mathbf{Hom}^\bullet_{(A,d)}(M,N) \simeq \mathbb{R}\,\mathbf{Hom}^\bullet_{H(A)}(M_r,N_r)\,,$$

where, in the right-hand side term, $H(A)$ is simply a graded algebra (with no differential) whose action is compatible with the differentials in M_r and N_r.

(1) Let $N := (A,d)$. Then $N_r = H(A)$ and we have the quasi-isomorphism

$$\mathbb{R}\,\mathbf{Hom}^\bullet_{(A,d)}(M,N) \simeq \mathbb{R}\,\mathbf{Hom}^\bullet_{H(A)}(M_r,H(A))\,. \tag{A.61}$$

Let $H(A) \to \mathcal{I}_\star$ be an injective resolution in $\mathrm{GM}(H(A))$, which we take of finite length since $H(A)$ is of finite homological dimension. The $H(A)$-dgm $\mathbf{Tot}_\oplus \mathcal{I}_\star$ is then K-injective in $\mathrm{DGM}(H(A))$.[25] Indeed, if (N,d) is acyclic, then $\mathbf{Hom}^\bullet((N,d),\mathbf{Tot}_\oplus \mathcal{I}_\star)$ is acyclic too, since its cohomology is the abutment of the spectral sequence associated with the decreasing $\star$-filtration of $\mathbf{Hom}^\bullet((N,d),\mathbf{Tot}_\oplus \mathcal{I}_\star)$ (regular since finite), whose first page is

$$\mathbb{E}_1^{p,q} = \mathbf{Hom}^\bullet_{H(A)}(\mathbf{h}^q(N,d),\mathcal{I}^p) = 0\,.$$

We can therefore write

$$\mathbb{R}\,\mathbf{Hom}^\bullet_{(A,d)}(M,(A,d)) = \mathbf{Hom}^\bullet_{H(A)}(M_r,\mathbf{Tot}_\oplus \mathcal{I}_\star)\,,$$

and the same $\star$-filtration gives a convergent spectral sequence with second page

$$\mathbb{E}_2^{p,q} = \mathbf{h}^p\,\mathbf{Hom}^\bullet_{H(A)}(H^q(M_r),\mathcal{I}_\star) = \mathbf{h}^p\,\mathbb{R}\,\mathbf{Hom}^\bullet_{H(A)}(H^q(M_r),H(A))$$

converging to

$$\mathbf{h}^{p+q}\,\mathbf{Hom}^\bullet_{H(A)}(M_r,\mathbf{Tot}_\oplus \mathcal{I}_\star) = \mathbf{h}^{p+q}\,\mathbb{R}\,\mathbf{Hom}^\bullet_{(A,d)}(M,(A,d))$$

as expected.

(2) Since $H(M_r) = H(M)$, the $H(A)$-dgm M_r is formal by Proposition A.3.3.1-(1), and we can exchange $M \leftrightarrow H(M)$ in A.61. The canonicity of the isomorphism when $H(M)$ is projective is due to Proposition A.3.3.1-(2). □

[25] Stacks Project [90] 24.25.

Appendix B
Sheaves of Differential Graded Algebras

The purpose of this appendix is to justify several auxiliary results used in various places in the book and which concern derived functors in categories of differential graded modules over sheaves of differential graded algebras, in particular for the sheaves of algebras of de Rham differential forms $\underline{\underline{\Omega}}_B$, and that of Alexander-Spanier cochains $\underline{\underline{\Omega}}_{B;\Bbbk}$. We prove the following statement which plays an central role in Relative Poincaré Duality (*cf*. Sect. 3.3.1, especially Proposition 3.3.2.1).

Theorem (B.9.1.3) *Let* (E, B, π, F) *be a fiber bundle of manifolds. There exist canonical isomorphisms in the derived category* $\mathcal{D}\mathrm{Vec}(\Bbbk)$*:*

$$\begin{array}{c} \mathbb{R}\,\mathbf{Hom}^\bullet_{\underline{\underline{\mathbb{R}}}_B}(\pi_!\underline{\underline{\Omega}}_E, \underline{\underline{\Omega}}_B) \\ \uparrow \\ \mathbb{R}\,\mathbf{Hom}^\bullet_{\mathcal{D}(\underline{\underline{\Omega}}_B,d)}(\pi_!\underline{\underline{\Omega}}_E, \underline{\underline{\Omega}}_B) \\ \downarrow \\ \mathbb{R}\,\mathbf{Hom}^\bullet_{\mathcal{D}(\Omega(B),d)}(\Omega_{\mathrm{cv}}(E), \Omega(B))\,. \end{array}$$

B.1 Mild Topological Spaces

Although the book concerns mainly manifolds or inductive limits of such, as are the classifying spaces, the results remain valid for more general topological spaces. Among the topological properties needed, the most crucial are the *separateness*, the *local contractibility*, and the *paracompactness and perfect normality*, all well-

A. Arabia, *Equivariant Poincaré Duality on G-Manifolds*, Lecture Notes in Mathematics 2288, https://doi.org/10.1007/978-3-030-70440-7

known properties of open subspaces of CW-complexes.[1],[2] The topological spaces in this appendix will therefore be locally contractible Hausdorff topological spaces such that all of its locally closed subspaces are paracompact. We will call such spaces *mild spaces*. Manifolds and, much more generally, open subspaces of CW-complexes are mild spaces.[3]

B.2 The Sheaf of Functions $\mathcal{O}_X$

We will consider the following sheaves of rings of functions on a mild space X:

- The sheaf $\mathcal{O}_X$ of continuous real valued functions;
- The sheaf $\mathcal{O}_X$ of differentiable real valued functions, when X is a manifold. This is the sheaf $\underline{\underline{\Omega}}_X^0$ of real de Rham differential forms of degree 0.
- The sheaf $\mathcal{O}_X$ the sheaf all functions (continuous or not) with values in a field $\mathbb{k}$. This is the sheaf $\underline{\underline{\Omega}}_{X,\mathbb{k}}^0$ of Alexander-Spanier cochains of degree 0, with coefficients in $\mathbb{k}$.

A statement concerning $\mathcal{O}_X$-(graded)-modules without any other precision, will be simultaneously valid in all these cases.

Conventions

- By '$\mathbb{k}$' we denote a field. Its meaning varies according to the context. In statements concerning de Rham complexes, it will be $\mathbb{k} := \mathbb{R}$, otherwise it is an arbitrary field. In this sense, the sheaf $\mathcal{O}_X$ is a sheaf of $\mathbb{k}$-vector spaces. The word '*sheaf*' will implicitly mean '*sheaf of* $\mathbb{k}$*-vector spaces*'.
- Rings and algebras have a multiplicative identity.
- Graded algebras are graded by the set of integers $\mathbb{Z}$ and are tacitly assumed anticommutative (graded commutative).
- In differential graded objects, differentials are of degree $+1$.
- Modules over rings and algebras are tacitly left modules.
- Acronyms for '*graded algebra*', '*graded module*', '*differential graded algebra*', '*differential graded module*' are respectively '*ga*', '*gm*', '*dga*', '*dgm*'.

---

[1]These properties are thoroughly studied in Lundell-Weingram book [74]. *Local contractibility* in Theorem 6.6 (p. 67). *Paracompactness and perfect normality* in Theorems 4.2 (p. 54) and 4.3 (p. 55), and the extension of these to all subspaces, in App. I, Theorem 10, p. 205.

[2]A topological space X is '*normal*' if for every pair of disjoint closed subspaces F_1, F_2, there exist disjoint neighborhoods $V_i \supseteq F_i$, and it is '*perfectly normal*' if, in addition, there exist continuous functions $f : B \to \mathbb{R}_{\geq 0}$ verifying $f|_{V_1} = 1$ and $f|_{V_2} = 0$.

[3]Notice that we excluded the properties of being *locally compact* and of being *countable at infinity*. The first, since we need our conclusions to be applied to classification spaces which are generally not locally compact. The second, simply because we do not need it. Recall however that a CW-complex is countable at infinity if and only if it is *locally finite*, which entails that it is *metrizable*, after Lundell-Weingram [74] Proposition 3.8, p. 52.

- The notations $\mathcal{F}(U)$ or $\Gamma(U;\mathcal{F})$ equivalently denote the sets of sections of a sheaf of $\Bbbk$-vector spaces $\mathcal{F} \in \mathrm{Sh}(X;\Bbbk)$, over subspace $U \subseteq X$.
- A graded sheaf $\mathcal{M}$ is a family of sheaves $\mathcal{M} := \{\mathcal{M}^i\}_{i\in\mathbb{Z}}$. The sheaf $\mathcal{M}^i$ is called the *i'th homogeneous component* of $\mathcal{M}$. It is usual to denote for a graded sheaf $\mathcal{M}$ and an open subspace $U \subseteq X$:

$$\mathcal{M}(U) = \Gamma(U;\mathcal{M}) := \bigoplus\nolimits_{i\in\mathbb{Z}} \mathcal{M}^i(U)\,,$$

  viewed as graded vector space.[4]
- As graded sheaves, the sheaves $\mathcal{O}_X$ are homogeneous of degree 0.
- An $\mathcal{O}_X$-graded algebra is a family $\mathcal{A} := \{\mathcal{A}^i\}_{i\in\mathbb{Z}}$ of $\mathcal{O}_X$-modules, together with a family of $\mathcal{O}_X$-bilinear morphisms $\{\cdot : \mathcal{A}^i \otimes \mathcal{A}^j \to \mathcal{A}^{i+j}\}_{i,k\in\mathbb{Z}}$, such that the triple $(\mathcal{A}, 0, +, \cdot)$ verifies the axioms of a algebras.
- If $\mathcal{A}$ is an $\mathcal{O}_X$-ga, then $\mathrm{GM}(\mathcal{A})$ denotes the category of $\mathcal{A}$-gm's. If $(\mathcal{A}, d)$ is an $\mathcal{O}_X$-dga, then $\mathrm{DGM}(A, d)$ denotes the category of $(\mathcal{A}, d)$-dgm's.

B.3 Global Lifting of Germs on $\mathcal{O}_X$-modules

The following proposition establishes that in a mild space X, the germs of sections of an $\mathcal{O}_X$-module are germs of global sections. We will see that despite the apparent shallowness of this property, its consequences are far-reaching.

Proposition B.3.1 *Let X be a mild space.*

1. *For every $x \in X$ and every open neighborhood $U \ni x$, there exists $f \in \mathcal{O}_X(X)$, such that the sets*

$$\{f = 1\} \subseteq \overline{\{f \neq 0\}} =: |f|$$

   are neighborhoods of x which are closed in X and contained in U.
2. *For every $\mathcal{O}_X$-module $\mathcal{M}$ and every $x \in X$, the map*

$$\mathcal{M}(X) \to \mathcal{M}_x$$

   which associates with $s \in \mathcal{M}(X)$ its germ $s_x \in \mathcal{M}_x$ at x, is a surjective map. In particular, for every $t \in \mathcal{M}_x$, there exists a morphism of $\mathcal{O}_X$-modules

$$\boxed{\xi_t : \mathcal{O}_X \to \mathcal{M}}$$

   verifying $(\xi_t(1))_x = t$.

[4] Notice that it would be a mistake to denote $\mathcal{M}$ as the sheaf *direct-sum-of-sheaves* $\bigoplus_{i\in\mathbb{Z}} \mathcal{M}^i$, since, while the natural map $\bigoplus_{i\in\mathbb{Z}} \Gamma(U;\mathcal{M}^i) \to \Gamma(U;\bigoplus_{i\in\mathbb{Z}} \mathcal{M}^i)$ is always injective, it is generally not surjective (*cf*. Godement [46] §2.7, p. 136, or Bredon [20] §I.5, p. 19.)

Proof (1) The sets $\{x\}$ and $X \smallsetminus U$ are disjoint closed subspaces of X. We then distinguish three cases:

a. *X is a mild space and $\mathcal{O}_X$ is the sheaf of continuous real valued functions*. Then, X being perfectly normal, there exists a continuous positive function $f_0 : X \to \mathbb{R}$ verifying $f_0(x) = 1$ and $f_0|_{X \smallsetminus U} = 0$, and we define $f := \sup\{1, 2\, f_0\}$, so that the sets $\{f = 1\}$ and $|f| = |f_0|$ are closed neighborhoods of x in X, contained in U.
b. *X is a manifold and $\mathcal{O}_X$ is the sheaf of differentiable real valued functions.* We can assume U relatively compact and contained in the domain of some chart of X, which leads us to prove (1) for $x \in U \subseteq \mathbb{R}^n$. Then, proceeding as in Godement [46] (p. 158), we '*regularize*' the continuous function f defined in case (a), by setting $\tilde{f}(y) := \int_{\mathbb{R}^n} f(y - z)\, \lambda(z)\, dz$, where λ is a positive differential function with compact support and integral equal to 1. Choosing λ with sufficiently small support, one then easily checks that the sets $\{\tilde{f} = 1\}$ and $|\tilde{f}|$ are again closed neighborhoods of x in X, contained in U.
c. *X is a mild space and $\mathcal{O}_X$ is the sheaf of all functions with values in* $\mathbb{k}$. For $x \in X$, we choose a continuous real valued function $g : X \to \mathbb{R}$, as in case (a), so that $Z := \{g = 1\}$ is a neighborhood of x which is closed in B and is contained in U. Then, the *indicator function* $f := \mathbf{1}_Z : X \to \mathbb{k}$, defined by $\mathbf{1}_Z(y) = 1$ if $y \in Z$, and $\mathbf{1}_Z(y) = 0$ otherwise, answers the question.

(2) Let $x \in X$. Given $s_x \in \mathcal{M}_x$, let $U \subseteq X$ be an open neighborhood of x such that s_x is the germ at x of a section $s \in \mathcal{M}(U)$. Then, if f is a function satisfying the conditions in (1) for the pair (x, U), the section $fs \in \mathcal{M}(U)$ verifies $(fs)_x = ((fs)|_{\{f=1\}})_x = s_x$, and, furthermore, since fs is zero in $X \smallsetminus |f|$, it extends by zero in a unique global section $\tilde{s} \in \Gamma(X; \mathcal{M})$ still verifying the equality $\tilde{s}_x = s_x$, as announced. □

Exercise B.3.2 Recall that a sheaf $\mathcal{F}$ on a topological space X is said to be *soft* if for every closed subspace $S \subseteq X$ the restriction map $\mathcal{F}(X) \to \mathcal{F}(S)$ is surjective. Prove that $\mathcal{O}_X$-modules are soft on a mild space X. (🕮, p. 357)

Comment B.3.3 *Softness*, *Φ-softness*, *c*-softness, *flabbiness*, *fineness*, etc., are properties related to different ways of lifting local sections to the whole space. A sheaf with one of these properties is $\Gamma_\Phi(X; \_)$-acyclic (Sect. A.2.4), something which is established following a standard procedure.[5] In the sequel, we will prove the same acyclicity results for $\mathcal{O}_X$-modules on mild spaces following a somewhat different approach.

---

[5]Look at '*soft sheaves*' in Bredon [20], Godement [46], Iversen [58], Kashiwara-Schapira [61], Weibel [95].

B.4 $\mathcal{O}_X$-Graded Algebras

Let $\mathcal{A}$ be an $\mathcal{O}_X$-ga and consider an $\mathcal{A}$-gm $\mathcal{M}$ endowed with the induced structure of $\mathcal{O}_X$-gm through the action of $\mathcal{A}$. One of the most relevant consequences of the existence of global lifts of the germs of $\mathcal{M}$, is that it can be realized as a quotient of a free $\mathcal{A}$-gm. More precisely, the morphism of sheaves

$$\xi := \sum_{x\in X;\, s\in\mathcal{M}_x} \xi_s : \bigoplus_{x\in X;\, s\in\mathcal{M}_x} \mathcal{A}[s] \twoheadrightarrow \mathcal{M}\,, \tag{B.1}$$

where s is homogeneous of degree $[s]$, and $\xi_s : \mathcal{A}[s] \to \mathcal{M}$ is the morphism defined in Proposition B.3.1-(2), is a surjection of $\mathcal{A}$-gm's, since it is so, by construction, at the stalks level at each homogeneous component.

The procedure can obviously be iterated on $\ker(\xi)$ since it is also an $\mathcal{A}$-gm. We get in this way a *free presentation* of $\mathcal{M}$ as $\mathcal{A}$-gm, i.e. a right exact sequence in the category $\mathrm{GM}(\mathcal{A})$:

$$\mathcal{A}^{\natural} \to \mathcal{A}^{\natural} \to \mathcal{M} \to 0\,, \tag{B.2}$$

where '$\mathcal{A}^{\natural}$' indicates a direct sum of $\mathcal{A}[\_]$'s, which it is not necessary to specify.

Terminology The existence of a free presentation (B.2) is referred to as the *quasi-coherence* (or *$\mathcal{A}$-quasi-coherence*) of the $\mathcal{A}$-gm $\mathcal{M}$.

Proposition B.4.1 *Let $\mathcal{A}$ be an $\mathcal{O}_X$-graded algebra on a mild space X.*

1. *An $\mathcal{A}$-gm is quasi-coherent.*
2. *If $\nu : \mathcal{A}^{\natural} \twoheadrightarrow \mathcal{M}$ is a surjective morphism of $\mathcal{O}_X$-gm's, then the following morphism of global section is surjective too:*[6]

$$\nu(X) : \Gamma(X;\mathcal{A}^{\natural}) \twoheadrightarrow \Gamma(X;\mathcal{M})\,.$$

3. *The functor*

$$\Gamma(X;\_) = \mathrm{GM}(\mathcal{A}) \to \mathrm{GM}(\Gamma(X;\mathcal{A})) \tag{B.3}$$

is exact.

In particular, if $\mathcal{A}^{\natural} \to \mathcal{A}^{\natural} \to \mathcal{M} \to 0$ is a presentation of an $\mathcal{A}$-gm $\mathcal{M}$, then the following induced sequence of $\Gamma(X;\mathcal{A})$-gm's is exact:

$$\Gamma(X;\mathcal{A}^{\natural}) \to \Gamma(X;\mathcal{A}^{\natural}) \to \Gamma(X;\mathcal{M}) \to 0\,. \tag{B.4}$$

[6] As recalled in the fn. ([4]), p. 303, the natural map $\Gamma(X;\mathcal{A})^{\natural} \to \Gamma(X;\mathcal{A}^{\natural})$ is injective but generally not surjective for infinite sums, contrary to the map between stalks $(\mathcal{A}_x)^{\natural} \to (\mathcal{A}^{\natural})_x$ which is always bijective. In particular, the natural morphism $\mathcal{A}\otimes(\bigoplus_{i\in\mathfrak{I}} \mathcal{M}_i) \to \bigoplus_{i\in\mathfrak{I}}(\mathcal{A}\otimes\mathcal{M}_i)$ is an isomorphism (*cf.* Godement [46] §2.7, p. 136, or Bredon [20] §I.5, p. 19.)

Proof (1) Already justified in the preliminary remarks.
(2) It suffices to restrict to homogeneous components. A section $s \in \mathcal{M}^{[s]}(X)$ determines a germ $s_x \in \mathcal{M}^{[s]}{}_x$ for every $x \in X$. Let $t_x \in (\mathcal{A}[s]^{\natural})_x$ be such that $\nu_x(t_x) = s_x$. Applying Proposition B.3.1-(2), we can choose, for each $x \in X$, a section $\tilde{t}_x \in \Gamma(X; \mathcal{A}[s]^{\natural})$ and an open neighborhood $V_x \ni x$ such that $\nu(V_x)(\tilde{t}_x) = s|_{V_x}$. Since X is paracompact, there exists a partition of unity $\{\phi_x\}_{x \in X} \subseteq \mathcal{O}_X(X)$ subordinate to the cover $\{V_x\}_{x \in X}$ of X. The sum $t := \sum_x \phi_x \tilde{t}_x$ is then well-defined in $\Gamma(X; \mathcal{A}[s]^{\natural})$, and verifies, by construction,

$$\nu(X)(t) = \sum\nolimits_x \phi_x \, \nu(X)(\tilde{t}_x) = \sum\nolimits_x \phi_x \, s = s\,.$$

(3) Since $\Gamma(X; \_)$ is left exact, we need only to prove it respects surjectivity. Given a surjective morphism of $\mathcal{A}$-gm's $\nu : \mathcal{M} \twoheadrightarrow \mathcal{N}$, we compose it with a surjection $\xi : \mathcal{A}^{\natural} \twoheadrightarrow \mathcal{M}$ (viz. (B.1)), and conclude, using (2), that $\Gamma(X; \nu \circ \xi)$, and hence $\Gamma(X; \nu)$, are surjective. The rest of (3) follows straightforwardly. □

B.5 Localization Functor for $\mathcal{O}_X$-GA's

B.5.1 The Isomorphism $\mathbf{Hom}^{\bullet}_{\mathcal{A}}(\mathcal{A}, \_) \simeq \Gamma(Y; \_)$

Let $\mathcal{A}$ be a sheaf of graded algebras on a topological space Y. Given $\mathcal{F} \in \mathrm{GM}(\mathcal{A})$, a homogeneous global section $s \in \mathcal{F}^{[s]}(Y)$ uniquely determines a morphism of $\mathcal{A}$-gm's of degree $[s]$, from $\mathcal{A}$ to $\mathcal{F}$, by setting, for every open subspace $U \subseteq Y$,

$$\mathcal{A}(U) \ni t \mapsto t\, s|_U \in \mathcal{F}(U)\,.$$

In this way we get a canonical map $\Gamma(Y; \mathcal{F}) \to \mathbf{Hom}^{[s]}_{\mathcal{A}}(\mathcal{A}; \mathcal{F})$ which is functorial in $\mathcal{F}$, and gives rise to a homomorphism of functors

$$\big(\Gamma(Y; \_) \to \mathbf{Hom}^{\bullet}_{\mathcal{A}}(\mathcal{A}; \_)\big) : \mathrm{GM}(\mathcal{A}) \rightsquigarrow \mathrm{GM}(\mathcal{A}(Y))\,, \tag{B.5}$$

which is easily seen to be an isomorphism with inverse the map which associates with a morphism $\nu : \mathcal{A} \mapsto \mathcal{F}$ of degree d, the section $\nu(Y)(1) \in \mathcal{F}^d(Y)$.

Comment B.5.1.1 Beware that although $\Gamma(Y; \_)$ and $\mathbf{Hom}^{\bullet}_{\mathcal{A}}(\mathcal{A}, \_)$ coincide in the category $\mathrm{GM}(\mathcal{A})$, they are not equal functors since they first is defined in the whole category $\mathrm{GM}(\Bbbk_X)$. In particular, one must be careful when comparing the derived functors $I\!R\,\Gamma(Y; \_)$ and $I\!R\,\mathbf{Hom}^{\bullet}_{\mathcal{A}}(\mathcal{A}; \_)$ (Theorem B.6.1.1).

B.5.2 *Right Adjoint to* $\mathbf{Hom}^\bullet_{\mathcal{A}}(\mathcal{A}, \_)$

B.5.2.1 Localization Presheaf Functor

Let $\mathcal{A}$ be a sheaf of graded algebras on a topological space Y.

Given $M \in \mathrm{GM}(\mathcal{A}(Y))$, the correspondence which assigns to an inclusion of open subspaces $V \subseteq U \subseteq Y$ the morphism

$$\mathcal{A}(U) \otimes_{\mathcal{A}(Y)} M \xrightarrow{\rho_V^U \otimes \mathrm{id}_M} \mathcal{A}(V) \otimes_{\mathcal{A}(Y)} M \tag{B.6}$$

where $\rho_V^U : \mathcal{A}(U) \to \mathcal{A}(V)$ is the restriction map, clearly defines a presheaf of $\mathcal{A}$-gm's on Y which we denote by $\mathcal{A} \underline{\otimes} M$. The construction is functorial on M. The resulting functor

$$\mathcal{A} \underline{\otimes} (\_) = \mathrm{GM}(\mathcal{A}(Y)) \rightsquigarrow \mathrm{Mod}_{\mathrm{presh.}}(\mathcal{A}) \tag{B.7}$$

is the *$\mathcal{A}$-localization presheaf functor*.

It verifies, by construction, the adjunction equality:

$$\mathbf{Hom}^\bullet_{\mathcal{A}}(\mathcal{A} \underline{\otimes} M, \mathcal{N}) = \mathbf{Hom}^\bullet_{\mathcal{A}(Y)}(M, \mathcal{N}(Y)), \tag{B.8}$$

since a morphism of presheaves $\nu : \mathcal{A}\underline{\otimes} M \to \mathcal{N}$ is, by (B.6), a family of morphisms $\{\nu(U) : \mathcal{A}(U) \otimes M \to \mathcal{N}(U)\}$ making commutative the diagrams

$$\begin{array}{ccc} M & \xrightarrow{\nu(Y)} & \mathcal{N}(Y) \\ \downarrow & \oplus & \downarrow \rho_U^Y \\ \mathcal{A}(U) \otimes M & \xrightarrow{\nu(U)} & \mathcal{N}(U) \end{array}$$

This means that one has the equalities

$$\nu(U)(a \otimes m) = a\, \nu(U)(1 \otimes m) = a\, \rho_U^Y(\nu(Y)(m)),$$

showing that ν is completely determined by $\nu(Y)$.

Definition B.5.2.1 Let $\mathcal{A}$ be a sheaf of graded algebras on a topological space Y. The *$\mathcal{A}$-localization functor*,

$$\mathcal{A} \otimes (\_) = \mathrm{GM}(\mathcal{A}(Y)) \rightsquigarrow \mathrm{GM}(\mathcal{A}). \tag{B.9}$$

is the composition of the localization presheaf functor $\mathcal{A} \underline{\otimes} (\_)$ (B.7), followed by the exact functor '*associated sheaf to a presheaf*'.

Proposition B.5.2.2 (Localization Functor) *The natural map*

$$\begin{array}{rcl} \mathbf{Hom}^\bullet_{\mathcal{A}}(\mathcal{A} \otimes M, \mathcal{N}) & \to & \mathbf{Hom}^\bullet_{\mathcal{A}(Y)}(M, \mathcal{N}(Y)) \\ \nu & \mapsto & \nu(Y) \end{array} \tag{B.10}$$

is an isomorphism, functorial in $M \in \mathrm{GM}(\mathcal{A}(Y))$ *and* $\mathcal{N} \in \mathrm{GM}(\mathcal{A})$.

The functors

$$\begin{cases} \Gamma(Y; \_) = \mathbf{Hom}^\bullet_{\mathcal{A}}(\mathcal{A}, \_) : \mathrm{GM}(\mathcal{A}) \rightsquigarrow \mathrm{GM}(\mathcal{A}(Y)) \\ \mathcal{A} \otimes (\_) : \mathrm{GM}(\mathcal{A}(Y)) \rightsquigarrow \mathrm{GM}(\mathcal{A}) \end{cases}$$

constitute a pair of adjoint functors.

Proof Everything follows from the adjunction equality (B.8) and the universal property of the functor '*associated sheaf to a presheaf*', by which the natural map $\mathbf{Hom}^\bullet(\mathcal{A} \otimes (\_), \_) \to \mathbf{Hom}^\bullet(\mathcal{A} \underline{\otimes} (\_), \_)$ is an isomorphism.

The map (B.10) is then the composition of equalities

$$\mathbf{Hom}^\bullet_{\mathcal{A}}(\mathcal{A} \otimes M, \mathcal{N}) \underset{(=)}{\longrightarrow} \mathbf{Hom}^\bullet_{\mathcal{A}}(\mathcal{A} \underline{\otimes} M, \mathcal{N}) = \mathbf{Hom}^\bullet_{\mathcal{A}(Y)}(M, \mathcal{N}(Y)) . \qquad \square$$

B.5.3 The Localization Functor for $\mathcal{O}_X$-GA's

We now come back to the case of a mild space X and a sheaf $\mathcal{A}$ of $\mathcal{O}_X$-ga's. Under these assumptions, we show that for every $\mathcal{A}$-gm $\mathcal{M}$, the stalks $\mathcal{M}_x$ for $x \in X$, are determined in terms of the $\mathcal{A}(X)$-gm $\mathcal{M}(X)$ alone.

Proposition B.5.3.1 *Let* $\mathcal{A}$ *be an* $\mathcal{O}_X$*-ga on a mild space* X.

For $x \in X$, *denote by* S_x *the multiplicative system of functions in* $\mathcal{O}_X(X)$ *equal to* 1 *in a neighborhood of* x.

1. *Let* $\mathcal{M} \in \mathrm{GM}(\mathcal{A})$. *For every* $x \in X$, *the germ map* $\mathcal{M}(X) \to \mathcal{M}_x$ *factors through the localization map* $\mathcal{M}(X) \to S_x^{-1}\mathcal{M}(X)$ *in an isomorphism of* $S_x^{-1}\mathcal{A}(X)$*-gm's*

$$\xi_x(\mathcal{M}) : S_x^{-1}\mathcal{M}(X) \to \mathcal{M}_x \,,$$

   which is functorial in $\mathcal{M} \in \mathrm{GM}(\mathcal{A})$.
2. *The localization functor*

$$\mathcal{A} \otimes (\_) : \mathrm{GM}(\mathcal{A}(X)) \rightsquigarrow \mathrm{GM}(\mathcal{A})$$

   is an exact functor.

3. *Let $\mathcal{M} \in \mathrm{GM}(\mathcal{A})$. The natural morphism of sheaves of $\mathcal{A}$-gm's*

$$\mathcal{A} \otimes \mathcal{M}(X) \to \mathcal{M}$$

is an isomorphism. More precisely, the localization functor $\mathcal{A} \otimes (\_)$ is a left inverse of the global section functor $\Gamma(X; \_)$.

Proof (1) The $\mathcal{A}(X)$-gm $\mathcal{M}_x$ is an $S_x^{-1}\mathcal{A}(B)$-gm so that we have a canonical factorization

$$\begin{array}{ccc} \mathcal{M}(B) & \longrightarrow & \mathcal{M}_x \\ & \searrow & \uparrow {\scriptstyle \xi_x(\mathcal{M})} \\ & & S_x^{-1}\mathcal{M}(B) \end{array}$$

where $\xi_x(\mathcal{M}) : S_x^{-1}\mathcal{M}(B) \to \mathcal{M}_x$ is a morphism of $S_x^{-1}\mathcal{A}(B)$-gm's.

The kernel of the canonical map $\mathcal{M}(B) \to S_x^{-1}\mathcal{M}(B)$ is the submodule of $m \in \mathcal{M}(B)$ annihilated by some $f \in S_x$, hence of m's which vanish on a neighborhood of x. But this is exactly the kernel of the map $\mathcal{M}(B) \to \mathcal{M}_x$, so that $\xi_x(\mathcal{M})$ is injective.

Up to this point we have not yet used the fact that $\mathcal{A}$ is an $\mathcal{O}_X$-module on a mild space. Indeed, these assumptions are needed only to establish the surjectivity of $\xi_x(\mathcal{M})$, which is then an obvious consequence of the surjectivity of the germ morphism $\mathcal{M}(B) \to \mathcal{M}_x$ stated in Proposition B.3.1-(2).

(2) Given an exact sequence of $\mathcal{A}(X)$-gm's $0 \to L \to M \to N \to 0$, the sequence of localized $\mathcal{A}$-gm's

$$0 \to \mathcal{A} \otimes L \to \mathcal{A} \otimes M \to \mathcal{A} \otimes N \to 0 \qquad \text{(B.11)}$$

is exact, if it is so at each of its stalks. But, we have seen in (1) that we have the identifications

$$(\mathcal{A} \otimes K)_x = \mathcal{A}_x \otimes K = S_x^{-1}K\,, \qquad \text{(B.12)}$$

which are functorial in $K \in \mathrm{GM}(\mathcal{A}(X))$. Therefore, (B.11) is exact if and only if the sequence $0 \to S_x^{-1}L \to S_x^{-1}M \to S_x^{-1}N \to 0$ is exact. And this is indeed the case since localization by S_x^{-1} is an exact functor.

(3) The morphism is induced by the morphism of presheaves $\mathcal{A} \underline{\otimes} \mathcal{M}(X) \to \mathcal{M}$, which associates with $a \otimes m \in \mathcal{A}(U) \otimes \mathcal{M}(X)$ the section $a\,\rho_U^X(m) \in \mathcal{M}(U)$. As usual, to see that it is an isomorphism, it suffices to check it at stalks level, and thus show that, for all $x \in X$, the map $\mathcal{A}_x \otimes \mathcal{M}(X) \to \mathcal{M}_x$ is an isomorphism. But then, equality (B.12) leads us again to the natural map $S_x^{-1}\mathcal{M}(X) \to \mathcal{M}_x$, which we already showed is an isomorphism in (1). □

B.6 Equivalences of Some Derived Functors in $\mathcal{D}\,\mathrm{GM}(\mathcal{A})$

B.6.1 An Equivalence of Categories

Proposition B.5.3.1-(3), stating that the functor $\Gamma(X;\_):\mathrm{GM}(\mathcal{A})\rightsquigarrow\mathrm{GM}(\mathcal{A}(X))$ admits a left inverse, has as first consequence that the image by $\Gamma(X;\_)$ of objects and morphisms constitutes a subcategory of $\mathrm{GM}(\mathcal{A}(X))$. The following theorem states important consequences of this fact.

Theorem B.6.1.1 *Let $\mathcal{A}$ be an $\mathcal{O}_X$-algebra on a mild space X.*

1. *The functor*

$$\Gamma(X;\_)=\mathbf{Hom}^\bullet_{\mathcal{A}}(\mathcal{A},\_):\mathrm{GM}(\mathcal{A})\rightsquigarrow\mathrm{GM}(\mathcal{A}(X))$$

   is exact and fully faithful. It is an equivalence of categories between $\mathrm{GM}(\mathcal{A})$ *and the* full essential image of $\Gamma(X;\_)$*, denoted by* $\Gamma(X;\mathrm{GM}(\mathcal{A}))$.[7]
2. *An injective $\mathcal{A}$-gm $\mathcal{I}$ is an injective sheaf of $\Bbbk$-vector spaces.*
3. *There is a canonical isomorphism of functors*

$$\big(I\!R\,\Gamma(X;\_)\to I\!R\,\mathbf{Hom}^\bullet_{\mathcal{A}}(\mathcal{A};\_)\big):\mathcal{D}\,\mathrm{GM}(\mathcal{A})\rightsquigarrow\mathcal{D}\,\mathrm{GM}(\mathcal{A}(X))\,.$$

4. *If $\mathcal{I}$ is an injective $\mathcal{A}$-gm, then $\mathcal{I}(X)$ is an injective $\mathcal{A}(X)$-gm.*
5. *For all $\mathcal{M},\mathcal{N}\in\mathrm{GM}(\mathcal{A})$, there is a canonical isomorphism in $\mathcal{D}\,\mathrm{GM}(\mathcal{A}(X))$*

$$I\!R\,\mathbf{Hom}^\bullet_{\mathcal{A}}(\mathcal{M},\mathcal{N})\to I\!R\,\mathbf{Hom}^\bullet_{\mathcal{A}(X)}(\mathcal{M}(X),\mathcal{N}(X))$$

   functorial in $\mathcal{M}$ and $\mathcal{N}$.

Proof (1) Exactness is stated in Proposition B.4.1-(3). The adjunction B.5.2.2-B.10 and the isomorphism $\mathcal{A}\otimes\mathcal{M}(X)\to\mathcal{M}$ in Proposition B.5.3.1-(3), give the isomorphisms:

$$\begin{aligned}\mathbf{Hom}^\bullet_{\mathcal{A}}(\mathcal{M},\mathcal{N})&\simeq\mathbf{Hom}^\bullet_{\mathcal{A}}(\mathcal{A}\otimes\mathcal{M}(X),\mathcal{N})\\&\simeq\mathbf{Hom}^\bullet_{\mathcal{A}(X)}(\mathcal{M}(X),\mathcal{N}(X))\,,\end{aligned}\tag{B.13}$$

proving fully faithfulness. The fact that $\Gamma(X;\_):\mathrm{GM}(\mathcal{A})\rightsquigarrow\Gamma(X;\mathrm{GM}(\mathcal{A}))$ is an equivalence of categories is then straightforward.
(2) The sheaf $\mathcal{A}$ considered with its natural structure of sheaf of $\Bbbk$-vector spaces is a *flat* $\underline{\Bbbk}_X$-gm. Indeed, flatness is a property of stalks, and $(\underline{\Bbbk}_X)_x$, being the field $\Bbbk$, the stalk $\mathcal{A}_x$ is trivially $(\underline{\Bbbk}_X)_x$-flat. In particular, the base change functor

$$\mathcal{A}\otimes_{\Bbbk}(\_):\mathrm{GM}(\underline{\Bbbk}_X)\rightsquigarrow\mathrm{GM}(\mathcal{A})\,,$$

[7] Given a functor between categories $F:C\rightsquigarrow C'$, the *full essential image of* F is the smallest full subcategory of C' containing $F(\mathrm{Ob}(C))$ and stable under isomorphisms in C'.

is exact. As a consequence, the functor

$$\mathbf{Hom}^{\bullet}_{\underline{\underline{\Bbbk}}_X}(\_, \mathcal{I}) : \mathrm{GM}(\underline{\underline{\Bbbk}}_X) \rightsquigarrow \mathrm{Vec}(\Bbbk)\,,$$

is the composition of the two exact functors $\mathcal{A} \otimes_{\Bbbk} (\_)$ and $\mathbf{Hom}^{\bullet}_{\mathcal{A}}(\_, \mathcal{I})$, since

$$\mathbf{Hom}^{\bullet}_{\underline{\underline{\Bbbk}}_X}(\_, \mathcal{I}) = \mathbf{Hom}^{\bullet}_{\mathcal{A}}(\mathcal{A} \otimes_{\Bbbk} (\_), \mathcal{I})\,,$$

and is therefore exact. Hence, the sheaf of vector spaces $\mathcal{I}$ is injective.
(3) let $\mathcal{M} \in \mathrm{GM}(\mathcal{A})$. An injective resolution $\mathcal{M} \to \mathcal{I}_{\star}$ in $\mathrm{GM}(\mathcal{A})$ is automatically an injective resolution in $\mathrm{GM}(\underline{\underline{\Bbbk}}_X)$ after (2), hence

$$I\!R\,\Gamma(X; \mathcal{M}) = \Gamma(X; \mathcal{I}_{\star}) =_1 \mathbf{Hom}^{\bullet}_{\mathcal{A}}(\mathcal{A}, \mathcal{I}_{\star}) = I\!R\,\mathbf{Hom}^{\bullet}_{\mathcal{A}}(\mathcal{A}; \mathcal{M})\,,$$

where $(=_1)$ was justified in Sect. B.5.1.
(4) In the same way as in (2), the adjunction equality B.5.2.2-(B.10)

$$\mathbf{Hom}^{\bullet}_{\mathcal{A}(X)}(\_, \mathcal{I}(X)) = \mathbf{Hom}^{\bullet}_{\mathcal{A}}(\mathcal{A} \otimes (\_), \mathcal{I})\,,$$

expresses the fact that the left-hand functor is the composition of the localization functor $\mathcal{A} \otimes (\_)$ which is exact (Proposition B.5.3.1-(1)), and the functor $\mathbf{Hom}^{\bullet}_{\mathcal{A}}(\_, \mathcal{I})$ which is also exact. Hence, the $\mathcal{A}(X)$-gm $\mathcal{I}(X)$ is injective.
(5) Let $\mathcal{N} \to \mathcal{I}_{\star}$ be an injective resolution in $\mathrm{GM}(\mathcal{A})$, then

$$\begin{aligned} I\!R\,\mathbf{Hom}^{\bullet}_{\mathcal{A}}(\mathcal{M}, \mathcal{M}) &=_1 \mathbf{Hom}^{\bullet}_{\mathcal{A}}(\mathcal{M}, \mathcal{I}_{\star}) \\ &=_2 \mathbf{Hom}^{\bullet}_{\mathcal{A}(X)}(\mathcal{M}(X), \mathcal{I}_{\star}(X)) \\ &=_3 I\!R\,\mathbf{Hom}^{\bullet}_{\mathcal{A}(X)}(\mathcal{M}(X), \mathcal{N}(X)) \end{aligned}$$

where $(=_1)$ is by definition of derived functor, $(=_2)$ is (B.13) and $(=_3)$ again by definition of derived functor, since $\mathcal{N}(X) \to \mathcal{I}_{\star}(X)$ is an injective resolution after (1) and (2). □

B.6.2 Family of Supports[8]

This is the name given to a collection Φ of subspaces on a topological space Y, verifying the following conditions:

Φ-1. every $S \in \Phi$ is a closed subset of Y;
Φ-2. if F is closed in Y and $F \subseteq S \in \Phi$, then $F \in \Phi$;
Φ-3. if $S_1, S_2 \in \Phi$, then $S_1 \cup S_2 \in \Phi$.

---

[8]Bredon [20] §I.6 Supports (p. 21), or Godement [46] §II.2.5 (p. 133) and §II.3.2 Espaces paracompacts (p. 150).

The family is called *paracompactifying* if, in addition,

Φ-4. every $S \in \Phi$ is a paracompact space;
Φ-5. every $S \in \Phi$ has a neighborhood in Y belonging to Φ.

If $\mathcal{F}$ is a sheaf of vector spaces on Y, the set

$$\Gamma_\Phi(Y;\mathcal{F}) := \big\{\sigma \in \Gamma(Y;\mathcal{F}) \mid |\sigma| \in \Phi\big\}. \tag{B.14}$$

is called *the set of global sections with support in* Φ.

Familiar equivalent notations for $\Gamma_\Phi(Y; \_)$, for particular families Φ, are

- $\Gamma(Y, \_)$, for the paracompactifying family of closed subspaces of Y,
- $\Gamma_c(Y, \_)$, for the paracompactifying family of compact subspaces of Y,
- $\Gamma_Z(Y, \_)$, for the family of closed subspaces of a closed subspace $Z \subseteq Y$.

The set $\Gamma_\Phi(Y;\mathcal{F})$ is vector subspace of $\Gamma(Y;\mathcal{F})$, and if $\nu : \mathcal{F} \to \mathcal{G}$ is a morphism of sheaves of vector spaces, then $\nu(Y)(\Gamma_\Phi(Y;\mathcal{F})) \subseteq \Gamma_\Phi(Y;\mathcal{G})$, and we have an induced homomorphism $\Gamma_\Phi(Y;\nu) : \Gamma_\Phi(Y;\mathcal{F}) \to \Gamma_\Phi(Y;\mathcal{G})$. In this way, we get a left exact functor

$$\Gamma_\Phi(Y; \_) : \mathrm{GM}(\underline{\mathbb{k}}_Y) \rightsquigarrow \mathrm{Vec}(\mathbb{k}),$$

its derived functor

$$I\!R\,\Gamma_\Phi(Y; \_) : \mathcal{D}^+\,\mathrm{GM}(\underline{\underline{\mathbb{k}}}_Y) \rightsquigarrow \mathcal{D}\mathrm{Vec}(\mathbb{k}),$$

and the corresponding cohomology functor $H_\Phi(Y; \_) := \boldsymbol{h}\,I\!R\,\Gamma_\Phi(Y; \_)$

$$H_\Phi(Y; \_) := \mathcal{D}^+\,\mathrm{GM}(\underline{\underline{\mathbb{k}}}_Y) \rightsquigarrow \mathrm{Vec}(\mathbb{k})^{\mathbb{N}}.$$

Definition B.6.2.1 A sheaf $\mathcal{I} \in \mathrm{GM}(\underline{\underline{\mathbb{k}}}_Y)$ is said to be $\Gamma_\Phi(Y; \_)$-acyclic is

$$H^i_\Phi(Y;\mathcal{I}) = 0, \quad \forall i > 0.$$

Recall that if $\mathcal{F} \to \mathcal{I}_\star$ is an *acyclic resolution* (Sect. A.2.4), then

$$I\!R\,\Gamma_\Phi(Y;\mathcal{F}) = \Gamma_\Phi(Y;\mathcal{I}_\star).$$

B.6.3 The functor Γ_Φ in GM($\mathcal{A}(X)$)

Let X be a mild space. We are interested in two cases

$\mathcal{O}$-1. Φ is a family of supports of X, and $\mathcal{O}_X$ is the sheaf $\underline{\underline{\Omega}}^0_{X,\mathbb{k}}$ of all functions with values in the field $\mathbb{k}$.

O-2. Φ is a paracompactifying family of X, and

a. O_X is the sheaf of continuous real valued functions;
b. X is a manifold and O_X is the sheaf of differentiable real valued functions.

Lemma B.6.3.1 *In both cases O-1 and O-2, the set of functions*

$$O_\Phi(X) := \Gamma_\Phi(X; O_X)$$

verifies the following properties.

1. *The set $O_\Phi(X)$ is an ideal in $O(X)$ such that Φ is the set of closed subsets $S \subseteq X$ verifying $S \subseteq \{\phi = 1\}$ for some $\phi \in O_\Phi(X)$, in symbols:*

$$S \in \Phi \Leftrightarrow \big(S \text{ is closed in } X \text{ and } (\exists \phi \in O_\Phi(X))(S \subseteq \{\phi = 1\})\big).$$

2. *For every O_X-gm $\mathcal{M}$, we have*

$$\Gamma_\Phi(X; \mathcal{M}) = \{m \in \mathcal{M}(X) \mid \exists \phi \in O_\Phi(X) \ \text{s.t.} \ \phi m = m\}. \tag{B.15}$$

Proof (1) The implication '$\Leftarrow$' is the condition B.6.2-(Φ-2), since $S \subseteq |\phi|$.
For the converse, let $S \in \Phi$. We distinguish the two cases:

Case O-1. Take $\phi := \mathbf{1}_S : X \to \{0, 1\} \subseteq \Bbbk$, the *indicator function* of S.
Case O-2. By B.6.2-(B.6.2), there exists a neighborhood $T \supseteq S$ in Φ. Denote by T° the interior of T in X. Since X is perfectly normal, there exists $\phi : X \to \mathbb{R}$ continuous (resp. differential) such that $S \subseteq \{\phi = 1\}$ and $|\phi| \subseteq T^\circ$. Hence, $|\phi| \subseteq T \in \Phi$ and $\phi \in O_\Phi(X)$, as needed.

(2) The inclusion '$\supseteq$' is clear since, if $m = \phi m$, then $|m| \subseteq |\phi| \in \Phi$. For the converse, we apply (1). If $m \in \mathcal{M}(X)$ verifies $|m| \in \Phi$, we choose $\phi \in O_\Phi(X)$ such that $|m| \subseteq \{\phi = 1\}$, in which case $\phi m = m$, since $\phi(x) = 1$, for all x such that $m_x \neq 0$. □

Definition B.6.3.2 Let $\mathcal{A}$ be an O_X-ga. Inspired by equality B.15, we define for $M \in \mathrm{GM}(\mathcal{A}(X))$,

$$\Gamma_\Phi(M) := \{m \in M \mid \exists \phi \in O_\Phi(X) \ \text{s.t.} \ \phi m = m\}. \tag{B.16}$$

This is a sub-$\mathcal{A}(X)$-gm of M. If $\nu : M \to N$ is a morphism of $\mathcal{A}(X)$-gm's, we have $\nu(\Gamma_\Phi M) \subseteq \Gamma_\Phi N$, and then denote by $\Gamma_\Phi \nu : \Gamma_\Phi M \to \Gamma_\Phi N$ the induced morphism. In this way, we get an additive functor

$$\Gamma_\Phi : \mathrm{GM}(\mathcal{A}(X)) \rightsquigarrow \mathrm{GM}(\mathcal{A}(X)). \tag{B.17}$$

Lemma B.6.3.3 *The functor Γ_Φ is exact.*

Proof Given an exact sequence of $\mathcal{A}(X)$-gm's $0 \to L \to M \to N \to 0$, it is immediate that the sequence $0 \to \Gamma_\Phi L \to \Gamma_\Phi M \to \Gamma_\Phi N$ is exact. We thus need only justify that if $\nu : M \twoheadrightarrow N$ is a surjective morphism of $\mathcal{A}(X)$-gm's, then $\Gamma_\Phi \nu$ is surjective. Indeed, an element $n \in \Gamma_\Phi N$ verifies $\phi\, n = n$ for some $\phi \in \mathcal{O}_\Phi(X)$, and if $n = \nu(m)$, we have $n = \phi\, n = \nu(\phi\, m)$. This leads us to check that $\phi\, m \in \Gamma_\Phi M$. But, since $|\phi| \in \Phi$, we can apply Lemma B.6.3.1-(1) and state that there exists $\phi' \in \mathcal{O}_\Phi(X)$ verifying $|\phi| \subseteq \{\phi' = 1\}$, in which case $\phi'\,\phi\, m = \phi\, m$, and $\phi\, m \in \Gamma_\Phi M$ by definition (B.16), as announced. □

Theorem B.6.3.4 (Acyclicity of $\mathcal{O}_X$-Modules) *Let X be a mild space.*

1. *In both cases B.6.3-(O-1,O-2), every $\mathcal{O}_X$-gm is $\Gamma_\Phi(X; \_)$-acyclic.*
2. *Let (E, X, π, F) be a fiber bundle of mild spaces.*
   a. *For every $i \in \mathbb{N}$, the $\mathcal{O}_X$-gm's $\pi_!\, \underline{\underline{\Omega}}^i_{E,\Bbbk}$ and $\pi_*\, \underline{\underline{\Omega}}^i_{E\,\Bbbk}$ are $\Gamma_\Phi(X; \_)$-acyclic for every family of supports Φ.*
   b. *There are canonical identifications in $\mathcal{D}(X; \Bbbk)$:*

$$\begin{cases} I\!R\, \Gamma_\Phi(X; \pi_!\, \underline{\underline{\Omega}}^i_{E,\Bbbk}) = \Gamma_\Phi(X; \pi_!\, \underline{\underline{\Omega}}^i_{E,\Bbbk}) \\ I\!R\, \Gamma_\Phi(X; \pi_*\, \underline{\underline{\Omega}}^i_{E,\Bbbk}) = \Gamma_\Phi(X; \pi_*\, \underline{\underline{\Omega}}^i_{E,\Bbbk}) \,. \end{cases}$$

3. *Let (E, X, π, F) be a fiber bundle of manifolds.*
   a. *For every $i \in \mathbb{N}$, the $\mathcal{O}_X$-gm's $\pi_!\, \underline{\underline{\Omega}}^i_E$ and $\pi_*\, \underline{\underline{\Omega}}^i_E$ are $\Gamma_\Phi(X; \_)$-acyclic for every paracompactifying family of supports Φ.*
   b. *There are canonical identifications in $\mathcal{D}(X; \mathbb{R})$:*

$$\begin{cases} I\!R\, \Gamma_\Phi(X; \pi_!\, \underline{\underline{\Omega}}^i_E) = \Gamma_\Phi(X; \pi_!\, \underline{\underline{\Omega}}^i_E) \\ I\!R\, \Gamma_\Phi(X; \pi_*\, \underline{\underline{\Omega}}^i_E) = \Gamma_\Phi(X; \pi_*\, \underline{\underline{\Omega}}^i_E) \end{cases}$$

Proof (1) On the category GM($\mathcal{A}$), the functor $\Gamma_\Phi(X; \_)$ coincides with the composition of the functors $\mathbf{Hom}^\bullet_{\mathcal{A}}(\mathcal{A}, \_)$ and Γ_Φ, both exact respectively by Theorem B.6.1.1-(1) and Lemma B.6.3.3. Therefore, likewise Theorem B.6.1.1-(2), if $\mathcal{M} \to \mathcal{I}_\star$ is an injective resolution in GM($\mathcal{A}$), we get $I\!R\, \Gamma_\Phi(X; \mathcal{M}) = \Gamma_\Phi(X; \mathcal{I}_\star) = \Gamma_\Phi(X; \mathcal{M})$ since injective graded modules in GM($\mathcal{A}$) are injective in GM($\underline{\Bbbk}_X$).

We therefore have, for all $i > 0$:

$$\boldsymbol{h}^i\, I\!R\, \Gamma_\Phi(X; \mathcal{M}) = \boldsymbol{h}^i\, \Gamma_\Phi \Gamma(X; \mathcal{I}_\star) =_1 \boldsymbol{h}^i\, \Gamma_\Phi \Gamma(X; \mathcal{M}) = \boldsymbol{h}^i\, \Gamma_\Phi(X; \mathcal{M}) = 0\,,$$

($=_1$) since $\Gamma(X; \_)$ is exact (Theorem B.6.1.1-(1)).
(2) The complexes $\pi_!\, \underline{\underline{\Omega}}_{E,\Bbbk}$ and $\pi_!\, \underline{\underline{\Omega}}_{E,\Bbbk}$ are graded modules over $\underline{\underline{\Omega}}_{X,\Bbbk}$ (Sect. 8.2), hence over over $\mathcal{O}_X := \underline{\underline{\Omega}}^0_{X,\Bbbk}$. We can then conclude, by (1), that they are $\Gamma_\Phi(X; \_)$-acyclic. The last part of (2), is a straightforward application of the general fact concerning complexes of acyclic objects of Proposition A.2.4.2.

(3) The complexes $\pi_! \underline{\underline{\Omega}}_E$ and $\pi_! \underline{\underline{\Omega}}_E$ are graded modules over $\underline{\underline{\Omega}}_y$ (Sect. 3.1.10.2). The proof is then same as for (2) with appropriate modifications. □

B.7 $\mathcal{O}_X$-Differential Graded Algebras

We now extend the validity of Theorems B.6.1.1 and B.6.3.4 about $\mathcal{O}_X$-graded algebras $\mathcal{A}$ and their modules, to the same statements about $\mathcal{O}_X$-dga's $(\mathcal{A}, d)$ and their dgm's. As the reader will notice, the approach is the same, with only heavier notations …

B.7.1 Localization Functor for $\mathcal{O}_X$-DGA's

The localization presheaf functor defined in B.5.2.1-(B.7) for any sheaf of graded algebras $\mathcal{A}$ on a topological space Y, i.e. the functor

$$\mathcal{A} \underline{\otimes} (\_) = \mathrm{GM}(\mathcal{A}(Y)) \rightsquigarrow \mathrm{Mod}_{\mathrm{presh.}}(\mathcal{A})$$

can be modified in order to incorporate differential structures. This is done in the restriction formula (B.6), where for $V \subseteq U \subseteq X$, we now define:

$$(\mathcal{A}(U), d) \otimes_{\mathcal{A}(X)} (M, d) \xrightarrow{\rho_V^U \otimes \mathrm{id}_M} (\mathcal{A}(V), d) \otimes_{\mathcal{A}(X)} (M, d) \tag{B.18}$$

where $\rho_V^U : \mathcal{A}(U) \to \mathcal{A}(V)$ is always the restriction map, but where the tensor product is now the tensor product *in categories of dgm's*.

We get in this way, exactly as in Sect. B.5.2.1 for $\mathcal{A}$-gm's, a presheaf of $(\mathcal{A}, d)$-dgm's on Y, which we denote by $(\mathcal{A}, d) \underline{\otimes} (M, d)$. The construction is functorial on (M, d), and the resulting functor

$$(\mathcal{A}, d) \underline{\otimes} (\_) = \mathrm{DGM}((\mathcal{A}, d)(X)) \rightsquigarrow \mathrm{DGM}_{\mathrm{presh.}}(\mathcal{A}, d) \tag{B.19}$$

is the $(\mathcal{A}, d)$*-localization presheaf functor*.

It verifies, by construction, the adjunction equality:

$$\mathbf{Hom}^{\bullet}_{\mathcal{A}}((\mathcal{A}, d) \underline{\otimes} (M, d), (\mathcal{N}, d)) = \mathbf{Hom}^{\bullet}_{\mathcal{A}(X)}((M, d), (\mathcal{N}, d)(Y)), \tag{B.20}$$

since a morphism of presheaves $\nu : (\mathcal{A}, d) \underline{\otimes} (M, d) \to (\mathcal{N}, d)$ is, by (B.18), a family of morphisms

$$\left\{\nu(U) : \mathcal{A}(U) \otimes_{\mathcal{A}(X)} (M, d) \to (\mathcal{N}, d)(U)\right\}_{U \subseteq X},$$

making commutative the diagrams

$$\begin{array}{ccc} (M,d) & \xrightarrow{\quad \nu(Y) \quad} & (\mathcal{N},d)(Y) \\ \downarrow & \oplus & \downarrow \rho_U^Y \\ (\mathcal{A},d)(U) \otimes_{\mathcal{A}(X)} (M,d) & \xrightarrow{\quad \nu(U) \quad} & (\mathcal{N},d)(U) \,, \end{array}$$

Which means that $\nu(U)(a \otimes m) = a\, \nu(U)(1 \otimes m) = a\, \rho_U^Y(\nu(Y)(m))$, showing that ν is uniquely determined by $\nu(Y)$.

Definition B.7.1.1 Let $(\mathcal{A}, d)$ be a sheaf of differential graded algebras on a topological space Y. The $(\mathcal{A}, d)$-*localization functor*,

$$(\mathcal{A}, d) \otimes (\_) = \mathrm{DGM}((\mathcal{A}, d)(Y)) \rightsquigarrow \mathrm{DGM}(\mathcal{A}, d)\,. \tag{B.21}$$

is the composition of the localization presheaf functor $(\mathcal{A}, d)\underline{\otimes}(\_)$ (B.19), followed by the exact functor '*associated sheaf to a presheaf*'.

The following is the analogue to Proposition B.5.2.2.

Proposition B.7.1.2 *The natural map*

$$\begin{array}{ccc} \mathbf{Hom}^\bullet_{\mathcal{A}}((\mathcal{A}, d) \otimes M, \mathcal{N}) & \to & \mathbf{Hom}^\bullet_{\mathcal{A}(Y)}(M, \mathcal{N}(Y)) \\ \nu & \mapsto & \nu(Y) \end{array} \tag{B.22}$$

is an isomorphism, functorial in $M \in \mathrm{DGM}((\mathcal{A}, d)(Y))$ *and* $\mathcal{N} \in \mathrm{DGM}(\mathcal{A}, d)$.

The functors

$$\begin{cases} \Gamma(Y; \_) = \mathbf{Hom}^\bullet_{\mathcal{A}}((\mathcal{A}, d), \_) : \mathrm{DGM}(\mathcal{A}, d) \rightsquigarrow \mathrm{DGM}((\mathcal{A}, d)(Y)) \\ (\mathcal{A}, d) \otimes (\_) : \mathrm{DGM}((\mathcal{A}, d)(Y)) \rightsquigarrow \mathrm{DGM}(\mathcal{A}, d) \end{cases}$$

constitute a pair of adjoint functors.

Proof Same as for Proposition B.5.2.2. □

B.8 The Localization Functor for $\mathcal{O}_X$-DGA's

As in Sect. B.5.3, we come back to the case of a mild space X and a $\mathcal{O}_X$-dga $(\mathcal{A}, d)$. Under these assumptions, we show that for every $(\mathcal{A}, d)$-dgm $(\mathcal{M}, d)$, the stalks $(\mathcal{M}, d)_x$ for $x \in X$, are determined in terms of the $(\mathcal{A}, d)(X)$-dgm $(\mathcal{M}, d)(X)$ alone. The following proposition is the exact analogue to Proposition B.5.3.1.

Proposition B.8.1 *Let* $(\mathcal{A}, d)$ *be an* $\mathcal{O}_X$*-dga on a mild space* X.

For $x \in X$, denote by S_x the multiplicative system of functions in $O_X(X)$ equal to 1 in a neighborhood of x.

1. *Let $(\mathcal{M}, d) \in \mathrm{DGM}(\mathcal{A}, d)$. For $x \in X$, the germ map $(\mathcal{M}, d)(X) \to (\mathcal{M}, d)_x$ is a morphism of $(\mathcal{A}, d)(X)$-dgm's which factors through the localization map $(\mathcal{M}, d)(X) \to S_x^{-1}(\mathcal{M}, d)(X)$ in a isomorphism of $(\mathcal{A}, d)(X)$-dgm's*

$$\xi_x(\mathcal{M}, d) : S_x^{-1}(\mathcal{M}, d)(X) \to (\mathcal{M}, d)_x ,$$

which is functorial in $(\mathcal{M}, d)$.

2. *The localization functor*

$$(\mathcal{A}, d) \otimes (\_) : \mathrm{DGM}((\mathcal{A}, d)(X)) \rightsquigarrow \mathrm{DGM}(\mathcal{A}, d)$$

is an exact functor, i.e. respects acyclicity.

3. *Let $(\mathcal{M}, d) \in \mathrm{DGM}(\mathcal{A}, d)$. The natural map of $(\mathcal{A}, d)$-modules*

$$(\mathcal{A}, d) \otimes (\mathcal{M}, d)(X) \to (\mathcal{M}, d)$$

is an isomorphism functorial in $(\mathcal{M}, d)$. In other words, the localization functor $(\mathcal{A}, d) \otimes (\_)$ is a left inverse of the global section functor $\Gamma(X; \_)$.

Proof Same as for Proposition B.5.3.1 with just two new details to make clear.

First, the extension of the dgm structure of an $(\mathcal{A}, d)(X)$-dgm M to the localized module $S_x^{-1}M$, which is given by the well-known Leibniz rule for fractions

$$d\Big(\frac{m}{s}\Big) = \frac{dm}{s} - \frac{m}{s^2}\,d(s) ,$$

for all $m \in M$ and $s \in S_x$.

Second, in (2), the definition of exact functors between dgm's categories, which requires that acyclic dgm's must remain acyclic. In our case, if (M, d) is acyclic, we have to show that the sheaf of dgm's $(\mathcal{A}, d) \otimes (M, d)$ is acyclic. This concerns only stalks, so that we are lead to check that $(\mathcal{A}, d)_x \otimes (M, d) = S_x^{-1}(M, d)$ is acyclic. A germ at $x \in X$ of a fraction $\frac{m}{s}$ coincides with $1_x \otimes m$ since $s_x = 1_x$. Hence, $d(\frac{m}{s})_x = 0$ if and only if $d\,m = 0$, in which case $m = d\,m'$ for some $m' \in M$ since (M, d) is acyclic, but then $d(1_x \otimes m') = 1_x \otimes m$. □

B.9 Equivalences of Derived Functors in $\mathcal{D}$DGM($\mathcal{A}$, d)

B.9.1 *K*-Injective Differential Graded Modules

In Sect. A.3, we reviewed the concept of the derived functor of an additive functor

$$F : \mathcal{K}\mathrm{DGM}(\mathcal{A}, d) \to \mathcal{K}\mathrm{DGM}(\mathcal{B}, d)$$

between homotopy categories of categories of dgm's over dga's on a ringed space.

The definition is based on the Existence Theorem A.3.1.1, which states that for every $\mathcal{M} \in \mathrm{DGM}(\mathcal{A}, d)$, there exists a quasi-isomorphism $\mathcal{M} \to \mathcal{I}$ where $\mathcal{I}$ is injective as $\mathcal{A}$-gm and is K-injective as $\mathcal{A}$-dgm. In that case the derived functor which we denote by

$$I\!R\, F : \mathcal{D}\mathrm{DGM}(\mathcal{A}, d) \to \mathcal{D}\mathrm{DGM}(\mathcal{B}, d)\,,$$

is the $(\mathcal{B}, d)$-dgm

$$I\!R\, F(\mathcal{M}) := F(\mathcal{I})\,. \tag{B.23}$$

In this definition, an $(\mathcal{A}, d)$-dgm $\mathcal{I}$ is said to be K-injective (Sect. A.3.1), if, for every *acyclic* $(\mathcal{A}, d)$-dgm $\mathcal{N}$, one has

$$\mathbf{Hom}^\bullet_{K(\mathcal{A},d)}(\mathcal{N}, \mathcal{I}) = 0\,,$$

where '$K(\mathcal{A}, d)$' is an abbreviation of the notation '$K\mathrm{DGM}(\mathcal{A}, d)$' of the homotopy category of the category of $\mathrm{DGM}(\mathcal{A}, d)$.

Proposition B.9.1.1 *Let $(\mathcal{A}, d)$ be an $\mathcal{O}_X$-dga on a mild space X.*

1. *For all $\mathcal{M}, \mathcal{N} \in \mathrm{DGM}(\mathcal{A}, d)$, the canonical map*

$$I\!R\, \Gamma\big(X; I\!R\, \underline{\underline{\boldsymbol{Hom}}}^\bullet_{\mathcal{D}(\mathcal{A},d)}(\mathcal{M}, \mathcal{N})\big) \to I\!R\, \mathbf{Hom}^\bullet_{\mathcal{D}(\mathcal{A},d)}(\mathcal{M}, \mathcal{N})$$

   is an isomorphism in $\mathcal{D}\mathrm{DGM}(\mathcal{A}(X), d)$.
2. *If the natural homomorphism of sheaves of algebras $\iota : \underline{\underline{\Bbbk}}_X \to (\mathcal{A}, d)$ is a quasi-isomorphism, the induced homomorphism of bifunctors*

$$\big(I\!R\, \underline{\underline{\boldsymbol{Hom}}}^\bullet_{\mathcal{D}(\mathcal{A},d)}(\_, \_) \to I\!R\, \underline{\underline{\boldsymbol{Hom}}}^\bullet_{\Bbbk}(\_, \_)\big) : \mathcal{D}\mathrm{DGM}(\mathcal{A}, d)^2 \rightsquigarrow \mathcal{D}\mathrm{DGM}(\underline{\underline{\Bbbk}}_X, d)\,,$$

   is an isomorphism.

Proof (1) Let $\mathcal{N} \to \mathcal{I}$ be a quasi-isomorphism with $\mathcal{I}$ injective as $\mathcal{A}$-gm, and K-injective as $(\mathcal{A}, d)$-dgm. We have, by (B.23),

$$I\!R\, \underline{\underline{\boldsymbol{Hom}}}^\bullet_{\mathcal{D}(\mathcal{A},d)}(\mathcal{M}, \mathcal{N}) = \underline{\underline{\boldsymbol{Hom}}}^\bullet_{\mathcal{A}}(\mathcal{M}, \mathcal{I})\,,$$

were we claim that $\underline{\underline{\boldsymbol{Hom}}}^\bullet_{\mathcal{A}}(\mathcal{M}, \mathcal{I})$ is a complex of injective sheaves.

Indeed, this is equivalent to the fact that the functor

$$\mathbf{Hom}_{\underline{\underline{\Bbbk}}_X}\big(\_, \underline{\underline{\boldsymbol{Hom}}}^\bullet_{\mathcal{A}}(\mathcal{M}, \mathcal{I})\big) : \mathrm{GM}(\underline{\underline{\Bbbk}}_X) \to \mathrm{GM}(\Bbbk)$$

is exact, which follows immediately from the well-known identification

$$\mathbf{Hom}_{\underline{\underline{\Bbbk}}_X}\big(\_, \underline{\underline{\boldsymbol{Hom}}}^\bullet_{\mathcal{A}}(\mathcal{M}, \mathcal{I})\big) = \mathbf{Hom}_{\mathcal{A}}\big(\_ \otimes_{\underline{\underline{\Bbbk}}_X} \mathcal{M}, \mathcal{I}\big)$$

which shows that the left-hand functor is the composition of $(\_)\otimes_{\underline{\underline{\mathbb{R}}}_X}\mathcal{I}$, exact since $\mathcal{I}$ is $\underline{\underline{\mathbb{R}}}_X$)-flat, followed by $\mathrm{Hom}_{\mathcal{A}}(\_,\mathcal{I})$ exact since $\mathcal{I}$ is an injective $\mathcal{A}$-gm.

We therefore have:

$$\begin{aligned}\mathbb{R}\,\Gamma\big(X;\mathbb{R}\,\underline{\underline{\boldsymbol{Hom}}}^{\bullet}_{\mathcal{D}(\mathcal{A},d)}(\mathcal{M},\mathcal{N})\big)&=\mathbb{R}\,\Gamma\big(X;\underline{\underline{\boldsymbol{Hom}}}^{\bullet}_{\mathcal{A}}(\mathcal{M},\mathcal{I})\big)\\&=_1\Gamma\big(X;\underline{\underline{\boldsymbol{Hom}}}^{\bullet}_{\mathcal{A}}(\mathcal{M},\mathcal{I})\big)\\&=\mathbf{Hom}^{\bullet}_{\mathcal{A}}(\mathcal{M},\mathcal{I})=\mathbb{R}\,\mathbf{Hom}^{\bullet}_{\mathcal{A}}(\mathcal{M},\mathcal{N})\end{aligned}$$

where $(=_1)$ is justified by the injectivity of $\underline{\underline{\boldsymbol{Hom}}}^{\bullet}_{\mathcal{A}}(\mathcal{M},\mathcal{I})$.
(2) Let $(\mathcal{I},d)\in\mathrm{DGM}(\mathcal{A},d)$ be K-injective and such that $\mathcal{I}$ is an injective $\mathcal{A}$-gm. Since $\mathcal{I}$ is an injective $\underline{\underline{\Bbbk}}_X$-gm after Theorem B.6.1.1-(2), we have a natural adjunction morphism of exact functors,

$$\mathbf{Hom}^{\bullet}_{\underline{\underline{\Bbbk}}_X}\big(\_,(\mathcal{I},d)\big)\to\mathbf{Hom}^{\bullet}_{\mathcal{A}}\big((\mathcal{A},d)\otimes_{\underline{\underline{\Bbbk}}_X}(\_),(\mathcal{I},d)\big)$$

which is an isomorphism of $\Bbbk$-dgm's since it respects differentials and since it is already an isomorphism at the underlying level of graded modules.

As a consequence, we get a natural isomorphism of bifunctors

$$\mathbb{R}\,\mathbf{Hom}^{\bullet}_{\underline{\underline{\Bbbk}}_X}(\_,\_)\to\mathbb{R}\,\mathbf{Hom}^{\bullet}_{\mathcal{D}(\mathcal{A},d)}\big((\mathcal{A},d)\otimes_{\underline{\underline{\Bbbk}}_X}(\_),\_\big)$$

which leads us to prove that the morphism of bifunctors on $\mathrm{DGM}(\mathcal{A})$

$$\mathbb{R}\,\mathbf{Hom}^{\bullet}_{(\mathcal{D}(\mathcal{A}),d)}(\_,\_)\to\mathbb{R}\,\mathbf{Hom}^{\bullet}_{(\mathcal{D}(\mathcal{A}),d)}\big(\mathcal{A}\otimes_{\underline{\underline{\Bbbk}}_X}(\_),\_\big),$$

induced by the morphisms of $\mathcal{A}$-dgm's

$$\mathcal{A}\otimes_{\underline{\underline{\Bbbk}}_X}(\mathcal{M},d)\to(\mathcal{M},d)\,,\quad a\otimes m\mapsto a\,m\,,$$

is an isomorphism. This is indeed the case, since it suffices to look at the level of stalks, where the morphism $(\mathcal{A},d)_x\otimes_k(\mathcal{F},d)_x\to(\mathcal{F},d)_x$ is easily seen to a quasi-isomorphism after the well-known Künneth's theorem. □

The following theorem extends the validity of Theorem B.6.1.1 to dgm's.

Theorem B.9.1.2 *Let $(\mathcal{A},d)$ be an $\mathcal{O}_X$-dga on a mild space X.*

1. *The global section functor*

$$\Gamma(X;\_)=:\mathrm{DGM}(\mathcal{A},d)\to\mathrm{DGM}\big(\mathcal{A}(X),d\big)$$

   is fully faithful.
2. *If $(\mathcal{I},d)$ is a K-injective object in* $\mathrm{DGM}(\mathcal{A},d)$, *then $(\mathcal{I},d)(X)$ is a K-injective object in* $\mathrm{DGM}(\mathcal{A}(C),d)$.

3. *For all* $\mathcal{M}, \mathcal{N} \in \mathrm{DGM}(\mathcal{A}, d)$, *the canonical map*

$$I\!R\,\mathbf{Hom}^{\bullet}_{\mathcal{D}(\mathcal{A},d)}(\mathcal{M}, \mathcal{N}) \to I\!R\,\mathbf{Hom}^{\bullet}_{\mathcal{D}(\mathcal{A}(X),d)}(\mathcal{M}(X), \mathcal{N}(X))$$

is an isomorphism in $\mathcal{D}\mathrm{DGM}(\mathcal{A}(X), d)$.

Proof Same as for Theorem B.6.1.1, with few new justifications needed.
(1) Immediate after Proposition B.8.1-(2,3)
(2) By the adjunction property (Proposition B.7.1.2), we can identify functors

$$\mathbf{Hom}^{\bullet}_{\mathcal{A}(X)}(\_, (\mathcal{I}, d)(X)) = \mathbf{Hom}^{\bullet}_{\mathcal{A}}((\mathcal{A}, d) \otimes (\_), (\mathcal{I}, d)), \tag{B.24}$$

and conclude that the left-hand functor in (B.24) respects acyclicity, since $\mathcal{A} \otimes (\_)$ do respects acyclicity by Proposition B.8.1-(2), and the same for $\mathbf{Hom}^{\bullet}_{\mathcal{A}}(\_, (\mathcal{I}, d))$ since $(\mathcal{I}, d)$ is K-injective. Hence, $(\mathcal{I}, d)(X)$ is K-injective in $\mathrm{DGM}((\mathcal{A}, d)(X))$.
(3) Follows straightforwardly by simply applying the definition of derived functors in $\mathrm{DGM}(\mathcal{A}, d)$ as recalled in Sect. A.3. let $\mathcal{N} \to \mathcal{I}$ be a quasi-isomorphism where $\mathcal{I}$ is an injective $\mathcal{A}$-gm and a K-injective $\mathcal{A}$-dgm. Then

$$\begin{aligned} I\!R\,\mathbf{Hom}^{\bullet}_{\mathcal{D}(\mathcal{A},d)}(\mathcal{M}, \mathcal{N}) &=_1 \mathbf{Hom}^{\bullet}_{\mathcal{A}}(\mathcal{M}, \mathcal{I}) =_2 \mathbf{Hom}^{\bullet}_{\mathcal{A}(X)}(\mathcal{M}(X), \mathcal{I}(X)) \\ &=_3 I\!R\,\mathbf{Hom}^{\bullet}_{\mathcal{D}(\mathcal{A}(X),d)}(\mathcal{M}(X), \mathcal{N}(X)) \end{aligned}$$

($=_1$) by definition of derived functor, ($=_2$) by (1), and ($=_3$) by definition of derived functor since $(\mathcal{I}, d)(X)$ is an injective $\mathcal{A}(X)$-gm after Theorem B.6.1.1-(4), and a K-injective $(\mathcal{A}, d)(X)$-dgm after (2). □

Corollary B.9.1.3 *1. Let* (E, X, π, F) *be a fiber bundle of* <u>*mild spaces*</u>.

a. The following morphisms induced by the augmentation $\epsilon : \underline{\underline{\Bbbk}}_X \to \underline{\underline{\Omega}}_{X;\Bbbk}$

$$\begin{array}{ccc} I\!R\,\underline{\underline{\mathbf{Hom}}}^{\bullet}_{\mathcal{D}(\underline{\underline{\Omega}}_{X,\Bbbk},d)}(\pi_!\underline{\underline{\Omega}}_{E,\Bbbk}, \underline{\underline{\Omega}}_{X,\Bbbk}) & \to & I\!R\,\underline{\underline{\mathbf{Hom}}}^{\bullet}_{\underline{\underline{\Bbbk}}_X}(\pi_!\underline{\underline{\Omega}}_{E,\Bbbk}, \underline{\underline{\Omega}}_{X,\Bbbk}) \\ & & \uparrow \\ & & I\!R\,\underline{\underline{\mathbf{Hom}}}^{\bullet}_{\underline{\underline{\Bbbk}}_X}(\pi_!\underline{\underline{\Omega}}_{E,\Bbbk}, \underline{\underline{\Bbbk}}_X) \end{array}$$

are isomorphisms in $\mathcal{D}(X; \Bbbk)$.

b. The morphisms

$$\begin{array}{c} I\!R\,\mathbf{Hom}^{\bullet}_{\underline{\underline{\Bbbk}}_X}(\pi_!\underline{\underline{\Omega}}_{E,\Bbbk}, \underline{\underline{\Bbbk}}_X) \\ \uparrow \\ I\!R\,\mathbf{Hom}^{\bullet}_{\mathcal{D}(\underline{\underline{\Omega}}_{X,\Bbbk},d)}(\pi_!\underline{\underline{\Omega}}_{E,\Bbbk}, \underline{\underline{\Omega}}_{X,\Bbbk}) \\ \downarrow \\ I\!R\,\mathbf{Hom}^{\bullet}_{\mathcal{D}(\Omega(X,\Bbbk),d)}(\Omega_{\mathrm{cv}}(E, \Bbbk), \Omega(X, \Bbbk)) \end{array}$$

are isomorphisms in $\mathcal{D}\mathrm{Vec}(\Bbbk)$.

2. *Let* (E, X, π, F) *be a fiber bundle of* *<u>manifolds</u>.*

   a. *The following morphisms induced by the augmentation* $\epsilon : \underline{\underline{\mathbb{R}}}_X \to \underline{\underline{\Omega}}_X$

$$\begin{array}{ccc} \mathbb{R}\,\underline{\underline{\boldsymbol{Hom}}}^\bullet_{\mathcal{D}(\underline{\underline{\Omega}}_X, d)}(\pi_!\underline{\underline{\Omega}}_E, \underline{\underline{\Omega}}_X) & \to & \mathbb{R}\,\underline{\underline{\boldsymbol{Hom}}}^\bullet_{\underline{\underline{\mathbb{R}}}_X}(\pi_!\underline{\underline{\Omega}}_E, \underline{\underline{\Omega}}_X) \\ & & \uparrow \\ & & \mathbb{R}\,\underline{\underline{\boldsymbol{Hom}}}^\bullet_{\underline{\underline{\mathbb{R}}}_X}(\pi_!\underline{\underline{\Omega}}_{E,\Bbbk}, \underline{\underline{\Bbbk}}_X) \end{array}$$

   are isomorphisms in $\mathcal{D}(X; \mathbb{R})$.

   b. *The morphisms*

$$\begin{array}{c} \mathbb{R}\,\mathbf{Hom}^\bullet_{\underline{\underline{\Bbbk}}_B}(\pi_!\underline{\underline{\Omega}}_E, \underline{\underline{\Omega}}_B) \\ \uparrow \\ \mathbb{R}\,\mathbf{Hom}^\bullet_{\mathcal{D}(\underline{\underline{\Omega}}_B, d)}(\pi_!\underline{\underline{\Omega}}_E, \underline{\underline{\Omega}}_B) \\ \downarrow \\ \mathbb{R}\,\mathbf{Hom}^\bullet_{\mathcal{D}(\Omega(B), d)}(\Omega_{\mathrm{cv}}(E), \Omega(B)) \end{array}$$

   are isomorphisms in $\mathcal{D}\mathrm{Vec}(\mathbb{R})$.

Proof Immediate consequences of Theorem B.9.1.1-(2) and Proposition B.9.1.2-(3), since X is locally contractible and the morphism $\underline{\underline{\Bbbk}}_X \to \underline{\underline{\Omega}}_{X,\Bbbk}$ (resp. $\underline{\underline{\mathbb{R}}}_X \to \underline{\underline{\Omega}}_X$ for manifolds) is a quasi-isomorphism. □

Appendix C
Cartan's Theorem for $\mathfrak{g}$-dg-Ideals

In Cartan [27] §4, given a $\mathfrak{g}$-dg-algebra $E := (E, d, \theta, \iota)$, an element $\omega \in E$, is called *basic* if it is killed by every $\mathfrak{g}$-interior product and every $\mathfrak{g}$-derivation, i.e.

$$\iota(X)(\omega) = \theta(X)(\omega) = 0\,, \quad \forall X \in \mathfrak{g}\,.$$

The same definition clearly makes sense for any $\mathfrak{g}$-*dg-ideal* $K \subseteq E$. This is the name we give to any dg-ideal $(K, d) \subseteq (E, d)$ stable under the $\mathfrak{g}$-interior products $\iota(X)$, for all $X \in \mathfrak{g}$. The fact that $\theta(X) = d \circ \iota(X) + \iota(X) \circ d$ (Sect. 4.2.3) warrants that such K's are also stable under the action of $\mathfrak{g}$-derivations. We can therefore consider the sub-E^{bas}-module K^{bas}.

When E is a $\mathfrak{g}$-dg-algebra admitting an algebraic connection $f : W(\mathfrak{g}) \to E$ and $K \subseteq E$ is a proper $\mathfrak{g}$-dg-ideal, the composition of f with the canonical surjection $E \twoheadrightarrow E' := E/K$ is a connection for E', which naturally leads us to consider the following diagram which extends diagram (4.2) in Sect. 4.1.1.1

$$\begin{array}{ccccc}
K & \xrightarrow{\ \mathfrak{i}\ } & W(\mathfrak{g}) \otimes K & \xrightarrow{\ \mathfrak{f}\ } & K \\
\downarrow & & \downarrow & & \downarrow \\
E & \xrightarrow{\ \mathfrak{i}\ } & W(\mathfrak{g}) \otimes E & \xrightarrow{\ \mathfrak{f}\ } & E \\
\downarrow & & \downarrow & & \downarrow \\
E' & \xrightarrow{\ \mathfrak{i}\ } & W(\mathfrak{g}) \otimes E' & \xrightarrow{\ \mathfrak{f}\ } & E'
\end{array} \tag{C.1}$$

where, we recall, $\mathfrak{i}(\omega) := 1 \otimes \omega$ and $\mathfrak{f}(\alpha \otimes \omega) := f(\alpha)\omega$.

A. Arabia, *Equivariant Poincaré Duality on G-Manifolds*, Lecture Notes in Mathematics 2288, https://doi.org/10.1007/978-3-030-70440-7

The problem with this diagram is that when applying the functor $(\_)^{\text{bas}}$, we get the diagram of Cartan-Weil morphisms

$$
\begin{array}{ccccc}
K^{\text{bas}} & \xrightarrow{\bar{\mathfrak{i}}} & (W(\mathfrak{g}) \otimes K)^{\text{bas}} & \xrightarrow{\bar{\mathfrak{f}}} & K^{\text{bas}} \\
\downarrow & & \downarrow & & \downarrow \\
E^{\text{bas}} & \underset{\text{q.i.}}{\xrightarrow{\bar{\mathfrak{i}}}} & (W(\mathfrak{g}) \otimes E)^{\text{bas}} & \underset{\text{q.i.}}{\xrightarrow{\bar{\mathfrak{f}}}} & E^{\text{bas}} \\
\downarrow & & \downarrow & & \downarrow \\
E^{\prime,\text{bas}} & \underset{\text{q.i.}}{\xrightarrow{\bar{\mathfrak{i}}}} & (W(\mathfrak{g}) \otimes E')^{\text{bas}} & \underset{\text{q.i.}}{\xrightarrow{\bar{\mathfrak{f}}}} & E^{\prime,\text{bas}}
\end{array}
\tag{C.2}
$$

where, although in the second and third rows, Cartan-Weil morphisms are isomorphisms after Cartan's theorem 4.1.1.1, we cannot conclude the same for the first row since columns are generally not exact.

The way to avoid this issue is to replace the cokernel of the inclusion $\kappa : K \subseteq E$ by its mapping cone. Indeed, the cone in question :

$$E' := \hat{c}(\kappa) := (E \oplus K[1], \Delta),$$

is equipped with a structure of $\mathfrak{g}$-dg-algebra by defining a multiplication by

$$(x, y) \cdot (x, y') := (xx', xy' + yx' + yy'),$$

and an action of $\mathfrak{g}$-interior products by

$$\iota(X)(x, y) := (\iota(X)(x), \iota(X)(y)), \quad \forall X \in \mathfrak{g}.$$

These settings entail that $\mathfrak{g}$-derivations are given by the formula

$$\theta(X)(x, y) := (\theta(X)(x), \theta(X)(y)), \quad \forall X \in \mathfrak{g},$$

and the verification of the axioms of $\mathfrak{g}$-dg-algebras in Sect. 4.2.3 is almost immediate.

More importantly, an algebraic connection $f : W(\mathfrak{g}) \to E$ composed with the inclusion $\iota : E \to E \oplus K[1]$ clearly gives an algebraic connection for E' so that Cartan's theorem applies to E'.

If we now reconsider diagram (C.2) under these new data, we see that while rows remain the same, the columns have become mapping cones since $\hat{c}(\kappa)^{\text{bas}} = \hat{c}(\kappa^{\text{bas}})$, which is easily checked. But then, applying the cohomology functor, we get the diagram of *exact* columns

$$\begin{array}{ccccc}
h^{i-1}(E^{\mathrm{bas}}) & \xrightarrow[\simeq]{\bar{\mathfrak{i}}} & h^{i-1}((W(\mathfrak{g})\otimes E)^{\mathrm{bas}}) & \xrightarrow[\simeq]{\bar{\mathfrak{f}}} & h^{i-1}(E^{\mathrm{bas}}) \\
\downarrow & & \downarrow & & \downarrow \\
h^{i-1}(E^{\prime,\mathrm{bas}}) & \xrightarrow[\simeq]{\bar{\mathfrak{i}}} & h^{i-1}((W(\mathfrak{g})\otimes E')^{\mathrm{bas}}) & \xrightarrow[\simeq]{\bar{\mathfrak{f}}} & h^{i-1}(E^{\prime,\mathrm{bas}}) \\
\downarrow & & \downarrow & & \downarrow \\
h^{i}(K^{\mathrm{bas}}) & \xrightarrow{\bar{\mathfrak{i}}} & h^{i}((W(\mathfrak{g})\otimes K)^{\mathrm{bas}}) & \xrightarrow{\bar{\mathfrak{f}}} & h^{i}(K^{\mathrm{bas}}) \\
\downarrow & & \downarrow & & \downarrow \\
h^{i}(E^{\mathrm{bas}}) & \xrightarrow[\simeq]{\bar{\mathfrak{i}}} & h^{i}((W(\mathfrak{g})\otimes E)^{\mathrm{bas}}) & \xrightarrow[\simeq]{\bar{\mathfrak{f}}} & h^{i}(E^{\mathrm{bas}}) \\
\downarrow & & \downarrow & & \downarrow \\
h^{i}(E^{\prime,\mathrm{bas}}) & \xrightarrow[\simeq]{\bar{\mathfrak{i}}} & h^{i}((W(\mathfrak{g})\otimes E')^{\mathrm{bas}}) & \xrightarrow[\simeq]{\bar{\mathfrak{f}}} & h^{i}(E^{\prime,\mathrm{bas}})
\end{array} \tag{C.3}$$

where we can apply the Five Lemma and conclude that Cartan-Weil morphisms in the third (middle) row are indeed isomorphisms. Hence, the following enhancement of Cartan's theorem.

C.1 Proposition *If $E \in \mathrm{DGA}(\mathfrak{g})$ admits algebraic connections, then, for every $\mathfrak{g}$-dg-ideal $K \subseteq E$, the Cartan-Weil morphism $\bar{\mathfrak{i}} : K^{\mathrm{bas}} \to (W(\mathfrak{g}) \otimes K)^{\mathrm{bas}}$ is a quasi-isomorphism. In particular, the Cartan-Weil morphism*

$$H(\bar{\mathfrak{f}}) : H\big((W(\mathfrak{g}) \otimes K)^{\mathrm{bas}}\big) \to H(K^{\mathrm{bas}})$$

is an isomorphism too, inverse of $H(\bar{\mathfrak{i}})$, hence independent of the connection.

Appendix D
Graded Ring of Fractions

Let $A := A^0 \oplus A^1 \oplus \cdots$ be a graded ring and denote by S the multiplicative system generated by the nonzero graded elements of A. The *graded ring of fractions of A* is, by definition, the ring $Q_A := S^{-1}A$.[1]

The following fact has been mentioned several times (Exercise 5.1.2.3 and Sect. 7.1).

D.1 Proposition *The graded ring Q_A is such that, for every A-graded module N, the tensor product $Q_A \otimes_A N$ is a flat injective A-graded module.*

Proof *$Q_A \otimes N$ is flat.* For any graded ideal I of A, one has the long exact sequence:

$$0 \to \mathbf{Tor}_1^A(Q_A, A/I) \longrightarrow Q_A \otimes I \longrightarrow Q_A \longrightarrow Q_A \otimes (A/I) \to 0 \qquad \text{(D.1)}$$

where A/I is a torsion graded A-module. The annihilators of the elements of A/I are graded ideals, generated, as such, by invertible elements of Q_A. Therefore

$$\mathbf{Tor}_1^A(Q_A, A/I) = 0, \quad \forall i \in \mathbb{N},$$

and we have from (D.1) the equality $Q_A \otimes I = Q_A$ from which, we deduce

$$Q_A \otimes I \otimes N = Q_A \otimes N$$

for any A-graded module N. The *ideal criterion of flatness* applies, and the A-graded module $Q_A \otimes N$ is flat.

---

[1]The ring Q_A is the zero ring is A has zero divisors, something that never occurs for $A := H_G$, since H_G is an integral domain.

A. Arabia, *Equivariant Poincaré Duality on G-Manifolds*, Lecture Notes in Mathematics 2288, https://doi.org/10.1007/978-3-030-70440-7

$Q_A \otimes N$ is injective. Let $\alpha : M_1 \subseteq M_2$ be a graded inclusion of graded A-modules. We must show that any morphism $\lambda : M_1 \to Q_A \otimes N$ of graded A-modules can be extended to M_2.

$$\begin{array}{ccc} M_1 & \xrightarrow[\subseteq]{\alpha} & M_2 \\ {\scriptstyle\lambda}\downarrow & \swarrow{\scriptstyle\lambda'} & \\ Q_A \otimes N & & \end{array}$$

Otherwise, the Zorn Lemma will lead us to assume that $M_2 \supsetneq M_1$ and that λ may not be further extended. In particular, $A \cdot m \cap M_1 \neq 0$ for any homogeneous $m \in M_2 \setminus M_1$, hence the quotient M_2/M_1 is a torsion module. One then has

$$Q_A \otimes M_1 = Q_A \otimes M_2 \,,$$

and a contradiction arises as a consequence of the diagram

$$\begin{array}{ccc} \mathrm{Homgr}_A(M_2, Q_A \otimes N) & \longrightarrow & \mathrm{Homgr}_A(M_1, Q_A \otimes N) \\ \cong\downarrow & & \cong\downarrow \\ \mathrm{Homgr}_{Q_A}(Q_A \otimes M_2, Q_A \otimes N) & \xrightarrow[(=)]{} & \mathrm{Homgr}_{Q_A}(Q_A \otimes M_1, Q_A \otimes N) \end{array}$$

where the horizontal arrows are induced by the inclusion $M_1 \subseteq M_2$ and the vertical arrows are the well-known canonical natural isomorphisms. □

Appendix E
Hints and Solutions to Exercises

Ch. 2. Nonequivariant Background

1. *Exercise 2.1.5.2–(3) (p. 12).* In both cases, $\oplus$ and Π, the coordinates λ_i of λ are linear forms determined by their action in the coordinate subspaces V^m. If all these linear forms vanish, then $\lambda = 0$ on each V^m, in which case $\lambda = 0$.

2. *Exercise 2.1.5.2 (p. 12).*

- In the case of the functor $\oplus$, we have to compare the two sets

$$\begin{cases} \mathrm{Hom}_{\Bbbk}(\oplus \boldsymbol{V}, \oplus \boldsymbol{W}) = \Pi_i \, \mathrm{Hom}_{\Bbbk}(V_i, \oplus \boldsymbol{W}) \\ \mathrm{Homgr}^*_{\Bbbk}(\boldsymbol{V}, \boldsymbol{W}) = \Pi_i \Pi_j \, \mathrm{Hom}_{\Bbbk}(V_i, W_j) = \Pi_i \, \mathrm{Hom}_{\Bbbk}(V_i, \Pi \boldsymbol{W}), \end{cases}$$

  which are clearly equal if and only if $\boldsymbol{V} = 0$ or $\oplus \boldsymbol{W} = \Pi \boldsymbol{W}$. Hence, $\Phi_{\boldsymbol{V},\boldsymbol{W}}$ is surjective if and only if $\boldsymbol{V} = 0$ or $\boldsymbol{W}$ is bounded.
- In the case of the functor Π, the sets to compare are

$$\begin{cases} \mathrm{Hom}_{\Bbbk}(\Pi \boldsymbol{V}, \Pi \boldsymbol{W}) = \Pi_j \, \mathrm{Hom}_{\Bbbk}(\Pi \boldsymbol{V}, W_j) \\ \mathrm{Homgr}^*_{\Bbbk}(\boldsymbol{V}, \boldsymbol{W}) = \Pi_i \Pi_j \, \mathrm{Hom}_{\Bbbk}(V_i, W_j) = \Pi_j \, \mathrm{Hom}_{\Bbbk}(\oplus \boldsymbol{V}, W_j), \end{cases}$$

  which are equal if and only if $\boldsymbol{W} = 0$ or $\oplus \boldsymbol{V} = \Pi \boldsymbol{V}$. Hence, $\Phi_{\boldsymbol{V},\boldsymbol{W}}$ is surjective if and only if $\boldsymbol{V}$ is bounded or $\boldsymbol{W} = 0$.
- In both cases, the announced equivalence (4a) and nonequivalence (4b) of categories is proved.

3. *Exercise 2.1.6.1 (p. 13).* Given a $d_i : X_i \to X_{i+1}$ in a split category Ab, we can assume that $X_i = Z_i \oplus B_{i+1}$, that $X_{i+1} = B_{i+1} \oplus N_{i+1}$ and that $d(z, b) = (b, 0)$. Then, $d_{i-1} : X_{i-1} \to X_i$, such that $d_i \circ d_{i-1} = 0$, factors through the inclusion $Z_i \subseteq X_i$, and we can assume that $X_{i-1} = Z_{i-1} \oplus B_i$, that $Z_i = B_i \oplus H_i$ and that

A. Arabia, *Equivariant Poincaré Duality on G-Manifolds*, Lecture Notes in Mathematics 2288, https://doi.org/10.1007/978-3-030-70440-7

$d_{i-1}(z, b) = (b, 0)$. As a consequence, the complex $(\boldsymbol{X}, \boldsymbol{d}) \in C(\mathrm{Ab})$ is isomorphic to a *split complex*:

$$\cdots \; X_{i-2} \xrightarrow{d_{i-2}} X_{i-1} \xrightarrow{d_{i-1}} X_i \xrightarrow{d_i} X_{i+1} \xrightarrow{d_{i+1}} X_{i-2} \; \cdots$$

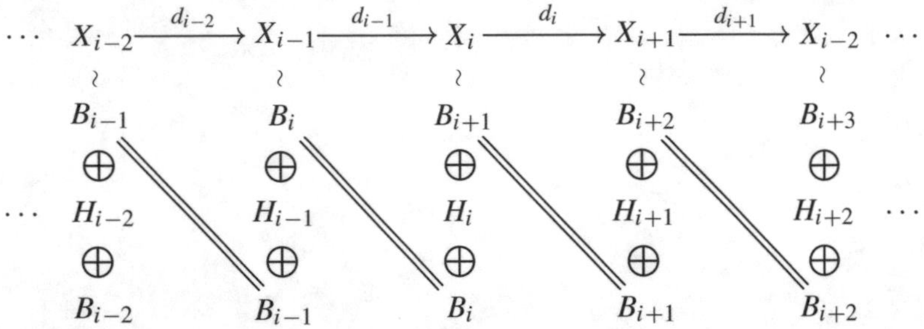

(E.1)

In which case, the cohomology of $\boldsymbol{h}(\boldsymbol{X}, \boldsymbol{d})$ is canonically isomorphic to the graded object $\{H_i\}_{i \in \mathbb{Z}} \in \mathrm{Ab}^{\mathbb{Z}}$.

Now, an additive functor $F : \mathrm{Ab} \to \mathrm{Ab}'$ applied to (E.1) respects the decompositions in direct sums, so that it tautologically satisfies the equality

$$\boldsymbol{h}(F(\boldsymbol{V}, \boldsymbol{d})) = \{F(H_i)\}) = F\boldsymbol{h}(\boldsymbol{V}, \boldsymbol{d})$$

proving the exactness of F.

Conversely, assume that every additive functor on Ab is exact. Then if $\iota : X \rightarrowtail Y$ is a monomorphism in Ab, we consider the short exact sequence

$$0 \to X \xrightarrow{\iota} Y \xrightarrow{\nu} K \to 0 \,,$$

where ν is a cokernel for ι, to which we apply the (exact) functor $\mathrm{Hom}_{\mathrm{Ab}}(K, \_)$. We obtain the short exact sequence

$$0 \to \mathrm{Hom}_{\mathrm{Ab}}(K, X \xrightarrow{\iota}) \mathrm{Hom}_{\mathrm{Ab}}(K, Y \xrightarrow{\nu}) \mathrm{Hom}_{\mathrm{Ab}}(K, K \to 0) \,.$$

Hence, there exists a morphism $\rho : K \to Y$ such that $\nu \circ \rho = \mathrm{id}_K$. But then the morphism $X \oplus K \to Y$, $(x, k) \mapsto x + \rho(k)$ is an isomorphism. We have thus shown that every subobject d'un object in Ab admits a complement. The category Ab is therefore a split category.

The *incomplete basis theorem* shows that $\mathrm{Vec}(\Bbbk)$ and $\mathrm{GV}(\Bbbk)$ are both split categories, and any additive functor defined in these categories is exact.

4. *Exercise 2.1.8.2 (p. 15).* We have

$$\begin{aligned} \boldsymbol{D}\big(\Psi(v'\otimes\boldsymbol{\lambda})\big)(v) &= \boldsymbol{d}'(\boldsymbol{\lambda}(v)v') - (-1)^{[v]+[\lambda]}\boldsymbol{\lambda}(\boldsymbol{d}v)v' \\ &= \boldsymbol{\lambda}(v)\boldsymbol{d}'(v') + (-1)^{[v]}(\boldsymbol{D\lambda})(v)v' \\ &= \Psi(\boldsymbol{d}'(v')\otimes\boldsymbol{\lambda})(v) + \Psi((-1)^{[v]}v'\otimes(\boldsymbol{D\lambda}))(v) \\ &= \Psi\big(\Delta(v'\otimes\boldsymbol{\lambda})\big)(v) \end{aligned}$$

5. *Exercise 2.1.8.3 (p. 15).* We treat only the case of the tensor product.
The map

$$\Xi : (\boldsymbol{V}\otimes\boldsymbol{W})[s+t] \to \boldsymbol{V}[s]\otimes\boldsymbol{W}[t],$$

which is the identity over each component $V^a \otimes W^b$, is an obvious isomorphism of graded vector spaces. Given $v \in \boldsymbol{V}$ homogeneous, the notation $v[s] \in \boldsymbol{V}[s]$ should recall that the degree of $v[s]$ is $[v] - s$. We then have:

$$\begin{aligned} \Xi\big(\Delta[s+t](v\otimes w)\big) &= (-1)^{s+t}\, \Xi\big(dv\otimes w + (-1)^{[v]}v\otimes dw\big) \\ &= d[s]v[s]\otimes(-1)^t\, w[t] + (-1)^{[v]+s}v[s]\otimes d[t]w[t] \end{aligned} \tag{E.2}$$

while

$$\Delta\big(\Xi(v\otimes w)\big) = d[s]v[s]\otimes w[t] + (-1)^{[v]-s}v[s]\otimes d[t]w[t], \tag{E.3}$$

which shows that Ξ does not respect the differentials. More precisely, the issue lies in the t-shift, otherwise, if $t = 0$, then Ξ is compatible, so that we can state that the map

$$\Xi_s : (\boldsymbol{V}\otimes\boldsymbol{W})[s] \to \boldsymbol{V}[s]\otimes\boldsymbol{W}, \quad (v\otimes w)[s] \mapsto v[s]\otimes w,$$

identifies both complexes.

It is now useful to recall that the following (anticommutative) transposition

$$\tau : \boldsymbol{V}\otimes\boldsymbol{W} \to \boldsymbol{W}\otimes\boldsymbol{V}, \quad \tau(v\otimes w) := (-1)^{[v][w]}\, w\otimes v,$$

dictated by the Koszul sign rule (Sect. 2.1.9), is an isomorphism of complexes.

Indeed, it is clearly bijective and it verifies:

$$\begin{aligned}\tau\big(\Delta(v\otimes x)\big) &= \tau\big(dv\otimes w + (-1)^{[v]} v\otimes dw\big)\\ &= (-1)^{([v]+1)[w]}\, w\otimes dv + (-1)^{[v]}(-1)^{[w+1][v]}\, dw\otimes v\\ &= (-1)^{[u][v]}\big(dw\otimes v + (-1)^{[w]}\, w\otimes dv\big) = \Delta\tau(v\otimes w)\,.\end{aligned}$$

We can now give an explicit isomorphism of complexes from $(\boldsymbol{V}\otimes\boldsymbol{W})[s+t]$ to $\boldsymbol{V}[s]\otimes\boldsymbol{W}[t]$ by simply composing the following isomorphisms:

$$\begin{aligned}(\boldsymbol{V}\otimes\boldsymbol{W})[s][t] &\xrightarrow[\simeq]{\Xi_s[t]} (\boldsymbol{V}[s]\otimes\boldsymbol{W})[t] \longrightarrow\\ &\xrightarrow[\simeq]{\tau[t]} (\boldsymbol{W}\otimes\boldsymbol{V}[s])[t] \longrightarrow\\ &\xrightarrow[\simeq]{\Xi_t} (\boldsymbol{W}[t]\otimes\boldsymbol{V}[s]) \longrightarrow\\ &\xrightarrow[\simeq]{\tau} (\boldsymbol{V}[s]\otimes\boldsymbol{W}[t])\,.\end{aligned}$$

An elementary computation shows that the resulting isomorphism associates:

$$v\otimes w \mapsto (-1)^{t([v]+s)}\, v[s]\otimes w[t]\,,$$

which is not obvious a priori.

6. *Exercise (fn. ([10]), p. 18) (p. 18).* Let $f : M\to N$ be proper. In the locally compact space N, a subset $A\subseteq N$ is closed if and only, for all $K\subseteq N$ compact, the set $A\cap K$ is compact. As a consequence, the map $f : M\to N$ is closed if and only if, for all $Y\subseteq M$ closed, and all $K\subseteq N$ compact, the set $f(Y)\cap K = f(Y\cap f^{-1}(K))$ is compact, which is always the case when f is proper, since then $f^{-1}(K)$, $Y\cap f^{-1}(K))$ and $f(Y\cap f^{-1}(K))$ are compact.

Conversely, suppose $f : M\to N$ closed with compact fibers. Let $K\subseteq N$ be compact. Given a family $\{Y_{\mathrm{i}}\}_{\mathrm{i}\in\mathfrak{J}}$ of nonempty closed subsets of $f^{-1}(K)$, which we assume to be stable by finite intersections, we necessary have $L := \bigcap_{\mathrm{i}} f(Y_{\mathrm{i}}) \neq \emptyset$, since otherwise, K being compact, some finite sub-intersection must be empty, which is not the case. But then, for $x\in L$, the fiber $f^{-1}(x)$ meets every Y_{i} in a nonempty set. As a consequence, $\bigcap_{\mathrm{i}}\big(Y_{\mathrm{i}}\cap f^{-1}(x)\big)\neq\emptyset$, since the fiber is compact. Therefore $\bigcap_{\mathrm{i}} Y_{\mathrm{i}}\neq\emptyset$. We have thus proved that the set $f^{-1}(K)$ is compact for every compact subspace $K\subseteq N$, i.e. that f is a proper map. Notice that we did not use the fact that the spaces are locally compact, but only that they are Hausdorff.

7. *Exercise 2.4.1.2 (p. 25).* To show that the Poincaré adjunctions are not surjective, let $\varphi : U\to W\subseteq\mathbb{R}^m$ $(m := d_M)$ be a chart of M and let $(f_n)_{n\in\mathbb{N}}$ be any sequence of nonzero functions in W with two by two disjoint supports and uniformly converging to the zero function.

Given $0 \leq i \leq n$, consider the family $\{\beta_{i,n}\}_{n\in\mathbb{N}} \subseteq \Omega^i_c(M)$ where

$$\beta_{i,n} := \varphi^*\big(f_n\, dx_1 \wedge \cdots \wedge dx_i\big)\,.$$

By construction, for all $\alpha \in \Omega^{m-i}(M)$ (resp. $\Omega^{m-i}_c(M)$) we have

$$\lim_{n\mapsto+\infty} \langle\, \alpha,\, \beta_{i,n}\,\rangle_M = 0\,. \tag{E.4}$$

On the other hand, since the family of i-forms $\{\beta_{i,n}\}_{n\in\mathbb{N}}$ is linearly free, it can be extended to a basis of $\Omega^i_c(M)$ (resp. $\Omega^i(M)$), so that there exist linear forms

$$\Lambda : \Omega^i_c(M) \to \mathbb{R} \quad \text{such that} \quad (\forall n \in \mathbb{N})\ \Lambda(\beta_{i,n}) = 1\,,$$

(resp. $\Lambda : \Omega^i(M) \to \mathbb{R}$). Such linear forms do not verify the condition (E.4), hence they are not of the form $\langle\, \alpha, \_\,\rangle_M$ with $\alpha \in \Omega^{m-i}(M)$ (resp. $\Omega^{m-i}_c(M)$).

8. *Exercise 2.4.2.1 (p. 30).*

(1) An orientation of M defines, on each connected component $C \in \Pi_0(M)$, an orientation $[C]$ and a nonzero class in $H^{d_M}_c(C)$, viz. the fundamental class $\zeta_{[C]}$. Conversely, let $\omega \in H^{d_M}_c(C) \smallsetminus \{0\}$. If $[C]$ is an orientation of C of fundamental class $\zeta_{[C]}$, we will have $\omega = \lambda(\omega, [C])\, \zeta_{[C]}$ for a unique scalar $\lambda(\omega, [C]) \in \mathbb{R} \smallsetminus \{0\}$. Changing $[C]$ by the opposite orientation $-[C]$, changes the sign of $\lambda(\omega, [C])$. As a consequence, there is a unique orientation of C, which we denote $[C]_\omega$, such that $\lambda(\omega, [C]_\omega) > 0$. The correspondences

$$[C] \mapsto \zeta_{[C]} \quad \text{and} \quad \omega \mapsto [C]_\omega\,,$$

are then clearly inverse of each other.

(2) When $|\Pi_0(M)| < +\infty$, we have

$$H^0(M) = \bigoplus_{C\in\Pi_0(M)} H^0(C) \quad \text{and} \quad H^{d_M}_c(M) = \bigoplus_{C\in\Pi_0(M)} H^{d_M}_c(C)\,,$$

and $\mathbf{1}_M = \sum_{C\in\Pi_0(M)} \mathbf{1}_C$, with $\mathbf{1}_C \in H^0(C)$. Therefore,

$$D_M(\mathbf{1}_M) = \sum_{C\in\Pi_0(M)} D_C(\mathbf{1}_C) = \sum_{C\in\Pi_0(M)} \textstyle\int_C \leftrightarrow \displaystyle\sum_{C\in\Pi_0(M)} \zeta_C \in H^{d_M}_c(M)$$

(3) When $|\Pi_0(M)| < \aleph_0$, we rather have

$$H^0(M) = \prod_{C\in\Pi_0(M)} H^0(C) \quad \text{and} \quad H^{d_M}_c(M) = \bigoplus_{C\in\Pi_0(M)} H^{d_M}_c(C)\,,$$

in which case $\mathbf{1}_M$ is the infinite family $\{\mathbf{1}_C \mid C \in \Pi_0(M)\}$, and the element

$$D_M(\mathbf{1}_M) = \Big\{ \int_C \;\Big|\; C \in \Pi_0(M)\Big\} \leftrightarrow \big\{\zeta_C \in H_c^{d_M}(C) \;\big|\; C \in \Pi_0(M)\big\}$$

can no more be represented by a single cohomology class in $H_c^{d_M}(M)$.

9. *Exercise 2.5.2.4–(1) (p. 33).* Fix a locally finite *good cover* $\mathcal{V} := \{V_i\}_{i\in\mathbb{N}}$ of N (fn. ([15]), p. 28) by trivializing open subsets for f. For each $m \in \mathbb{N}$, let $\mathcal{W}_m := \{V_0, \ldots, V_m\}$ and let $W_m := \bigcup \mathcal{W}_m$. The manifold $f^{-1}W_m$ is of finite type. Indeed, thanks to Leray, one knows that $H(f^{-1}W_m)$ is the abutment of a spectral sequence of second page $\mathbb{E}_2^{p,q} = H^p(W_m; \mathcal{H}^q)$, where $\mathcal{H}^p$ is the local system $\mathcal{H}^q\, \mathbb{R} f_* \underline{\mathbb{R}}_M$ of germs $H^q(F)$, where F denotes the fiber of f. To show that $\dim(H(f^{-1}W_m)) < +\infty$, it will then suffice to show that $\dim(\mathbb{E}_2) < +\infty$, and since q is bounded, only that $\dim(H(W_m; \mathcal{H}^q)) < +\infty$, for each q. For that, one uses the fact that $\mathcal{W}_m$ is a good cover of W_m, and that, therefore, one can compute $H(W_m; \mathcal{H}^q)$ using the Čech complex $\check{C}(\mathcal{W}_m; \mathcal{H}^q)$, [1] which is easily seen to be finite dimensional.

10. *Exercise 2.5.2.4–(2) (p. 33).* If M is orientable, then $M = \tilde{M}$ and $\dim(H_c(M)) = \dim(H(M))$ after Poincaré duality.

If M is not orientable, the orientation manifold $\tilde{M}$ is connected and it is the total space of a 2-fold covering $\pi : \tilde{M} \to M$ with $\mathrm{Aut}(\pi) = \mathbb{Z}/2\mathbb{Z}$. In particular, one has $M = \tilde{M}/\langle\tau\rangle$ as manifolds.

The generator τ of $\mathrm{Aut}(\pi)$ is an orientation-reversing involutive diffeomorphism $\tau : M \to M$. Its action by pullback, also denoted by 'τ', gives rise to the canonical decompositions

$$\begin{cases} H(\tilde{M}) = H(M)^+ \oplus H(M)^- \\ H_c(\tilde{M}) = H_c(M)^+ \oplus H_c(M)^- \end{cases} \tag{E.5}$$

where $(\_)^+$ denotes the vector subspaces of τ-invariants, i.e. $\tau(x) = x$, while $(\_)^-$ denotes the space of τ-anti-invariants, i.e. $\tau(x) = -x$. The key fact is that subspaces with same invariance $\epsilon \in \{+, -\}$ are orthogonal under the Poincaré pairing $\langle\cdot,\cdot\rangle_{\tilde{M}}$. Indeed, if $\alpha \in H(\tilde{M})^\epsilon$, and $\beta \in H_c(\tilde{M})^\epsilon$, then

$$\langle\alpha,\beta\rangle_{\tilde{M}} = \langle\tau\alpha,\tau\beta\rangle_{\tilde{M}} = \int_{\tilde{M}} \tau\alpha\wedge\tau\beta = \int_{\tilde{M}} \tau(\alpha\wedge\beta) = -\int_{\tilde{M}} \alpha\wedge\beta = -\langle\alpha,\beta\rangle_{\tilde{M}},$$

since τ reverses orientation, which immediately implies that

$$\langle\alpha,\beta\rangle_{\tilde{M}} = 0.$$

[1] See Godement [46], sec. 5.3–4, Theorem 5.4.1 and its Corollary pages 212–213.

As a consequence, the Poincaré duality map $D_{\tilde{M}}$ (Theorem 2.4.1.3–(2.29)) establishes an isomorphism

$$D_{\tilde{M}} : H(\tilde{M})^{-} \underset{\simeq}{\longrightarrow} (H_c(\tilde{M})^{+})^{\vee},$$

which, introduced in the decomposition (E.5), gives rise to a canonical isomorphism of vector spaces

$$H(\tilde{M}) \simeq H(\tilde{M})^{+} \oplus (H_c(\tilde{M})^{+})^{\vee}. \tag{E.6}$$

On the other hand, since $M \sim \tilde{M}/\operatorname{Aut}(\pi)$ and since π is proper, the pullback map π^* identifies $H(M) \leftrightarrow H(\tilde{M})^{+}$ and $H_c(M) \leftrightarrow H_c(\tilde{M})^{+}$, so that the decomposition (E.6), can be best seen as the decomposition

$$\boxed{H(\tilde{M}) \simeq H(M) \oplus H_c(M)^{\vee}}$$

whence the claim.

11. *Exercise 2.6.1.1–(1) (p. 38).* We have

$$\begin{aligned}
I\!D'_M(d\,\beta)(\alpha) &= \int_M \alpha \wedge d\,\beta = (-1)^{[\alpha]+1} \int_M d\,\alpha \wedge \beta \\
&= (-1)^{[\alpha]+1}\, I\!D'_M(\beta)(d\,\alpha) = (-1)^{[\alpha]+1}(-1)^{d_M-[\beta]+1}(\boldsymbol{D}I\!D'_M(\beta))(\alpha) \\
&= (-\boldsymbol{D}I\!D'_M(\beta))(\alpha)\,.
\end{aligned}$$

Hence, $I\!D'_M : (\Omega(M)[d_M], d) \to (\Omega_c(M)^{\vee}, -\boldsymbol{D})$ is a morphism of complexes.

12. *Exercise 2.6.1.1–(2) (p. 38).* The differential in $(\Omega_c(M)[d_M], d)$ is d, while in $(\Omega(M), d)[d_M]$ is $(-1)^{dM}d$, which immediately shows that ι is not a morphism of complexes, contrary to ϵ. For $I\!D_M$ and $I\!D_{M'}$, we know already that they are morphisms of complexes. For Ξ, we have

$$\begin{aligned}
\Xi(-\boldsymbol{D}\lambda) &= \Xi\big((-1)^{[\lambda]}\,\lambda\circ d\big) = (-1)^{[\lambda]}(-1)^{[\lambda]+1+d_M}\,\lambda\circ d\circ\iota \\
\boldsymbol{D}(\Xi(\lambda)) &= \boldsymbol{D}\big((-1)^{[\lambda]+d_M}\,\lambda\circ\iota\big) = (-1)^{[\lambda]+d_M}(-1)^{[\lambda]+1}\lambda\circ\iota\circ d
\end{aligned}$$

and Ξ is also a morphism of complexes.

To check the commutativity of the diagram, let $\beta, \beta' \in \Omega_c(M)$ be homogeneous. Then,

$$\begin{aligned}
I\!D_M(\epsilon(\omega))(\omega') &= (-1)^{[\omega]d_M} \int_M \omega\wedge\omega' = (-1)^{[\omega]} \int_M \omega'\wedge\omega \\
&= (-1)^{[\omega]}\, I\!D'(\omega)(\omega') = \Xi(I\!D'(\omega))(\omega')\,.
\end{aligned}$$

13. *Exercise 2.6.2.3 (p. 42).* Simple application of Proposition 2.5.1.1 on adjoint operators.

Ch. 3. Relative Poincaré Duality

14. *Exercise 3.1.1.1 (p. 50).* Since the projection map π is open and since E' is connected, the space $B' := \pi(E')$ is an open connected submanifold of B, hence of some well-defined dimension $d_{B'}$. On the other hand, from general topology, for all $b \in \overline{B}'$, the connected components of the fibers F_b are contained in the connected components of E, hence the fact that $F_b' := F_b \cap E'$ is union of connected components of F_b. The triviality of π above small open balls $\mathbb{B}_b$ centered at $b \in \overline{B}'$ then warrants that F_b' is of dimension $d_{E'} - d_{B'}$, and, furthermore, that the fibers F_x' are diffeomorphic for all $x \in \mathbb{B}_b$, hence for all $x \in \overline{B}'$, since $\overline{B}'$ is connected. Notice that we have also proved that B' is a closed subspace of B. The map $\pi' : E' \to B'$ is therefore a locally trivial fibration of manifolds onto a connected component of B.

15. *Exercise 3.1.4.1 (p. 52).*

(1) Let $\pi \in \mathrm{Mor}_{\mathcal{C}}(X, W)$ and let $X' := X \times_{(\pi, \mathrm{id}_W)} W$. The map

$$\begin{array}{rccl} \Xi(Y) : & \mathrm{Mor}_{\mathcal{C}}(Y, X) & \to & \mathrm{Mor}_{\mathcal{C}}(Y, X') \\ & \eta & \mapsto & (\eta, \pi \circ \eta) \end{array}, \qquad \begin{array}{ccccc} & & X & & \\ & \overset{\eta}{\nearrow} & & \overset{\pi}{\searrow} & \\ Y & & & & W \\ & \underset{\pi\circ\eta}{\searrow} & & /\!/ & \\ & & W & & \end{array}$$

is a natural bijection for all $Y \in \mathcal{C}$. The morphism $(\mathrm{id}_X, \pi) : X \to X'$ corresponds to $\mathrm{id}_X : X \to X$ through the isomorphism $\Xi(X)$, and it has as left inverse the morphism $\psi := \Xi(X')^{-1}(\mathrm{id}_{X'}) : X' \to X$ since the diagram

$$\begin{array}{ccc} \mathrm{Mor}_{\mathcal{C}}(X, X) & \xrightarrow{\mathrm{Mor}(\psi, X)} & \mathrm{Mor}_{\mathcal{C}}(X', X) \\ {\scriptstyle \Xi(X)}\downarrow\simeq & & \simeq\downarrow{\scriptstyle \Xi(X')} \\ \mathrm{Mor}_{\mathcal{C}}(X, X') & \xrightarrow{\mathrm{Mor}(\psi, X')} & \mathrm{Mor}_{\mathcal{C}}(X', X') \end{array} \qquad \begin{array}{ccc} \mathrm{id}_X & \overset{\psi^*}{\longmapsto} & \psi \\ {\scriptstyle \Xi(X)}\downarrow\simeq & & \simeq\downarrow{\scriptstyle \Xi(X')} \\ (\mathrm{id}_X, \pi) & \overset{\psi^*}{\longmapsto} & \mathrm{id}_{X'} \end{array}$$

is commutative. A symmetric argument shows that ψ is a right inverse to (id_X, π), hence that (id_X, π) is an isomorphism.

(2) If the notation $(X_1 \times_W W) \times_Z X_2$ denotes $(X_1 \times_W W) \times_{(\nu, \pi_2)} X_2$, we set $\pi_1 := \nu \circ (\mathrm{id}_X, \pi)$, in which case we have a commutative diagram

$$\begin{array}{ccc} X & \underset{\simeq}{\overset{(\mathrm{id}_X, \pi)}{\longmapsto}} & X \times_W W \\ {\scriptstyle \pi_1}\downarrow & & \downarrow{\scriptstyle \nu} \\ Z & = & Z \end{array}$$

with horizontal isomorphisms, which implies that the induced morphism

$$(X_1 \times_W W) \times_{(\nu,\pi_2)} X_2 \to X_1 \times_{(\pi_1,\pi_2)} X_2$$

is an isomorphism.

16. *Exercise 3.1.4.3 (p. 54).* We give the answer for $H := \{e\}$ and leave the general case to the reader. We thus consider the diagram

$$\begin{array}{ccc} X & \xrightarrow{\nu_X} & X/G \\ {\scriptstyle f}\downarrow & \oplus & \downarrow{\scriptstyle f_G} \\ Y & \xrightarrow{\nu_Y} & Y/G. \end{array}$$

Assume the diagram Cartesian. Given $x \in X$, such that $f(x) = f(k\cdot x)$, the two elements x and $k\cdot x$ must coincide after Proposition 3.1.4.2–(1) since they verify simultaneously $f(x) = f(k\cdot x)$ and $\nu_X(x) = \nu_X(k\cdot x)$. Hence, $\mathrm{Stab}_G(x) \subseteq \mathrm{Stab}_G(f(x))$. The opposite inclusion being obvious, the map

$$f : G\cdot x \to G\cdot f(x) \tag{E.7}$$

is bijective. Conversely, assume the maps (E.7) bijective for all $x \in X$. Then

$$w : X \to \big(Y \times_{(\nu_Y, f_G)} (X/G)\big), \quad w(x) := (f(x), G\cdot x),$$

is surjective since, given $y \in Y$ and $G\cdot x \in X/G$ such that $y \in f(G\cdot x)$, there always exist $k \in G$ verifying $y = f(k\cdot x)$, in which case $w(k\cdot x) = (y, G\cdot x)$. But such k is unique because (E.7) is bijective. The map w is therefore a continuous bijection. It is also an open map. Indeed since f is continuous, an open subspace $U \subseteq X$ is the union of open subspaces of the form $f^{-1}(V)$ where $V \subseteq Y$ is open. It therefore suffices to show that

$$w\big(f^{-1}(V)\big) = V \times_{(\nu_Y, f_G)} f_G^{-1}\big(\nu_Y(V)\big) \tag{E.8}$$

is open in $Y \times_{(\nu_Y, f_G)} (X/G)$. But this is clear since ν_Y is an open map, in which case the right-hand side of (E.8) is the induced fiber product of the two open subspaces $V \subseteq Y$ and $f_{\gg}^{-1}\big(\nu_Y(V)\big) \subseteq X/G$.

17. *Counterexample to a fiber product of manifolds 3.1.6 (p. 56).* Given two maps of manifolds $p_i : M_i \to N$, assume that their fiber product exists in the category of manifolds, denote it by $(M_1 \times_N M_2)_{\mathrm{diff}}$. Denote by $(M_1 \times_N M_2)_{\mathrm{top}}$ the fiber product in the category of topological spaces. By the universal property of fiber product in Top, we have a natural continuous map

$$\xi : (M_1 \times_N M_2)_{\mathrm{diff}} \to (M_1 \times_N M_2)_{\mathrm{top}} \subseteq M_1 \times M_2\,.$$

The universal property applied to 0-dimensional manifolds, for which continuous and differentiable maps are the same, implies immediately that ξ is a bijection.

Now, suppose that for $(x_1, x_2) \in (M_1 \times_N M_2)_{\text{top}}$ the map p_1 is étale on x_1, i.e. that there exists an open neighborhood $U_{x_1} \ni x_1$ such that $V_1 := p_1(U_{x_1})$ is open in N and such that $p : U_1 \to V_1$ is a diffeomorphism. Then, the subset

$$W(x_1, x_2) := \big(U_1 \times p_2^{-1}(V_1)\big) \cap \big(M_1 \times_N M_2\big)$$

is an open neighborhood of $(x_1, x_2) \in (M_1 \times_N M_2)_{\text{top}}$ and is clearly the graph of the differentiable map $x_2' \mapsto p_1^{-1}(p_2(x_2'))$ for $x_2' \in p_2^{-1}(V_1) \subseteq M_2$. Hence, $W(x_1, x_2)$ is a locally closed differentiable submanifold of $M_1 \times M_2$ of dimension $\dim(M_2)$. Since $W(x_1, x_2)$ is also open in $(M_1 \times_N M_2)_{\text{diff}}$, this gives a constructive proof of the existence of this manifold at a neighborhood of (x_1, x_2).

We thereby see that to find a counterexample to a fiber product of manifolds, we need a case were, for $(x_1, x_2) \in M_1 \times_N M_2$, neither x_i is étale for p_i. For example $p_i : \mathbb{R} \to \mathbb{R}$, $p_i(t) = t^2$, and $x_i = 0$. In that case

$$(\mathbb{R} \times_{\mathbb{R}} \mathbb{R})_{\text{top}} = \{(t, t)\} \cup \{(t, -t)\} \subseteq \mathbb{R} \times \mathbb{R}\,.$$

Since the paths $\gamma_\pm : t \mapsto (t, \pm t) \in (\mathbb{R} \times_{\mathbb{R}} \mathbb{R})_{\text{top}}$ are differentiable in $\mathbb{R} \times \mathbb{R}$, they must lift to $(\mathbb{R} \times_{\mathbb{R}} \mathbb{R})_{\text{diff}}$, if this manifold exist. But this manifold should be a curve after the previous paragraph, and the sets $\{\gamma_\pm(\mathbb{R})\}$ must therefore coincide in $(\mathbb{R} \times_{\mathbb{R}} \mathbb{R})_{\text{diff}}$, but this is impossible since they are different in $\mathbb{R} \times \mathbb{R}$.

We have thus proved that the fiber product of $p_i : \mathbb{R} \to \mathbb{R}$, $p_i(t) = t^2$, does not exist in the category of manifolds.

18. *Exercise 3.1.7.4(1) (p. 62).* We can assume, after Exercise 3.1.1.1, that E, B and F are equidimensional, respectively of dimensions m, n and d.

- If B is orientable and $\omega_B \in \Omega^n(B)$ is nowhere vanishing, then $\pi^*\omega_B \wedge \omega_\pi$ is nowhere vanishing. Indeed, for $x \in F_b$, fix a family $\{\xi_i, \dots, \xi_n\}$ of vector fields on some trivializing open neighborhood U of b such that $\{\xi_1(b), \dots, \xi_n(b)\}$ spans $T_b B$, and fix a family $\{\zeta_1, \dots, \zeta_d\}$ of vector fields on F_b such that $\{\zeta_i(x), \dots, \zeta_n(x)\}$ spans $T_x F_b$. Extend the vector fields to the whole $\pi^{-1}(U)$ using some trivializing diffeomorphism $\varphi : \pi^{-1}(U) \to U \times F_b$. Then, we have

$$(\pi^*\omega_B \wedge \omega_\pi)(\xi_1, \dots, \xi_n, \zeta_1, \dots, \zeta_d)(x) = \omega_B(\xi_1, \dots, \xi_n)(b)\, \omega_\pi\,|_{F_b}(\zeta_1, \dots, \zeta_d)(x) \neq 0\,.$$

  The differential form $\pi^*\omega_B \wedge \omega_\pi$ is therefore nowhere vanishing and of highest degree m, hence E is orientable.
- Assume E orientable and choose $\omega_E \in \Omega^m(E)$ nowhere vanishing. Fix an open cover $\mathscr{U} := \{U_\mathfrak{i}\}_{\mathfrak{i} \in \mathfrak{I}}$ of B by trivializing subspaces for π, where each $U_\mathfrak{i}$ is diffeomorphic to $\mathbb{R}^n$, hence is orientable. For each $\mathfrak{i} \in \mathfrak{I}$, we can then choose a nowhere vanishing form $\omega_U \in \Omega^n(U)$ such that $\pi^*\omega_U \wedge \omega_\pi$ (which

is nowhere vanishing after the previous paragraph) and ω_E define the same orientation on $\pi^{-1}(U_{\mathrm{i}})$. Then, for any partition of unity $\{\phi_{\mathrm{i}}\}_{\mathrm{i}\in\mathfrak{I}}$ subordinate to $\mathcal{U}$, the differential form $\sum_{\mathrm{i}} \phi_{\mathrm{i}}\, \omega_{U_{\mathrm{i}}} \in \Omega^n(B)$ is nowhere vanishing since, by construction, for every $b \in B$, the differential forms $\omega_{U_{\mathrm{i}}}$ such that $b \in U_{\mathrm{i}}$, define the same orientation on a neighborhood of b. Hence B is orientable.

19. *Exercise 3.1.7.4(1) (p. 62).* The Möbius Strip is naturally fibered over its central circle $\mathbb{S}^1$ with fibers diffeomorphic to $\mathbb{R}$. More generally, the normal bundle to any non orientable submanifold N of any Euclidean space $\mathbb{R}^m$.

20. *Exercise 3.1.9.3-(1) (p. 64).* In the commutative diagram

$$\begin{array}{ccc} X' & \xrightarrow{\;g\;} & X \\ {\scriptstyle \pi'}\big\downarrow & \oplus & \big\downarrow{\scriptstyle \pi} \\ B' & \xrightarrow{\;\bar{g}\;} & B \end{array}$$

we have, for all $P \subseteq X$ and all $K \subseteq B'$,

$$g^{-1}(P) \cap \pi'^{-1}(K) = g^{-1}\big(P \cap \pi^{-1}(\bar{g}(K))\big)\,.$$

Consequently, if g and $p : P \to Z$ are proper maps, then the right-hand term in the last equality is always compact, which proves that $g^{-1}(P)$ is π'-proper.

Counterexamples to the converse are easily found applying Proposition 3.1.9.2-(1), for example any direct product

$$\begin{array}{ccc} Y \times K & \xrightarrow{\;p_2\;} & K \\ {\scriptstyle p_1}\big\downarrow & \oplus & \big\downarrow \\ Y & \xrightarrow{\qquad} & \{\bullet\} \end{array}$$

with K compact and Y noncompact.

21. *Exercise 3.1.9.3-(2) (p. 64).* To see that p_2 is open, it suffices to show that for all $(x_1, x_2) \in X_1 \times_B X_2$ and for all pair of open neighborhoods $x_i \in V_{x_i} \subseteq X_i$, the set W_{x_2} of elements $x_2' \in V_{x_2}$ verifying $(V_{x_1} \times \{x_2'\}) \cap (X_1 \times_B X_2) \neq \emptyset$, is a neighborhood of x_2. In the present case, where π_1 is open, the set $\pi_1(V_{x_1})$ is open in B and the set $W_{x_2} := \pi_2^{-1}(\pi_1(V_{x_1})) \cap V_{x_2}$ fulfills the requirements.

As counterexample for the 'closed' statement, take $\pi_i := \mathbb{R} \to \{\bullet\}$. The maps π_i are closed and $\mathbb{R} \times_{\{\bullet\}} \mathbb{R} = \mathbb{R} \times \mathbb{R}$, but it is well-known that the projection maps $p_i : \mathbb{R}^2 \to \mathbb{R}$ are not closed. The hyperbola $H := \{x_1 x_2 = 1\}$ is closed in $\mathbb{R}^2$ while its projections $p_i(H) := \mathbb{R} \setminus \{0\}$ are not closed in $\mathbb{R}$.

22. *Exercise 3.1.9.3-(3) (p. 64).* In a locally compact space Y, a subset $A \subseteq Y$ is closed if and only, for all $K \subseteq Y$ compact, the set $A \cap K$ is compact. As a consequence, a map $f : W \to Y$ is closed if and only if, for all $A \subseteq W$ closed, and all $K \subseteq Y$ compact, the set $f(A) \cap K = f(A \cap f^{-1}(K))$ is compact, but this is always the case if f is proper map. We can thus conclude that a proper map is universally closed, since, by Proposition 3.1.9.2-(1), if $\pi_1 : X \to B$ is proper, then for all $\pi_2 : X_2 \to B$ the map $p_2 : X_1 \times_B X_2 \to X_2$ is proper, hence closed.

Conversely, let $\pi_1 : X \to B$ be universally closed and let $K \subseteq B$ be compact. Then, for $\pi_2 : K \subseteq B$ the inclusion map, the map $p_2 : X_1 \times_B K \to K$ is universally closed, and since $K \to \{\bullet\}$ is universally closed also, their composition, i.e. the constant map $X_1 \times_B K \to \{\bullet\}$, is universally closed. But a constant map on a locally closed space $W \to \{\bullet\}$ is universally closed (if and) only if W is compact. Indeed, denote by $\widehat{W}$ the Alexandroff compactification of W. The map $p_2 : W \times_{\{\bullet\}} \widehat{W} \to \widehat{W}$ is a closed map and the image of the diagonal $\Delta_W \subseteq W \times \widehat{W}$, which is closed since the spaces are Hausdorff, is also closed, hence compact. But this image is homeomorphic to Δ_W which, in turn, is homeomorphic to W. We have thus proved that $X_1 \times_B K$ is compact. We can now conclude that π_1 is proper. Indeed, for $K' \subseteq B$ compact, the subspace $K := K' \cap \pi_1(X)$ is also compact since $\pi_1(X)$ is closed. But then, $\pi_1^{-1}(K') = \pi_1^{-1}(K) = X_1 \times_B K$, which we already showed it is compact.

23. *Exercise 3.1.9.3-(4) (p. 65).* Applying Proposition 3.1.9.2-(2), we get the factorization $\varphi = p_2 \circ \xi$ through the fiber product:

$$
\begin{array}{ccc}
\pi^{-1}(U) & \xrightarrow{\ \varphi\ } & F \\
{\scriptstyle\pi}\downarrow & \square & \downarrow \\
U & \longrightarrow & \{\bullet\}
\end{array}
\qquad\Bigg\|\qquad
\begin{array}{ccccc}
\pi^{-1}(U) & \xrightarrow{\ \xi\ } & U \times F & \xrightarrow{\ p_2\ } & F \\
{\scriptstyle\pi}\downarrow & & {\scriptstyle p_1}\downarrow & \square & \downarrow \\
U & = & U & \longrightarrow & \{\bullet\}
\end{array}
$$

where for ξ, a priori a continuous bijection, to be a homeomorphism it is necessary an sufficient that it be proper, i.e. that φ^{-1} preserves properness after Proposition 3.1.9.2-(2a).

In the category of manifolds, to see that ξ is a diffeomorphism it suffices to show that it is locally invertible, i.e. étale, which is clearly equivalent to ask that the restrictions of φ to fibers are all étales, since the first coordinate π already provides the full horizontal tangent spaces.

24. *Exercise 3.1.10.1 (p. 65).* We have $|\phi\,\beta| \subseteq \pi^{-1}(|\phi|) \cap |\beta|$, so that if $\beta \in \Omega_c(E)$, then $\phi\,\beta \in \Omega_c(E)$. Conversely, by Urysohn's lemma, given $K \subseteq B$ compact, there exists $\phi \in \Omega_c^0(B)$ such that $K \subseteq |\phi|$. Whence $\pi^{-1}(K) \cap |\beta|$ is compact since closed and contained in $|\phi\,\beta|$.

25. *Construction of a Thom form 3.2.3 (p. 81).* Given an oriented fiber bundle of manifolds (E, B, π, M) with connected fiber M of dimension d_M, we show that there always exists a differential form $\zeta_\pi \in \Omega_{cv}^{d_M}(E)$ such that

$$
\int_M \pi^*(\alpha) \wedge \zeta_\pi = \alpha\,, \qquad \forall \alpha \in \Omega(B)\,.
$$

Let $\mathscr{U} := \{U_i\}_{i \in \mathfrak{I}}$ be a trivializing cover of B, and fix some oriented family of trivializations $\{\Phi_i : \pi^{-1}(U_i) \to U_i \times M\}_{i \in \mathfrak{I}}$. We have the Cartesian diagram of oriented fiber bundles of fiber M

$$\begin{array}{ccccc} \pi^{-1}(U_i) & \xrightarrow[\simeq]{\Phi_i} & U_i \times M & \xrightarrow{p_2} & M \\ {\scriptstyle\pi}\downarrow & \square & \downarrow{\scriptstyle p_1} & & \\ U_i & = & U_i & & \end{array}$$

i

and, after Proposition 3.2.1.3, a commutative diagram

$$\begin{array}{ccc} \Omega_{cv}(U_i \times M)[d_M] & \xrightarrow[\simeq]{\Phi_i^*} & \Omega_{cv}(\pi^{-1}(U_i))[d_M] \\ {\scriptstyle\int_M}\downarrow & & \downarrow{\scriptstyle\int_M} \\ \Omega(U_i) & = & \Omega(U_i)\,. \end{array}$$

Hence, if we define

$$\zeta_{\pi,i} := \Phi_i^*(p_2^*(\zeta_M)) \in \Omega_{cv}^{d_M}(\pi^{-1}(U_i))\,,$$

where $\zeta_M \in \Omega_c^{d_M}(M)$ is a differential form representing ζ_M, we get the equality

$$\int_M \pi^*(\alpha_i) \wedge \zeta_{\pi,i} = \int_M \pi^*(\alpha_i) \wedge p_2^*(\zeta_M) = \alpha_i\,, \quad \forall \alpha_i \in \Omega(U_i). \tag{E.9}$$

Now, taking a partition of unity $\{\phi_i\}_{i\in\mathfrak{I}}$ subordinate do $\mathscr{U}$, we define

$$\zeta_\pi := \sum_i \pi^*(\phi_i)\,\zeta_{\pi,i} \in \Omega_{cv}^{d_M}(E)\,.$$

By (E.9), we then have, for all $\alpha \in \Omega(B)$,

$$\begin{aligned} \int_M \pi^*(\alpha) \wedge \zeta_\pi &= \int_M \pi^*(\alpha) \wedge \sum_i \pi^*(\phi_i)\,\zeta_{\pi,i} \\ &= \sum_i \int_M \pi^*(\phi_i\,\alpha) \wedge \zeta_{\pi,i} = \sum_i \phi_i\,\alpha = \alpha\,. \end{aligned}$$

26. *Exercise 3.7.1-(1) (p. 104)*. The sequence is a complex since, the support $|\beta|$ of $\beta \in \Omega_c(U)$, being compact disjoint to F, there exists an open neighborhood $W \supseteq F$ disjoint to $|\beta|$, hence $R_W^M(\beta) = 0$.

The injectivity of $j_!$ and the fact that $\ker(R_{\mathscr{F}}^M) \subseteq \Omega_c(U)$ are clear.

To see that $R_{\mathscr{F}}^M$ is onto, let now $W \supseteq F$ be relatively compact, and let $\{\phi_1, \phi_2\}$ be a partition of unity relative to the open cover $\{W, M \smallsetminus F\}$. Notice that $|\phi_1|$ is compact since closed in $\overline{W}$ which is compact. Then, for any open subspace $V \supseteq F$

and any $\alpha \in \Omega(V)$, we have $\phi_1\alpha \in \Omega_c(M)$ and $R^M_{W\cap V}(\phi_1\alpha) = R^V_{W\cap V}(\alpha)$. Hence, the surjectivity of $R^M_{\mathscr{F}}$.

27. *Exercise 3.7.1-(1) (p. 104).* The *Tubular Neighborhood Theorem* states that every submanifold N in M has an open neighborhood V diffeomorphic to the normal bundle of N in M (Sect. 3.2.3.1). Since the statement is also true for any open neighborhood $V \supseteq F$ in place of M, the collection $\mathscr{F}_T$ of tubular neighborhoods of F is cofinal in $\mathscr{F}$, in which case the natural map

$$\varinjlim_{\mathscr{F}} \Omega(V) \to \varinjlim_{\mathscr{F}_T} \Omega(V)$$

is an isomorphism.

We can now easily conclude since all the restrictions R^V_V are quasi-isomorphic and that a filtrant inductive limit of quasi-isomorphisms is a quasi-isomorphism.

28. *Exercise 3.7.1-(1) (p. 104).* Since the composition $R^{\mathscr{F}}_F \circ R^M_{\mathscr{F}}$ is the restriction morphism $i^* : \Omega_c(M) \to \Omega(F)$, we get the short exact sequence of complexes:

$$0 \to \Omega_c(U) \xrightarrow{j_!} \Omega_c(M) \xrightarrow{i^*} \Omega(F) \to 0$$

inducing the long exact sequence 3.85.

29. *Exercise 3.7.1-(1) (p. 105).* By the general definition of the connecting morphism corresponding to the short exact sequence

$$0 \to \Omega_c(U) \xrightarrow{j_!} \Omega_c(M) \xrightarrow{i^*} \Omega(F) \to 0,$$

we have to lift $\omega \in \Omega(F)$ to $\Omega_c(M)$. For this, we multiply $\pi^*\omega \in \Omega(\mathbb{B}_{2\epsilon})$ by a Urysohn function $\phi : M \to \mathbb{R} \geq 0$ of compact support equal to 1 on a neighborhood of F. The differential form $\phi\pi^*\omega$ has now compact support and verifies $i^*(\phi\pi^*\omega) = \omega$. When ω the snake diagram

$$\begin{array}{ccccccc}
0 \to & \Omega_c(U) & \xrightarrow{j_!} & \Omega_c(M) & \xrightarrow{i^*} & \Omega(F) \to 0 \\
 & & & \phi\pi^*\omega & \longmapsto & \omega \\
 & & & \downarrow d & & \downarrow d \\
 & d\phi\wedge\pi^*\omega & = & d\phi\wedge\pi^*\omega & \longmapsto & 0
\end{array}$$

gives the connecting morphism

$$c : H(F) \longrightarrow H_c(U)[1], \quad \omega \mapsto d\phi \wedge \pi^*\omega\,.$$

Looking closely, we see that in fact $d\phi \wedge \pi^*\omega \in \Omega_c(\mathbb{B}_{2\epsilon} \setminus F)$, where we have a Cartesian diagram of fibrations

$$\begin{array}{ccc} \mathbb{B}_{2\epsilon} \setminus F & \longrightarrow & \mathbb{R} \times \mathbb{S}_\epsilon \\ {\scriptstyle p}\downarrow & & {\scriptstyle p_2}\downarrow \\ \mathbb{S}_\epsilon & = & \mathbb{S}_\epsilon \end{array}$$

so that we can apply the study of the zero section of a vector bundle in Sect. 3.6.5. The Gysin map

$$\sigma_! : H(\mathbb{S}_\epsilon) \to H_c(\mathbb{B}_{2\epsilon} \setminus F)$$

is therefore an isomorphism with inverse the morphism $p_!$ of integration along fibers (*cf*. 3.6.2). Hence,

$$\begin{aligned} d\phi \wedge \pi^*\omega &= \sigma_!\big(p_!(d\phi \wedge \pi^*\omega)\big) = \sigma_!\big(p_!(d\phi \wedge p^*(\pi'^*\omega))\big) \\ &= \sigma_!\big(p_!(d\phi) \wedge \pi'^*\omega)\big) = \sigma_!\big(\pi'^*\omega)\big) \end{aligned}$$

after the projection formula (2.6.2.1-(2)).

To finish, we have still to pushforward $d\phi \wedge \pi^*\omega$ from $\Omega_c(\mathbb{B}_{2\epsilon} \setminus F)$ to U, which is obviously done through the extension by zero morphism $j_{\epsilon !}$.

30. *Exercise 3.7.1-(2) (p. 105).* This is almost tautological. By duality, we have the long exact sequence

$$\cdots \to H_c(F)^\vee \xrightarrow{i^{*\vee}} H_c(M)^\vee \xrightarrow{j_!^\vee} H_c(U)^\vee \xrightarrow{c^\vee} H_c(F)^\vee \to \cdots$$

that we can link through Poincaré adjunctions,

$$\begin{array}{ccccccccc} H(F)[d_F] & \xrightarrow{i_*} & H(M)[d_M] & \xrightarrow{j^*} & H(U)[d_M] & \xrightarrow[{[1]}]{\delta} & H(F)[d_F][1] & \longrightarrow \\ {\scriptstyle \mathbb{D}_F}\downarrow\simeq & \text{(I)} & {\scriptstyle \mathbb{D}_M}\downarrow\simeq & \text{(II)} & {\scriptstyle \mathbb{D}_U}\downarrow\simeq & & {\scriptstyle \mathbb{D}_F[1]}\downarrow\simeq & \\ H_c(F)^\vee & \xrightarrow{i^{*\vee}} & H_c(M)^\vee & \xrightarrow{j_!^\vee} & H_c(U)^\vee & \xrightarrow[{[1]}]{c^\vee} & H_c(F)^\vee[1] & \longrightarrow \end{array}$$

Where i_* is the Gysin morphism *in cohomology* corresponding to the closed embedding $i : N \subseteq M$. The subdiagrams (I) and (II) are commutative thanks to the exchanges of Poincaré adjoints $j_! \leftrightarrow j^*$ and $i^* \leftrightarrow i_*$ (see 2.5.1.1-(4)). The connecting morphism δ is then the left Poincaré adjoint to c.

31. *Exercise 3.7.1-(2) (p. 106).* The connecting morphism δ is the left Poincaré adjoint to c which was decomposed in (1) as

$$H(F) \xrightarrow{\pi^*} H_c(\mathbb{S}_\epsilon) \xrightarrow{\sigma_![-d_{\mathbb{S}_\epsilon}]} H_c(\mathbb{B}_{2\epsilon} \smallsetminus F)[1] \xrightarrow{j_{\epsilon!}[1]} H_c(U)[1]$$

$$c : H(F) \to H_c(U)[1]$$

By the theorem of adjoint morphisms 2.5.1.1-(4), we see that δ is obtained by successively applying

- the adjoint to the zero extension $j_{\epsilon!}$, i.e. the restriction j_ϵ^*
- the adjoint to Thom isomorphism $\sigma_!$, i.e. the restriction i^*,
- the adjoint of the pullback π^*, i.e. the integration along fibers $\pi_!$.

Hence, the announced formula

$$H(U) \ni \alpha \mapsto \delta(\alpha) = \int_{\mathbb{S}_\epsilon} \alpha|_{\mathbb{S}_\epsilon} \in H(F)\,.$$

32. *Exercise 3.7.2-(1) (p. 106).* Both result by adjunction. The first:

$$\int_M \mathrm{Gr}(f)^*(\delta_*(1)) = \int_M \mathrm{Gr}(f)^*(\delta_*(1)) \cup 1 = \int_{M\times M} \delta_*(1) \cup \mathrm{Gr}(f)_*(1)\,.$$

The second:

$$\begin{aligned}\int_{M\times M} \delta_*(1) \cup \mathrm{Gr}(f)_*(1) &= (-1)^{d_M.d_M} \int_{M\times M} \mathrm{Gr}(f)_*(1) \cup \delta_*(1) \\ &= (-1)^{d_M} \int_M \delta^*\big(\mathrm{Gr}(f)_*(1)\big)\,.\end{aligned}$$

33. *Exercise 3.7.2-(2) (p. 106).* If f has no fixed points, then the map $\mathrm{Gr}(f)$ factors through the open subspace $(M \times M) \smallsetminus \Delta_M$:

$$M \xrightarrow{\mathrm{Gr}'(f)} (M \times M) \smallsetminus \Delta_M \xrightarrow{j} M \times M$$

$$\mathrm{Gr}(f) : M \to M \times M$$

where $\mathrm{Gr}'(f)$ denotes the restriction of $\mathrm{Gr}(f)$ and j is the open inclusion. But then

$$\mathrm{Gr}(f)_* = \mathrm{Gr}(f)_! = j_! \circ \mathrm{Gr}'(f)_!\,,$$

where $j_!$ is the extension by zero, in which case $\delta^* \circ j_! = 0$ and

$$\delta^* \circ \mathrm{Gr}(f)_* = \mathrm{Gr}(f)_! = (\delta^* \circ j_!) \circ \mathrm{Gr}'(f)_! = 0\,.$$

Hence, $\Lambda_f = 0$, after the last equality in 3.89.

34. *Exercise 3.7.2-(3) (p. 106).* Since $\delta_*(1) \in H^{d_M}(M \times M)$ and that by Künneth's theorem we have $H(M \times M) = H(M) \otimes H(M)$, the decomposition of $\delta_*(1)$ in the basis $\{e_i \otimes e'_j\}$ looks like:

$$\delta_*(1) = \sum\nolimits_{[e_i]+[e_j]=d_M} x_{i,j}\; e_i \otimes e'_j\,.$$

The coefficients $x_{a,b}$ are given by the formula

$$x_{a,b} = (-1)^{[e_b][e_a]} \int_{M\times M} (e'_a \otimes e_b) \cup \delta_*(1)\,,$$

which gives, by adjunction,

$$\begin{aligned} x_{a,b} &= (-1)^{[e_b][e_a]} \int_M \delta^*(e'_a \otimes e_b) \\ &= (-1)^{[e_b][e_a]} \int_M e'_a \cup e_b = \{\, (-1)^{[e_a]} \text{ if } a = b, \mathbf{0} \text{ if } a \neq b. \end{aligned}$$

Hence, the announced formula:

$$\delta_*(1) = \sum\nolimits_{i\in I} (-1)^{\deg(e_i)}\; e_i \otimes e'_i\,.$$

Therefore,

$$\int_M \delta_*(1)|_\Delta = \sum_{0\le k\le d_M} \Big((-1)^k \sum_{[e_i]=k} e_i \cup e'_i \Big) = \sum_{0\le k\le d_M} (-1)^k \dim\big(H^k(M)\big)\,.$$

35. *Exercise 3.7.2-(4) (p. 106).* After 3.7.2-(3), we have

$$\begin{aligned} \Lambda(f) := \int_M \mathrm{Gr}(f)^* \delta_*(1) &= \sum_{0\le k\le d_M} \Big((-1)^k \sum_{[e_i]=k} f^*(e_i) \cup e'_i \Big) \\ &= \sum\nolimits_{0\le k\le d_M} (-1)^k \mathrm{Tr}\big(f^* : H^k(M) \to H^k(M)\big)\,. \end{aligned}$$

The fixed points criterion results then from 3.7.2-(2).

36. *Exercise 3.7.2.1 (p. 107).* 1) By the Lefschetz Fixed point theorem 3.7.2-(3), we have

$$\int_M \mathrm{Eu}(\Delta_M) = \chi_M\,.$$

2) $j : M'_\epsilon \to M_\epsilon$ is an open embedding of fiber bundles of base M and the orientations of the fibers of π induce an orientation of the fibers of π'.
(2) The fibers of π' are clearly homeomorphic to $\mathbb{R}^d_M \smallsetminus \{0\}$.
(2) The sequence in question contains the exact subsequence:

$$\begin{array}{c} \to H^{d_M-1}(M) \overset{c}{\twoheadrightarrow} H^{d_M}_{\mathrm{cv}}(M'_\epsilon) \underset{0}{\overset{j_!}{\to}} H^{d_M}_{\mathrm{cv}}(M_\epsilon) \overset{\rho}{\hookrightarrow} H^{d_M}(M) \to \\ \qquad\qquad\qquad\qquad\qquad \wr\| \\ \qquad\qquad\qquad\qquad\qquad \mathbb{R} \ni \Phi_\pi \longmapsto \mathrm{Eu}(\Delta_M) \end{array}$$

where $\pi : M_\epsilon \to M$ is a vector bundle, so that $\delta_! : H^0(B) \to H^{d_M}_{\mathrm{cv}}(M_\epsilon)$ is the Thom isomorphism (3.6.5.1). The restriction map ρ is therefore necessarily injective, since $\rho(\delta_!(1)) = \mathrm{Eu}(\Delta_M) \neq 0$, after (1). As a consequence, $j_! = 0$ and the connecting morphism c is surjective.
(2) After (2), it suffices to take any manifold M such that $\chi_M \neq 0$ and $H^{d_M-1}(M) = 0$. For example, an even dimensional sphere $M := \mathbb{S}^{2m}$.

Ch. 4. Equivariant Background

37. *Exercise 4.2.2.1 (p. 120).* (3)⇒(1)⇒(2) is obvious. (2)⇒(1). Let $\mathcal{A}$ be the set whose members are the sets A of simple submodules $S \subseteq V$, such that the sums $\sum A := \sum_{S\in A} S$ are direct sums. Notice that $\mathcal{A}$ is nonempty as it contains the singletons $\{S\}$. We endow $\mathcal{A}$ with the partial order of set inclusion.

We claim that $(\mathcal{A}, \subseteq)$ is an inductive poset. Indeed, let $C := \{A_\mathfrak{a}\}_{\mathfrak{a}\in(\mathfrak{A},\leq)}$ be a totally ordered subset in $(\mathcal{A}, \subseteq)$. Express the union $\bigcup_{\mathfrak{a}\in\mathfrak{A}} A_\mathfrak{a}$ as the family of its elements $\{S_\mathfrak{b}\}_{\mathfrak{b}\in\mathfrak{B}}$. If we have $\sum_{\mathfrak{b}\in\mathfrak{B}} s_\mathfrak{b} = 0$ with almost all $s_\mathfrak{b} \in S_\mathfrak{b}$ equal to zero, then the same vanishing expression would be verified in some $A_\mathfrak{a}$, for $\mathfrak{a} \in (\mathfrak{A}, \leq)$ big enough, in which case $s_\mathfrak{b} = 0$ for all $\mathfrak{b} \in \mathfrak{B}$. As a consequence, the set $\bigcup_{\mathfrak{a}\in\mathfrak{A}} A_\mathfrak{a}$ is a member of $\mathcal{A}$ and is also an upper bound for C.

We can therefore apply the Zorn Lemma, and state that there exists a maximal element $A \in (\mathcal{A}, \subseteq)$. We claim that the sum $\sum A$ contains every simple submodule $S \subseteq V$. Indeed, if $S \not\subseteq \sum A$ then the sum $S + \sum A$ is direct, and $A \sqcup \{S\}$ is a member of $(\mathcal{A}, \subseteq)$ strictly greater than A, which is not possible.

We have thus proved that if V is a sum of simple submodules, then there exists a set A of simple submodules of V, such that the sum $\sum A$ is a direct sum containing every simple submodule of V. Hence $V = \bigoplus A$, and V is semisimple.

(1,2)⇒(3,4). If V is a sum of simple modules, then the same holds for any quotient $\nu : V \to Q$, and Q is semisimple after (2⇒1). Moreover, in proving the semisimplicity of Q, we consider the poset $(\mathcal{A}(\nu), \subseteq)$, whose members are the sets A of simple submodules $S \subseteq V$ such that $\nu(S) \subseteq Q$ is simple, and such that the sum $\sum_{S\in A} \nu(S)$ is a direct sum in Q. This condition automatically implies that $\sum A := \sum_{S\in A} S$ is a direct sum in V, and, therefore, that the restriction $\nu : \sum A \to \nu(\sum A)$ is an isomorphism. In particular, if A is maximal in $(\mathcal{A}(\nu), \subseteq)$, then $\sum A$ is a semisimple complement to $\ker(\nu)$, which proves (4).

(4)⇒(1). By induction on $\dim V$. If $\dim V \leq 1$ semisimplicity is obvious. Otherwise, the module V is either simple, or it has a proper submodule $0 \subsetneq W \subsetneq V$

with some complement W'. But then, since both, W and W', verify (4) and are of strictly smaller dimensions than V, we can conclude that they are semisimple, hence, that V is semisimple.

38. *Exercise 4.2.5.2 (p. 122).* Denote by $\nu = M \to M/N$ the canonical projection. Given $M' \subseteq M$ such that $N \subseteq M'$ and that M'/N is finite dimensional, let $\varphi : M'/N \to M$ lift the inclusion $M'/N \subseteq M/N$. The induced morphism $\nu : \varphi(M'/N) \to M'/N$ is then isomorphism, and we have $\varphi(M'/N) \oplus N = M'$. Conversely, let V be finite dimensional and let $\varphi : V \to M/N$ be given. Since N is of finite codimension in $M' := \nu^{-1}(\varphi(V))$, it admits a complementary $\mathfrak{g}$-submodule $H \subseteq M'$ by hypothesis. But then the restriction of ν induces an isomorphism $\nu_H : H \to \varphi(V)$, and $\nu_H^{-1} \circ \varphi : V \to M$ lifts φ.

39. *Basic Elements in Cartan-Weil Complexes (Sect. 4.3.2) (p. 125).*
To show that the map

$$\Xi' : \big(\Lambda(\mathfrak{g}) \otimes S(\mathfrak{g}^\vee) \otimes C\big)^{\mathrm{hor}} \to \frac{\Lambda(\mathfrak{g}) \otimes S(\mathfrak{g}^\vee) \otimes C}{\Lambda^+(\mathfrak{g}) \otimes S(\mathfrak{g}^\vee) \otimes C} = S(\mathfrak{g}^\vee) \otimes C$$

is a θ-equivariant isomorphism we can disregard the $S(\mathfrak{g}^\vee)$ factor (since interior products are $S(\mathfrak{g}^\vee)$-linear), which reduces us to simply showing

$$\begin{cases} \text{(i) } (\Lambda(\mathfrak{g}^\vee) \otimes C)^{\mathrm{hor}} \cap (\Lambda^+(\mathfrak{g}^\vee) \otimes C) = 0\,; \\ \text{(ii) } (\Lambda(\mathfrak{g}^\vee) \otimes C)^{\mathrm{hor}} + (\Lambda^+(\mathfrak{g}^\vee) \otimes C) = \Lambda(\mathfrak{g}^\vee) \otimes C\,. \\ \text{(iii) } (\Lambda(\mathfrak{g}^\vee) \otimes C)^{\mathrm{hor}} \text{ is } \theta\text{-stable.} \end{cases}$$

(i) For $d \in \mathbb{Z}$, defining $C_d := \sum_{i \le d} C^i$, one gets an increasing $\star$-filtration

$$(\Lambda(\mathfrak{g}^\vee) \otimes C)_\star := \big(\cdots \subseteq (\Lambda(\mathfrak{g}^\vee) \otimes C)_d \subseteq (\Lambda(\mathfrak{g}^\vee) \otimes C)_{d+1} \subseteq \cdots\big),$$

by graded subspaces which are stable by interior products $\iota(X)$, for all $X \in \mathfrak{g}$. The action of $\iota(X)$ on $\mathrm{Gr}^\star\big((\Lambda(\mathfrak{g}^\vee) \otimes C)_\star\big)$ is then given by the map

$$\iota(X) \otimes 1 : \Lambda(\mathfrak{g}^\vee) \otimes C \to \Lambda(\mathfrak{g}^\vee) \otimes C\,,$$

so that

$$\mathrm{Gr}^\star\big((\Lambda(\mathfrak{g}^\vee) \otimes C)_\star\big)^{\mathrm{hor}} = \Lambda(\mathfrak{g}^\vee)^{\mathrm{hor}} \otimes C\,.$$

But, elementary reasons give the equalities

$$\Lambda(\mathfrak{g}^\vee)^{\mathrm{hor}} = \bigcap\nolimits_{X \in \mathfrak{g}} \ker(\iota(X) : \Lambda(\mathfrak{g}^\vee) \to \Lambda(\mathfrak{g}^\vee)) = \bigcap\nolimits_{X \in \mathfrak{g}} \Lambda(X^\perp) = \Lambda(\{0\})\,,$$

where $X^\perp := \{\lambda \in \mathfrak{g}^\vee \mid \lambda(X) = 0\}$, which lead us to conclude that

$$(\Lambda^+(\mathfrak{g}^\vee) \otimes C)^{\mathrm{hor}} = 0\,,$$

since the $\star$-filtration is regular.

(ii) We need only check the inclusion

$$1 \otimes C \subseteq (\Lambda(\mathfrak{g}^\vee) \otimes C)^{\mathrm{hor}} + (\Lambda^+(\mathfrak{g}^\vee) \otimes C)\,,$$

which is obvious after the equality

$$1 \otimes \omega = \Big(1 \otimes \omega - \sum\nolimits_i e^i \otimes \iota(e_i)\omega\Big) + \Big(\sum\nolimits_i e^i \otimes \iota(e_i)\omega\Big)\,,$$

where $\{e_i\}$ is a basis of $\mathfrak{g}$, of dual basis $\{e^i\}$, and where the first term in the r.h.s. is clearly horizontal, as it is an obvious zero of the interior products $\iota(e_j)$.

(iii) If $\varpi \in (\Lambda(\mathfrak{g}^\vee) \otimes C)^{\mathrm{hor}}$, then, for all $X, X' \in \mathfrak{g}$, we have

$$\iota(X')\theta(X)(\varpi) = \big(\iota(X')\theta(X) - \theta(X)\iota(X')\big)(\varpi) \underset{1}{=} \iota([X', X])(\varpi) = 0\,,$$

where $(=_1)$ is condition 4.13-(iii) for $\mathfrak{g}$-dgm's (in Sect. 4.2.3). We have thus proved that $\theta(X)(\varpi) \in (\Lambda(\mathfrak{g}^\vee) \otimes C)^{\mathrm{hor}}$, for all $X \in \mathfrak{g}$, as announced.

40. *Exercise 4.3.4.2 (p. 128).* The statement results by proving that, given two G-modules $N \subseteq M$, we have $N|M$ as G-modules, if and only if we have $N|M$ as G_0-modules.

Recall that the subgroup G_0 is an open normal subgroup of G, and the quotient $W := G/G_0$ is a finite group since G is compact. Recall also that on the category of $\mathbb{R}[G]$-modules, we have natural identifications of bifunctors:

$$\mathrm{Hom}_G(\_,\_) = \mathrm{Hom}_{\mathbb{R}}(\_,\_)^G = \big(\mathrm{Hom}_{\mathbb{R}}(\_,\_)^{G_0}\big)^W = \mathrm{Hom}_{G_0}(\_,\_)^W\,.$$

Furthermore, given a G-module V and a G_0-module V', we have canonical isomorphisms of functors from $\mathrm{Mod}(G)$ to $\mathrm{Vec}(\mathbb{R})$:

$$\begin{cases} \text{(i)} \qquad \mathrm{Hom}_G(V,\_) \simeq \mathrm{Hom}_{G_0}(V,\_)^W\,, \\ \text{(ii)}\ \mathrm{Hom}_G(\mathbb{R}[G] \otimes_{G_0} V',\_) \simeq \mathrm{Hom}_{G_0}(V',\_)\,. \end{cases} \tag{E.10}$$

– Assume that $N|M$ as G_0-modules. Let V be a finite dimensional G-module. Since V is also a G_0-module, the map $\mathrm{Hom}_{G_0}(V, M) \to \mathrm{Hom}_{G_0}(V, M/N)$ is surjective, but then $\mathrm{Hom}_{G_0}(V, M)^W \to \mathrm{Hom}_{G_0}(V, M/N)^W$ is surjective too, since the functor W-invariants $(\_)^W$ is exact on the category $\mathrm{Mod}(\mathbb{R}[W])$. Whence we have $N|M$ as G-modules, by isomorphism (i)-(E.10).

– Assume that $N|M$ as G-modules. If V' is a finite dimensional G_0-module, then $V := \Bbbk[G] \otimes_{G_0} V'$ is a finite dimensional G-module (since W is finite), the map $\mathrm{Hom}_G(V, M) \to \mathrm{Hom}_G(V, M/N)$ is surjective, and we can conclude that $N|M$ as G-modules, by isomorphism (ii)-(E.10).

41. *Exercise 4.7.2.1 (p. 150).* For every space X, the set $\Omega^d(X; \Bbbk)$ consists of all the set-theoretic maps $f : X^{d+1} \to \Bbbk$, and for $Z \subseteq X$, the restriction morphism $\Omega^d(X; \Bbbk) \to \Omega^d(Z; \Bbbk)$ is simply the restriction of maps from X^d to Z^d. If $X = \bigcup \uparrow_{n\in\mathbb{N}} Z_n$ is an increasing cover, any d-tuple in X^d, belongs to Z_n^d for some $n \in \mathbb{N}$, so that $X^d = \bigcup \uparrow_{n\in\mathbb{N}} Z_n^d$. The bijectivity of (i,ii) in (4.52) then immediately follows. For (iii), let $M = \bigcup \uparrow_{k\in\mathbb{N}} K_k$ be an increasing cover of M by compact subspaces. The subspaces $K_{n,G}(n) \subseteq M_G(n)$ are compact and the *indicator function* $\mathbf{1}_n$ of $K_{n,G}(n)^d$ belongs to $\Omega^d_{\mathrm{cv}}(M(n); \Bbbk)$. The family $\{\mathbf{1}_n\}_{n\in\mathbb{N}}$ is a projective family whose limit is the constant map $\mathbf{1}_\infty : M_G^d \to \{1\}$, but $\mathbf{1}_\infty \in \Omega^d_{\mathrm{cv}}(M_G; \Bbbk)$ if and only if M is compact.

42. *Exercise 4.10.1.3 (p. 174).*
(1,2) These are straightforward applications of Theorem 4.3.3.1-(2). The filtration is given by the subcomplexes $K_m = \big(S^{\geq m}(\mathfrak{g}^\vee) \otimes \hat{c}(\alpha)\big)^{\mathfrak{g}}$ for all $m \in \mathbb{N}$. The rest of the argument is the same as in the proof of 4.3.3.1-(2).
(3) The hint refers to the fact that "*the spectral sequence of the mapping cone of a morphism of filtered complexes is the mapping cone of the induced morphism on the corresponding spectral sequences*". The exercise is then immediate.

43. *Proof of Proposition 5.1.2.2-(3) (p. 178).* Let $A := S(\mathfrak{g}^\vee)^{\mathfrak{g}} = \mathbb{R}[X_1, \ldots, X_r]$ and let $W := G/G_0$. Following Hilbert's Syzygy Theorem, for every A^W-graded module M, the module $A \otimes_{A^W} M$ admits a resolution of A-graded modules

$$0 \to L_r \to \cdots \to L_1 \to L_0 \to A \otimes_{A^W} M \to 0\,, \tag{E.11}$$

where the L_i's are *free* A-modules. The submodules $(L_i)^W$ are then *free* A^W-modules and, applying the exact functor $(\_)^W$ to (E.11), we conclude that

$$\text{dim-proj}_{A^W}\big((A \otimes_{A^W} M)^W\big) \leq r\,.$$

On the other hand, $(A \otimes_{A^W} M)^W \simeq M$ since the action of W on M is trivial. Hence, $\text{dim-proj}_{A^W}(M) \leq r$ for all $M \in \mathrm{Mod}(A^W)$, and $\dim_{\mathfrak{h}}(A^W) \leq \dim_{\mathfrak{h}}(A)$.

Ch. 6. Equivariant Euler Classes

44. *Exercise 6.5.2.3 (p. 229).* Use the projection formula for Gysin morphisms.

45. *Exercise 6.5.4.3 (p. 233).* A maximal torus $T \subseteq \mathrm{SO}(3)$ is the group of rotations around a one dimensional vector subspace $L \subseteq \mathbb{R}^3$. Hence, although $\dim(\mathbb{R}^3)^G = 0$, we have $\dim(\mathbb{R}^3)^T = 1$, and then $\mathrm{Eu}_G(0, \mathbb{R}^3) = \mathrm{Eu}_T(0, \mathbb{R}^3) = 0$, after 6.5.4.2–(3).

Ch. 7. Localization

46. *Exercise 7.2.2 (p. 236).* Let M be an H_G-gm. Show that the canonical map $M \to Q_G \otimes_{H_G} M$ is injective if and only if M is torsion-free. Show that if M is also of finite type, then the natural map $\mathbf{Hom}^\bullet_{H_G}(M, H_G) \to \mathbf{Hom}^\bullet_{H_G}(M, Q_G)$ induces an isomorphism

$$Q_G \otimes_{H_G} \mathbf{Hom}^\bullet_{H_G}(M, H_G) \simeq \mathbf{Hom}^\bullet_{Q_G}(Q_G \otimes_{H_G} M, Q_G).$$

Apply 7.2.1.

47. *Exercise 7.4.1.1–(3) (p. 237).* Let $\lambda : V \to Q_G$ be H_G-linear. If V is torsion and $v \in V$, we have $P.v = 0$ for some $P \in H_G \setminus \{0\}$. But then $P.\lambda(v) = \lambda(P.v) = 0$, in which case $\lambda(v) = 0$, because P is invertible in Q_G. Since this is true for all $v \in V$, we conclude that $\lambda = 0$.

Conversely, if $v \in V$ is not torsion, we have $H_G \simeq H_G \cdot v \subseteq V$, and we claim that the linear map $\lambda : H_G \cdot v \to H_G$, $P \cdot v \mapsto P$, can be extended to the whole of V. Indeed, applying the localization functor $Q_G \otimes_{H_G} (\_)$ to the inclusion $\iota : H_G \cdot v \rightarrowtail V$, we get a commutative diagram

$$\begin{array}{ccc} H_G \cdot v & \xrightarrow{\iota} & V \\ {\scriptstyle j_v}\downarrow & & \downarrow{\scriptstyle j_V} \\ Q_G = Q_G \otimes_{H_G} H_G \cdot v & \xrightarrow{\alpha} & Q_G \otimes_{H_G} V \end{array} \quad (\rho : Q_G \otimes_{H_G} V \to Q_G \otimes_{H_G} H_G \cdot v) \tag{E.12}$$

where $j_V(u) := 1 \otimes u$, $j_v(P \cdot v) = P \otimes v$, and α is induced by ι. The morphism α is injective, since the localization functor is exact (Q_G is a flat H_G-module). But, as injection between Q_G-vector spaces, α admits a Q_G-linear retraction ρ, i.e. $\rho \circ \alpha = \mathrm{id}$. Then, the H_G-linear map $\rho \circ \iota_V : V \to Q_G$ is nonzero because it extends j_v. We have therefore proved that if V in not a torsion module, then $\mathrm{Hom}_{H_G}(V, Q_G) \neq 0$

48. *Exercise 7.4.1.2–1 (p. 237).* For $P \in H_G$, let $W(P) := H_G/(H_G \cdot P)$ and take $V := \bigoplus_{P \in H_G} W(P)$.

49. *On the General Slice Theorem 7.5.1 (p. 238).* The G-equivariant maps

$$G \times_{G_x} S(x) \xrightarrow{p} G/G_x \xrightarrow{c} \{\bullet\}$$

where $p([g, m]) := [g]$ and c is the constant map, give rise to maps of Borel constructions

$$\mathbb{E}G \times_G (G \times_{G_x} S(x)) \xrightarrow{\overline{p}} \mathbb{E}G \times_G G/G_x \xrightarrow{\overline{c}} \mathbb{E}G \times_G \{\bullet\}$$

where $\mathbb{E}G\times_G(G\times_{G_x}S(x)) = V(x)_G = S(x)_{G_x}$. Here the space G/G_x is compact, which implies that for a subspace $P \subseteq V(x)_G$ being $(\bar{c}\circ\overline{p})$-proper is equivalent to being $\overline{p}$-proper. Consequently, $\Omega_{\mathrm{cv}}(V(x)_G;\Bbbk) = \Omega_{\mathrm{cv}}(S(x)_{G_x};\Bbbk)$ and the equality $H_{G,\mathrm{c}}(V_x) = H_{G_x,\mathrm{c}}(S(x))$ is proved.

50. *Exercise 7.5.2.2 (p. 239).* Use the slice theorem. If $x \notin M^T$, show that the slice $S(x)$ is a strict submanifold of M stable under G_x and that $O_G(G\cdot S(x)) = O_{G_x}(S(x))$, then conclude by induction on $\dim(M)$. Otherwise, if $x \in M^T$, linearize the action as in Proposition 6.5.4.2 and conclude showing that there is a one-to-one correspondence between isotropy groups in the T-space T_xM and subsets of the set of nonzero weights of the linear representation of T on T_xM.

Ch. A. Appendix: Basics on Derived Categories

51. *Exercise A.1.4.2 (p. 270).* The equivalence (1)⇔(2) is tautological. The equivalence (2)⇔(3) comes from the fact that, since F is additive, it respects the mapping cone. We can then apply 3-(3b) and conclude.

52. *Exercise A.1.5.1 (p. 270).* Given $\alpha \in \mathrm{Mor}_{\mathcal{C}(\mathrm{Ab})}(A,B)$ and $h : A[1] \to B$, let $\alpha' = \alpha + h\,d + d\,h$. The map

$$L(h) : (\hat{c}(\alpha),\Delta) \to (\hat{c}(\alpha'),\Delta')\,,\qquad L(h)(b,a) := (b - h(a), a)\,. \tag{E.13}$$

verifies

$$\begin{aligned} L(h)\big(\Delta(b,a)\big) &= L(h)\big(db + \alpha(a), -da\big)\big) = (db + \alpha(a) + h(da), -da) \\ \Delta'\big(L(h)(b,a)\big) &= \Delta'\big(b - h(a), a\big) = (db - dh(a) + \alpha(a) + hd(a) + dh(a), -da) \\ &= (db + \alpha(a) + h(da), -da) \end{aligned}$$

and (E.13) is a morphism of complexes.

For the same reasons, the map $L(-h) : (\hat{c}(\alpha'),\Delta') \to (\hat{c}(\alpha),\Delta)$ is a morphism of complexes, and we can conclude since we have the obvious equalities

$$L(-h)\circ L(h) = \mathrm{id}_{\hat{c}(\alpha)} \quad\text{and}\quad L(h)\circ L(-h) = \mathrm{id}_{\hat{c}(\alpha')}\,.$$

The fact that mapping cones are canonically defined in the homotopy category $\mathcal{K}(\mathrm{Ab})$ is then a straightforward consequence of the fact that $L(h)$ is uniquely determined by the homotopy h.

53. *Exercise A.1.5.3-(1) (p. 272).* After Proposition A.1.5.2, applying the functor $\mathrm{Mor}_{\mathcal{K}}(C,\_)$ to the mapping cone $A \xrightarrow{\alpha} B \xrightarrow{\iota} \hat{c}(\alpha) \xrightarrow{p} A[1]$ we obtain the long exact sequence

$$\mathrm{Mor}_{\mathcal{K}}(C,\hat{c}(\alpha)[-1]) \xrightarrow{p_*} \mathrm{Mor}_{\mathcal{K}}(C,A) \xhookrightarrow{\alpha_*} \mathrm{Mor}_{\mathcal{K}}(C,B) \xrightarrow{\iota_*} \mathrm{Mor}_{\mathcal{K}}(C,\hat{c}(\alpha)) \xrightarrow{p_*}$$

where α_* is injective since α is assumed to be a monomorphism. This immediately implies that $p_* = 0$, which, in the particular case where $C = \hat{c}(\alpha)$, gives us $0 = p_*(\mathrm{id}_{\hat{c}(\alpha)}) = p$, and the fact that $\iota_* : \mathrm{Mor}_{\mathcal{K}}(\hat{c}(\alpha), B) \to \mathrm{Mor}_{\mathcal{K}}(\hat{c}(\alpha), \hat{c}(\alpha))$ is surjective, hence the existence of $\sigma \in \mathrm{Mor}_{\mathcal{K}}(\hat{c}(\alpha), B)$ such that $\iota \circ \sigma = \mathrm{id}_{\hat{c}(\alpha)}$, which is the announced section of ι. We have

$$\underbrace{\hat{c}(\alpha) \xrightarrow{\ \rho\ } B \xrightarrow{\ \iota\ } \hat{c}(\alpha)}_{\mathrm{id}}\,,$$

which allows to conclude that $i^* : \mathrm{Mor}_{\mathcal{K}}(\hat{c}(\alpha), C) \to \mathrm{Mor}_{\mathcal{K}}(B, C)$ is injective, hence, that ι is an epimorphism in $\mathcal{K}(\mathrm{Ab})$.

By a symmetric argument, replacing α is mono, by ι is epi, and $\mathrm{Mor}_{\mathcal{K}}(C, \_)$ by $\mathrm{Mor}_{\mathcal{K}}(\_, C)$, one shows that there exists $\rho : B \to A$ such that $\mathrm{id}_A = \rho \circ \alpha$, which is the announced retraction of α.

When the category $\mathcal{K}(\mathrm{Ab})$ is abelian we can further look at the kernel of ρ and the cokernels of ι, and show by standard arguments that the sequence

$$0 \longrightarrow A \underset{\rho}{\overset{\alpha}{\rightleftarrows}} B \underset{\sigma}{\overset{\iota}{\rightleftarrows}} \hat{c}(\alpha) \longrightarrow 0\,,$$

is a split sequence. The abelian category $\mathcal{K}(\mathrm{Ab})$ is therefore split. Now, if $\alpha : X \to Y$ is a morphism in Ab, the morphism $\alpha[0] : X[0] \to Y[0]$ is split in $\mathcal{K}(\mathrm{Ab})$, which means that we have direct decompositions $X[0] = K \oplus L$ and $Y[0] = L \oplus M$ in $\mathcal{K}(\mathrm{Ab})$, through which $\alpha[0]$ reads as $(\iota \circ p)(k, l) = (l, 0)$. All things preserved by the cohomology functor

$$\begin{array}{ccc} X[0] & \xrightarrow{\alpha[0]} & Y[0] \\ \simeq\updownarrow & & \updownarrow\simeq \\ K \oplus L & \xrightarrow{\iota \circ p} & L \oplus M \end{array} \quad \xrightarrow{\ \mathbf{h}\ } \quad \begin{array}{ccc} X & \xrightarrow{\qquad\alpha\qquad} & Y[0] \\ \simeq\updownarrow & & \updownarrow\simeq \\ \mathbf{h}(K) \oplus \mathbf{h}(L) & \xrightarrow{\iota \circ p} & \mathbf{h}(L) \oplus \mathbf{h}(M) \end{array}$$

the morphism α is split in Ab, and this ends the proof that Ab is split.

54. *Exercise A.1.5.3-2 (p. 272).* If Ab is a split category every complex $C \in C(\mathrm{Ab})$ is homotopy-equivalent to its cohomology (*cf.* answer of Exercise 2.1.6.1, p. 329) and the fully faithful functor $\mathrm{Ab}^{\mathbb{Z}} \to \mathcal{K}(\mathrm{Ab})$ which associates with a graded object the corresponding complex with 0 differential, is then essentially surjective, hence is an equivalence of categories.

55. *Exercise A.1.5.7-(1) (p. 273).* The map $\epsilon : \hat{c}(\alpha) \to C$ is a morphism of complexes since $\epsilon(\Delta(x, y)) = \epsilon(dx + \alpha(y), -dy) = \beta(dx) = d(\epsilon(x, y))$. If $\beta(x) \in C$ is a cocycle, we have $dx = y$ for some cocycle $y \in A$. Hence $(x, -y)$ is a cocycle in $\hat{c}(\alpha)$ and $\epsilon(x, -y) = \beta(x)$. But then, in cohomology, we get

$$c\big(\mathbf{h}(\epsilon)\overline{(x, -y)}\big) = \bar{y} = -\mathbf{h}(p)\big(\overline{(x, -y)}\big)\,,$$

where $c : \boldsymbol{h}(C) \to \boldsymbol{h}(A[1])$ is the connecting morphism. We therefore have a commutative diagram

$$\begin{array}{ccccccccc}
\boldsymbol{h}(A) & \xrightarrow{\boldsymbol{h}(\alpha)} & \boldsymbol{h}(B) & \xrightarrow{\boldsymbol{h}(\iota)} & \hat{c}(\alpha) & \xrightarrow{\boldsymbol{h}(p)} & \boldsymbol{h}(A)[1] & \xrightarrow{\boldsymbol{h}(\alpha)[1]} & \boldsymbol{h}(B)[1] \\
\| & & \| & & \downarrow{\scriptstyle \boldsymbol{h}(\epsilon)} & & \| & & \| \\
\boldsymbol{h}(A) & \xrightarrow{\boldsymbol{h}(\alpha)} & \boldsymbol{h}(B) & \xrightarrow{\boldsymbol{h}(\beta)} & \boldsymbol{h}(C) & \xrightarrow{-c} & \boldsymbol{h}(A)[1] & \xrightarrow{\boldsymbol{h}(\alpha)[1]} & \boldsymbol{h}(B)[1]
\end{array}$$

where we apply the Five Lemma to conclude that $\boldsymbol{h}(\epsilon)$ is an isomorphism.

In $C(\mathrm{Mod}(\mathbb{Z}))$ we can take $E := \big(0 \to \mathbb{Z}[0] \xrightarrow{\times 2} \mathbb{Z}[0] \to (\mathbb{Z}/2\mathbb{Z})[0] \to 0\big)$. Then $\hat{c}(\times 2) = \mathbb{Z} \oplus \mathbb{Z}$ is a non-torsion module and $\mathrm{Homgr}_{\mathbb{Z}}\big((\mathbb{Z}/2\mathbb{Z})[0], \hat{c}(\times 2)\big) = 0$.

56. *Exercise A.1.5.7-(2) (p. 273).* In a triangle $T := (A[0], B[0], C[0], p[0], q[0], \gamma)$, any morphism $\gamma : C[0] \to A[1]$ vanishes since the complexes are concentrated in different degrees. Hence $\gamma = 0$. But then, if T is an exact triangle, the arguments in Exercise A.1.5.3-(1) show that the exact sequence E is slit in Ab. Conversely, if E is split, it is isomorphic to $E' := (0 \to A \xrightarrow{\iota} A \oplus C \xrightarrow{p} C \to 0)$, and denoting by $\zeta : C[-1] \to A$ the zero morphism, we have $\hat{c}(\zeta) = A \oplus C$ and the triangle $(A[0], (A \oplus C)[0], C[0], \iota[0], p[0], \zeta)$ is exact since up to a rotation it coincides with the mapping cone of ζ, i.e. with $(C[-1], A[0], \hat{c}(\zeta), \zeta, \iota[0], p[1])$.

57. *Exercise A.1.6.1-(1) (p. 276).* 1) The complexes $C(m)$ are all exact hence acyclic. The morphism of complexes $f := (0, 0, 0) : C(m) \to C(n)$ is therefore a quasi-isomorphism.

Given a morphism of complexes $f := (f_0, f_1, f_2) : C(m) \to C(m)$, consider the following notations

$$\begin{array}{ccccccccc}
0 & \longrightarrow & (m) & \longrightarrow & \mathbb{Z} & \xrightarrow{\nu_m} & \mathbb{Z}/(m) & \longrightarrow & 0 \\
 & & \downarrow{\scriptstyle f_0} & \swarrow{\scriptstyle h_1} & \downarrow{\scriptstyle f_1} & \swarrow{\scriptstyle h_2} & \downarrow{\scriptstyle f_2} & & \\
0 & \longrightarrow & (m) & \longrightarrow & \mathbb{Z} & \xrightarrow{\nu_m} & \mathbb{Z}/(m) & \longrightarrow & 0
\end{array}$$

For $m > 0$ and $f_2 = 0$, the morphism f cannot be homotopic to the identity, since then $\mathrm{id}_{\mathbb{Z}/(m)} = \nu_m \circ h_2$, which is impossible because $h_2 = 0$. As a consequence, a morphism of complexes $g : C(m) \to C(0)$ does not accept any homotopic inverse. Indeed, if $g' : C(0) \to C(m)$ is such, then $f := g' \circ g$ would be homotopic to the identity while $f_2 = g'_2 \circ g_2 = 0$, because $g_2 = 0$. We have thus proved that C_0 is not homotopic to any other $C(m)$.

For $m > 0$ and $n > 0$, write $m = m' \gcd(m, n)$ and $n = n' \gcd(m, n)$. Then, in a morphism of complexes $g : C(m) \to C(n)$, the map g_1 is the multiplication by a multiple of n', and the same for a homotopic inverse $g' : C(n) \to C(m)$. As a consequence, if $f := g' \circ g$, the map f_1 is the multiplication by a multiple of $n'm'$, but then, as $f \sim \mathrm{id}$, we would have $1 - \lambda n'm' \in \mathrm{im}(h_1) \subseteq (m)$, which implies that $m' = 1$. A symmetric argument shows that $n' = 1$. Hence $m = n$.

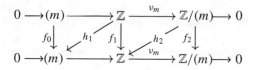

58. *Exercise A.1.6.1-(2) (p. 276).* Given a monomorphism $\iota : X \to Y$ in Ab, let $Y \twoheadrightarrow K := \operatorname{coker}(\iota)$ denote its cokernel. The short exact sequence:

$$0 \longrightarrow X \overset{\iota}{\rightarrowtail} Y \overset{\nu}{\twoheadrightarrow} K \longrightarrow 0$$

is an acyclic complex and if it is homotopic to 0, any homotopy

$$\begin{array}{ccccccccc} 0 & \longrightarrow & X & \overset{\iota}{\rightarrowtail} & Y & \overset{\nu}{\twoheadrightarrow} & K & \longrightarrow & 0 \\ & \swarrow^{h_0} & \| & \swarrow^{h_1} & \| & \swarrow^{h_2} & \| & \swarrow^{h_3} & \\ 0 & \longrightarrow & X & \underset{\iota}{\rightarrowtail} & Y & \underset{\nu}{\twoheadrightarrow} & K & \longrightarrow & 0 \end{array}$$

uses a section $h_2 : K \to Y$ of $\nu : Y \to K$, i.e. $\nu \circ h_2 = \mathrm{id}_K$. In that case, the morphism

$$X \oplus K \to Y, \quad (x, k) \mapsto x + h_2(k),$$

is an isomorphism with inverse

$$Y \to X \oplus K, \quad y \mapsto \big(y - h_2 \circ \nu(y), \nu(y)\big).$$

Now, any morphism $\alpha : X \to Y$ in Ab factors through its image $\operatorname{im}(\alpha)$, so that we get two short exact sequences:

$$0 \to \ker(\alpha) \to X \to \operatorname{im}(\alpha) \to 0 \quad \text{and} \quad 0 \to \operatorname{im}(\alpha) \to Y \to \operatorname{coker}(\alpha) \to 0,$$

and, after what we have just seen, two direct decompositions

$$X \sim \ker(\alpha) \oplus \operatorname{im}(\alpha) \quad \text{and} \quad Y \sim \operatorname{im}(\alpha) \oplus \operatorname{coker}(\alpha),$$

through which α clearly reads as $(a, b) \mapsto (b, 0)$. Hence α is a split morphism.

Conversely, if Ab is a split category, any acyclic complex of Ab is isomorphic to a complex of the form:

$$C := \big(\cdots \xrightarrow{d_{i-1}} A_{i-1} \oplus A_i \xrightarrow{d_i} A_i \oplus A_{i+1} \xrightarrow{d_{i+1}} A_{i+1} \oplus A_{i+2} \xrightarrow{d_{i+2}} \cdots\big),$$

where $d_i(a, b) = (b, 0)$. But then, the morphisms

$$h_{i+1} : A_i \oplus A_{i+1} \to A_{i-1} \oplus A_i, \quad h_i(x, y) = (0, y),$$

constitute an obvious homotopy $\mathrm{id}_C \sim 0$.

59. *Proof of Proposition A.1.6.3 (p. 277).* We have to check the conditions ($\mathcal{S}$-1) and ($\mathcal{S}$-2) of multiplicative families of morphisms (Sect. A.1.6.1).

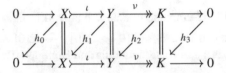

($\mathcal{S}$-1) Given $s \in \mathcal{S}$, we have to complete in $\mathcal{K}(\mathrm{Ab})$ a diagram $D := \left(\begin{array}{ccc} & & X \\ & & \downarrow s \\ Y & \cdot t \to & Z \end{array}\right)$ to a commutative diagram $\left(\begin{array}{ccc} \bullet & -t' \to & X \\ s' \downarrow & & \downarrow s \\ Y & -t \to & Z \end{array}\right)$.

We begin by completing D, in $C(\mathrm{Ab})$, to the commutative diagram

$$\begin{array}{ccccccccc}
 & & X & \xrightarrow{\iota_1 \circ s} & \hat{c}(t) & \xrightarrow[(3)]{\iota_3} & \hat{c}(\iota_1 \circ s) & \xrightarrow{t'} & X[1] \\
 & D & \downarrow s & & \| & & (4) \downarrow u & \mathrm{I} & \downarrow s[1] \\
Y & -t \to & Z & \xrightarrow[(1)]{\iota_1} & \hat{c}(t) & \xrightarrow[(2)]{\iota_2} & \hat{c}(\iota_1) & -p_3 \to & Z[1]
\end{array} \tag{E.14}$$

by successively constructing the cones $\hat{c}(t)$, $\hat{c}(\iota_1)$ and $\hat{c}(\iota_1 \circ s)$ (Sect. A.1.3). The natural morphism u closing the diagram is then a quasi-isomorphism.

By the definition A.1.3-(A.4), one has $\hat{c}(\iota_1) = Z \oplus Y[1] \oplus Z[1]$ with differential

$$D(z, y', z') = \big(d\,z + t(y') + z', -d\,y', -d\,z'\big)\,.$$

Hence, a commutative diagram of complexes in $C(\mathrm{Ab})$

$$\begin{array}{ccc}
\hat{c}(\iota_1 \circ s) & \xrightarrow{t'} & X[1] \\
(4) \downarrow u & \mathrm{I} & \downarrow s[1] \\
\hat{c}(\iota_1) = Z \oplus Y[1] \oplus Z[1] & -p_3 \to & Z[1] \\
(5)\ u' \uparrow\ \downarrow p_2 & & \| \\
Y[1] & -t[1] \to & Z[1]
\end{array}$$

where $u'(y') = (0, -y', t(y'))$. Now, the key point here is that although it is easy to see that u' is a quasi-isomorphism (exercise), it cannot be composed with u, in $C(\mathrm{Ab})$, since it goes in the wrong direction. There is nothing to do about this since simple counterexamples show that $\mathcal{S}$ is not a multiplicative collection of morphisms in $C(\mathrm{Ab})$ (*cf*. p. 277). Nonetheless, u' has a *homotopic* inverse, *viz.* the projection $p_2 : \hat{c}(\iota_1) \to Y[1]$, $p_2(z, y', z') := y'$, and while the morphisms $t[1] \circ p_2$ and p_3 do not coincide in $C(\mathrm{Ab})$, they do coincide in $\mathcal{K}(\mathrm{Ab})$ since they are homotopic (exercise). We can therefore set $s' := p_2 \circ u$ in $\mathcal{K}(\mathrm{Ab})$ to complete the diagram D as needed.

The statement ($\mathcal{S}$-1) for shapes $\left(\begin{array}{cc} \bullet & -t \to \bullet \\ s \downarrow & \\ \bullet & \end{array}\right)$ follow by the same ideas.

($\mathcal{S}$-2) We limit ourselves to showing which steps to follow, leaving the details as exercises. We have to show that in $\mathcal{K}(\mathrm{Ab})$ if $s \circ u = 0$ for some $s \in \mathcal{S}$, then there is $s' \in \mathcal{S}$ such that $u \circ s' = 0$.

We complete the diagram $\begin{pmatrix} X \dashrightarrow^{u} Y \\ \downarrow s \\ Z \end{pmatrix}$ adding cones, in order to construct the following diagram in $C(\mathrm{Ab})$

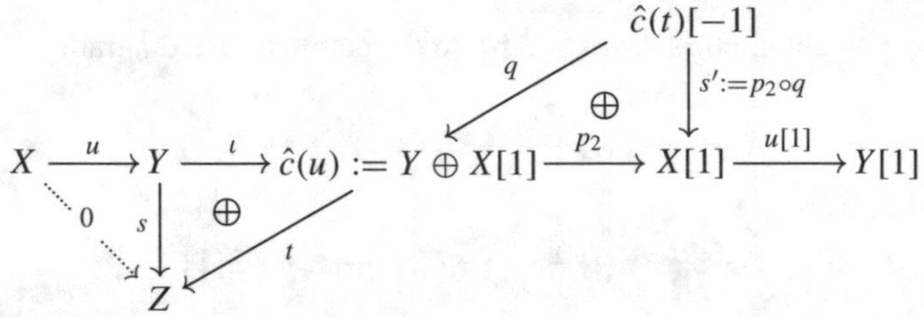

The morphism $t : Y \oplus X[1] \to Z$ is defined as

$$t(y, x') := y + h(x'),$$

where $h : X[1] \to Z$ is a homotopy to zero for the morphism $s \circ u : X \to Z$. The other arrows are those of the corresponding mapping cones.

- Show that t is a well-defined morphism of complexes.
- Show that $t \circ \iota$ is homotopic to s.
- Show that $s' := p_2 \circ q$ is a quasi-isomorphism such that $u \circ s' = 0$.

The converse statement $(u \circ s = 0) \Rightarrow (s' \circ u = 0)$, follows by the same ideas.

60. *Counterexample in A.1.6.3 (p. 277).* Let $X \neq 0$ and assume there is some commutative diagram $\begin{pmatrix} Y & \longrightarrow & 0 \\ s\downarrow & & \downarrow \\ X & \xrightarrow{\iota} & \hat{c}(\mathrm{id}_X) \end{pmatrix}$, where s a quasi-isomorphism. Then $Y \neq 0$ and $s \neq 0$. On the other hand, since $\hat{c}(\mathrm{id}_X) = X \oplus X[1]$ and $\iota(x) = (x, 0)$, we have $\iota \circ s \neq 0$, contrary to the vanishing of the composition $Y \to 0 \to \hat{c}(\mathrm{id}_X)$.

In $\mathcal{K}(\mathrm{Ab})$ the diagram $\begin{pmatrix} X & \longrightarrow & 0 \\ \mathrm{id}_X\downarrow & & \downarrow \\ X & \xrightarrow{\iota} & \hat{c}(\mathrm{id}_X) \end{pmatrix}$ is commutative, i.e. the morphism $\iota : X \to \hat{c}(\mathrm{id}_X) := X \oplus X[1]$, $\iota(x) := (x, 0)$, is homotopic to zero.

Indeed, if we set

$$h : X \to \hat{c}(\mathrm{id}_X)[-1] := X[-1] \oplus X, \quad h(x) = (0, x),$$

we get the diagram:

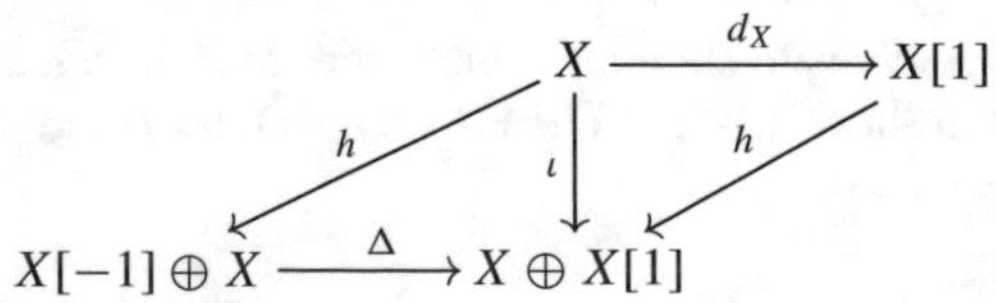

where $\Delta(x, x') := (d_X\, x + x', -d_X\, x')$. Whence,

$$(h \circ d_X + \Delta \circ h)(x) = \quad (0, d_X\, x) + \Delta(0, x) = \quad (0, d_X\, x) + (x, -d_X\, x) = \iota(x)\,,$$

and $\iota = 0$ in $\mathcal{K}$(Ab) as expected.

61. *Exercise A.1.6.6 (p. 278).* Given a path representing a morphism α in $\mathcal{D}$(Ab)

$$(C, d) \xleftarrow{s_1} \bullet \xrightarrow{t_1} \bullet \xleftarrow{s} \bullet \xrightarrow{t} \cdots, \tag{E.15}$$

the exchange property ($\mathcal{S}$-2) for multiplicative families (Sect. A.1.6.1) tells us that there exist $s_2 \in \mathcal{S}$ en t_2, such that $t_1 \circ s_1^{-1} = s_2^{-1} \circ t_2$. The path (E.15) is then equivalent by A.1.6.4-(A.19) to the path

$$(C, d) \xrightarrow{t_2} \bullet \xleftarrow{s_2} \bullet \xleftarrow{s} \bullet \xrightarrow{t} \cdots$$

where we can compose s and s_2 in a single quasi-isomorphism $s_2 \circ s$, hence decreasing the number of steps in the path representing α in $\mathcal{D}$(Ab). An obvious induction then finishes the proof.

62. *Exercise A.2.3.6 (p. 285).*
1) For $X \in$ Ob(Ab), to say that $\boldsymbol{\epsilon} : X[0] \to \boldsymbol{I} := (0 \to I_0 \to I_1 \to I_2 \to \cdots)$ is a quasi-isomorphism is equivalent to say that the complex

$$0 \to X \xrightarrow{\epsilon_0} I_0 \to I_1 \to I_2 \to \cdots$$

is acyclic. Therefore, if F is left exact, the sequence

$$0 \to F(X) \xrightarrow{F(\epsilon_0)} F(I_0) \to F(I_1) \to F(I_2) \to \cdots$$

is left exact, or, in other terms, the induced morphism $\epsilon_X : F(X) \to H^0(F(\boldsymbol{I}))$ is an isomorphism. As a consequence, if in addition $\boldsymbol{I}$ is a complex of injective objects, then $F(\boldsymbol{I}) = \mathbb{R}\, F(X[0])$ and we get the announced natural isomorphism

$$\epsilon : F(X) \to \mathbb{R}^0\, F(X[0])\,.$$

The same arguments apply, reversing arrows, for right exact functors.
2) No other changes that the source of the functor $\mathbb{R}\, F$ which is $\mathcal{D}^-$(Ab), and the source of the functor $\mathbb{L} F$ which is $\mathcal{D}^+$(Ab).

63. *Exercise B.3.2 (p. 304).* Let $s \in \Gamma(S; \mathcal{F})$. After Proposition B.3.1-(2), we know that for each $x \in S$, there exists an open subspace $V_x \subseteq X$, neighborhood of x, and a global section $s_{V_x} \in \Gamma(X; \mathcal{F})$, such that $(s_{V_x})_t = s_t$, for all $t \in S \cap V_x$. In that case, and since the open subspace $\bigcup_{x \in S} V_x$ is paracompact, there exists a partition of unity $\{\phi_x\}_{x \in S}$ subordinate to $\{V_x\}_{x \in S}$. Then $\tilde{s} := \sum_{x \in S} \phi_x\, s_{V_x}$ is a well-defined global section such that $\tilde{s}_x = s_x$ for all $x \in S$.

References

*B*ooks are in slanted characters.

1. A. Alekseev, E. Meinrenken, Equivariant cohomology and the Maurer-Cartan equation. Duke Math. J. **130**(3), 479–521 (2005)
2. C. Allday, V. Puppe, *Cohomological methods in transformation groups*. Cambridge Studies in Advanced Mathematics, vol. 32 (Cambridge University, Cambridge, 1993).
3. C. Allday, M. Franz, V. Puppe, Equivariant cohomology, syzygies and orbit structure. Trans. Am. Math. Soc. **366**(12), 6567–6589 (2014)
4. C. Allday, M. Franz, V. Puppe, Equivariant Poincaré-Alexander-Lefschetz duality and the Cohen-Macaulay property. Algebr. Geom. Topol. **14**(3), 1339–1375 (2014)
5. C. Allday, M. Franz, V. Puppe, Syzygies in equivariant cohomology in positive characteristic. arXiv:2007.00496v2 (2020)
6. M. André, Cohomologie des algèbres différentielles où opère une algèbre de Lie. Ph.D. Thesis 1962, advisor Claude C. Chevalley. Tôhoku Math. J. (2) **14**(3), 263–311 (1962)
7. M.F. Atiyah, R. Bott, The moment map and equivariant cohomology. Topology **23**(1), 1–28 (1984).
8. M.F. Atiyah, G.B. Segal, The index of elliptic operators II. Ann. Math. (2) **87**, 531–545 (1968)
9. N. Berline, M. Vergne, Classes caractéristiques équivariantes. Formule de localisation en cohomologie équivariante. C. R. Acad. Sci. Paris Sér. I Math. **295**(9), 539–541 (1982). http://gallica.bnf.fr/ark:/12148/bpt6k62356694/f77
10. N. Berline, M. Vergne, Zéros d'un champ de vecteurs et classes caractéristiques équivariantes. Duke Math. J. **50**(2), 539–549, (1983)
11. N. Berline, M. Vergne, Fourier transforms of orbits of the coadjoint representation, in *Representation Theory of Reductive Groups* (Park City, Utah, 1982), pp. 53–67. Prog. Math. vol. 40. Birkhäuser (1983)
12. N. Berline, E. Getzler, M. Vergne, *Heat kernels and Dirac operators*. Corrected reprint of the 1992 original. Grundlehren Text Editions (Springer, New York, 2004)
13. A. Borel, Sur La Cohomologie des Espaces Fibres Principaux et des Espaces Homogenes de Groupes de Lie Compacts. Ann. Math. Second Series **57**(1), 115-207 (1953)
14. A. Borel, Nouvelle démonstration d'un théorème de P. A. Smith. Comment. Math. Helv. **29**, 27–39 (1955)
15. A. Borel, in *Seminar on Transformation Groups*, ed. by G. Bredon, E.E. Floyd, D. Montgomery, R. Palais. Annals of Mathematics Studies, vol. 46 (Princeton University, Princeton, 1960)
16. A. Borel et al. *Intersection Cohomology* (Bern, 1983). Prog. Math. **50**. Swiss Sem., Birkhäuser Boston, Boston, MA, (1984). Especially chapter V: Sheaf Theoretic Intersection cohomology, pp. 47–182

A. Arabia, *Equivariant Poincaré Duality on G-Manifolds*, Lecture Notes in Mathematics 2288, https://doi.org/10.1007/978-3-030-70440-7

17. R. Bott, Vector fields and characteristic numbers. Michigan Math. J. **14**, 231–244 (1967)
18. R. Bott, L.W. Tu, *Differential forms in algebraic topology*, in *Graduate Texts in Mathematics*, vol. 82 (Springer, New York, 1982)
19. N. Bourbaki, *Algebra I. Chapters 1–3*. Elements of Mathematics (Berlin) (Springer, Berlin, 1998)
20. G.E. Bredon, *Sheaf theory*, 2nd edn. Graduate Texts in Mathematics, vol. 170 (Springer, New York, 1997)
21. M. Brion, Equivariant Chow groups for torus actions. Transform. Groups **2**(3), 225–267 (1997)
22. M. Brion, Equivariant cohomology and equivariant intersection theory, in *Representation Theories and Algebraic Geometry (Montreal, PQ, 1997)*, pp. 1–37 (Kluwer Acad. Publication, Dordrecht, 1998)
23. M. Brion, Poincaré duality and equivariant (co)homology. Michigan Math. J. **48**, 77–92 (2000)
24. J.-L. Brylinski, Equivariant intersection cohomology, in *Kazhdan-Lusztig theory and related topics (Chicago, IL, 1989)*, pp. 5–32. Contemp. Math. vol. 139. Am. Math. Soc. Providence, RI (1992)
25. H. Cartan, Cohomologie réelle d'un espace fibré principal différentiable. I: notions d'algèbre différentielle, algèbre de Weil d'un groupe de Lie. Séminaire Henri Cartan, vol. 2 (1949–1950), Talk no. 19, pp. 1–10 (May 15, 1950).
26. H. Cartan, Cohomologie réelle d'un espace fibré principal différentiable. II: transgression dans un groupe de Lie et dans un espace fibré principal ; recherche de la cohomologie de l'espace de base. Séminaire Henri Cartan, tome 2 (1949-1950), exp. no 20, pp. 1–11 (May 23 and June 19, 1950)
27. H. Cartan, Notions d'algèbre différentielle; application aux groupes de Lie et aux variétés où opère un groupe de Lie. Colloque de topologie (espaces fibrés), Bruxelles, 1950, pp. 15–27. Georges Thone, Liège; (Masson et Cie., Paris, 1951)
28. H. Cartan, La transgression dans un groupe de Lie et dans un espace fibré principal. (French) Colloque de topologie (espaces fibrés), Bruxelles (1950), pp. 57–71. Georges Thone, Liège; Masson et Cie., Paris, 1951
29. C. Chevalley, Review of Cartan's articles [27, 28], in *Mathematical Reviews* (1952).
30. C. Chevalley, Invariants of finite groups generated by reflections. Am. J. Math. **77**, 778–782 (1955)
31. N. Chriss, V. Ginzburg, *Representation theory and complex geometry* (Birkhäuser Boston, Inc., Boston, MA, 1997)
32. *Colloque de Topologie (Espaces Fibrés), tenu á Bruxelles du 5 au 8 juin 1950*. Publications du Centre Belge de Recherches Mathématiques (Liège, Thone, et Paris, Masson, 1951)
33. P.E. Conner, On the action of the circle group. Michigan Math. J. **4**, 241–247 (1957)
34. S.R. Costenoble, S. Waner, Equivariant ordinary homology and cohomology. Lecture Notes Math. **2178** (2016)
35. P. Deligne, Cohomology á support propre et construction du foncteur $f^!$, Appendix to: Residues and Duality, Lecture Notes in Math., vol. 20 (Springer, Heidelberg, 1966), pp. 404–421. MR 36:5145
36. P. Deligne, P. Griffiths, J. Morgan, D. Sullivan, Real homotopy theory of Kähler manifolds. Invent. Math. **29**(3), 245–274 (1975)
37. A. Dold, Partitions of unity in the theory of fibrations. Ann. Math. (2) **78**, 223–255 (1963)
38. J.J. Duistermaat, G.J. Heckman, On the variation in the cohomology of the symplectic form of the reduced phase space. Invent. Math. 69(2), 259–268 (1982). Addendum to: "On the variation in the cohomology of the symplectic form of the reduced phase space." Invent. Math. **72**(1), 153–158 (1983)
39. D. Edidin, W. Graham, Equivariant intersection theory. Invent. Math. **131**(3), 595–634 (1998)
40. R. Fossum, H.-B. Foxby, The category of graded modules. Math. Scand. **35**, 288–300 (1974)
41. M. Franz, Big polygon spaces. Int. Math. Res. Not. **2015**, 13379–13405 (2015)
42. M. Franz, Syzygies in equivariant cohomology for non-abelian Lie groups, in *Configuration spaces (Cortona, 2014)* (eds.) by F. Callegaro et al. Springer INdAM Ser., vol. 14 (Springer, Cham, 2016)

43. M. Franz, V. Puppe, Freeness of equivariant cohomology and mutants of compactified representations, in *Toric Topology (Osaka, 2006)* (eds.) by M. Harada et al. Contemp. Math. vol, 460 (AMS, Providence, 2008), pp. 87–98
44. V.A. Ginzburg, Equivariant cohomology and Kähler geometry. Funct. Anal. Appl. **21**, 271–283 (1987)
45. A.M. Gleason, Spaces with a compact Lie group of transformations. Proc. Am. Math. Soc. **1**, 35–43 (1950)
46. R. Godement, *Topologie algébrique et théorie des faisceaux*. Troisième édition revue et corrigée. Actualités Scientifiques et Industrielles, No. 1252.(Hermann, Paris, 1973)
47. M. Goresky, R. Kottwitz, R. MacPherson, Equivariant cohomology, Koszul duality, and the localization theorem. Invent. Math. **131**(1), 25–83 (1998)
48. D.H. Gottlieb, Fibre bundles and the Euler characteristic. J. Differ. Geom. **10**, 39–48, (1975)
49. A. Grothendieck, Sur quelques points d'algèbre homologique. Tôhoku Math. J. (2) **9**, 119–221 (1957)
50. V.W. Guillemin, S. Sternberg, *Supersymmetry and equivariant de Rham theory*. With an appendix containing two reprints by Henri Cartan. Mathematics Past and Present (Springer, Berlin, 1999)
51. R. Hartshorne, *Residues and duality*. Lecture notes of a seminar on the work of A. Grothendieck, given at Harvard 1963/64. With an appendix by P. Deligne. Lecture Notes in Mathematics, vol. 20 (Springer, Berlin, 1966)
52. R. Hartshorne, in *Local cohomology* (ed.) by A. Grothendieck (Harvard University, Fall, 1961). Lecture Notes in Mathematics, vol. 41 (Springer, Berlin, 1967)
53. A. Hatcher, Algebraic topology, in *Algebraic Topology* (Cambridge University, Cambridge, 2002). https://pi.math.cornell.edu/~hatcher/AT/AT.pdf (2001)
54. A. Hatcher, *Vector Bundles and K-Theory*. Version 2.2. https://pi.math.cornell.edu/~hatcher/VBKT/VB.pdf (2017)
55. W. Hsiang, *Cohomology theory of topological transformation groups*. Ergebnisse der Mathematik und ihrer Grenzgebiete, Band, vol. 85 (1975)
56. D. Husemöller, Fibre bundles, in *Fibre Bundles*, 3rd edn. Graduate Texts in Mathematics, vol. 20 (Springer, New York, 1994)
57. D. Husemöller, M. Joachim, B. Jurčo, M. Schottenloher, The Milnor construction: homotopy classification of principal bundles, in *Basic Bundle Theory and K-Cohomology Invariants*. Lecture Notes in Physics, vol. 726, pp. 75–81 (2008). https://doi.org/10.1007/978-3-540-74956-1_8
58. B. Iversen, *Cohomology of Sheaves* (Springer, Berlin, 1986)
59. N. Jacobson, *Basic algebra. II*, 2nd edn. (W.H. Freeman and Company, New York, 1989)
60. R. Joshua, Vanishing of odd-dimensional intersection cohomology. Math. Z. **195**(2), 239–253 (1987)
61. M. Kashiwara, P. Schapira, Sheaves on manifolds, in *Grundlehren der Mathematischen Wissenschaften*, vol. 292 (Springer, Berlin, 1994)
62. M. Kashiwara, P. Schapira, Categories and sheaves, in *Grundlehren der Mathematischen Wissenschaften*, vol. 332 (Springer, Berlin, 2006)
63. K. Kawakubo, *The theory of transformation groups* (The Clarendon Press, Oxford University Press, New York, 1991)
64. B. Keller, Deriving DG categories. Ann. Sci. école Norm. Sup. (4) **27**(1), 63–102 (1994)
65. B. Keller, Derived categories and their uses, in *Handbook of Algebra* (ed.) by M. Hazewinkel, vol. 1 (Elsevier, Amsterdam, 1996)
66. A.W. Knapp, *Lie groups beyond an introduction*. Prog. Math. 140, 2nd ed. (Birkhaüser, Basel, 2005)
67. B. Kostant, Lie group representations on polynomial rings. Bull. Am. Math. Soc. **69**, 518–526 (1963)
68. B. Kostant, Lie group representations on polynomial rings. Am. J. Math. **85**, 327–404 (1963)
69. J.L. Koszul, Sur certains groupes de transformations de Lie. (French) Géométrie différentielle. *Colloques Internationaux du Centre National de la Recherche Scientifique, Strasbourg*, pp.

137–141 (1953), in *Selected papers of J.-L. Koszul*. (French) Series in Pure Mathematics, vol. 17, pp. 79–83 (World Scientific Publishing Co., Inc., River Edge, NJ, 1994)
70. S. Kumar, A remark on universal connections. Math. Ann. **260**(4), 453–462 (1982)
71. T.Y. Lam, Lectures on modules and rings, in *Graduate Texts in Mathematics*, vol. 189 (Springer, New York, 1999)
72. S. Lang, Fundamentals of differential geometry, in *Graduate Texts in Mathematics*, vol. 191 (Springer, New York, 1999)
73. J. Leray, *Sur l'homologie des groupes de Lie, des espaces homogènes et des espaces fibrés principaux (French)*. (Colloque de topologie (espace fibrés), Bruxelles, 1950), pp. 101–115
74. A. Lundell, S. Weingram, The topology of CW-complexes, in *The University Series in Higher Mathematics* (Van Nostrand Reinhold Co., New York, 1969)
75. J. Milnor, Construction of Universal Bundles, II. Ann. Math. **63**(3), 430–436 (1956)
76. J. Milnor, J.D. Stasheff, *Characteristic Classes*. Annals of Mathematics Studies, vol. 76 (Princeton University, Princeton, 1974)
77. D. Montgomery, C.T. Yang, The existence of a slice. Ann. Math. (2) **65**, 108–116 (1957)
78. G.D. Mostow, Equivariant embeddings in Euclidean space. Ann. Math. (2) **65**, 432–446 (1957)
79. R.S. Palais, Imbedding of compact, differentiable transformation groups in orthogonal representations. J. Math. Mech. **6**, 673–678 (1957)
80. D. Quillen, The spectrum of an equivariant cohomology ring. I, II. Ann. Math. (2) **94**, 549–572 (1971); ibid. (2) **94**, 573–602 (1971)
81. O. Randal-Williams, On D. Else question: Is there a kind of Poincare duality for Borel equivariant cohomology? MathOverflow (2018). https://mathoverflow.net/q/310533
82. G. Segal, Equivariant K-theory, in *Instutite Hautes études Science Publication Mathematical*, vol. 34 (1968), pp. 129–151
83. M. Spivak, *A Comprehensive Introduction to Differential Geometry*, vol. I, 3rd edn. Publish or Perish (1999)
84. N. Steenrod, *The Topology of Fibre Bundles* (Princeton University, Princeton, 1951)
85. D. Sullivan, Infinitesimal computations in topology. Inst. Hautes études Sci. Publ. Math. **47**(1977), 269–331 (1978)
86. The Stacks Project, in *More on Algebra* (2020). https://stacks.math.columbia.edu/download/more-algebra.pdf
87. The Stacks Project, in *Differential Graded Algebra* (2020). https://stacks.math.columbia.edu/download/dga.pdf
88. The Stacks Project. Section 15.67 Hom complexes (2020). https://stacks.math.columbia.edu/tag/0A8H
89. The Stacks Project. Section 22.19 Injective modules and differential graded algebras (2020). https://stacks.math.columbia.edu/tag/0FQD
90. The Stacks Project. *Chapter 24: Differential Graded Sheaves. §24.25 K-injective differential graded modules. §24.26 The derived category. §24.30 Equivalences of derived categories.* https://stacks.math.columbia.edu/tag/0FQS (2020)
91. L. Tu, *Introductory Lectures on Equivariant Cohomology*. Annals of Mathematics Studies (Princeton University, Princeton, 2020)
92. J.-L. Verdier, Dualité dans la cohomologie des espaces localement compacts. Séminaire N. Bourbaki, exp. no 300, pp. 337–349 (1966)
93. J.-L. Verdier, *Des catégories dérivées des catégories abéliennes*. Astérisque, tome, vol. 239 (1996)
94. T. Wedhorn, *Manifolds, Sheaves, and Cohomology*. Springer Studium Mathematik—Master (Springer Spektrum, Wiesbaden, 2016)
95. Ch.A. Weibel, An introduction to homological algebra, in *Cambridge Studies in Advanced Mathematics*, vol. 38 (Cambridge University, Cambridge, 1994)
96. A. Weil, Œuvres scientifiques. *Collected papers*, vol. II (1951–1964). Reprint of the 1979 original (Springer, Berlin, 2009)
97. E. Witten, Supersymmetry and Morse theory. J. Differ. Geom. **17**(1982/4), 661–692 (1983)

Glossary

Chapter 2.	***Nonequivariant Background***	
$\Bbbk$	an arbitrary field	9
$\mathbb{R}, \mathbb{C}$	the fields of real and complex numbers	9
$\mathrm{Vec}(\Bbbk)$	the category of $\Bbbk$-vector spaces and $\Bbbk$-linear maps	9
$V^\vee$	alias for $\mathrm{Hom}_\Bbbk(V, \Bbbk)$, dual of the vector space V	9
$\mathrm{Homgr}^d_\Bbbk(\_, \_)$	space of graded homomorphisms of degree d	10
$\mathrm{Homgr}^*_\Bbbk(\_, \_)$	graded space of graded homomorphisms	10
$\mathrm{GV}(\Bbbk)$	category of graded $\Bbbk$-vector spaces	10
$\mathrm{Vec}(\Bbbk)^\mathbb{Z}$	alias for $\mathrm{GV}(\Bbbk)$	10
$\mathrm{GV}^+(\Bbbk)$, $\mathrm{GV}^-(\Bbbk)$, $\mathrm{GV}^b(\Bbbk), \ldots$	subcategories of bounded graded vector spaces	10
$\mathrm{Endgr}^*_\Bbbk(\_)$	algebra of graded endomorphisms of a graded space	11
$\mathfrak{g}$	Lie algebra of a Lie group G	11
$S(\mathfrak{g}^\vee)^\mathfrak{g}$	algebra of $\mathfrak{g}$-invariant polynomials functions over $\mathfrak{g}$	11
Ω_G	alias for $S(\mathfrak{g}^\vee)^\mathfrak{g}$	11
$\mathrm{DGV}(\Bbbk)$	the category of complexes of $\Bbbk$-vector spaces	13
$C(\mathrm{Vec}(\Bbbk))$	alias for $\mathrm{DGV}(\Bbbk)$	13
$\boldsymbol{h}(\boldsymbol{V}, \boldsymbol{d})$	cohomology of $\boldsymbol{V} \in \mathrm{DGV}(\Bbbk)$	13
$\boldsymbol{h}$	$: \mathrm{DGV}(\Bbbk) \rightsquigarrow \mathrm{GV}(\Bbbk)$, the cohomology functor	13
$[\_]$	shift functors	13
$\mathbf{Hom}^\bullet_\Bbbk(\_, \_)$	functor on $\mathrm{DGV}(\Bbbk)$	14
$(\_ \otimes_\Bbbk \_)^\bullet$	functors on $\mathrm{DGV}(\Bbbk)$	14
$(\_)^\vee$	alias for $\mathbf{Hom}^\bullet_\Bbbk(\_, \Bbbk[0])$, duality functor	16
$(\boldsymbol{V}, \boldsymbol{d})^\vee$	$:= (V^\vee, \boldsymbol{D})$, dual complex	17
Man	category of manifolds and differentiable maps	18
$\mathrm{Man}_{\mathrm{pr}}$	category of manifolds and differentiable proper maps	18
$\Omega(\_)$, $\Omega_c(\_)$,	$: \mathrm{Man} \rightsquigarrow \mathrm{DGV}(\mathbb{R})$ de Rham differential forms functors	18

A. Arabia, *Equivariant Poincaré Duality on G-Manifolds*, Lecture Notes in Mathematics 2288, https://doi.org/10.1007/978-3-030-70440-7

$H(\_),\ H_{\mathrm{c}}(\_)$	$:\mathrm{Man} \rightsquigarrow \mathrm{Mod}^{\mathbb{N}}(\mathbb{R})$ de Rham cohomology functors	18
G-Man	category of G-manifolds	19
G-$\mathrm{Man}_{\mathrm{pr}}$	category of G-manifolds and proper maps	19
$[M]$	orientation of M	20
$\int_{[M]}$	$\Omega_{\mathrm{c}}^{d_M}(M) \to \mathbb{R}$, integration operation	21
$\wedge$	exterior product of differential forms	23
$\langle\cdot,\cdot\rangle_M$	Poincaré pairing	23
$\cup$	cup-product of cohomology classes	23
$\mathbb{D}_M$	left Poincaré adjunction for complexes	23
D_M	left Poincaré adjunction in cohomology	24
$\zeta_{[M]}$	the fundamental class of the oriented manifold $(M,[M])$	30
$\mathbb{D}'_M$	$:\big(\Omega_{\mathrm{c}}(M)[d_M], d\big) \hookrightarrow \big(\Omega(M)^\vee, -\boldsymbol{D}\big)$, right Poincaré adjunction in complexes	37
D'_M	right Poincaré adjunction in cohomology	38
$f_!$	$: H_{\mathrm{c}}(M)[d_M] \to H_{\mathrm{c}}(N)[d_N]$, Gysin morphism	39
$(\_)_!$	$: \mathrm{Man}^{\mathrm{or}} \rightsquigarrow \mathrm{GV}(\mathbb{R})$, Gysin functor	40
f_*	$: H(M)[d_M] \to H(N)[d_N]$, Gysin morphism	45
$(\_)_*$	$: \mathrm{Man}^{\mathrm{or}}_{\mathrm{pr}} \rightsquigarrow \mathrm{GV}(\mathbb{R})$, Gysin functor for proper maps	46
Chapter 3.	***Poincaré Duality Relative to a Base Space***	
$X_1 \times_{(\pi_1,\pi_2)} X_2$	fiber product of two maps $\pi_i : X_i \to (\pi_1,\pi_2)$	51
$\square$	inside a diagram, it indicates a Cartesian sub-diagram	53
$\bar{h}^{-1}(\_) : \mathrm{Top}_B \rightsquigarrow \mathrm{Top}_{B'}$, *pullback* or *base change* functor		54
$\Omega_{\mathrm{cv}}(E)$	complex of compactly supported differential forms	65
$\pi_!\,\underline{\underline{\Omega}}_E \subseteq \pi_*\,\underline{\underline{\Omega}}_E$	sheaves on B of $\underline{\underline{\Omega}}_B$-differential graded modules	68
$\mathcal{H}(\pi_!) := \mathcal{H}(\pi_!\,\underline{\underline{\Omega}}_E)$	graded sheaf cohomology of $\pi_!\,\underline{\underline{\Omega}}_E$	68
$\mathcal{H}(\pi_*) := \mathcal{H}(\pi_*\,\underline{\underline{\Omega}}_E)$	graded sheaf cohomology of $\pi_*\,\underline{\underline{\Omega}}_E$	68
$\mathscr{F} \boxtimes \mathscr{G} \in \mathrm{Sh}(X \times Y)$	external tensor product of $\mathscr{F} \in \mathrm{Sh}(X)$ and $\mathscr{G} \in \mathrm{Sh}(Y)$	70
$\mathbb{D}_{U,V}$	relative left Poincaré adjunction for complexes	73
$\int_M : \Omega_{\mathrm{cv}}(E)[d_M] \to \Omega(B)$, integration along the fiber		75
$\mathbb{D}_{B,M}$	relative left Poincaré adjunction for complexes	77
$\langle\cdot,\cdot\rangle_{B,M}$	Poincaré pairing relative to B	77
$\int_M : \pi_!\,\underline{\underline{\Omega}}_E[d_M] \to \underline{\underline{\Omega}}_B$, sheafification of integration along fibers		78
$\underline{\underline{\mathbb{D}}}_{B,M}$	$\pi_*\underline{\underline{\Omega}}_E \to \underline{\boldsymbol{Hom}}^{\bullet}_{\underline{\underline{\Omega}}_B}\big(\pi_!\underline{\underline{\Omega}}_E, \underline{\underline{\Omega}}_B\big)$, sheafification of the left Poincaré adjunction	83
$(\_)_*$	$: \mathrm{Fib}^{\mathrm{or}}_{B,\mathrm{pr}} \rightsquigarrow \mathcal{D}(\mathrm{DGM}(\Omega(B), d))$, Relative Gysin functor for proper maps	96
$(\_)_!$	$: \mathrm{Fib}^{\mathrm{or}}_{B} \rightsquigarrow \mathcal{D}(\mathrm{DGM}(\Omega(B), d))$, Relative Gysin functor for arbitrary maps	96
$j_!$	$: \Omega_{\mathrm{c}}(U) \to \Omega_{\mathrm{c}}(M)$, extension by zero for $j : U \subseteq M$ open	100
$\mathrm{Gr}(f)$	$: M \to M \times M$, the graph map $x \mapsto (f(x), x)$	106
$L(f)$	$\in H^{d_M(M)}$, the Lefschetz class	106

$\Lambda(f)$	the Lefschetz number	106
Chapter 4.	***Equivariant Background***	
$d_{\mathfrak{g}}$	differential of the Cartan complex	112
$\Omega_G(M)$	the Cartan complex of the G-manifold M	113
X_G	alias for $\mathbb{E}G \times_G X$, the Borel Construction of a G-space X	114
$K_G(X)$	equivariant K-theory of G-space X	115
$V^{\mathfrak{g}}$	subspace of $\mathfrak{g}$-invariant elements of a $\mathfrak{g}$-module V	120
$\mathfrak{g}$-dgm/dga	alias for $\mathfrak{g}$-differential graded module/algebra	120
$C := (C, d, \theta, \iota)$	a $\mathfrak{g}$-differential graded module/algebra	120
DGM($\mathfrak{g}$)	category of differential graded $\mathfrak{g}$-modules over $\mathbb{R}$	121
DGM($\mathfrak{g}$)	category of differential graded $\mathfrak{g}$-modules over $\mathbb{R}$	121
$(\_)\vert(\_)$	$\mathfrak{g}$-split inclusion of $\mathfrak{g}$-modules	122
$S(\mathfrak{g}^\vee)$	the ring of real polynomial functions on $\mathfrak{g}$	124
$C_{\mathfrak{g}}$	the Cartan Complex associated with a $\mathfrak{g}$-dgm C	126
$d_{\mathfrak{g}}$	the Cartan Differential	126
$H_{\mathfrak{g}}(C)$	alias for $\boldsymbol{h}(C_{\mathfrak{g}}, d_{\mathfrak{g}})$, $\mathfrak{g}$-equivariant cohomology of C	126
$\mathbf{Y}(\_)$	fundamental vector field on a G-manifold associated with $Y \in \mathfrak{g}$	129
ι	$: \mathfrak{g} \to \mathrm{Endgr}^{-1}_{\mathrm{GV}(\Bbbk)}(\Omega(M))$, *interior product representation*	130
$\mathcal{L}(\mathbf{Y})$	*Lie derivative* with respect to the vector field $\mathbf{Y}$	130
θ	$: \mathfrak{g} \to \mathrm{Endgr}_{\mathrm{GV}(\Bbbk)}(\Omega(M))$, *Lie derivative representation*	130
$(\Omega(M), d, \theta, \iota)$	$\mathfrak{g}$-dga of differential forms	130
$(\Omega_{\mathrm{c}}(M), d, \theta, \iota)$	$\mathfrak{g}$-dga of differential forms compactly supported	130
$(\Omega_G(\_), d_G)$	Cartan complex functor	130
$(\Omega_{G,\mathrm{c}}(\_), d_G)$	Cartan complex functors for compact supports	130
$H_G(\_),\ H_{G,\mathrm{c}}(\_)$	G-equivariant cohomology functors	130
$\mathbb{E} \times_G X$	quotient of $\mathbb{E} \times X$ under the diagonal action of G	144
$\mathbb{E}G$	universal fiber bundle for G	146
$\mathbb{E}G(n)$	n-th approximation of universal fiber bundle for G	146
$\mathbb{B}G$	$:= \mathbb{E}G/G$, classifying space for G	146
$\mathbb{B}G(n)$	n-th approximation of classifying space for G	146
$O(r)$	group of orthogonal $(r \times r)$-matrices	147
$(\_)_G$	$(\mathbb{E}G \times M)/G$, Borel construction functor	148
$H_{G,N}(M)$	local equivariant (de Rham) cohomology	171
$(\Omega_{G,N}(M), d_G)$	local Cartan complex of M relative to N	172
Chapter 5.	***Equivariant Poincaré Duality***	
$\boldsymbol{A}$	graded $\Bbbk$-algebra	175
$\mathrm{Homgr}^*_{\boldsymbol{A}}(\_, \_)$	graded space of graded homomorphisms of $\boldsymbol{A}$-gm's	175
GM($\boldsymbol{A}$)	category of $\boldsymbol{A}$-graded modules	175
DGM(Ω_G)	category of Ω_G-differential graded modules	179

$\mathbf{Hom}^\bullet_{\Omega_G}(\_,\_)$	$: \mathrm{DGM}(\Omega_G) \times \mathrm{DGM}(\Omega_G) \rightsquigarrow \mathrm{DGM}(\Omega_G)$	180
$(\_ \otimes_{\Omega_G} \_)^\bullet$	$: \mathrm{DGM}(\Omega_G) \times \mathrm{DGM}(\Omega_G) \rightsquigarrow \mathrm{DGM}(\Omega_G)$	180
$\mathbf{Hom}^\bullet_{\Omega_G}(\_, \Omega_G)$	$: \mathrm{DGM}(\Omega_G) \rightsquigarrow \mathrm{DGM}(\Omega_G)$, duality functor	180
o	$: \mathrm{DGM}(\Omega_G) \rightsquigarrow \mathrm{GM}(\Omega_G)$, forgetful functor	180
$\mathbf{Tot}_\oplus, \mathbf{Tot}_\Pi$	$: C(\mathrm{GM}(\Omega_G)) \rightsquigarrow \mathrm{DGM}(\Omega_G)$	183
$\mathbf{Hom}^\bullet_{\Omega_G}(\_,\_)$	$: C(\mathrm{GM}(\Omega_G)) \times C(\mathrm{GM}(\Omega_G)) \rightsquigarrow C(\mathrm{GM}(\Omega_G))$	184
$(\_ \otimes_{\Omega_G} \_)^\bullet$	$: C(\mathrm{GM}(\Omega_G)) \times C(\mathrm{GM}(\Omega_G)) \rightsquigarrow C(\mathrm{GM}(\Omega_G))$	186
$\mathbb{R}F, \mathbb{L}F$	$: \mathrm{GM}(\Omega_G) \rightsquigarrow \mathcal{K}(\mathrm{Ab})$, right and left derived functors	187
$C^{\star,\natural}(\mathrm{Ab})$	alias for $C^\star(C^\natural(\mathrm{Ab}))$	188
$\mathbb{R}\,\mathbf{Hom}^\bullet_{\Omega_G}(\_,\_)$	$: \mathcal{D}(\mathrm{GM}(\Omega_G)) \times \mathcal{D}^+(\mathrm{GM}(\Omega_G)) \rightsquigarrow \mathcal{D}(\mathrm{GM}(\Omega_G))$	191
$(\_) \otimes^{\mathbb{L}}_{\Omega_G} (\_)$	$: \mathcal{D}(\mathrm{GM}(\Omega_G)) \times \mathcal{D}^-(\mathrm{GM}(\Omega_G)) \rightsquigarrow \mathcal{D}(\mathrm{GM}(\Omega_G))$	191
$\mathbf{Ext}^{i,*}_{\Omega_G}(\_,\_)$	$: \mathcal{D}(\mathrm{GM}(\Omega_G)) \times \mathcal{D}^+(\mathrm{GM}(\Omega_G)) \rightsquigarrow \mathrm{GM}^*(\Omega_G)$	191
$\mathbf{Tor}^*_{\Omega_G,i}(\_,\_)$	$: \mathcal{D}(\mathrm{GM}(\Omega_G)) \times \mathcal{D}^+(\mathrm{GM}(\Omega_G)) \rightsquigarrow \mathrm{GM}^*(\Omega_G)$	191
$\mathbb{R}\,\mathbf{Hom}^\bullet_{\Omega_G}(\_, \Omega_G)$	$: \mathcal{D}(\mathrm{GM}(\Omega_G)) \rightsquigarrow \mathcal{D}(\mathrm{GM}(\Omega_G))$, derived duality functor	192
$\mathbb{R}\,\mathbf{Hom}^\bullet_{\Omega_G}(\_, \Omega_G)$	$: \mathcal{D}(\mathrm{DGM}(\Omega_G)) \rightsquigarrow \mathcal{D}(\mathrm{DGM}(\Omega_G))$, derived duality functor	192
$\mathbf{Ext}^{i,*}_{\Omega_G}(\_, \Omega_G)$	$: \mathcal{D}(\mathrm{DGM}(\Omega_G)) \rightsquigarrow \mathrm{GM}^*(\Omega_G)$	192
$\langle\cdot,\cdot\rangle_{M,G}$	$: \Omega_G(M) \times \Omega_{G,\mathrm{c}}(M) \to \Omega_G$, equivariant Poincaré pairing	201
$\langle\cdot,\cdot\rangle_{M,G}$	$: H_G(M) \times H_{G,\mathrm{c}}(M) \to H_G$, *idem* in cohomology	201
$\mathbb{D}_{G,M}$	$: \Omega_G(M) \to \mathbf{Hom}^\bullet_{\Omega_G}(\Omega_{G,\mathrm{c}}(M), \Omega_G)$, equivariant Poincaré left adjunction	201
$\mathbb{D}'_{G,M}$	$: \Omega_{G,\mathrm{c}}(M) \to \mathbf{Hom}^\bullet_{\Omega_G}(\Omega_G(M), \Omega_G)$, equivariant Poincaré right adjunction	201
$D_{G,M}$	$: H_G(M)[d_M] \to \mathbf{Hom}^\bullet_{H_G}(H_{G,\mathrm{c}}(M), H_G)$, equivariant Poincaré duality morphism is cohomology	204
Chapter 6.	***Equivariant Gysin Morphism***	
$f_!$	$: H_{G,\mathrm{c}}(M)[d_M] \to H_{G,\mathrm{c}}(N)[d_N]$, equivariant Gysin morphism	212
$(\_)_!$	G-$\mathrm{Man}^{\mathrm{or}} \rightsquigarrow \mathrm{GM}(H_G)$, equivariant Gysin functor	213
f_*	$: H_G(M)[d_M] \to H_G(N)[d_N]$, equivariant Gysin morphism	214
$(\_)_*$	G-$\mathrm{Man}^{\mathrm{or}}_{\mathrm{pr}} \rightsquigarrow \mathrm{GM}(H_G)$, equivariant Gysin functor	214
Res^G_H	group restriction morphism	217
Chapter 6.	***Equivariant Euler Classes***	
$\Phi(N, M)$	$i_*(1) \in H^{d_M - d_N}(M)$, equivariant Thom class of $(N \subseteq M)$	229
$\mathrm{Eu}(N, M)$	$i^* i_*(1) = \Phi(N, M)$, equivariant Euler class of $(N \subseteq M)$	229
Chapter 7.	***Localization***	
$Q_G \otimes_{\Omega_G} (\_)$	$: \mathrm{GM}(\Omega_G) \rightsquigarrow \mathrm{Vec}(Q_G)$, localization functor	235
$D_{G,M}$	$: Q \otimes H_G(M)[d_M] \to \mathbf{Hom}^\bullet_{Q_G}(Q \otimes H_{G,\mathrm{c}}(M), Q_G)$, localized equivariant Poincaré duality	236

$D_{G,M}$	$: \mathcal{Q} \otimes H_G(M)[d_M] \to \mathbf{Hom}^\bullet_{\mathcal{Q}_G}\big(\mathcal{Q} \otimes H_{G,c}(M), \mathcal{Q}_G\big)$, localized equivariant Poincaré duality	236
$f_!$	$: \mathcal{Q} \otimes H_{G,c}(M) \to \mathcal{Q} \otimes H_{G,c}(N)$, localized Gysin morphism	236
f_*	$: \mathcal{Q} \otimes H_G(M) \to \mathcal{Q} \otimes H_G(N)$, localized Gysin morphism	236
Chapter 8.	***Changing the Coefficients Field***	
$\mathbb{k}$	field of arbitrary characteristic	245
$\underline{\underline{\omega}}^\bullet_{(\_)}(\mathbb{k})$	dualizing complex on the space (_)	246
$\mathrm{Sh}(X) := \mathrm{Sh}(X; \mathbb{k})$	category of sheaves of $\mathbb{k}$-vector spaces on X	246
$C(X) := C(X; \mathbb{k})$	category of complexes of sheaves on X	246
$\mathcal{K}(X) := \mathcal{K}(X; \mathbb{k})$	homotopy category of $C(X)$	246
$\mathcal{D}(X) := \mathcal{D}(X; \mathbb{k})$	derived category of $C(X)$	246
$\mathbb{B}, \mathbb{B}G$	classifying space of the compact Lie group G	247
$f_*, f_!$	$: \mathrm{Sh}(X) \rightsquigarrow \mathrm{Sh}(X')$, direct images	247
$\mathbb{R} f_*, \mathbb{R} f_!$	$: \mathcal{D}^+(X) \rightsquigarrow \mathcal{D}^+(X')$, derived direct images	247
$\underline{\underline{\mathbb{k}}}_X$	constant sheaf on X with fiber $\mathbb{k}$	247
$\underline{\underline{\Omega}}(X; \mathbb{k})$	sheaf of Alexander-Spanier cochains with coefficients in $\mathbb{k}$	248
Φ	paracompactifying family of supports	248
$\underline{\underline{\Omega}}_G$	$:= \underline{\underline{\Omega}}(\mathbb{B})$, sheafification of Ω_G	249
$\underline{\underline{\Omega}}_G(X)$	$:= \pi_* \underline{\underline{\Omega}}(X_G)$, sheafification of $\Omega_G(M)$	249
$\underline{\underline{\Omega}}_{G,c}(X)$	$:= \pi_* \underline{\underline{\Omega}}(X_G)$, sheafification of $\Omega_{G,c}(M)$	249
$\mathrm{DGM}(\underline{\underline{\Omega}}_G)$	category of sheaves of differential graded modules over $\underline{\underline{\Omega}}_G$	249
$\mathbb{R}\,\underline{\underline{\mathbf{Hom}}}^\bullet_{\underline{\underline{\Omega}}_G}(\_, \underline{\underline{\Omega}}_G)$	duality functor in $\mathrm{DGM}(\underline{\underline{\Omega}}_G)$	250
$\underline{\underline{\mathbb{D}}}_{G,M}$	left Poincaré adjunction in $\mathcal{D}\,\mathrm{DGM}(\underline{\underline{\Omega}}_G)$	256
$D_{G,M}$	left Poincaré adjunction in $\mathcal{D}(\mathrm{DGM}(\Omega_G))$	256
$\underline{\underline{\mathbb{D}}}'_{G,M}$	right Poincaré adjunction in $\mathcal{D}\,\mathrm{DGM}(\underline{\underline{\Omega}}_G)$	256
$f_!$	$H_{G,c}(M; \mathbb{k})[d_M] \to H_{G,c}(N; \mathbb{k})[d_N]$, Gysin morphism over $\mathbb{k}$	262
f_*	$H_{G,c}(M; \mathbb{k})[d_M] \to H_{G,c}(N; \mathbb{k})[d_N]$, Gysin morphism over $\mathbb{k}$	263
Appendix A.	***Basics on Derived Categories***	
Ab	abelian category	266
$C(\mathrm{Ab})$	category of complexes on Ab	266
$\boldsymbol{h} : C(A) \rightsquigarrow \mathrm{Ab}^{\mathbb{Z}}$	cohomology functor	266
$[m]$	$: C(\mathrm{Ab}) \to C(\mathrm{Ab})$, shift functor	266
$\hat{c}(\alpha)$	cone of a morphism of complexes α	268
$\mathrm{Hot}_{C(\mathrm{Ab})}(\_, \_)$	the group of homotopy classes of complex morphisms	269
$\mathcal{K}(\mathrm{Ab})$	homotopy category of $C(\mathrm{Ab})$	270
$\mathcal{D}(\mathrm{Ab})$	derived category of $C(\mathrm{Ab})$	275
$\mathcal{S}$	multiplicative collection of morphisms in a category $\mathcal{C}$	276
$\mathcal{Q} : \mathcal{C} \rightsquigarrow \mathcal{C}[\mathcal{S}^{-1}]$	localization of the category $\mathcal{C}$ by $\mathcal{S}$	276
$\mathbb{R} F$	$: \mathcal{D}^+(\mathrm{Ab}) \rightsquigarrow \mathcal{D}^+(\mathrm{Ab}')$, right derived functor of $\boldsymbol{F}$	283

$\mathbb{L}F$	$: \mathcal{D}^-(\mathrm{Ab}) \rightsquigarrow \mathcal{D}^-(\mathrm{Ab}')$, left derived functor of $\boldsymbol{F}$	283
$\mathrm{Hot}_{\mathrm{DGM}(\Omega_G)}(\_, \_)$	the group of homotopy classes of Ω_G-dgm's	288
$\mathcal{D}(\mathrm{DGM}(\Omega_G))$	$:= \mathcal{K}(\mathrm{DGM}(\Omega_G))[\mathcal{S}^{-1}]$, derived category of $\mathrm{DGM}(\Omega_G)$	289
Appendix B.	***Sheaves of Differential Graded Algebras***	
$\mathcal{A} \underline{\otimes} (\_)$	$\mathcal{A}$-localization presheaf functor	307
$\mathcal{A} \otimes (\_)$	$\mathcal{A}$-localization functor	307
$S_x \subseteq O_X(X)$	set of functions equal to 1 in a neighborhood of x	308
Φ	(paracompactifying) family of supports	311
$\Gamma_\Phi(Y; \_)$	functor of global sections with supports in Φ	312
Γ_Φ	functor on $\mathrm{GM}(\mathcal{A}(X))$	313
$(\mathcal{A}, d) \underline{\otimes} (\_)$	$(\mathcal{A}, d)$-localization presheaf functor	315
$(\mathcal{A}, d) \otimes (\_)$	$(\mathcal{A}, d)$-localization functor	316
$S^{-1}(M, d)$	localization of an (A, d)-dgm by $S \subseteq A$.	317

Index

Underlined page numbers refers to definitions.

A. Arabia, *Equivariant Poincaré Duality on G-Manifolds*, Lecture Notes in Mathematics 2288, https://doi.org/10.1007/978-3-030-70440-7

LECTURE NOTES IN MATHEMATICS

Springer

Editorial Policy

1. Lecture Notes aim to report new developments in all areas of mathematics and their applications – quickly, informally and at a high level. Mathematical texts analysing new developments in modelling and numerical simulation are welcome.

   Manuscripts should be reasonably self-contained and rounded off. Thus they may, and often will, present not only results of the author but also related work by other people. They may be based on specialised lecture courses. Furthermore, the manuscripts should provide sufficient motivation, examples and applications. This clearly distinguishes Lecture Notes from journal articles or technical reports which normally are very concise. Articles intended for a journal but too long to be accepted by most journals, usually do not have this "lecture notes" character. For similar reasons it is unusual for doctoral theses to be accepted for the Lecture Notes series, though habilitation theses may be appropriate.

2. Besides monographs, multi-author manuscripts resulting from SUMMER SCHOOLS or similar INTENSIVE COURSES are welcome, provided their objective was held to present an active mathematical topic to an audience at the beginning or intermediate graduate level (a list of participants should be provided).

   The resulting manuscript should not be just a collection of course notes, but should require advance planning and coordination among the main lecturers. The subject matter should dictate the structure of the book. This structure should be motivated and explained in a scientific introduction, and the notation, references, index and formulation of results should be, if possible, unified by the editors. Each contribution should have an abstract and an introduction referring to the other contributions. In other words, more preparatory work must go into a multi-authored volume than simply assembling a disparate collection of papers, communicated at the event.

3. Manuscripts should be submitted either online at www.editorialmanager.com/lnm to Springer's mathematics editorial in Heidelberg, or electronically to one of the series editors. Authors should be aware that incomplete or insufficiently close-to-final manuscripts almost always result in longer refereeing times and nevertheless unclear referees' recommendations, making further refereeing of a final draft necessary. The strict minimum amount of material that will be considered should include a detailed outline describing the planned contents of each chapter, a bibliography and several sample chapters. Parallel submission of a manuscript to another publisher while under consideration for LNM is not acceptable and can lead to rejection.

4. In general, **monographs** will be sent out to at least 2 external referees for evaluation.

   A final decision to publish can be made only on the basis of the complete manuscript, however a refereeing process leading to a preliminary decision can be based on a pre-final or incomplete manuscript.

   Volume Editors of **multi-author works** are expected to arrange for the refereeing, to the usual scientific standards, of the individual contributions. If the resulting reports can be forwarded to the LNM Editorial Board, this is very helpful. If no reports are forwarded or if other questions remain unclear in respect of homogeneity etc, the series editors may wish to consult external referees for an overall evaluation of the volume.

5. Manuscripts should in general be submitted in English. Final manuscripts should contain at least 100 pages of mathematical text and should always include

   – a table of contents;
   – an informative introduction, with adequate motivation and perhaps some historical remarks: it should be accessible to a reader not intimately familiar with the topic treated;
   – a subject index: as a rule this is genuinely helpful for the reader.
   – For evaluation purposes, manuscripts should be submitted as pdf files.

6. Careful preparation of the manuscripts will help keep production time short besides ensuring satisfactory appearance of the finished book in print and online. After acceptance of the manuscript authors will be asked to prepare the final LaTeX source files (see LaTeX templates online: https://www.springer.com/gb/authors-editors/book-authors-editors/manuscriptpreparation/5636) plus the corresponding pdf- or zipped ps-file. The LaTeX source files are essential for producing the full-text online version of the book, see http://link.springer.com/bookseries/304 for the existing online volumes of LNM). The technical production of a Lecture Notes volume takes approximately 12 weeks. Additional instructions, if necessary, are available on request from lnm@springer.com.

7. Authors receive a total of 30 free copies of their volume and free access to their book on SpringerLink, but no royalties. They are entitled to a discount of 33.3 % on the price of Springer books purchased for their personal use, if ordering directly from Springer.

8. Commitment to publish is made by a *Publishing Agreement*; contributing authors of multiauthor books are requested to sign a *Consent to Publish form*. Springer-Verlag registers the copyright for each volume. Authors are free to reuse material contained in their LNM volumes in later publications: a brief written (or e-mail) request for formal permission is sufficient.

Addresses:
Professor Jean-Michel Morel, CMLA, École Normale Supérieure de Cachan, France
E-mail: moreljeanmichel@gmail.com

Professor Bernard Teissier, Equipe Géométrie et Dynamique,
Institut de Mathématiques de Jussieu – Paris Rive Gauche, Paris, France
E-mail: bernard.teissier@imj-prg.fr

Springer: Ute McCrory, Mathematics, Heidelberg, Germany,
E-mail: lnm@springer.com

Printed in the United States
by Baker & Taylor Publisher Services